2nd International Conference on Civil Engineering and Materials Science (ICCEMS 2017)

IOP Conference Series: Materials Science and Engineering
Volume 216

Seoul, South Korea
26 – 28 May 2017

ISBN: 978-1-5108-4417-9
ISSN: 1757-8981

Printed from e-media with permission by:

Curran Associates, Inc.
57 Morehouse Lane
Red Hook, NY 12571

proceedings.com

Some format issues inherent in the e-media version may also appear in this print version.

Printed by Curran Associates, Inc. (2017)

Institute of Physics
Dirac House, Temple Back
Bristol BS1 6BE UK

Phone: 44 1 17 929 7481
Fax: 44 1 17 920 0979

techtracking@iop.org

Additional copies of this publication are available from:

Curran Associates, Inc.
57 Morehouse Lane
Red Hook, NY 12571 USA
Phone: 845-758-0400
Fax: 845-758-2633
Email: curran@proceedings.com
Web: www.proceedings.com

Table of contents

Volume 216

The 2nd International Conference on Civil Engineering and Materials Science
26–28 May 2017, Seoul, Korea

Accepted papers received: 23 June 2017
Published online: 2 July 2017

Preface

Design and Development of New Environmental Materials

Performance Testing and Verification of Smart Materials

Performance Analysis and Mechanical Simulation of Steel Structures

Design and Performance Optimization of Concrete Structures

Building Materials Analysis: Methods and Tests

Structural Engineering

Geotechnical Engineering

Hydraulic Engineering and Application

Road and Bridge Engineering

Architectural Design and Engineering Management

IOP Conf. Series: Materials Science and Engineering **216** (2017) 011001 doi:10.1088/1757-899X/216/1/011001

Preface

It is our great pleasure to welcome you to 2017 the Second International Conference on Civil Engineering and Materials Science (ICCEMS 2017), which will be held in Seoul, Korea during May 26-28, 2017.

This conference provides opportunities from different areas delegates to exchange new ideas and application experiences face to face, to establish business or research relations and to find global partners for future collaboration. The major goal and feature of the conference is to bring academic scientists, engineers, industry researchers together to exchange and share their experiences and research results, and discuss the practical challenges encountered and the solutions adopted. Professors from different areas are invited to deliver keynote speeches regarding latest information in their respective expertise areas.

This proceedings presents a selection from papers submitted to the conference from universities, research institutes and industries. The papers selected depended on their quality and their relevancy to the conference. The volume tends to present to the readers the recent advances in the field of civil engineering and materials science and various related areas, such as design and development of new environmental materials, performance testing and verification of smart materials, performance analysis and mechanical simulation of steel structures, design and performance optimization of concrete structures, building materials analysis, structural engineering, geotechnical engineering, hydraulic engineering and application, road and bridge engineering, architectural design and engineering management etc.

We would like to thank all the authors who have contributed to this volume and also to reviewers, speakers, chairpersons, sponsors and all the conference participants for their support to ICCEMS 2017.

ICCEMS 2017 Organizing Committee
May 20, 2017

IOP Conf. Series: Materials Science and Engineering **216** (2017) 011002 doi:10.1088/1757-899X/216/1/011002

Conference Committee

International Advisory Committees
Prof. George C. Manos, Aristotle University of Thessaloniki-GREECE, Greece
Prof. C. W. Lim, City University of Hong Kong, Hong Kong

Conference Chair
Prof. Kyoung Sun Moon, Yale University School of Architecture, USA

Conference Program Co-Chairs
Prof. Teik-Cheng Lim, SIM University, Singapore
Prof. Chien-Yuan Chen, National Chiayi University, Taiwan

Technical Committees
Dr. Jaroslaw Jakubowicz, Poznan University of Technology, Poland

Dr. Grzegorz ADAMEK, Poznan University of Technology, Poland

Dr. Ramadhansyah Putra Jaya, Universiti Teknologi Malaysia, Malaysia

Dr. Haryati Yaacob, Universiti Teknologi Malaysia, Malaysia

Dr. Norhidayah Abdul Hassan, Universiti Teknologi Malaysia, Malaysia

Dr. Mohammad Arif Kamal, Aligarh Muslim University, India

Dr. Tomasz Trapko, Wroclaw University of Technology, Poland

Dr. Sadaf Qasim, NED university of Engineering and Technology, Pakista

Dr. Sevil Yücel, Yildiz Technical University, Turkey

Dr. Kong Fah Tee, University of Greenwich, UK

Dr. Shantha Kumari Muniyandi, Tunku Abdul Rahman University College, Malaysia

Dr. Andrew Boyd, McGill University, Canada

Dr. Omer Mughieda, Abu Dhabi University, United Arab Emirates

Dr. Murat Gunduz, Qatar University, Qatar

Dr. Guoxin Dai, Chongqing University, China

Dr. Shang Huaishuai, Qingdao University of technology, China

Dr. Michał Musiał, Wroclaw University of Technology, Poland

Dr. Bo Yang, Chongqing University, China

Dr. Jianhui Si, Xi'an University of Technology, China

Dr. Qingbin Li, Tsinghua University, China

Dr. Pınar Terzioğlu, Mugla Sıtkı Kocman University, Turkey

Peer review statement

All papers published in this volume of *IOP Conference Series: Materials Science and Engineering* have been peer reviewed through processes administered by the proceedings Editors. Reviews were conducted by expert referees to the professional and scientific standards expected of a proceedings journal published by IOP Publishing.

IOP Conf. Series: Materials Science and Engineering **216** (2017) 012001 doi:10.1088/1757-899X/216/1/012001

Ammonia Adsorption of Four Thailand White Charcoals for Air Purification Application

Nuchanaporn Pijarn[1], Tinnphat Sribuarai[1], Manipa Butsee[1], Kanokwan Buakul[1], Hasan Seng[2], Pongtep Phonprasert[1], Thanate Kitisriworaphan[1], Prasertsil Atthameth[3]
[1]Division of Environmental Technology, Faculty of Science and Technology, Bangkokthonburi University, Bangkok 10170, Thailand
[2]Pentacle Pollutech Co., Ltd., Bangsrithong, Bangkruai, Nonthaburi 11130, Thailand
[3]Faculty of Engineering, Srinakharinwirot University, Nakhon Nayok 26120, Thailand

Email: nuchtennis@gmail.com

Abstract. The goal of this work was to study the efficiency of ammonia adsorption of four white charcoals include Bamboo, Coconut spathe, Coconut shell and Miscellaneous woods. The ammonia 20 mL was used as the odour sample for adsorption, 0.1 grams of each white charcoal was used as adsorbent and adsorption time were studied in 20, 40, 60, 80, and 100 min. The gravimetric method was used as method for study ammonia adsorption. The results show all white charcoals have saturated adsorption in range 40-60 min. The sequent of ammonia adsorption with white charcoal from descending were Miscellaneous > Coconut spathe > Coconut shell > Bamboo. A statistical test with Paired Samples T-Test, the result of four white charcoal to absorb the odour of ammonia in the all adsorption times have difference statistically significant at the 95% confidence.

1. Introduction

Thailand is agricultural country many where have wastes of wood from many activities such as miscellaneous woods from construction building, bamboo from basketwork and coconut spathe and coconut shell from food production process [1]. These are a big problem because of no enough storage places [2]. Maximizing the value of these materials is very important [3]. Nowadays, there are a lot of pollution types and intensified even more in everyday. All pollutions are problem on humans directly. Especially, the odour problem is one of the important issues that affect the ecological environment and also affects on respiratory tract of humans [4]. Currently, there are two kinds of charcoal include the ordinary "black charcoal" and "white charcoal". These charcoals are made by different technical method of burning [5]. Although, they are produced in a basically similar manner, the quality of the charcoal differs totally depending on how the fire is extinguished. The method for producing white charcoal is hard compared to ordinary black charcoal [6]. This technology is applied mostly in Japan, and some parts of China and Korea. Nowadays, the world admires the charcoal-making techniques of Japan and applies it in the field of apply science [7-9]. White charcoals are produced from natural raw materials waste such as wood, bamboo, coconut that the waste from the basketwork or residue waste wood for other applications [10-15]. White charcoal is a good of materials for adsorbing an odour because it has a high adsorption capacity. Many researchers reported about adsorption gases such as ethylene [16], hydrogen [17] and some work studied adsorption of acetic acid vapour [18].
Therefore, in the study is the investigation on the efficiency of four white charcoals, (bamboo, coconut spathe, coconut shell and miscellaneous wood) for ammonia absorption to reduce these wastes wood and remove odour. This study can be used as a model for adsorption odour in environment and air purification in the future work.

2. Materials and Methods

2.1. Materials

Analytical grade of ammonia solution 25% was purchased from Merck KGaA, Darmstadt, Germany. The wastes of miscellaneous woods were collected from construction site, the bamboo was collected from basketwork, and coconut shell and coconut spathe was collected from the fresh market in Prachinburi province, Thailand. Each materials wood was carbonized at >1000 °C for more than three hours to change as white charcoal.

2.2. Method

Ammonia adsorption in this work were studied by Gravimetric method using recording the initial weight of white charcoal before testing compared to the final weight of white charcoal after testing. For ammonia adsorption test, theses step were followed this following; (1) carefully weighting 0.1 g of white charcoal before testing, (2) preparing ammonia solution 20 mL into bottle test and record starting weight before testing (3) testing ammonia adsorption in various adsorption times for 20, 40, 60, 80, 100 min., (4) recording weight after adsorption testing of all samples, (5) calculating of percentage of adsorption follows equation 1, (6) observing microstructure of all white charcoal using scanning electron microscope (SEM, JSM 5410 JEOL, Japan), (7) measuring of specific surface area using A Flowsorb II 2300, Micromeritics, USA, and (8) comparing and discussing results.

$$\%Adsorption = \frac{W_A - W_B}{W_B} x100 \tag{1}$$

Where; W_A is weight after adsorption testing, and W_B is weight before adsorption testing

3. Results and Discussion

Microstructure of four white charcoal were observed using SEM with magnifications of 35X. The pore of white charcoal could be observed but the fine structure requires higher than this magnification.The SEM images revealed pore wide smaller than 1 mm as shown in Figure 1.

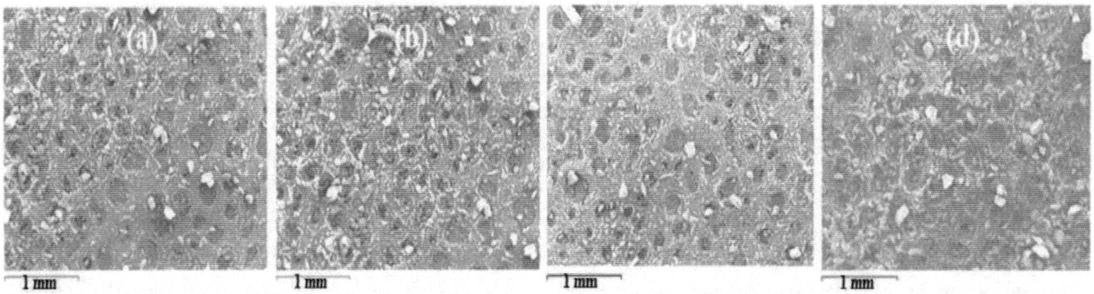

Figure 1.SEM of (a) Bamboo, (b) Coconut spathe, (c) Coconut shell, and (d) Miscellaneous wood

Table 1. Percentage of ammonia adsorption of four white charcoals in various adsorptiontimes

Time (min)	Ammonia Adsorption (%)			
	Bamboo	**Coconut spathe**	**Coconut Shell**	**Miscellaneous**
20	112.02±7.82	129.22±7.27	118.11±16.94	136.96±9.57
40	145.76±10.55	141.14±6.78	140.92±9.52	145.14±23.26
60	121.84±6.96	136.25±17.86	127.57±14.43	149.50±15.17
80	108.13±11.09	130.19±24.19	125.18±4.30	153.91±10.78
100	92.43±8.70	120.53±27.12	121.14±7.65	154.42±15.31

IOP Conf. Series: Materials Science and Engineering **216** (2017) 012001 doi:10.1088/1757-899X/216/1/012001

Average	**118.04±9.20**	**131.47±16.64**	**121.14±10.57**	**148.12±15.31**
Surface area (m^2/g)	**3,189**	**3,230**	**3,205**	**3,280**

The 0.1 g of bamboo white charcoal powder was used as adsorbent for ammonia adsorption in 20, 40, 60, 80 and 100 min. Percentage of ammonia adsorption of four white charcoals in various adsorption times are concluded in Table 1. This table found that a white charcoal from bamboo, coconut spathe, coconut shell and Miscellaneous had ammonia adsorption around 118 %, 131 %, 121 % and 148 %, respectively. These results related with specific surface area of white charcoal in Table 1. The results show that all of white charcoal had specific surface area more than 3100 m^2/g.

All white charcoals powder was used as adsorbent for ammonia adsorption in 20, 40, 60, 80 and 100 min. Figure 2(a) was the line graph with standard error of percentage ammonia adsorption all white charcoal. It was clear that percentage of ammonia adsorption more than 100 % of its weight almost all. At adsorption time in 40 min was highest adsorption at adsorption of bamboo, coconut spathe and coconut shell. After that the percentage continuously decreased with increasing of adsorption time. This may because this time is saturated adsorptive of these white charcoals [19]. However, miscellaneous white charcoal continuously increased with increasing of adsorption time. This may be caused by high specific surface area. When concern only average percentage of adsorption, it is indicated that miscellaneous wood has highest percentage of ammonia adsorption. The sequent of ammonia adsorption with white charcoal from descending were Miscellaneous > Coconut spathe > Coconut shell > Bamboo follow Figure 2(b). These results may be caused by density, pore volume and specific surface area of each white charcoal [20] and related specific surface area from this study.

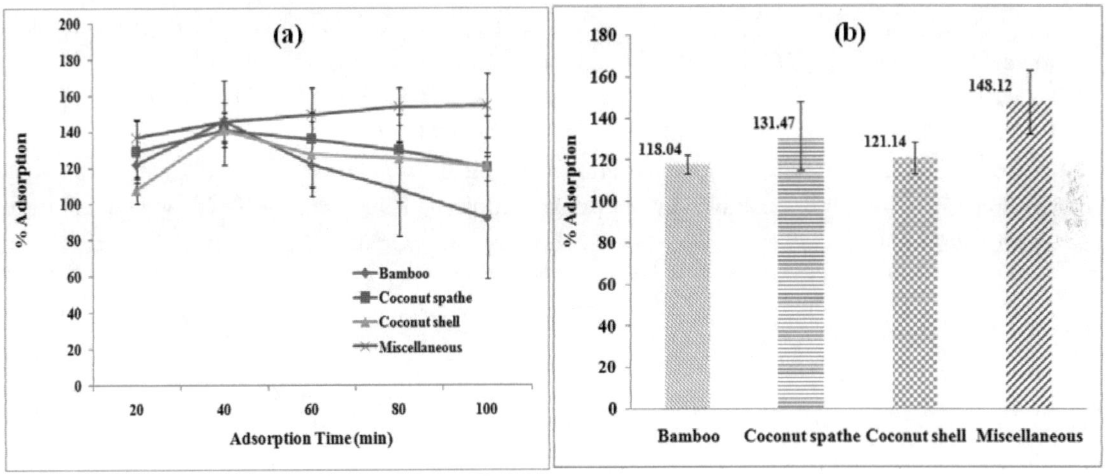

Figure 2. Ammonia adsorption of four white charcoals (a) in various times and (b) in average value

4. Summary

It can be conclude that every white charcoal have adsorption nearly the same value at adsorption time 40 min. Miscellaneous wood white charcoal has highest percentage of ammonia adsorption. The sequent of ammonia adsorption with white charcoal from descending were Miscellaneous > Coconut spathe > Coconut shell > Bamboo. These results can used as model for odor removal and can reduced these waste wood from environmental of Thailand because of using the waste wood from wicker (Bamboo), cutting wood for building house (Miscellaneous wood), and cooking (Coconut spathe and Coconut shell).

5. Acknowledgement
This research was financial supported by Bangkokthonburi University. The authors thank the staff of equipment center from National Metal and Materials Technology for sample characterization.

6. References
[1] Nirunnoot T and Porncharoen R 2016 *J. Art Architect.Naresuan Univ.***7** 14
[2] Shinde L M, Naktode P L and Saklecha P P 2015 *Int. J. Comput. Appl.***0975 – 8887** 254
[3] Kamyingyong P 2016 *Thesis Master of Business* Bangkok University
[4] Oiamo T H and Luginaah I N 2015 *J. Baxter: Soc. Sci. Med.***146** 19
[5] Miura I 1931 *Ind. Eng. Chem.***23** 631
[6] Wartluft J L 1984 *White S VITA*, Virginia 22209 USA
[7] Elsayed Y, Dalibalta S and Abu-Farha N 2016 *Sci. Total Environ.***569–570** 262
[8] Bui H H, Wang L, Tran K-Q and Skreiberg Ø 2016 *Fuel Process. Technol.***152** 207
[9] Pingree M R A, DeLuca E E, Schwartz D T and DeLuca T H 2016 *Geoderma***283** 68
[10] Patel V R, Patel D, Varia N and Patel R N 2017 *Energy***119** 834
[11] Wang L, Chen S S, Tsang D C W, Poon C S and Shih K 2016 *Constr. Build. Mater.***125** 316
[12] Nasser R A, Salem M Z M, Al-Mefarrej H A and Aref I M 2016 *Cement Concrete Comp.* **72** 246
[13] Vaidya A A, Gaugler M and Smith D A 2016 *Carbohyd. Polym.***136** 1238
[14] Won J M and Song J Y 2007 *J. Korea Tech. Assoc. Pulp Pap. Ind.***39** 64
[15] Ratajczak E, Bidzińska G, Szostak A and Herbeć M 2015 *Conserv. Recycling***97** 93
[16] Dodić S N, Vasiljević T Z, Marić R M,. Kosanović AJ R, Dodić J M and Popov S D 2012 *Renew. Sust.Energ. Rev.***16** (5) 2355
[17] Miao H Y, Chen G R, Chen D Y, Lue J T and Yu M S 2010 *EPJ Appl. Phys.***52** 21101
[18] Ikuo A, Satoshi I, Jun M and Tomoko F 2000 *Sci. Ind.***74** 106
[19] Griessacher T, Antrekowitsch J and Steinlechner S 2012 *Biomass Bioenerg.* **39** 139
[20] Chia C H, Joseph S D, Rawal A, Linser R, Hook J M and Munroe P 2014 *J. Anal. Appl. Pyrol.***109** 215

IOP Conf. Series: Materials Science and Engineering **216** (2017) 012002 doi:10.1088/1757-899X/216/1/012002

Effect of White Charcoal on COD Reduction in Wastewater Treatment

Nuchanaporn Pijarn[1], Manipa Butsee[1], Kanokwan Buakul[1], Hasan Seng[2], Tinnphat Sribuarai[1], Pongtep Phonprasert[1], Kla Taneeto[1], Prasertsil Atthameth[3]
[1]Division of Environmental Technology, Faculty of Science and Technology, Bangkokthonburi University, Bangkok 10170, Thailand
[2]Pentacle Pollutech Co., Ltd., Bangsrithong, Bangkruai, Nonthaburi 11130, Thailand
[3]Faculty of Engineering, Srinakharinwirot University, Nakhon Nayok 26120, Thailand

Email: nuchtennis@gmail.com

Abstract. The objective of this study is to compare the COD reduction in wastewater between using coconut shell and coconut spathe white charcoal from Khlong Wat NongPra-Ong, Krathumbaen, SamutSakhon province, Thailand. The waste water samples were collected using composite sampling method. The experimental section can be divided into 2 parts. The first part was study the optimum of COD adsorption time using both white charcoals. The second part was study the optimum amount of white charcoal for chemical oxygen demand (COD) reduction. The pre-treatment of wastewater was examined in parameters include temperature, alkalinity (pH), conductivity, turbidity, suspended solid (SS), total dissolved solid (TDS), and COD. The results show that both white charcoals can reduce COD of wastewater. The pH of pre-treatment wastewater had pH 9 but post-treatment wastewaters using both white charcoals have pH 8. The COD of pre-treatment wastewater had COD as 258 mg/L but post-treatment wastewater using coconut shell white charcoal had COD steady at 40 mg/L in 30 min and the amount of white charcoals 4 g. The COD of post-treatment wastewater using coconut spathe white charcoal had COD steady at 71 mg/L in 30 min and the amount of white charcoals 4 g. Therefore comparison of COD reduction between coconut shell white charcoal versus coconut spathe white charcoal found that the coconut shell white charcoal had efficiency for COD reduction better than coconut spathe white charcoal.

1. Introduction

Water is crucial to the life of plants and animals. Living things on earth need to use and drink water. Especially, humans used water in consumption and in many activities such as fisheries, agriculture, industry, and transportation. In the past, the water is not pollution due to natural balance and it can be refresh itself because0 it has a small volume of rotation contamination from chemicals or pollutants. Despite, nowadays, when the growth of social community changes from the agricultural to industrial community, the water pollution is one big problem to solve because natural could not refresh itself instantly. There are four main techniques to treat wastewater that will depend on how the appropriate use of wastewater includes chemical process, biological process, physical process and physical-chemical process. The chemical process is a way to treat wastewater by separating different substances or contaminants in treated water. The contaminated by adding various chemicals to react, however, this method has limitation because when adding chemicals to the wastewater which impact on the environment and high costs for chemicals [1-4]. The biological process is the principle of using microorganisms to decompose the organic material changed to carbon dioxide and ammonia. The limitation of this process is selection of aerobic bacteria or anaerobic bacteria suitable for wastewater [5-8]. The physical process is a simple wastewater treatment that separates insoluble solids. The disadvantage of this technique is the cost of expensive equipment for separation. There are many different methods of separation such as screening, skimming and trapping [9-10]. The physical-chemical process have two ways; carbon adsorption method uses charcoal powder or carbon as a substance adsorbent, and ion exchange is based on the principle of exchange of charge between pollutants in wastewater with medium containing both positive and negative charges. This is an economical way because natural materials readily available [11-14].

Therefore, in this work, the wastewater from Khlong Wat NongPra-Ong, the surface wastewater, from community in SamutSakhon province, was a source of water samples. All samples were treated with white charcoal from coconut shells and coconut spathe for applying on wastewater treatment in the surface water source in the future.

2. Materials and Methods

Coconut shell and coconut spathe is the wastes were collected from the fresh market in Prachinburi province. Coconut shell and coconut spathe were burned at more than 1000 ^{o}C for change to white charcoal. Then, both white charcoal were grinded and used as adsorbent for adsorption of COD. This experiment can be divided into two parts. The first part was study adsorption time of 2 g white charcoal for COD reduction at 10, 20, 30, 40, 50, 60, 70, 80, 90 and 100 min. The second part was study the optimum amount of whitecharcoal 1, 2, 3, 4, 5, 6, 7, 8, 9, and 10 g for COD reduction.

3. Results and Discussion

3.1 Pre-treatment water properties

Water properties of pre-treatment water sample were observed and measured included colour, odour, temperature, pH, conductivity, turbidity, SS and TDS. The results show that pH as 9, conductivity as 947μS/L, turbidity as28.03 NTU, SS as 158.72 mg/L, TDS as 472 mg/L, and COD as 258 mg/L.

3.2 Study adsorption time and the optimum amount of white charcoal for COD reduction

The adsorption time were studied at 10, 20, 30, 40, 50, 60, 70, 80, 90 and 100 min, and using 2 g of white charcoal per water sample 100 mL for COD reduction. The adsorptions of COD were measured using spectrophotometer. The results show that the best adsorption times were steadies at 30 min in both white charcoals follow as Figure 1(a). These results may cause by limitation of pore volume and specific surface area of white charcoals [15]. The COD of pre-treatment sample was 258 mg/L. The results indicated that the coconut shell white charcoal can reduce the COD of wastewater from 258 mg/L to 40 mg/L in adsorption time at 30 min, while the coconut spathe can reduce the COD of wastewater from 258 mg/L to 74 mg/L in adsorption time at 30 min. It is cleared that the coconut shell white charcoal can reduce COD better than the coconut spathe white charcoal.The optimum amount of white charcoal for COD reduction were studied using spectrophotometer at adsorption time 30 min, and various the amount of white charcoals as 1, 2, 3, 4, 5, 6, 7, 8, 9 and10 g in 100 mL wastewater. Figure 1(b) show of study the optimum amount of both white charcoal were4 gand indicated that coconut shell white charcoal can reduce the COD from 258 mg/L to 19 mg/L while coconut spathe can reduce the COD from 258 mg/L to 71 mg/L. These results were related with previous report [16]. It is confirmed that the coconut shell white charcoal can reduce COD better than the coconut spathe white charcoal the same amount of white charcoal.

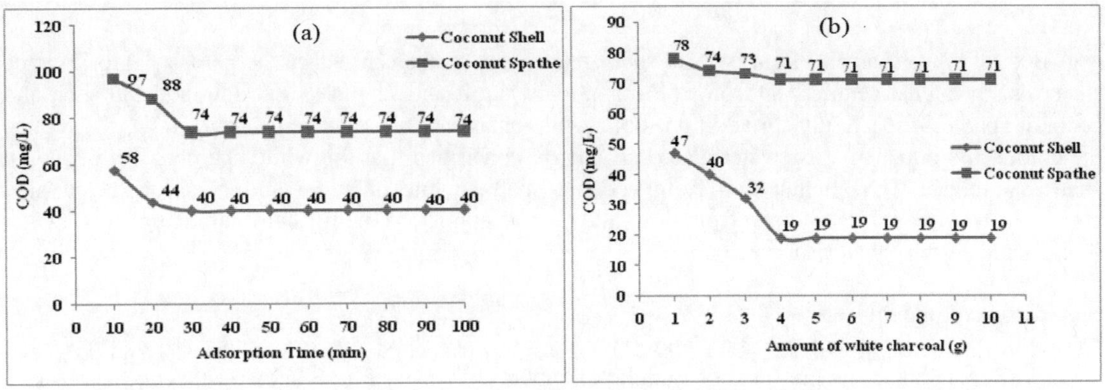

Figure 1. COD of water sample in: (a) v arious adsorption times and (b) various the amount of white charcoal (COD of pre-treatment sample as 258 mg/L)

3.3 Comparison of pre-treatment and post treatment wastewater with white charcoal
The pH of pre-treatment wastewater was 9 but pH of post-treatment as 8 in using both of white charcoal as shown in Figure 2(a). This is standard range of pH (7-9) of surface water follow Notification of the National Environmental Board No. 8 (BE 2537) issued under the National Environment Quality Promotion and Conservation Act, BE 2535 (1992), defining water quality standards in surface water sources (Category 3) of Thailand [17]. Figure 2(b) show the COD of pre-treatment wastewater as 258 mg/L, while post-treatment for 30 min using 4 g of coconut shell and 4 g of coconut spathe found that the COD of post-treatment with coconut shell and coconut spathe as 19 mg/L and 71 mg/L, respectively.

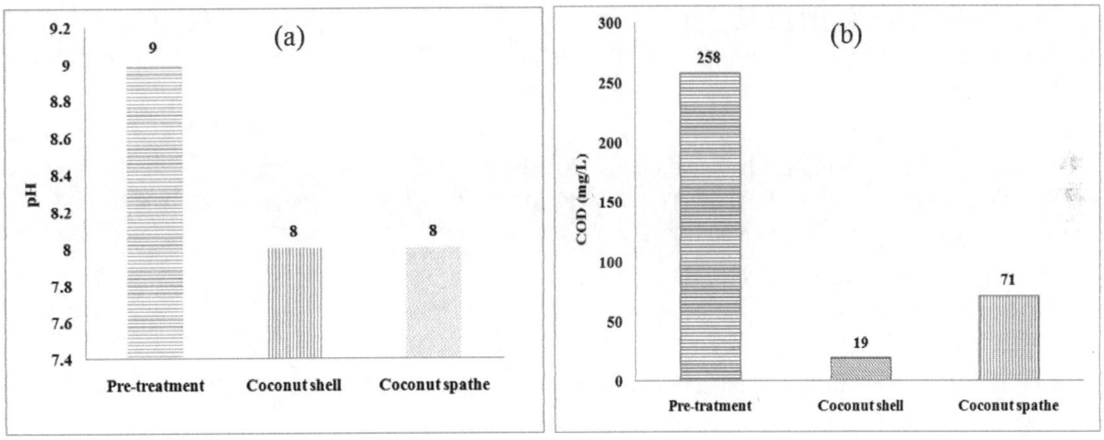

Figure 2. Comparison of pre-treatment and post-treatment wastewater with both white charcoals: (a) pH of wastewater, (b) COD of wastewater

These results were calculated in percentage of the COD reduction found that 4 g of coconut white charcoal can reduce the COD as 92.64%. This is percentage of COD removal more than previous report which used activated charcoal can remove only 83% [18]. However, 4 g of coconut spathe white charcoal can reduce the COD as 72.48%.

4. Summary
This work can be concluded that the saturated adsorption time of white charcoal was 30 min for using both of white charcoal from Prachinburi province. The optimum amount of both white charcoals for COD reduction of wastewater from Klong Wat NongPra-Ong, SamutSakhon province in Thailand was

4 g. Both of white charcoals from coconut shell and coconut spathe can reduce the pH from 9 to 8. White charcoals 4 g can reduce the COD from 258 mg/L to 19 mg/L and 71 mg/L for using coconut shell and coconut spathe, respectively. In addition, both white charcoals not only reduce the COD, but reduce the colour and odour of wastewater also. It is clearly confirmed that the white charcoal from coconut shell can reduce COD better than from coconut spathe white charcoal and better than activated charcoal.The knowledge obtained from this study can be applied to the further treatment of wastewater in the actual surface water source.

5. Acknowledgement

This work was supported by Bangkokthonburi University. The authors thank Assist. Prof. Dr. BungornBenjaathikul, the president of Bangkokthonburi University for all help.

6. References

[1] MoreiraF C, SolerJ, AlpenduradaM F, BoaventuraR A R, BrillasE, and VilarVJ P 2016 *Water Res.* **105** 251

[2] SahuO2016*Int. J. Sust. Built Environ.* **5** **(2)**288

[3] TiwariA andSahuO 2017 *Water Resour. Indust.***17**19

[4] BoluarteI AR, AndersenM, PramanikB K, ChangC-Y, BagshawS, FaragoL, JegatheesanV and ShuL 2016 *Int. Biodeter. Biodegr.***113** 44

[5] Ioannou-TtofaL and et.al. 2017*Water Res.***114** pp.1-13

[6] TrapidoM, TennoT, GoiA, DulovaN, KattelE, KlausonD, KleinK, TennoT andViisimaaM2017*J Water Process Eng.***16** 277

[7] GrandclémentC, SeyssiecqI, PiramA, Wong-Wah-ChungP, VanotG, TiliacosN, RocheN and DoumenqP*Water Res.***111** 297

[8] HuangC, LuoM-T, ChenX-F, XiongL, LiX-M and ChenX-D2017*Bioresource Technol.* In Press, Corrected Proof, Available online 20 February 2017

[9] KimuraK, HonokiD and SatoT2017*Sep. Purif. Technol.*2017 In Press, Accepted Manuscript

[10] ChenW-T, MaJ, ZhangY, GaiC and QianW2014*Bioresource Technol.***169**816

[11] LiuJ, OuH-S, WeiC-H, WuH-Z, HeJ-Z and LuD-H 2016 *J. Water Process Eng.* **10**98

[12] WuJ, YanG, ZhouG and XuT2014*Chem. Eng. J.***258** 450

[13] GhimpusanM, NechiforG, NechiforA-C, DimaS-O andPasseriP 2016 *J. Environ. Manage.* In Press, Corrected Proof

[14] GeorgeJ S, RamosA and ShipleyH J 2015 *J. Environ. Chem. Eng.* **3** **(2)** 969

[15] Thines R K, MubarakN M, NizamuddinS, SahuJ N, AbdullahE C and GanesanP 2017 *J. Taiwan Institute Chem. Eng.***72** 116

[16] ParandeA K, SivashanmugamA, BeulahH and PalaniswamyN 2009 *J Hazard. Mater.***168(2–3)** 800

[17] https://www.env.go.jp/earth/coop/oemjc/thai/e/thaie3. (B.E. 2535, A.D. 1992)

[18] Heibati B, et.al. 2015 *J. Mol. Liq.***208**, 99–105.

Evaluation of Aluminium Dross as Adsorbent for Removal of Carcinogenic Congo Red Dye in Wastewater

Mohamad Zulfika Hazielim b Zakaria[1], Nur Syuhada Ahmad Zauzi[2], Rubiyah Baini[3], Norsuzailina Mohamed Sutan[4], Md Rezaur Rahman[5],

[1,2,3,5] Department of Chemical Engineering and Energy Sustainability, Faculty of Engineering, Universiti Malaysia Sarawak Malaysia
[4] Department of Civil Engineering, Faculty of Engineering, Universiti Malaysia Sarawak, Malaysia

Email: aznsyuhada@unimas.my; zmzhazielim@unimas.my;

Abstract. In this study, aluminium dross waste generated from aluminium smelting industries was employed as adsorbent in removing of congo red dye in aqueous solution. The raw aluminium dross as adsorbent was characterized using Scanning Electron Microscope (SEM), Brunauer-Emmett-Teller (BET) for surface area and X-Ray Fluorescence (XRF) Spectroscopy. Adsorption experiments were carried out by batch system at different adsorbent mass, pH, and initial dye concentration. The results showed that the per cent removal of dye increased as adsorbent mass increased. It was found that 0.4 gram of adsorbent can remove approximately 100 % of dye at pH 9 for dye concentration 20 and 40 ppm. Therefore, it can be concluded that raw aluminium dross without undergone any treatment can be effectively used for the adsorption of congo red in textile wastewater related industries.

1. Introduction

Dyes have been recognised as an issue that usually contaminates soil and groundwater. There are a lot of industries that apply dyes in their products such as textiles, foods, leather, plastics, paints, cosmetics and etc. The list is endless. It is approximately 10,000 of different types of commercial dyes and pigments that available in market and it is about 700,000 tonnes of dyes are produced every year throughout the world [1]. Discharged of these compounds to the environment not just only visible to the human, it also could contribute high organic loading and ecotoxicity of water bodies. These compounds that are visible in water bodies could absorb and reflect the sunlight that can retard the photosynthesis process. Some of these dyes are very toxic, carcinogenic or mutagenic to the human [1]-[3].

In the removal of these dyes, the most challenging are anionic dyes due to the bright coloured, very reactive in water and show acidic properties. Congo red is the most common dyes that can be found in rubber, plastic, textile and paper industries. Congo red has a benzidine based anionic disazo dye which has the structural stability that makes it hard to biodegrade. There are several methods have been reported for the removal of Congo red pollution such as adsorption with activated carbon, chemical oxidation, reverse osmosis and biological treatment but among these methods, adsorption is the best

IOP Conf. Series: Materials Science and Engineering **216** (2017) 012003 doi:10.1088/1757-899X/216/1/012003

way to treat this pollution [3],[4]. This method usually being used in the removal of toxic dyes from waste water due to the high efficiency, simple and less cost compared to other treating process. Activated carbon has been extensively used in the adsorption of Congo red but due to the cost of the material itself, it become unfavourable for the water treatment [5]. Therefore, researchers had discovered several adsorbents such as fly ash and bentonite and etc. that act as an adsorbent in the removal of Congo red which is more cheaper than activated carbon [5],[6] and [7]. Hence, other possible material which has ability to adsorb the congo red from waste water is still need to be explored.

Aluminium dross is a by-product produced from aluminium smelting industries and consists of metal, salt oxides, and other non-metallic substances [8]. Globally, it was estimated 5 million tonnes of aluminium dross generated every year [9]. The consumption of aluminium dross is rising as it produced from the melting of aluminium scrap for production of can, tin, aluminium siding and other related aluminium products. However, the disposal of dross as a waste is a burden to the main manufacturer industries due to its negative side effects to the ecosystem, surface and ground water [8],[9]. Waste of aluminium dross has been applied in other application such as low quality refractory brick, catalyst, and as filler in to improve stiffness, abrasion resistance and to control micro cracking [10]-[12]. On the other hand, there are still large amount of it being dumped at the landfill without treated, this is due to its high cost and difficult to transport [13].

Therefore, in this study, aluminium dross is introduced as an adsorbent in the removal of Congo red in the aqueous solution. In the laboratory scale, it may seem do not have affected to the aluminium dross produced throughout the world, but if it applied to the industry, the usage of this waste may contribute to the management of aluminium dross disposal. In fact, aluminium dross also could be reducing since this waste can be converted into the useful product.

2. Research Methodology

2.1 Raw materials

Aluminium dross (AD) was obtained from aluminium smelting industry, Press Metal Company located in Samalaju Industrial Park, Bintulu, Sarawak, Malaysia. The properties of AD were investigated using Scanning Electron Microscope (SEM), Brunauer-Emmett-Teller (BET) for surface area (Quantachrome Instruments) and X-Ray Fluorescence (Bruker S4 EXPLORER X-Ray Fluorescence) Spectroscopy for chemical composition of AD. The surface area of dross is 10.06 m2/g.

2.2 Preparation of congo red dye solution

1g of Congo Red dye was dissolved in 1 L of distilled water to produce 1000 mg/L of stock solution. Four different concentrations 20, 40, 70 and 100 ppm were prepared by diluting the stock solution.

2.3 Adsorption Studies

The study was performed in batch process. The effect of adsorbent mass, pH and initial concentration of congo red solution were carried out to investigate the adsorption efficiency of AD. The adsorbent mass varied in the range of 0.1 to 1.0 gram, and pH was observed at pH 6, 8 and 9. The initial concentrations of congo red solution were 20, 40, 70 and 100 ppm. The mixture was stirred using orbital shaker at 200 rpm at room temperature until 180 minutes. After that, the final concentration of congo red was analysed using UV-Vis Spectrometer (Shimadzu, UV-1800) at 500 nm wavelength absorbance. The amount of congo red adsorbed by aluminium dross was calculated according to equation 1:

$$q = \frac{C_o - C_e}{m} V \quad (1)$$

q (mg/g) = amount of congo red adsorbed per unit gram of aluminium dross; C_o and C_e (mg/l) is the initial and final concentration of congo red; V (l) is the solution volume and m (g) is the dosage of aluminium dross.

3. Results and Discussion

3.1 Scanning Electron Microscopic (SEM) image of aluminium dross
Figure 1 shows the microstructure of raw aluminium dross and the dross has irregular shape, and there is no voids or cavities observed in the image indicates that, the particles are well dispersed.

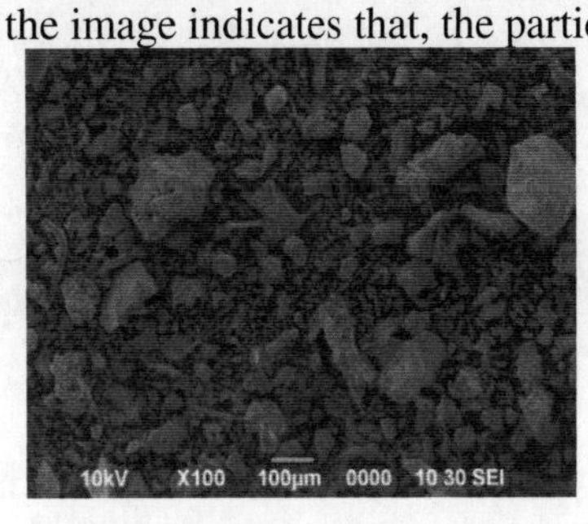

Figure 1. SEM image of Aluminium Dross

3.2 Chemical composition of aluminium dross
The chemical composition of aluminium dross has the highest content of Al_2O_3 with 89.80%, followed by SiO_2 with 1.20% and small amounts in between of 0.01 to 0.2% of Fe_2O_3, MgO, SO_3, CaO, Cr_2O_3, TiO_2 and Na_2O.

3.3 Effect of adsorbent mass, pH and concentration on the adsorption of congo red
Figure 1, 2 show the percentage of congo red removal by aluminum dross at different adsorbent mass for the dye concentrations of 20, 40, 70 and 100 ppm and for pH 6, 8 and 9, respectively. In general, the removal curves are single, smooth and continuous leading to saturation.

For the effect of adsorbent mass, all curves showed the increment in the percentage removal of dye as adsorbent mass increased. This is due to increase in adsorption surface area of the absorbent and the availability of adsorption sites [2], [14]-[15]. Besides, it was observed that 1.0 g has the highest per cent of removal, approximately 80% and above for all dye concentration for pH 6, 8 and 9. On the other hand, 0.4 g was a good enough to remove all congo red dye, approximately 100% from solution for 20 and 40 ppm of dye concentration at pH 9.

As the pH of the system increased the per cent dye removal increased. The per cent removal almost has the same values for lower adsorbent mass, masses from 0.1 to 0.4 g for 40, 70 and 100 ppm. The highest percentage of adsorption occurred at pH 9 for all different dye concentrations and at different adsorbent mass. Thus the highest efficiency of dye removal was achieved when carried out with aluminum dross at pH 9.

The removal of dye depends on the concentration of dye, Figure 2 (a,b,c), exhibited the same curve trends which is when dye concentration value increased, per cent dye removal decreased at same adsorbent mass. This can be explained by the limited number of available active sites on the surface of adsorbent to accommodate higher concentration of congo red [16].

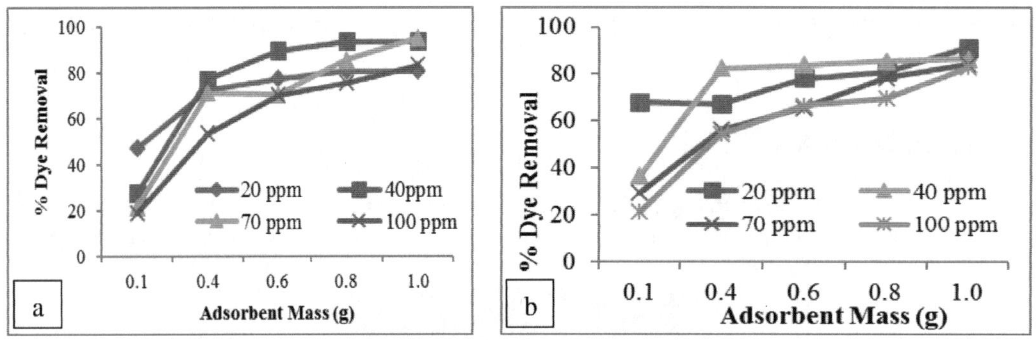

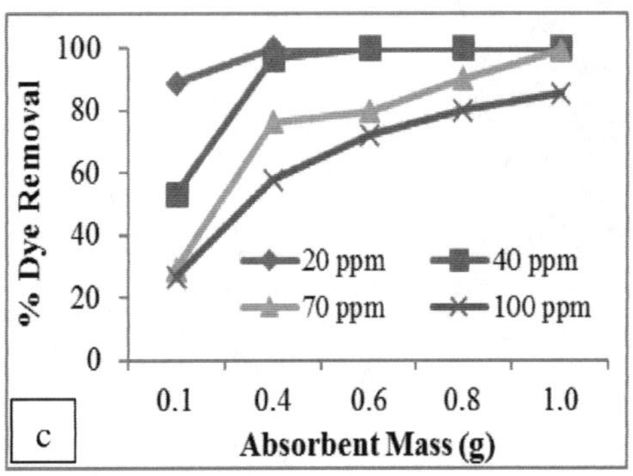

Figure 2. Per cent dye removal as function of adsorbent mass at different dye concentration for (a) pH 6, (b) pH 8 and (c) pH 9.

4. Conclusion

From the results presented, complete removal of dye can be achieved by using appropriate adsorbent mass and pH for the wastewater. Overall, 0.1 g of adsorbent mass provides significant impact on the per cent dye removal when it was carried out at low dye concentration, 20 ppm and at higher pH which is pH 9. In contrast, waste water with lower pH, at higher dye concentration, need more amount of adsorbent mass to successfully removing dye from waste water. Thus, it can be concluded that aluminium dross without having any treatment can be effectively used for the adsorption of congo red in wastewater industries. The kinetic and physicochemical of AD can be further explored for better understanding on the performance of AD as adsorbent in wastewater treatment.

5. Acknowledgement

The authors acknowledge the research grant provided by the UNIMAS under Ministry of Higher Education Fund grant (FRGS/SG06(03)/1288/2015(05).

6. References

[1] Smaranda C, Gavrilescu M and Bulgariu D 2011 Studies on sorption of congo red from aqueous solution onto soil. *Int. J. Environ. Res.* 5(1) 177–88.

[2] Raymundo A, Zanarotto R, Belisário M, Pereira M D D, Ribeiro J N and Ribeiro A V F N 2010 Evaluation of Sugar-Cane Bagasse as Bioadsorbent in the Textile Wastewater Treatment Contaminated with Carcinogenic Congo Red Dye. *Braz. arch. biol. technol.* 53 931-38.

[3] Tor A and Cengeloglu Y 2006 Removal of congo red from aqueous solution by adsorption onto acid activated red mud. *J Hazard Mater.* 138(2) 409-15.
[4] Lan S, Guo N, Liu L, Wu X, Li L and Gan S 2013 Facile preparation of hierarchical hollow structure gamma alumina and a study of its adsorption capacity. *Appl. Surf. Sci.* 283 1032–40.
[5] Acemioglu B 2004 Adsorption of Congo red from aqueous solution onto calcium-rich fly ash. *J. Colloid Interface Sci.* 274 371–9
[6] Mane V S and Babu P V V 2013 Kinetic and equilibrium studies on the removal of Congo red from aqueous solution using Eucalyptus wood (Eucalyptus globulus) saw dust. *J. Taiwan Inst Chem Eng.* 4(1) 81–8.
[7] Toor M, Jin B, Dai S and Vimonses V 2015 Activating natural bentonite as a cost-effective adsorbent for removal of Congo-red in wastewater. *J. Ind. Eng. Chem.* 21 653–61.
[8] Adeosun S O, Sekunowa O I, Taiwo O O, Ayoola W A and Mahado A 2014 Physical and mechanical properties of aluminium drosss. *J. Adv. Mater.* 3 6-10
[9] Murayama N, Maekawa I, Ushiro H, Miyoshi T, Shibata J and Valix M 2012 Synthesis of various layered double hydroxides using aluminum dross generated in aluminum recycling process. *Int. J. Miner. Process.* 110–111 46–52.
[10] Bajare D, Korjakins A and Kazjonovs J 2007 APPLICATION OF ALUMINIUM DROSS AND GLASS WASTE FOR PRODUCTION OF EXPANDED CLAY AGGREGATE, *3rd International Conference CIVIL ENGINEERING`11.* 27–31.
[11] Das B R, Dash B, Tripathy B C, Bhattacharya I N and Das S C 2007 Production of g -alumina from waste aluminium dross. *Miner. Eng.* 20 252–8.
[12] Murayama N, Okajima N, Yamaoka S, Yamamoto H, and Shibata J 2006 Hydrothermal synthesis of AlPO 4 -5 type zeolitic materials by using aluminum dross as a raw material. *J. Eur Ceram. Soc.* 26 459–62.
[13] Jian-ping H, Jun W and Hai-yan C 2015 Process of aluminum dross recycling and life cycle assessment for Al-Si alloys and brown fused alumina. *Trans. Nonferrous Met. Soc. China.* 20(11) 2155–61.
[14] Bhat A, Megeri G B, Thomas C, Bhargava H, Jeevitha C, Chandrashekar S, and Madhu G M 2015 Adsorption and optimization studies of lead from aqueous solution using g -Alumina. *Biochem. Pharmacol.* 3(1) 30–9.
[15] Namasivayam C and Kavitha D 2002 Removal of Congo Red from water by adsorption onto activated carbon prepared from coir pith, agricultural solid waste. *Dyes Pigm.* 54 47-58.
[16] Jolly P and Manas K D 2014 Efficient adsorption of congo red dye from aqueous solution using green synthesized coinage nanoparticles coated activated carbon beads. *Appl Nanosci.* 4 967-78.

IOP Conf. Series: Materials Science and Engineering **216** (2017) 012004 doi:10.1088/1757-899X/216/1/012004

Silver Recovery and Power Generation from Ammonia Chelated Silver Solution in a Bio-Electrochemical Reactor

N. A. D. Ho and S. Babel*
School of Biochemical Engineering and Technology, Sirindhorn International Institute of Technology, Thammasat University, P.O. Box 22, Pathum Thani 12121, Thailand

* Email: sandhya@siit.tu.ac.th

Abstract. Silver has valuable features and limited availability, and thus recovery from wastewater or aqueous solutions plays an important role in environmental protection and economic profits. In this study, silver recovery along with power generation and COD removal were investigated in a bio-electrochemical system (BES). The BES comprised of an anode and a cathode chamber which were separated by a cation exchange membrane to prevent the cross-over of electrolytes. During the biological oxidation of acetate as an electron donor in the anode chamber, the reduction of ammonia chelated silver ions as electron acceptors in the cathode side occurred spontaneously. Results showed that a silver recovery of 99% and COD removal efficiency of 60% were achieved at the initial silver concentration of 1,000 mg/L after 48 hours of operation. The power generation improved 4.66%, from 3,618 to 3,795 mW/m^3, by adding $NaNO_3$ of 850 mg/L to the catholyte containing 2,000 mg/L of silver ions. Deposits on the cathode surface were characterized using scanning electron microscope (SEM) and energy dispersive X-ray (EDX). Metallic silver with dendritic structures and high purity were detected. This study demonstrated that BES technology can be employed to recover silver from complex chelating solution, produce electricity, and treat wastewater.

1. Introduction

Silver ions in the form of complexes and chelating agents usually exist in the effluents of silver plating plant and photographic processes. The toxicity of silver to freshwater and marine organisms is an environmental concern, thus its release into the environment is strictly regulated. Silver also has valuable features and limited availability. Therefore, for both economic and environmental aspects, it is important to recover silver from wastewater or aqueous solutions.

Recently, the bio-electrochemical system (BES) technology has been demonstrated as an attractive approach for recovery of metals [1-3]. Basically, BES uses organic waste as the electron donor during the biological conversion in the anode chamber to supply for the reduction of metal ions as the electron acceptors in the cathode chamber. Recovery of silver from aqueous solution using the BES technology has been studied by several researchers [4-6]. However, the above studies mostly employed simple free silver ions (Ag^+) in acidic solutions (e.g. $AgNO_3$) as the model catholyte. Meanwhile, the practical waste solutions may be alkaline due to the existence of complex compositions such as chelating agents like ammonia, cyanide and EDTA [7]. Thus, the aqueous solutions containing silver in complex forms need to be investigated further.

In this study, the recovery of silver was conducted in a BES reactor using ammonia chelated silver solution as the synthetic wastewater in the cathode chamber. Acetate served as the substrate for microbial activities in the anode chamber. The BES performance was characterized by recovery efficiency, power generation, and substrate removal. The addition of sodium nitrate ($NaNO_3$) to the catholyte, which may affect the ionic conductivity of solution and power generation, was also evaluated.

2. Materials and Methods

2.1. BES reactor fabrication
A cubic-shape BES reactor consisting of an anode and a cathode chamber (1.0 L working volume) was fabricated. Two chambers were separated by a cation exchange membrane (CMI-7000S, Membrane International, Inc. USA) (figure. 1). The membrane (264 cm^2 surface area) was pre-treated by immersing in 5% NaCl solution for 12 hours to allow for the hydration and expansion. Carbon brush (15 cm x 6 cm) prepared by twisting carbon fiber (ThaiCarbonfiber Co., Thailand) with Ti wire of 1.0 mm in diameter (Prolog Titanium Co., Ltd), served as the anode, while the cathode was a graphite plate (90 cm^2, Fujian, China). Before making the electrical connection with Ti wire, both of electrodes were pre-conditioned according to previous studies [6, 8].

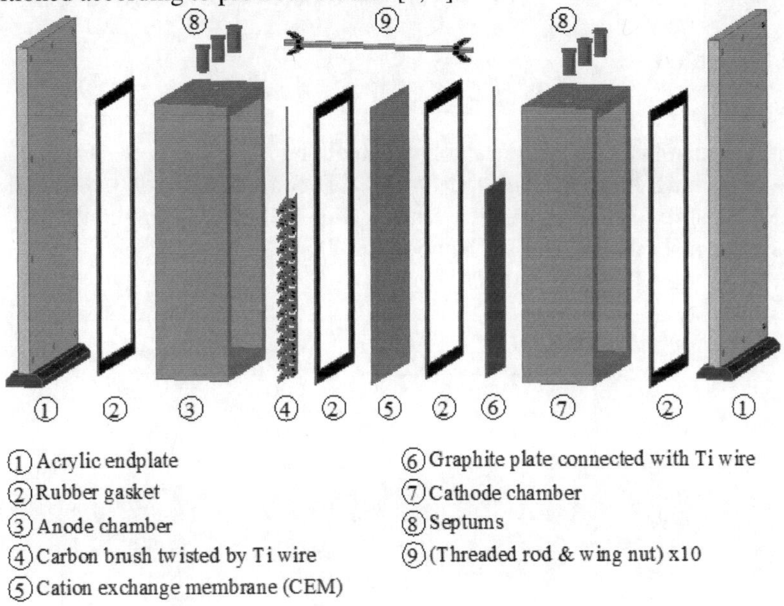

Figure 1. Schematic drawing of BES reactor.

2.2. Start-up and Operation
Anaerobic sludge (100 mL) from the digester of a brewery wastewater treatment plant (Pathum Thani, Thailand) was used as an inoculum in the anode chamber. The anolyte (900 mL) included (per liter): CH_3COONa (1.28 g) as electron donor, Na_2HPO_4 (3.55 g), KH_2PO_4 (3.4 g), NH_4Cl (0.31 g), yeast extract (0.2 g). During the inoculation stage, the cathode chamber was filled with phosphate buffer solution (1.0 L, pH 7.0) and purged continuously with air (80 mL/min). The anode and cathode were connected through an electrical circuit made by Ti wire and 1,000 Ω resistance. The system was considered to be successfully acclimated when measured cell voltage was stable after several feeding cycles.

Air was then stopped in the cathode chamber, and the catholyte was replaced by the ammonia chelated silver solution. The chelated solution was prepared by dissolving Ag_2SO_4 at different Ag^+ concentrations (1,000-2,000 mg/L) in the excess amount of 30% NH_3OH solution along with de-ionized water. Silver existed in the form of the chelated agent as $Ag(NH_3)_2^+$. The catholyte pH was around 10.2. The anolyte was refilled with fresh solution each time the catholyte was replaced. Two chambers were purged with N_2 (80 ml/min) for 15 min to remove the effect of dissolved oxygen and then sealed with septa during the operation. A 5 Ω resistance was connected to maximize the current transferred from the anode to the cathode. The effect of ionic conductivity of solution on power generation was investigated by adding a different amount of $NaNO_3$ to the catholyte containing 2,000 mg/L of silver ions.

2.3. Analytical methods and calculation

The cell voltage E (mV) was monitored through a data logger (Grant Instruments, Cambridge Co., Ltd) every 10 min and analyzed by SquirrelView software 2010 series. Power density P (mW/m^3) was calculated as $P = 1000E^2/R_{ext}V$, in which R_{ext} (Ω) is the external resistance, V (m^3) is the working volume of anode chamber. The COD measurement was based on the Closed Reflux, Titrimetric method. The silver concentration (mg/L) was measured by Inductively Coupled Plasma (ICP) Spectrometer (Optima 8000, PerkinElmer, USA). Before analysis, all aqueous samples were filtered through syringe filters NY 0.45 µm (Allpure). Deposits scraped from the graphite plate surface were characterized for morphology and chemical composition by using SEM (S-3400N, Hitachi, USA) and EDX (Metek, Apollo XP 2060, USA). All analytical procedures were performed in triplicate.

3. Results and Discussion

3.1. Silver recovery and COD removal

The experiment was conducted with the catholyte containing 1000 mg/L of silver ions as $Ag(NH_3)_2^+$ and the anolyte containing acetate as substrate with COD concentration of 1,000 mg/L. In the anode chamber, the biological oxidation of acetate occurred (Eq. 1), while the ammonia chelated silver ions was reduced spontaneously in the cathode chamber (Eq. 2).

$$2HCO_3^- + 9H^+ + 8e^- \leftrightarrow CH_3COO^- + 4H_2O, \qquad E^0 = -0.290\ V. \qquad (1)$$

$$Ag(NH_3)_2^+ + e^- \leftrightarrow Ag^0 + 2NH_3, \qquad E^0 = +0.373\ V. \qquad (2)$$

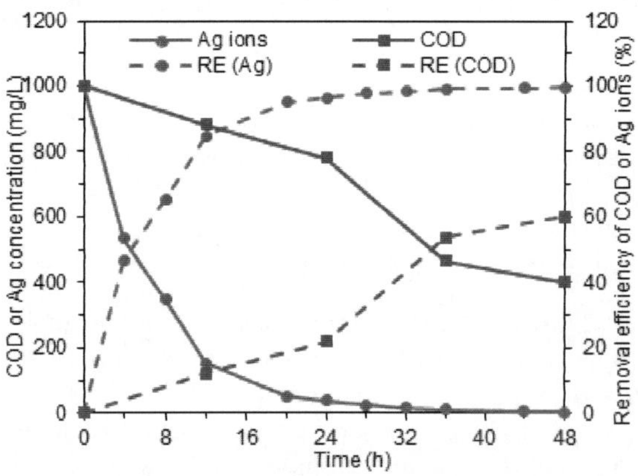

Figure 2. Chelated silver ions and COD removal by time.

During the experiment, COD and silver concentration in the anolyte and catholyte were monitored every 4 hours. Figure 2 showed that the silver concentration decreased significantly to below 5 mg/L, indicating > 99% chelated silver ions were removed from the catholyte after 48 hours of operation. Compared with single- form silver (i.e. $AgNO_3$, E^0 = 0.799 V) [4-6], the chelated silver ion had a slower reduction rate due to its lower redox potential (E^0 = 0.373 V). In this study, when silver diffusion from the cathode to anode chamber through CEM was ignored, it can be inferred that chelated silver ions were reduced to form deposits on the cathode surface.

This finding suggests that high silver recovery of 99% could be obtained in this study. In contrast, the COD concentration in the anolyte decreased slowly. After 48 hours, the remaining COD concentration was 400 mg/L, corresponding to the COD removal efficiency of 60% (figure. 2). It can be attributed to the biologically uncompleted conversion of organic matter in the anode chamber, which relied on the activated ability of microbial community. Higher COD removal can be achieved when the operation time is extended.

3.2. Effect of $NaNO_3$ on power generation

The ammonia chelated silver solution containing 2,000 mg/L of silver ions were used as a catholyte in the experiment to investigate the performance of power generation. The catholyte was then added $NaNO_3$ 425 mg/L and 850 mg/L to test its effect on the power generation and silver recovery. In order to characterize the fuel cell of reactor, the polarization curve analysis was also conducted when the external resistance was varied in a range of 10,000 - 5Ω.

Figure 3 shows that the cell voltage developed rapidly during the first 8 hours and decreased gradually thereafter due to the shortage of substrate in the anode chamber and the reduction of silver ions in the cathode chamber. The highest cell voltage was found when $NaNO_3$ of 850 mg/L was added to the catholyte containing 2000 mg/L of silver. It can be ascribed that the addition of $NaNO_3$ could decrease the internal resistance of reactor by improving the ionic conductivity of the catholyte. It was confirmed by the polarization and power curves (figure 4). The maximum power density (P_{max}) increased by 4.66%, from 3,618 to 3,795 mW/m^3, with the addition of $NaNO_3$ of 850 mg/l to the catholyte.

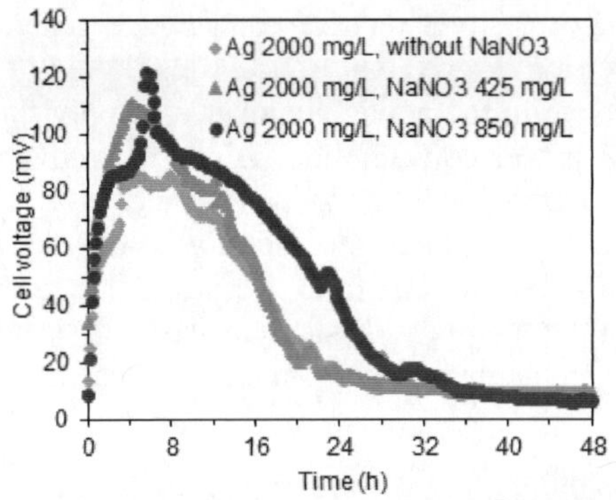

Figure 3. Cell voltage progression in reactor by time with different catholyte

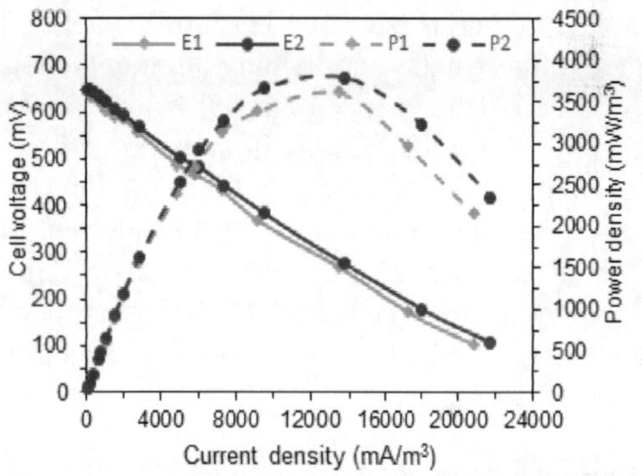

Figure 4. Polarization curves (E1 and E2), power curves (P1 and P2) without and with an addition of $NaNO_3$ 850 mg/L to the catholyte, respectively.

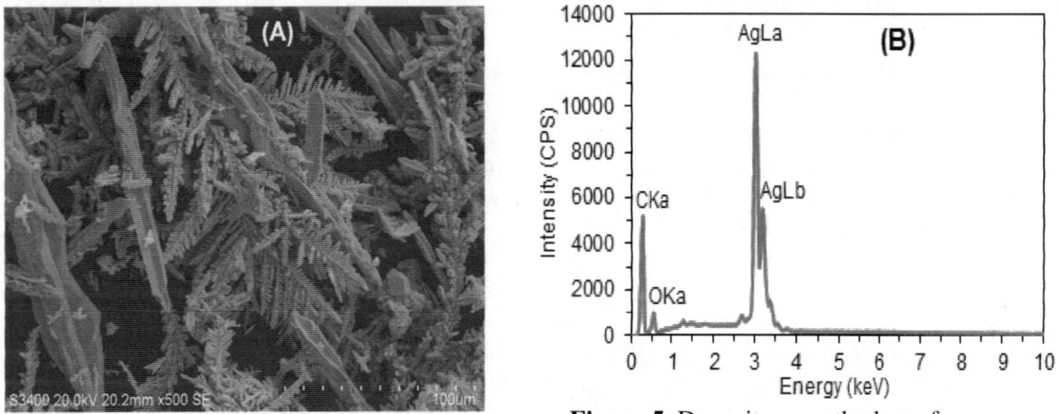

Figure 5. Deposits on cathode surface were characterized by (A) SEM and (B) EDX

During experiments, the reduction of silver ions in catholyte was not affected by the addition of $NaNO_3$. Results showed that the silver removal was over 99% in all of the cases. Similar results were found by other researchers when KNO_3 was used to improve the ionic conductivity of solution for recovery of copper (Cu) and electricity production using a microbial fuel cell (MFC) [9].

3.3. Characterization of deposits on the cathode surface

After 58 days of operation, the BES reactor was emptied. The deposits on graphite cathode surface were scrapped and air dried at room temperature before characterizing by SEM and EDX analysis. Silver with dendritic structures was detected at a magnification of 500 SE (figure 5A). These dendrites have long trunks and short branches which look like a fishbone. This result was also verified by the EDX spectrum (figure 5B), in which the sharp peak at an energy level of 2.98 keV was found, indicating the deposits comprised of pure silver. It was reported that dendritic structures are often produced in electrochemical deposition, especially in fine powder deposition [10]. Other researchers also showed that silver deposits in compact layer structure were found at cathode surface of a MFC used to recover silver from ammonia chelated solution [7].

4. Conclusions

This study demonstrated the feasibility of BES technology for silver recovery and power generation from alkaline solution containing ammonia chelated silver ions $Ag(NH_3)_2^+$ of 1,000-2,000 mg/L. The BES based on the energy obtained from oxidation of acetate in the anode chamber to supply for the reduction of $Ag(NH_3)_2^+$ ions in cathode without external power consumption. High Ag recovery (> 99%) were achieved after 48 hours of operation. The addition of $NaNO_3$ to the catholyte could improve the ionic conductivity of the solution, thus increased the power generation with fewer effects on the silver removal. Pure metallic silver deposits with dendritic structures were found at cathode surface by using SEM and EDX technique. Other silver complexes such as silver thiosulfate $[Ag(S_2O_3)_2]^{3-}$ which normally exists in effluent of photographic processing, should be investigated in further studies.

5. Acknowledgement

The authors would like to acknowledge the financial grant of an Excellent Foreign Student (EFS) doctoral scholarship from Sirindhorn International Institute of Technology (SIIT), Thammasat University (Thailand) to the first author. Technical support by the National Metal and Materials Technology Center, Thailand Science Park (Thailand) are also acknowledged

6. References

[1] Mathuriya A S and Yakhmi J 2014, Microbial fuel cells to recover heavy metals, *Environ. Chem. Lett.* **12**(4) 483-94.

[2] Nancharaiah Y, Mohan S V and Lens P 2015, Metals removal and recovery in bioelectrochemical systems: A review, *Bioresour. Technol.* **195** 102-14.

[3] Wang H and Ren Z J 2014, Bioelectrochemical metal recovery from wastewater: a review, *Water Resour.* **66** 219-32.

[4] Choi C and Cui Y 2012, Recovery of silver from wastewater coupled with power generation using a microbial fuel cell, *Bioresour. Technol.* **107** 522-25.

[5] Lim B, Lu H, Choi C and Liu Z 2015, Recovery of silver metal and electric power generation using a microbial fuel cell, *Desalin. Water Treat.* **54**(13) 3675-81.

[6] Tao H-C, Gao Z-Y, Ding H, Xu N and Wu W-M 2012, Recovery of silver from silver (I)-containing solutions in bioelectrochemical reactors, *Bioresour. Technol.* **111** 92-97.

[7] Wang Y-H, Wang B-S, Pan B, Chen Q-Y and Yan W 2013, Electricity production from a bio-electrochemical cell for silver recovery in alkaline media, *Appli. Energ.* **112** 1337-41.

[8] Feng Y, Yang Q, Wang X and Logan B E 2010, Treatment of carbon fiber brush anodes for improving power generation in air–cathode microbial fuel cells, *J. Power Source.* **195**(7) 1841-44.
[9] Wang Z, Lim B-S, Lu H, Fan J and Choi C-S 2010, Cathodic reduction of Cu^{2+} and electric power generation using a microbial fuel cell, *Bull. Korean Chem. Soc.* **31**(7) 2025-30.
[10] Liu W, Yang T, Li C, Che P and Han Y 2015, Regulating silver morphology via electrochemical reaction, *CrystEngComm.* **17**(31) 6014-22.

IOP Conf. Series: Materials Science and Engineering **216** (2017) 012005 doi:10.1088/1757-899X/216/1/012005

Apiaceae Family Plants as Low-Cost Adsorbents for the Removal of Lead Ion from Water Environment

W. Boontham and S. Babel*

School of Biochemical Engineering and Technology, Sirindhorn International Institute of Technology, Thammasat University, P.O. Box 22, Pathum Thani 12121, Thailand

* Email: sandhya@siit.tu.ac.th

Abstract. Adsorbents prepared from the three selected plants from Apiaceae famaily commonly known as parsley, coriander and culantro were observed to remove lead from aqueous solutions. Batch experiments were conducted to study the effect of dosage, pH, contact time and agitation speed at 10 mg L^{-1} initial Pb(II) concentration. Results revealed that three selected plants showed high adsorption capacity for removal of lead from aqueous solutions. The maximum biosorption of Pb^{2+} was found to be more than 97% with 1.0 g/l dosage for all three adsorbents under optimum pH of 3-5. The adsorption equilibrium was established after about 1 hr. The equilibrium adsorption capacity of parsley and coriander were found to fit well with the Langmuir isotherm whereas the Freundlich isotherm was better fit for culantro. The studies showed that the adsorbents can be used for removing lead ions from contaminated waters.

1. Introduction

Heavy metal pollution is an environmental problem of worldwide concern, among which lead (Pb) is one of the most common pollutants found in industrial effluents and is significantly toxic to human beings and ecological environments even at low concentrations [1]. It has the ability to accumulate in living organisms. Assimilation in the human body of relatively small amounts of Pb over a long period of time can lead to malfunctioning of certain organs and chronic toxicity.

According to World Health Organization (WHO) [2], the permissible limit of lead concentration in drinking water is 0.01 mg L^{-1} and according to Thailand Pollution Control Department (PCD) [3] is not more than 0.05 mg dm-3. Some of the methods for removal of heavy metals include chemical precipitation, ion exchange, electrodialysis, reverse osmosis and carbon adsorption. But these methods are often ineffective and/or very expensive for low concentrations of metals.

Since the last decade, biosorption or sorption of contaminants by sorbents of natural origin has gained important credibility due to its good performance and low cost of these complexing materials. Due to high uptake capacity and very cost effective source of raw materials, biosorption is a progression towards a perspective method [4].

In the present study, parsley (Petroselinum crispum), coriander (Coriandrum sativum), and culantro (Eryngium foetidum) plants from Apiaceae family were employed to investigate the biosorption of lead(II) from aqueous solution. Effect of different parameters such as dosage, pH, contact time and agitation speed on the removal process are studied. The experimental data obtained have been evaluated and fitted using adsorption isotherms.

2. Materials and Methods

2.1. Sample preparation

Coriander, culantro and parsley were bought from the local vegetable market Taladthai, Pathum Thani, Thailand. The plants were washed several times and dried vegetable under the sunlight. The dried materials were cut down, crushed and washed thoroughly with deionized water to remove any adhering dirt. Next, they were dried in an oven at 90 °C overnight. The dried plants were grounded and sieved through 250 mesh sieve size. The powdered plants were stored in an air tight container for further applications. Pb^{2+} ion solution was prepared by dissolving $Pb(NO_3)_2$ in deionized water. The pH adjustments were done by adding 0.1 M NaOH and/or 0.1 M HNO_3 solutions.

2.2. Adsorption studies

All the experiments were carried out in Erlenmeyer flasks with 50 ml test solution at room temperature (25±2 °C). The effect of adsorbent dose (0.1, 0.5, 1, 2, 4 gm L^{-1}), pH (2-10), contact time (10, 20, 30, 40, 50, 60, 120, 180, 360 min) and agitation speed (100, 150, 200, 250, 300 rpm) were investigated at 10 mg L^{-1} initial Pb(II) concentration. At the end of desired contact time, the samples were filtered by using syringe filters. Filtrates were then analyzed for residual lead concentration by using an Inductively Coupled Plasma (ICP) Spectrometer (Optima 8000, PerkinElmer, USA).

2.3. Data analysis

The adsorption at equilibrium, q_e (mg g^{-1}), was calculated as in equation (1);

$$q_e = \frac{Co - Ce}{m} xV \quad (1)$$

Where, C_o and C_e are the initial and the final (equilibrium) Pb(II) concentration (mg L^{-1}) respectively; V is the adsorbate volume (L); and m is the mass of adsorbent used (g).
The percentage of lead ion removal was calculated as in equation (2);

$$\text{Removal (\%)} = \frac{Co - Ct}{Co} x100 \quad (2)$$

Where, C_0 and C_t are the initial and the final lead concentration (mg L^{-1}) at time t, respectively.

3. Results and Discussion

3.1. Effect of variable parameters

3.1.1. Effect of dosage.

The effect of adsorbent concentration on the lead removal efficiency is presented in figure 1. Parsley, coriander and culantro were found to have maximum removal efficiency of 96 %, 94 % and 97 % under pH 5, respectively, for 2.0 g/l dose at 10 mg L^{-1} initial Pb(II) concentration. This is because the number of adsorption sites or surface area increases with the weight of adsorbent and hence results in a higher percent of lead removal at a high dose. However, the percent removal of three absorbents slightly decrease for 4.0 g/l dose. This is due to the fact that at higher adsorbent dose, the solution ion concentration drops to a lower value and the system reaches equilibrium [5].

3.1.2. Effect of pH.

From figure 2, it could be seen that for all the three adsorbents, lead adsorption decreased along with the increase in pH. Maximum adsorption of Pb^{2+} in parsley, coriander and culantro were attained at pH 3 with 97 % removal. The increase in lead removal at low pH of 2 and 3 can be explained on the basis of a decrease in competition between proton and the lead cations for the same functional groups and by the decrease in positive charge at pH of 4 afterward the adsorbents which result in a lower electrostatic repulsion between the lead cations and the surface [6].

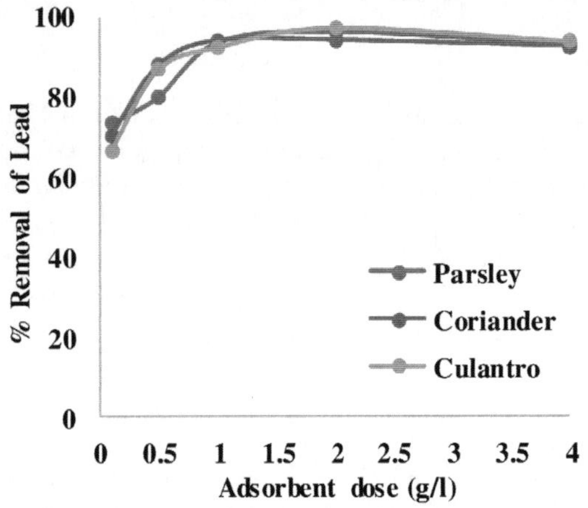

Figure 1. Effect of adsorbent dose on the sorption of Pb^{2+} (contact time = 2h at 170 rpm).

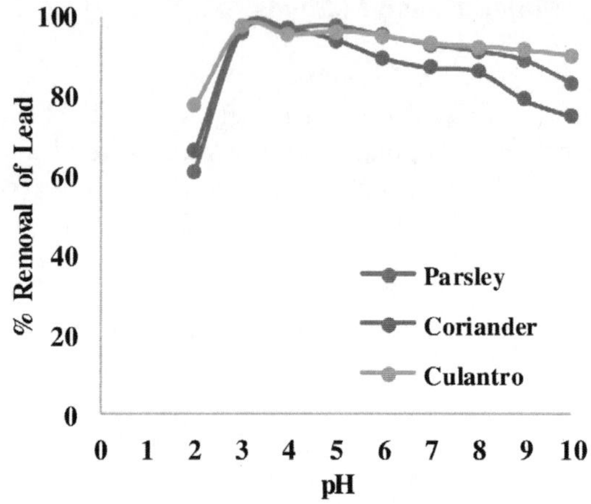

Figure 2. Effect of adsorbent dose on the sorption of Pb^{2+} (contact time = 2h at 170 rpm).

3.1.3. Effect of contact time.
Figure 3 shows a slowly increased adsorption rate of lead at the beginning until 60 min of contact time, thereafter, the adsorption rate became practically constant. The variation in the extent of adsorption may be due to the fact that initially all sites on the adsorbent surface were vacant and the solute concentration gradient was relatively high [6].

3.1.4. Effect of agitation speed.
The effect of agitation speed on lead removal by the given adsorbents is presented in figure 4. The speed of the reactor was increased, lead removal efficiency increased. It was observed that for parsley, coriander and culantro, maximum removal of 97%, 95% and 94%, respectively, were observed at 250 rpm. The removal was almost constant thereafter. However, the effective removal of Pb^{2+} ions at 150 rpm were not much different from at 250 rpm.

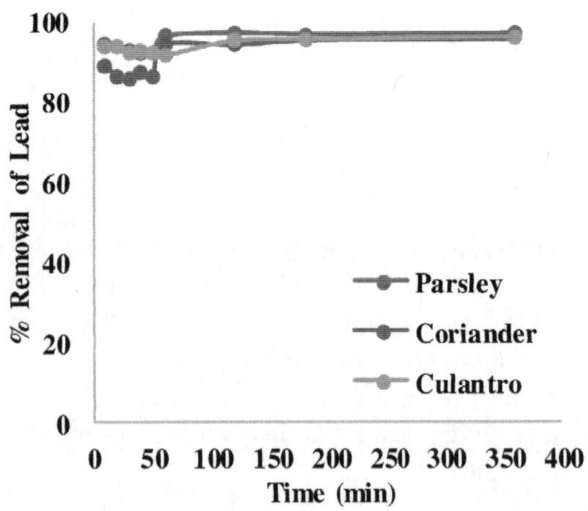

Figure 3. Effect of contact time on the sorption of Pb^{2+} (pH 5.0, at 170 rpm).

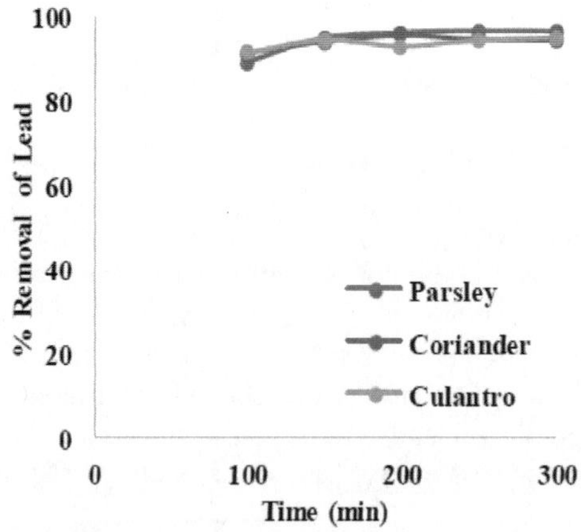

Figure 4. Effect of agitation speed on on the sorption of Pb^{2+} (pH 5.0, contact time = 2h).

Table 1 shows comparison of the adsorption efficiency of selected adsorbents with other plant materials. The removal capabilities for lead of low cost adsorbents depend on its initial concentration along with other factors as shown in table 1. Results from this study showed Pb(II) removal of more than 97% for all three species and are comparable to other studies.

Table 1. Comparison of the maximum adsorption capacity of lead (II) ion by various adsorbents from plant origin.

Adsorbent	pH	Initial concentration	Dosages	Contact time	Pb^{2+} Removal %	Reference
Soursop seeds	5	2.5 mg L^{-1}	1 gm	120 min	40.6	[2]
Banana Stalk	8	20 mg L^{-1}	0.90 gm	152 min	96.41	
Coffee grounds	Neutral	0.5 mg L^{-1}	0.5 mg L^{-1}	1 hr	87.2	
Rice husk and Maize cobs	2.5-6.5	25 mg L^{-1}	1.5 gm	120 min	98.5	
Parsley	5	10 mg L^{-1}	1 gm L^{-1}	2 hr	97.88	This study
Coriander	3	10 mg L^{-1}	1 gm L^{-1}	2 hr	97.62	
Culantro	3	10 mg L^{-1}	1 gm L^{-1}	2 hr	97.58	

3.2. Adsorption Isotherms

3.2.1. Langmuir Adsorption Isotherm.
Langmuir isotherm is based on the monolayer sorption of Pb(II) on the surface of carbon sites and is represented linearly by the equation (3);

$$(C_e/q_e) = (1/Q_0 b) + (C_e/Q_0) \quad (3)$$

A plot of C_e/q_e against Ce (figure 5) gave a straight line graph with a slope $1/Q_0$ and intercept of 1/bQ0. Values of Q0 and b are calculated from the graph and reported in table 2. Q_o and b are Langmuir constants related to adsorption capacity and energy of adsorption, respectively.

3.2.2. Freundlich Adsorption Isotherm.
Freundlich isotherm describes the heterogeneous surface energies by multilayer sorption and is expressed by the equation (4):

$$\log q_e = \log K_F + (1/n) \log C_e \quad (4)$$

where, K_F and n are the Freundlich constants incorporating the factors affecting the adsorption capacity and the degree of non-linearity between the solute concentration in the solution and the amount adsorbed at equilibrium respectively. Plots of log q_e against log C_e (figure 6) gave linear graphs with high R^2. Value of n > 1 indicates that the adsorption is favourable.

Table 2. Calculated Langmuir and Freundlich isotherm parameters.

Adsorbents	Langmuir constants			Freundlich constants		
	b	Q_0	R^2	N	K_F	R^2
Parsley	1.022	45.657	0.996	2.462	19.984	0.859
Coriander	0.392	53.550	0.993	1.875	14.601	0.957
Culantro	0.411	56.198	0.955	1.882	15.906	0.971

Comparing the R^2 values of the two isotherms (table 2), the adsorption of parsley and coriander fitted well with the Langmuir model suggesting that the adsorption of Pb(II) onto these adsorbents is a mono-layer process and the adsorption behaviour is homogeneous rather than heterogeneous whereas culantro fitted well with Freundlich model since R^2 from Langmuir model was not close to unity, it could be conceivable that there might be some heterogeneity in the system [7].

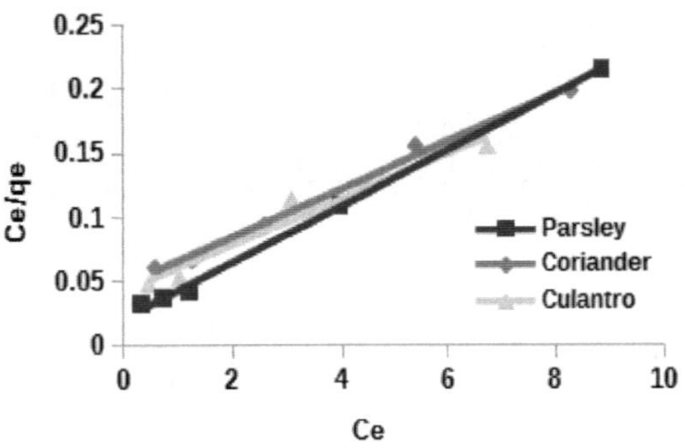

Figure 5. Langmuir isotherm of lead on adsorbents.

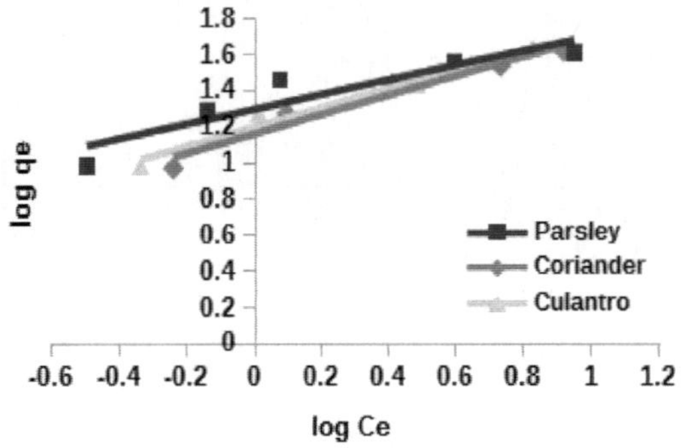

Figure 6. Freundlich isotherm of lead on adsorbents.

4. Conclusions

The present investigation shows that parsley, coriander and culantro can be employed as a potential low-cost adsorbents for the removal of Pb^{2+} ions from contaminated water because it can prepare without chemical and physical treatments. The Pb^{2+} adsorption is found to be greatly dependent on the adsorbent mass and initial pH of the solution. Optimum lead removal was observed at 1.0 g/l dosage, pH value between 3-5, 150 rpm of agitation speed with 1 hr contact time. The equilibrium data were well described for parsley and coriander by the Langmuir isotherm model whereas culantro fitted well to Freundlich isotherm model. These results show that adsorbents which have a very low economical value may be used effectively for removal of Pb^{2+} ions from aqueous systems for environmental protection purpose.

5. Acknowledgement

The authors would like to acknowledge the financial grant of TAIST-Tokyo Tech master degree scholarship from Sirindhorn International Institute of Technology (SIIT), Thammasat University, Thailand to the first author.

6. References

[1] Yoshita A, Lu J L, Ye J H and Liang Y R 2009, Sorption of lead from aqueous solutions by spent tea leaf, African Journal of Biotechnology, *Afr. J. Biotechnol.* **8**(10) 2212-17.

[2] Bachale S, Sharma S, Sharma A and Verma S 2016, Removal of lead(II) from aqueous solution using low cost adsorbent: A review, *Int. J. Appl. Res.* **2**(10) 523-27.

[3] Information on http://www.pcd.go.th/info_serv/en_reg_std_water01.html

[4] Jayaram K and Prasad M N V 2009, Removal of Pb(II) from aqueous solution by seed powder of *Prosopis juliflora* DC., *J. Hazard. Mater.* **169** 991–97.

[5] Amarasinghe B M W P K and Williams R A 2007, Tea waste as a low cost adsorbent for the removal of Cu and Pb from wastewater, *Chem. Eng. J.* **132** 299-309.

[6] Azouaou N, Belmedani M, Mokaddem H and Sadaoui Z 2013, Adsorption of Lead from Aqueous Solution onto Untreated Orange Barks, *Chem. Eng. Trans.* **32** 55-60.

[7] Khalid A, Kazmi M A, Habib M, Jabeen S and Shahzad K 2015, Kinetic & Equilibrium Modelling of Copper Biosorption, *J. Fac. Eng. Technol.* **22**(1).

IOP Conf. Series: Materials Science and Engineering **216** (2017) 012006 doi:10.1088/1757-899X/216/1/012006

Characterization of High-Energy Ball-Milled and Hot-Pressed Nanocrystalline Tantalum

J Jakubowicz[1,2], G Adamek[1] and M Sopata[1]
[1]Poznan University of Technology, Institute of Material Science and Engineering
Jana Pawla II, 24, 61-138 Poznan, Poland

[2]E-mail: jaroslaw.jakubowicz@put.poznan.pl;

Abstract. The paper shows a comparison of nanocrystalline and microcrystalline tantalum sinters. The nanocrystalline tantalum was made using high-energy ball milling. The sinters were made using hot pressing with high frequency induction heating. The structure and microstructure of the powders and sinters were investigated. After 48 hour milling, the microcrystalline tantalum transformed into nanocrystalline powder. The hot pressing resulted in a formation of bulk tantalum with ultrafine grains and hardness as high as 1067 HV. The nanostructure supports the densification process at lower sintering temperature in comparison to microcrystalline tantalum. The average crystallite size in nanocrystalline bulk materials reached 170 nm.

1. Introduction

Tantalum is a very promising material for different technical applications [1, 2]. Tantalum is a refractory material used as base for alloys used in air and space, chemical, energy and biomedical applications. However, the main downside of tantalum is its high cost and relatively hard processing because of its high melting point. Tantalum can be composed with other alloying elements such as Ti [3] as well as other refractory elements [4-6] to form alloys of very high resistance to temperature and corrosion. The Ta is relatively plastic but its combination with other alloying elements results in strengthening through solid solutioning, precipitating or dispersing [4]. The formation of nanostructure results in significant changes in the properties of materials [7, 8]. A reduction of grain size results in an increase in the grain boundaries. These microstructural elements cause an increase in the material strength following the grain refinement. Nanocrystalline powders can be successfully used for preparation of bulk materials [9]. The powder metallurgy processes are typical processes leading to their densification [10]. The nanocrystalline powders should be consolidated at lighter pressing and sintering conditions, i.e. temperature, pressure and time in comparison to microcrystalline counterparts. Thus, conventional powder metallurgy followed by pressureless sintering is not appropriate due to high temperature, long time and no pressure during sintering. The hot pressing, on the other hand, can minimize grain growth, thus preserving the nanostructure after the process, providing high mechanical properties.

In this work the formation of bulk nanocrystalline tantalum sinters was studied and compared to the microcrystalline ones. The authors used hot pressing in order to reduce the grain growth during consolidation.

2. Experimental details

In this work, nanocrystalline Ta sinters were made starting from high-energy ball milling followed by hot pressing. The tantalum powder of the mesh size of 325 (<44 μm) and the purity of 99.97% was purchased at Alfa Aesar (Germany). During all the processing stages, the powder was kept in high purity inert atmosphere (argon). The powder handling was conducted in a Labmaster 130 glove box (MBraun, Germany). The milling procedure was done in a hardened steel vial mounted inside the SPEX 8000M shaker Mixer/Mill (Spex SamplePrep, USA). The mill was run for 48h at a room temperature with the frequency of 875 cycles/minute. 5g of Ta powder and 7 hardened steel balls were placed inside the vial. The ball to powder weight ratio was 3. The obtained powders were then axially hot-pressed at 10^{-2} Pa vacuum using graphite die and tungsten stamps. The pressing temperature was established experimentally and the process was done at 1500°C and 1000°C for microcrystalline and nanocrystalline powders, respectively. The pressure was 50 MPa and the sintering time at a constant temperature was 5s after 120s and 100s temperature rising to 1500°C and 1000°C, respectively. The BN spray was applied to prevent the powder material sticking to the die and ensure easier stamp moving. The structure analysis was controlled by Empyrean XRD (Panalytical, The Netherlands). The investigation was done using CuKα radiation at the voltage of 45 mV and the anode current of 40 mA. The angular 2θ range was 30-120 deg., the step size was 0.0167 deg. and the time per step was 12.54 s/step. The crystallite size was estimated using the Scherrer method. The microstructure was analyzed using Vega 5135 SEM (Tescan, Czech Republic). Additionally, Q-Scope 250 AFM (Quesant, USA) working in tapping mode was applied for the microstructure and grain size analysis. The surface was scanned by a SSS-NCLR Nanosensors probe. Based on the AFM images, a particle size dimension distribution histogram was developed. For microscopic observations, the bulk samples were grinded on a sandpaper, polished in a diamond suspension and finally etched in a H_2SO_4 + HNO_3 +HF mixture to reveal the grain boundaries. Vickers hardness HV3.0 was measured using a Nexus hardness tester (Innovatest).

3. Results and discussion

Microcrystalline tantalum (figure 1a) was intensely milled for 48h (figure 1b). The high-energy ball-milling resulted in a reduction of the crystallite size to below 20 nm, nanostructure formation and significant material amorphization (figure 1b). Intense milling resulted in a strong deformation and stress introduction resulting in an increase of the lattice constant and the volume of the unit cell (figure 2). During HEBM, the initial Ta particles (figure 3a) undergo not only a reduction of the crystallite size, but also cold welding and plastic deformation, which can result in a formation of relatively large particles (figure 3b). These particles, however, are composed of nanocrystals (figure 4a). The strain hardening is the main factor leading to the size reduction of final powders and the crystallites. The particles after HEBM stuck together and formed larger agglomerates (figure 3b).

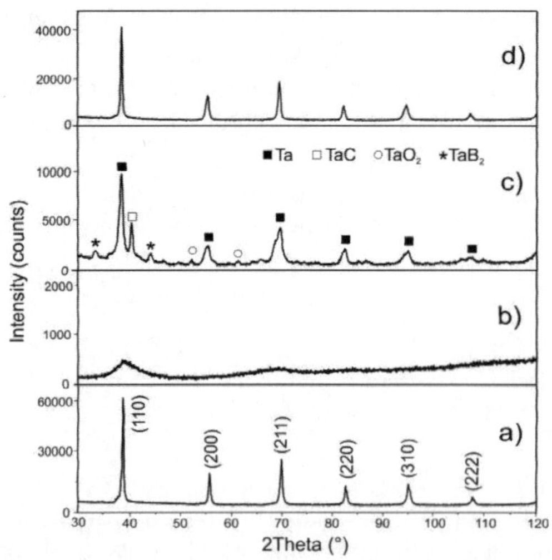

Figure 1. XRD spectra of tantalum: microcrystalline 325 mesh powder (a), nanocrystalline powder obtained after 48h of HEBM (b), nanocrystalline sinter (c) and microcrystalline sinter (d)

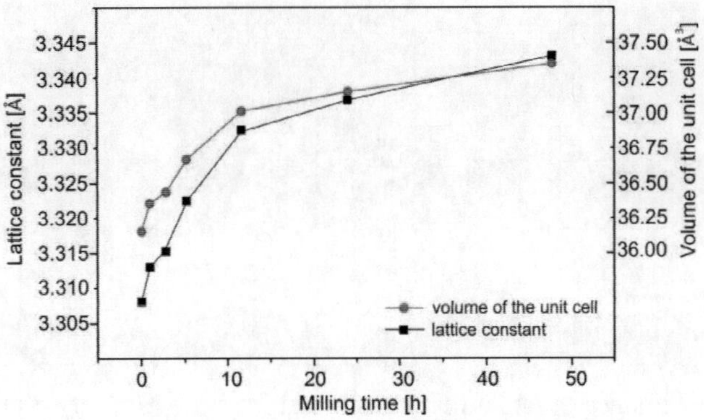

Figure 2. Lattice constant and the volume of the unit cell changes during HEBM

The hot-pressing process was done in a very short time and at a relatively low temperature. For nanocrystalline tantalum we have found that the temperature of 1500°C is too high in contrast to microcrystalline tantalum, which is why, for nanocrystalline material, the pressing was done at 1000°C. Due to the high reactivity of the nanocrystalline powder, the material reacts with graphite die and boron nitride covering the die and the stamp. As a result, small amounts of additional hard precipitations (TaC, TaB_2) appear in the structure of the compacts as well as TaO_2 due to high reactivity with oxygen (figure 1c). For comparison, the parent, much bigger microcrystalline powder does react with the above-mentioned die materials (figure 1d). The hot-pressing process results in a limited grain growth in comparison to conventional powder metallurgy. Low temperature and short time at a given temperature lead to inhibiting the grain growth of the compact (figure 4b). The compacting leads to the obtainment of uniform grains in the case of nanocrystalline material of a spherical shape (figure 4b), while in the case of microcrystalline materials it is typical polyhedral grains (figure 4c). In both cases, for nanocrystalline and microcrystalline materials, an over 95% density was obtained without significant pores (figure 3c,d and figure 4b,c). Small pits observed in figure 3c,d are a result of the polished sample etching in order to reveal the grain boundaries.

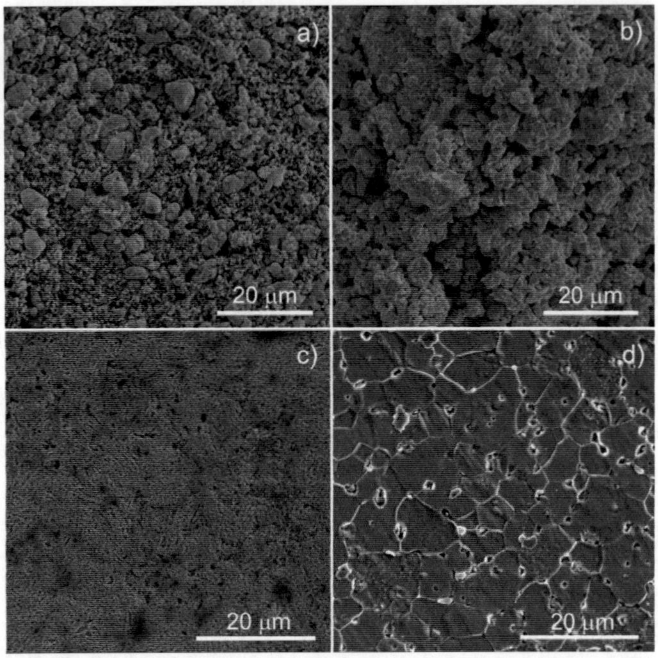

Figure 3. SEM images of tantalum: microcrystalline powder (a), nanocrystalline powder after 48h of HEBM (b), nanocrystalline sinter (c) and microcrystalline sinter (d)

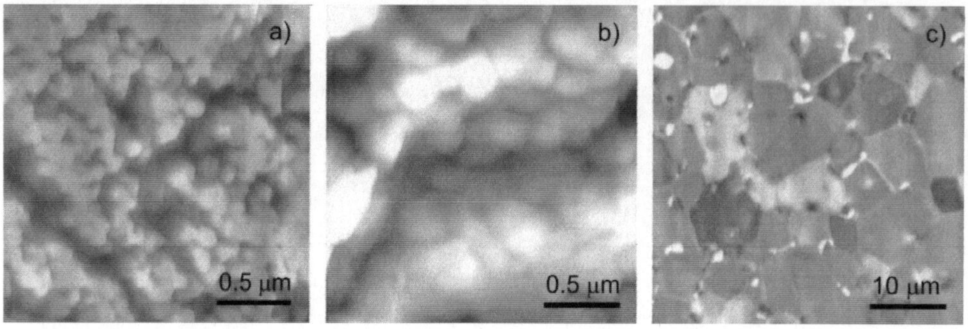

Figure 4. AFM images of tantalum: nanocrystalline powder after 48h of HEBM (a), nanocrystalline sinter (b) and microcrystalline sinter (c)

The grain size distribution was calculated taking into account the AFM data. The average grain size after 48h of HEBM reaches 70 nm (Figure 5a), while after sintering, it increases up to 170 nm (figure 5b). For comparison, when the microcrystalline particles were sintered, the average grain size reached 4.3 μm (figure 5c). In all cases, the grains have a wide range of distribution.

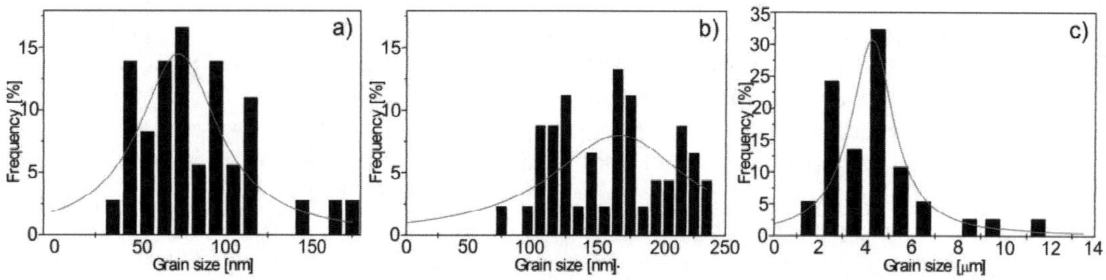

Figure 5. Grain size distribution of the: nanocrystalline powder after 48h of HEBM (a), nanocrystalline sinter (b) and microcrystalline sinter (c)

In the case of both, nanocrystalline and microcrystalline hot-pressed materials hardness was measured using the Vickers tester. The microcrystalline material had a hardness of 357 ± 6.6 HV and the nanocrystalline material due to small grains and hard precipitations reached 1067 ± 78 HV. For comparison, the hardness was also measured on commercial bulk sintered tantalum, which reached the value of 101 ± 2.2 HV.

4. Conclusions

This work discussed the formation of hot-pressed nanocrystalline tantalum in comparison to the microcrystalline ones. HEBM results in a reduction of the crystallite size up to the average value of approx. 70 nm. The hot pressing of the nanocrystalline tantalum results in a limited grain growth and an average crystal size of approx. 170 nm. For comparison, microcrystalline powders were compacted, giving standard microcrystalline sinters with crystal size of approx. 4.3 μm. The nanocrystalline material showed additional hard precipitations, that derived from the graphite die and lubricant, forming during a reaction with the nanocrystalline material. The hardness of the ultrafine tantalum is very high and reaches 1067 HV.

This work constitutes preliminary studies for further research of nanocrystalline tantalum alloys and composites prepared by mechanical alloying followed by densification process.

5. Acknowledgements
The work has been financed by National Science Centre, Poland under project identification DEC-2015/19/B/ST5/02595.

6. References
[1] Buckman R W 2000 *JOM* **52** 40
[2] Silva R A, Walls M, Rondot B, Cunha Belo M and Guidoin R 2002 *J. Mater. Sci. Mat. Med.* **13** 495
[3] Souza K A and Robin A 2007 *Mat. Chem. Phys.* **103** 351
[4] Cardonne S M, Kumar P, Michaluk C A and Schwartz H D 1995 *Int. J. Refract. Met. Hard Mat.* **13** 187
[5] Robin A and Rosa J L 2000 *Int. J. Refract. Met. Hard Mat.* **18** 13
[6] Schuster B E, Ligda J P, Pan Z L and Wei Q 2011 *JOM* **63** 27
[7] Tang Y, Bringa E M and Meyers M A 2013 *Mat. Sci. Eng.* A **580** 414
[8] Ligda J, Scotto D'antuono D, Taheri M L, Schuster B E and Wei Q 2016 *JOM* **68** 2832
[9] Maglia F, Tredici I G and Anselmi-Tamburini U 2013 *J. Euro. Ceramic Soc.* **33** 1045
[10] Morales R, Aune R E, Seetharaman S and Grinder O 2003 *JOM* **55** 20

IOP Conf. Series: Materials Science and Engineering **216** (2017) 012007 doi:10.1088/1757-899X/216/1/012007

Improving the Performance of Electrically Activated NiTi Shape Memory Actuators by Pre-Aging

Christian Rathmann1, Benjamin Fleczok1, Dennis Otibar1, Bernd Kuhlenkötter
Ruhr-University of Bochum, Universitätsstraße 150, 44801 Bochum, Germany

Email: rathmann@lps.rub.de; fleczok@lps.rub.de; otibar@lps.rub.de; kuhlenkoetter@lps.rub.de;

Abstract. Shape memory alloys possess an array of unique functional properties which are influenced by a complex interaction of different factors. Due to thermal sensitivity, slight changes in temperature may cause the properties to change significantly. This poses a huge challenge especially for the use of shape memory alloys as actuators. The displacement is the key performance indicator, which has to be of equal or better quality compared to conventional actuators. One problem of shape memory alloys is the change in functional fatigue in the first cycles, which makes it rather difficult to design the actuator. Therefore, the reduction of this shakedown effect is crucial. For this reason, this paper investigates the effect of electrical heat treatment as a method for pre-aging. This topic has so far been little investigated so that the investigations focus on identifying important factors and effects by using the design of experiments.

1. Introduction

Shape memory alloys (SMA) have a great reversible elongation with a high specific working capacity but also limited dynamics. [1] They have the ability to regain their original shape after a seemingly plastic deformation by heating above a certain temperature. The shape-memory effect was first discovered by [2] in 1932 on cadmium-gold alloys and first described by Vernon in 1941. [3] The most commercially important alloys currently available are shape memory alloys based on NiTi (also Nitinol) discovered by Buehler and Wang in 1963 at the Naval Ordnance Laboratory. [3]
The great application potential of shape memory alloys, apart from the medical technology, is still not fully used and SME actuator applications are almost exclusively produced in small and medium series. [4] [5] The main reasons for this are technical aspects such as the limited temperature range of the phase transformation, [6] limited dynamics [3] and challenging SMA manufacturing processes [7] as well as the lack of development guidelines. One reason for their limited use in the context of product design is their behavior under thermal cycling. SMAs show a decrease in the displacement in the first cycles (also known as shakedown) through irreversible plastic lattice deformation and martensite stabilization, [8] which makes the actuator design difficult. Therefore, this paper tries to investigate and meet this challenge.
The paper is organized as follows: The first part provides background information of shape memory alloys and their application fields as well as it describes the shape memory effect. Furthermore, the paper discusses the effect of functional and structural fatigue regarding important actuator performance indicators. Electrical heat treatment is introduced as it is chosen as a method for pretreatment to reduce the shakedown effect. Once the experimental design, which uses the design of experiments, is delivered, the choice of factors and response is thoroughly discussed. This paper

presents the possibility of reducing the shakedown effect of SMA actuators by pretreatment and eludes to future investigations in this field.

1.1. Shape memory alloys and actuators

SMAs belong to the group of active material, also called smart materials. The shape memory effect (SME) is based on a change in crystal structure and microstructure, by which many physical properties change. More precisely, the SME results from a reversible martensitic phase transformation between the cubic high temperature phase austenite and the monoclinic low temperature phase martensite. [9] The phase transformation happens without diffusion and occurs by shear movements without a change in shape, [10] which can be initiated thermally or mechanically. Basically three SMEs can be distinguished, the one-way effect, the two-way effect and pseudoelasticity.

In case of the one-way effect, the material can remember the imprinted form of the high-temperature phase by heating, but not by cooling. [11] It is used for connection, fastening and sealing elements. [10] The first commercial application was a shrink sleeve in the F-14 Tomcat of the U.S Navy for pipe connection in 1971. [12]

In contrast to the one-way effect, the material can remember also a form in the low-temperature phase in case of the two-way effect. [13] There are two types of two-way effects, but only extrinsic technical relevance. [14] The extrinsic two-way effect can be understood as an automatically repeated one-way effect, [15] which requires an external resetting force like a spring.

For the alloys which are based on the two previously mentioned effects, a distinction for applications can be made between the possibility of thermal and electrical activation. [16] Thermally activated SMAs are used as autarchic, self-regulating actuators, e.g. in heating and air conditioning. [17] This type of actuator reacts to a change of the ambient temperature. With electrically activated SMAs, the change in temperature is caused by joule heating. [16, 18] SMA actuators which use this activation are used as positioning elements for a wide range of applications. For this purpose, SMA wires are particularly suitable, since these have a high electrical resistance. [19]

In case of pseudoelasticity, the phase transformation is initiated mechanically. All SMEs have a hysteresis. For pseudoelastic elements, this can be used for damping. [17] [20] The most famous pseudoelastic applications are highly flexible stents, eyeglass frames and tooth tensioners.

1.2. Performance indicators of shape memory actuators with regard to fatigue

A repeated or cyclic use of the shape memory effects of a material leads to fatigue, which limits the functionality and lifetime. [21] The fatigue behavior of SMAs is very complex and multiclausal. Important factors in this context include external factors such as: time, temperature, voltage, load and number of cycles, but also internal factors such as alloying elements and composition, the nature of the phase transformation, and the lattice structure. [8] Fatigue in case of SMAs can be distinguished between functional and structural fatigue. [5] [21] [22] Particularly with regard to safety aspects, the fatigue has to be kept in mind when the components are designed. This study investigates the influence of mechanical and thermal stress on this fatigue.

Structural fatigue is understood as the behavior of materials to fail like any material, caused by cracking and crack growth. [8] [21] [23] Functional fatigue covers the changes in mechanical, thermal and other properties by thermomechanical cycles, [4] [21] which is an essential criterion for assessing the stability of the SME. [24] A challenge in the use of SMA actuators is that they first have a shakedown effect before they reach a largely constant operating point. Eggeler et al. demonstrated this effect in a study. [22]

To gain repeatable behavior, which is important for design products using SMAs, it is suggested to use relatively low stress instead of high stress in which case the shakedown would pose a bigger problem. [25] Major findings regarding the shakedown effect are the overload of SMA actuators, which accelerates functional fatigue. [25] Another approach is to limit the displacement of the SMA actuator, which reduces functional fatigue. [26] The authors do not consider possible pretreatment to reduce the shakedown effect prior to use even if it is stated that conditioned SMA wires (actuators) are preconditioned to eliminate the shakedown effect at moderate tensile loads. [25]

1.3. Heat treatment for shape memory alloys by resistance-based heat treatment

The SME is impressed by thermomechanical treatment. Semi-finished SMA products do not have any shape memory properties after the forming. [27] The required shape and functional properties can be adjusted by a separate heat treatment. [28] In addition to the widespread thermal heat treatment by means of a heating furnace, there is also the possibility of a resistance-based or electrical heat treatment based on Joule's heat. The generated heat is the product of the electrical voltage in the electrical conductor and the current. Advantages of this type of heat treatment compared to the thermal heat treatment are the simpler process or construction, and thus a higher process control, lower process time, but also lower energy consumption. [29] [30] Further, partial heat treatment is possible. [27], [31] and [32] have carried out important work on electrical heat treatment.

Like the conventional thermal heat treatment shaping, it is possible to adjust the phase transformation temperatures or to minimize functional fatigue. NiTi typically have a specific electrical resistance of 80 μΩcm in the martensite phase and 100 μΩcm in the austenite phase, which is comparatively high and thus leads to easy and rapid heating by current. [31] A danger using this type of heat treatment is that it can lead to a slight overheating and associated irreparable damage.

2. Material and methods

2.1. Experimental design

The design of experiments (DoE) is used for this study to identify significant factors and interactions between factors and response. DoE offers a systematical approach with various tools based on statistics, in which several factors are simultaneously changed. Despite the simultaneous change of several factors at a time, clear assignment is possible and thus a disadvantage of the one-factor-at-a-time method, the misinterpretation is not possible. [33] The advantage of this method is higher information quality in contrast to the one-factor-at-a-time experiment, but at the same time less effort. [34] [35]

Based on literature, research and experience, the factors and response are selected. For the response, our analysis focuses on the loss in displacement until stabilization (Δs), stabilized displacement (sstabilized), the number of cycles until stabilization (Δc) and lifetime with respect to cycles till crack (c). A visualization is provided in Fig. 1. The selection of investigated factors is based on a cause-effect diagram and further a factor-response-matrix through which the assumed influence regarding the response is evaluated. Based on this, the investigated factors are the type of heating, current, frequency, time and stress during heating. In Table 1, all factors and their levels are listed. A 52 full factorial design is used for the investigation, by which also interactions are determinable. The investigation covers 32 samples.

Table 1. factors and levels for the investigation

factors	unit	level	
type of heating	-	ramp	step
current	A	0.5	1
frequency	Hz	1	50
time	s	5	30
stress	MPa	275	400

IOP Conf. Series: Materials Science and Engineering **216** (2017) 012007 doi:10.1088/1757-899X/216/1/012007

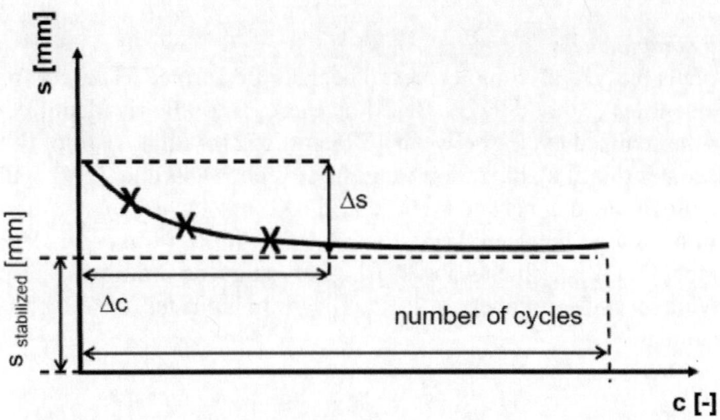

Figure 1. illustration of displacement and cycles as response.

2.2. Experimental setup and material

Due to their practical importance for actuator applications, only SMA wires with a diameter of 150 μm and an alloy composition of 54.8wt% Ni are considered as samples. The samples are crimped at the ends to form eyelets which are used to attach them to the test stand.

A programmable power supply is used for the electrical heat treatment. The sample is clamped in a holder and is loaded by a weight, which is guided by air bearing via a deflecting roller. Crocodile clamps are used for the contacting.

After the heat treatment, the samples are cyclically stressed until they are ruptured. For this purpose, the samples are clamped in two vertical hooks and are loaded by a weight equally to 400 MPa stress. Electrical activation takes place via crocodile clamps, with an amperage of 0.630 A for one second, followed by eight seconds of cooling time. Fig. 2 schematically summarizes the experimental setup, including the components and essential test parameters.

Preliminary tests have also been carried out in which five reference samples have cyclically been activated without electrical heat treatment to be able to compare the influence later on.

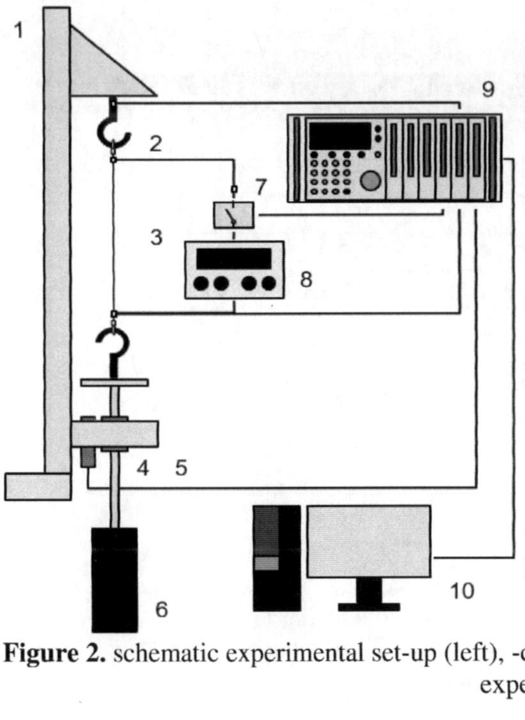

N#	component
1	test frame
2	mounting hook
3	SMA-sample
4	ultrasonic sensor
5	plain bearing
6	weight
7	electronic relay
8	laboratory power supply
9	measuring amplifier
10	measuring computer

parameter	unit	value
diameter	mm	0.15
length	mm	~ 100
connection	-	crimping
stress	MPa (g)	400 (710)
voltage	V	~ 3.2
current	mA	630
heating time	s	1
cooling time	s	4
measuring frequency	Hz	50

Figure 2. schematic experimental set-up (left), -components (center) and -parameters of the pre-aging experiments.

3. Results

3.1. Plausibility and completeness

As a first step, a plausibility and completeness check is performed. The criteria of the check for plausibility are the quantiles (2.5 and 97.5). Based on these, an outlier is identified and excluded. The response ranges for the realized cycles between 1306 and 5521 with a mean of 3908. The percentage of the loss in displacement until stabilization ranges between 13.49 and 35.91 with a mean of 21.62. The cycles until stabilization are between 160 and 1100 and have a mean of 543. The stabilized displacement is within a range between 4.051 to 4.684 % with a mean of 4.23 %. Compared to the reference samples, the DoE-experiments have a higher range regarding the responses. The higher range and lower average cycle number until stabilization indicates an effect of the electric heat treatment for pretreatment.

3.2. Effect screening

The first step to analyze the experimental results is a screening, which is illustrated in Table 2. Effects are considered significant regarding a response if the individual p-value is below 0.05. This is equal to a type 1 error (or α -error) of 5%. In Table 2 only values meeting these criteria are filled in. Additionally, the half-normal probability plot of the contrast is used for graphical interpretation. The current and interactions are the main influencing factors for cycles till crack. In case of the loss in displacement, the type of heating is another important factor.
The time, frequency and stress during heating in the investigated range of factors have no significant impact on the responses. The third-order interaction of current * type of heating * frequency slightly misses the criteria based on the p-value. However, in a graphical interpretation using a half-normal probability plot it becomes highly significant. For the response of cycle until stabilization interactions are exclusively relevant, which is one major indicator and objective of the investigation. One reason for this can be the determination of the number of cycles which can only be done by using graphical methods like a regression analysis and looking up the values from the diagrams. This negatively influences the accuracy and hence the accessability of factors regarding this response. However, it is certain that the current and type of heating are highly relevant; the parameters of these have to be further investigated. The response of stabilized displacement has similar significances as the cycle and loss in displacement until stabilization.

Table 2. effect screening based on p-value (α=5%).

factor	c	Δs	Δc	$s_{stabilized}$
current	0.0260	0.0048		0.0215
type of heating		0.0283		
current*time		0.0262		0.0438
current*type of heating		0.0016		
current*type of heating*frequency	0.0035	0.0698	0.0088	0.0039
current*type of heating*stress	0.0127		0.0564	0.0305
current*time*frequency	0.0391			
type of heating*frequency*time		0.0429		0.0855
type of heating*frequency*stress				
type of heating*frequency*stress*current			0.0569	

3.3. Discussion

The analysis of the effect of current by linear regression to evaluate the qualitative impact reveals that a higher current leads to a lower number of cycles till crack and to a higher loss in displacement until stabilization but also to a higher displacement after stabilization. If we group the impacts by the type of heating, these effects even increase. An explanation for this is that higher temperatures in heat

treatment and under cycling accelerate fatigue. Due to the fact that the type of heat treatment is a categorical factor, only differences can be recognized (see Table 2). This indicates using step as the type of heating has a tendency for higher loss in displacement until stabilization compared to using ramp.
Regarding the optima within the investigated parameters for the various responses, a distinction must be made between the number of cycles and the reduction of shakedown. For reducing shakedown, other factors, particularly the type of heating, are relevant compared to maximizing the number of cycles. When one response is optimized, interactions may probably not have been considered. The most important difference when optimizing a single response is the type of heating, as can be seen in Table 3.

Table 3. Overview of the optima for the response and their respective factors.

optima	factors					response			
	type of heating	current	fre-quency	time	stress	c	Δs	Δc	$s_{stabilized}$
c	ramp	0.5	50	5	275	5.566	23.71	793	4.0536
Δs, Δc, $s_{stabilized}$	step	0.5	50	5	275	2.942	14.81	270	4.4519
overall	step	0.5	50	5	275	2.942	14.81	270	4.4519

One reason for this is the fact that a longer lifetime quantified by the number of realized cycles leads to a higher number of cycles until stabilization is reached. If the cycles up to the stabilization are set in relation to the maximum lifetime, it is noticeable that a pre-treatment reduces the shakedown disproportionately.
For the design of SMA actuators, the following conclusions can be drawn from the experiments, which are summarized in Fig.3. For applications which have fewer cycles, the reduction of the shakedown effect by pretreatment can be achieved by at least one third up to two thirds. On the other hand, if a long lifetime, i.e. a high number of cycles is required, the result is a slower stabilization and therefore a higher number of cycles till the SME stabilizes. Furthermore, loss in displacement is higher and the stabilized displacement lower. Therefore, pretreatment is not promising in this case.

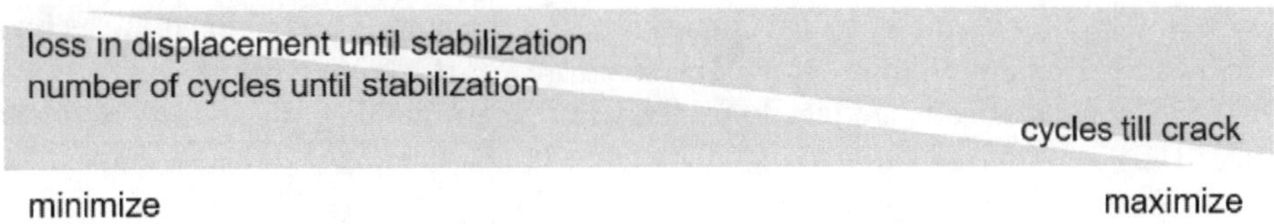

Figure 3. qualitative dependency of the response regarding optimization.

4. Conclusion and outlook

Even though it is not possible to completely eliminate the shakedown at the beginning of the lifetime of an SMA actuator, with the identified factors, it is possible to reduce the shakedown effect of NiTi SMA wires by electrical heat treatment in a certain range. The study has therefore been able to demonstrate the principal suitability of a pretreatment by electrical heat treatment to obtain stabilized SMA actuator behavior. This makes the material more manageable, which is necessary when designing SMA-based products. The pretreatment is thus an alternative approach to investigations of [25] and [26]. The main factor regarding the pretreatment is the electrical current, but also the type of heat treatment and the profile of current supply which was also indicated by [36] for regular cycling.

4.1. Outlook

Further investigations should be performed with larger sample size to further minimize random effects. The low significance of the factors and high relevance of interactions of the third and higher level suggest unconscious effect blending. For this reason, further investigations are necessary. Promising factors in this context are parameters in use or parameters of the pre-aging procedure using electrical heat treatment, e.g. stress. Alternative experimental designs like response surface plans can be used to investigate the nonlinear behavior of various factors especially the electrical current and interactions.

Acknowledgment
The authors acknowledge the funding of the FLAAI project through the aviation research program LuFo V-1 and the FGL-Recycling project through the central innovation program for small and medium-sized enterprises (zentrales Innovationsprogramm Mittelstand) both funded by the Federal Ministry of Economic Affairs and Energy.

5. References

[1] Tzou HS, Lee HJ and Arnold SM (2004) Smart Materials, Precision Sensors/Actuators, Smart Structures, and Structronic Systems. Mechanics of Advanced Materials and Structures 11(4-5): 367–393.
[2] Ölander A (1932) An electrochemical investigation of solid Cadmium-Gold alloys. Journal of the American Chemical Society 54(10): 3819–3833.
[3] Mohd Jani J, Leary M, Subic A and Gibson MA (2014) A review of shape memory alloy research, applications and opportunities. Materials & Design 56: 1078–1113.
[4] Gümpel P (2004) Shape Memory Alloy: applications in mechanical engineering, medical engineering and actuation. Renningen: expert.
[5] Mertmann M (2006) Fatigue in Nitinol Actuators. In: Borgmann H (ed.) Proceedings of the Actuator 2006. Bremen, Deutschland: Messe Bremen.
[6] Butera F. and Miyazaki S (2010) Recent Development of Shape Memory Alloys and Engineering Actuator Application. In: Borgmann H (ed.) Proceedings of the Actuator 2010. Bremen, Deutschland: Messe Bremen, pp. 206–211.
[7] Gugel, H. 2010. "Laser Welding of Similar Type and Unlike Kind Combinations of Material with Nickel-Titanium Shape Memory Alloys. Dissertation," Ruhr-University Bochum, Bochum.
[8] Wayman CM and Otsuka K (1999) Shape memory materials. Cambridge, New York: Cambridge University Press.
[9] Kaack M (2002) Elastic properties of NiTi-Shape Memory Alloys. Dissertation, Ruhr-University Bochum. Bochum
[10] Kristen M (1994) Studies for electrical control of shape memory drives in handling technology. Braunschweig: Mechanics Center oft the Technical University Braunschweig
[11] Bäker M (2014) Functional materials: Physical basics and principles. Wiesbaden: Springer Vieweg.
[12] Lagoudas DC (2007) Introduction to modeling and engineering applications of shape memory alloys. New York, London: Springer.
[13] Liu Y and Kohl M (2003) Shape memory micro valves with high energy density. Science reports. FZKA. 6934, research centre Kalsruhe. Kalsruhe.
[14] Steinhilper W, Sauer B and Albers A (2012) Princibles of machine elements for drive tasks. Berlin: Springer Vieweg
[15] Mertmann M. (1997) NiTi shape memory alloys for actuators of gripper technology: Functional characteristics and optimization. Düsseldorf: VDI Verlag.
[16] Janocha H (2007) Adaptronics and smart structures: Basics, materials, design, and applications. Berlin, New York: Springer
[17] Lygin K (2014) A method for developing environmental activated SMA-actuators with low hysteresis on the example of heating and air conditioning. München: Shaker
[18] Musolff A (2005) Shape memory alloys – experimental study and design of adaptive structures, TU Berlin. Berlin.
[19] Oelschläger L and Mandel D (2013) Keep the suitable form. Mechatronics: 20–24
[20] Treppmann D (1997) Thermo-mechanical treatment of NiTi – including solutions for quality assurance and standardization of shape memory alloys. Düsseldorf: VDI Verlag.
[21] Wagner MF X. (2005) A contribution of structural and functional fatigue of wires and springs made of NiTi memory alloys. Univ., Dissertation, Ruhr-University Bochum.

[22] Eggeler G, Hornbogen E, Yawny A, Heckmann A and Wagner MF X. (2004) Structural and functional fatigue of NiTi shape memory alloys. Materials Science and Engineering: A 378(1-2): 24–33.
[23] Rahim M (2013) Investigation on influence of carbon and oxygen impurities on the fatigue behavior of pseudoelastic NiTi shape memory alloys. Dissertation, Ruhr-University Bochum. Bochum
[24] Bauer B (2010) Form memory alloys in drive technology: Actuator in gearboxes and couplings; fatigue notch factor. Dissertation, Ruhr-University Bochum. Bochum.
[25] Churchill CB and Shaw JA (2008) Shakedown response of conditioned shape memory alloy wire. In: Dapino MJ and Ounaies Z (eds) The 15th International Symposium on: Smart Structures and Materials & Nondestructive Evaluation and Health Monitoring: SPIE, 69291F.
[26] Kim W, Barnes BM, Luntz JE and Brei DE (2010) A design method for shape memory alloy actuators accounting for cyclinc shakedown with constrained allowable strain. In: Proceedings of the SMASIS 2010 Conference on Smart Materials, Adaptive Structures and Intelligent Systems. New York: ASME Digital Collection.
[27] Pilch J, Heller L and Šittner P (2009) Final thermomechanical treatment of thin NiTi filaments for textile applications by electric current. In: ESOMAT 2009 - 8th European Symposium on Martensitic Transformations.
[28] Elahinia MH (2016) Shape memory alloy actuators: Design, fabrication, and experimental evaluation. Chichester: Wiley.
[29] Wang CC, Huang WM, Ding Z, Zhao Y, Purnawali H, Zheng LX, et al. (2012) Rubber-like shape memory polymeric materials with repeatable thermal-assisted healing function. Smart Materials and Structures 21(11): 115010.
[30] Pilch J, Heller L and Šittner P (2010) Heat treatment of thin NiTi filaments by electric current. Available at: http://energia.moebius.com.br/oo/Energia/Nitinol/0107Pilc0110Heat.pdf (accessed 3 September 2015)
[31] Wang ZG, Zu XT, Feng XD, Zhu S, Deng J and Wang LM (2004) Effect of electrothermal annealing on the transformation behavior of TiNi shape memory alloy and two-way shape memory spring actuated by direct electrical current. Physica B: Condensed Matter 349(1-4): 365–370.
[32] Delobelle V, Chagnon G, Favier D and Alonso T (2016) Study of electropulse heat treatment of cold worked NiTi wire: From uniform to localised tensile behaviour. Journal of Materials Processing Technology 227: 244–250.
[33] Siebertz K, van Bebber D and Hochkirchen T (2010) Statistical experimental design: Design of Experiments (DoE). Heidelberg: Springer.
[34] Kleppmann W (2013) Pocketbook for the design of experiments: Optimizing products and processes. Munich: Hanser.
[35] Lunau S (2012) Six Sigma+Lean Toolset: Mindset for successful implementation of improvement projects. Berlin, Heidelberg: Springer.
[36] Casati R, Passaretti F and Tuissi A (2011) Effect of electrical heating conditions on functional fatigue of thin NiTi wire for shape memory actuators. Procedia Engineering 10: 3423–3428

IOP Conf. Series: Materials Science and Engineering **216** (2017) 012008 doi:10.1088/1757-899X/216/1/012008

Impact of Different Electrical Time-Based Activations on NiTi Shape Memory Alloys

Benjamin Fleczok, Christian Rathmann, Dennis Otibar, Antonia Weirich, Bernd Kuhlenkötter
Ruhr-University Bochum, Universitätsstraße 150, 44801 Bochum, Germany

Email: fleczok@lps.rub.de; rathmann@lps.rub.de; otibar@lps.rub.de; weirich@lps.rub.de; kuhlenkoetter@lps.rub.de;

Abstract. The use of NiTi shape-memory alloys (SMA) in actuators bears significant advantages for designing robust, simple and lightweight applications. The SMA effect is based on a phase transformation of the atomic lattice in response to stress, strain and temperature. The resulting crystallographic configurations lead to a complex behavior revealing different electrical and mechanical characteristics. In view of the impact of thermo-mechanical cyclization on the operational lifetime, this paper investigates the influences of different types of electrical activation. For this purpose, six current curves with six samples each are compared to a reference activation with regard to the operational lifetime. The chosen time of activation is 1 second in accordance with an industrially relevant cycle of technical actuators. Based on the results of these investigations, recommendations of the activation type shall be developed for the operational lifetime of NiTi-SMA.

1. Introduction

Smart materials (SM) like shape memory alloys provide many advantages for the design of cross-industry technical products such as their energy saving potential (in Germany) of about 16TWh/a as well as their emission saving potential of about 8t CO_2/a [1]. Furthermore, they offer a great potential for developing lightweight and miniaturized actuators and sensors, while also reducing the complexity of technical actuator assemblies [2, 3]. Despite these advantages, the impact of shape memory alloys on the global market is still very low. Smart material applications for motors and actuators have a market value of USD 26 billion with a compound annual growth rate (CAGR) of 10.2% (between 2014 and 2019) instead of the predicted 15.8% (between 2011 and 2016) [4, 5]. This 5% reduction of the expected CAGR in the new study clearly shows the difficulty of smart materials to grow in the global market. Comparing the advantages of SMAs to their relatively low market value unveils certain problems with the developing process of SMA actuators. This is based on the complex relationship between stress, strain and temperature and, as a result, the electrical and mechanical characteristics that limit the handling for engineers and scientists [6].

Due to their fast and time-controlled activation with simultaneously usable sensory features, electrically activated NiTi SMA will have a huge significance in the future. [7] This is particularly true for cyber physical systems [8] as a result of the increasing amount of sensory-recorded information in technical systems and the possibility to keep the complexity of technical systems low. [9, 10]

In regards to electrically activated SMAs based on NiTi, many investigations focus on medical pseudo-elastic applications [11, 12, 13], sensory features [14, 15, 16] as well as thermal heat treatments. [17, 18] Furthermore, electrical activation were researched in connection with the positioning control and activation control. [19, 20] This has become particularly important for many

IOP Conf. Series: Materials Science and Engineering **216** (2017) 012008 doi:10.1088/1757-899X/216/1/012008

technical applications. However, the biggest application field for actuators in technical systems is the simple time-controlled actuation. This is based on the historical evolution of mechatronics (relay-circuit, Programmable Logic Controller [plc]), which realized only two positions/ functions (0 and 1). [21] SMA actuators are a good substitute for an electromotor for a variety of applications, such as a door locking device and can be directly integrated with existing electrical control systems. [22] Unfortunately, this time controlled activation has hardly been scientifically researched yet. [23, 24, 25] An approach to lower the market entrance barriers for SMA applications has been realized by [26] in accordance to [27] by using a cross-impact matrix of SMAs in order to analyze the influences among each other. One of the essential findings is that the operational lifetime as well as the activation frequency are interdependent. Due to the fact that in the thermo-mechanical cyclization the type of energy input has as well a substantial influence on the operational lifetime [28, 29, 30], this paper shall especially investigate the type of current within the determined and relevant time-controlled cycle time for technical actuators. The cycle time is set to (up to) 1s. [21] Based on this aim, the following four parameters are used as the main focus of this investigation:

1. the activation time
2. activation by means of current
3. course of the activation
4. operational lifetime until the rupture:

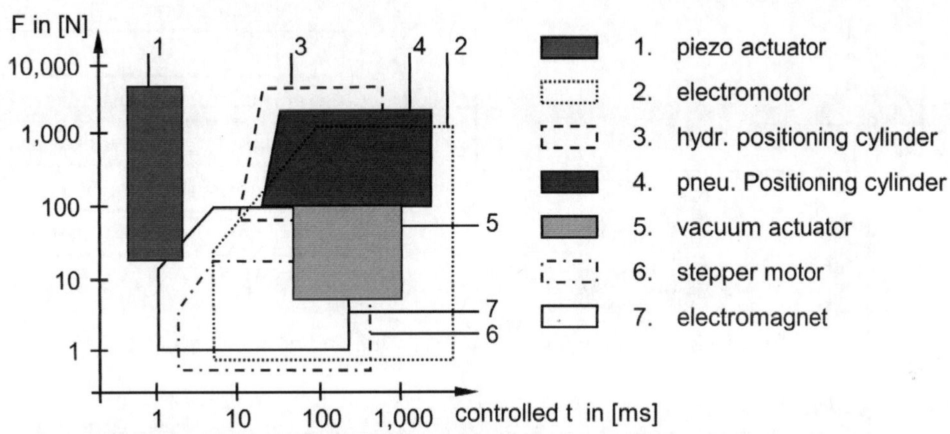

Figure 1. force – actuation time (controlled) for common actuators [21].

2. Experimental Setups, Material and Methods

2.1. Introduction

Ten reference tests are run in order to make a valid statement about the type of current that has the greatest impact on the operational lifetime. For this purpose, conventional rectangular current curves are used for the activation. [24, 28, 31] This and the following activation parameters of the reference were chosen according to a control strategy in which reaching the austenite finish point represents the end of the cycle. [16, 24] These preliminary tests serve to determine a standard deviation, which represent the calculation base for the required individual tests. Different curves for electrical activation are then selected and researched in the following test series.

2.2. Material and Experimental Setup

All presented tests make use of the economically significant and commercially available binary nickel-titanium wires by SAES Getters Company. The wires of Smartflex® SF90 (54.8wt% Ni) exhibit an austenite finish temperature of 90°C. Activated wire lengths of 100mm with a wire thickness of 0.3mm are used in all tests. All wires of this test series are cyclically activated for 1s and cooled down for 9s in a room temperature of 23°C (stationary air) with a mechanical load of 400MPa. Moreover, the tests prescribe that only the samples with a contraction of 5% are considered as completely activated and will be taken into account. The test samples are cycled as long as the wire rupture takes place.

The testing facility consists of five vertical pillars in order to perform five parallel tests. The wires will be activated by a programmable laboratory power supply. For this purpose, the samples are provided with a crimped loop at the end of the wire and are hooked into the designated positions that are presented in figure 2.

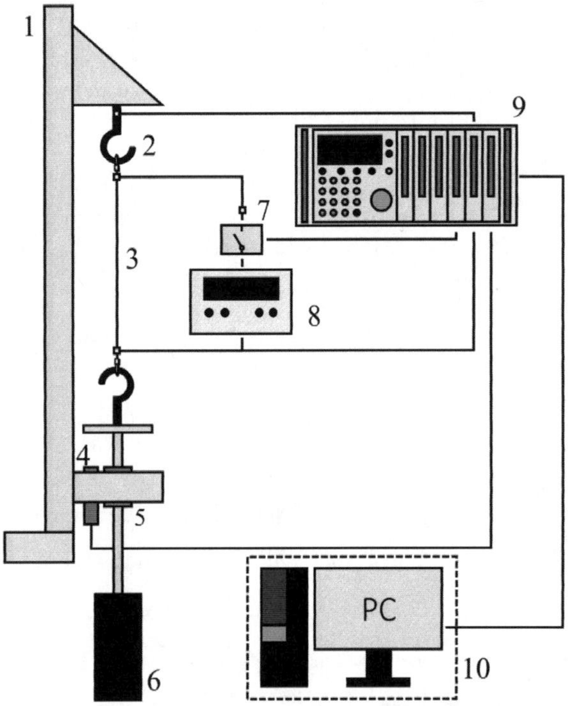

N#	component
1	test frame
2	mounting hook
3	SMA-sample
4	ultrasonic sensor
5	plain bearing
6	weight
7	electronic relay
8	laboratory power supply
9	measuring amplifier
10	measuring computer

Figure 2. schematic experimental set-up and components.

2.3. Design of experiment

This investigation is based on a method which uses the "standard deviation" of $\sigma = 476.1$ cyles (taken from an already performed sample) and the "technological relevance" $\Delta\mu$ to calculate the lowest experimental effort. [32, 33] With this minimal requirement, it is possible to find a statistically relevant distinction with a probability of 90%. The total amount of the measured values can be calculated with the following equation:

Scope of Experiment $$N = 60 \cdot \left(\frac{\sigma}{\Delta\mu}\right)^2 \quad \text{(Eq. 1a)}$$

The technological relevance $\Delta\mu$ amounts to approximately 580 cycles and has been derived by the experiences of performed durability tests of electrically activated SM actuators. By means of both, the values of the standard deviation and the technological relevance, the total amount 'N' of the required measured tests are calculated as part of the next step:

$$N = 60 \cdot \left(\frac{467.1}{580}\right)^2 = 38.91 \quad \text{(Eq. 1b)}$$

2.4. Tests

39 measurements are required to investigate seven types of current in relation to the operational lifetime. These 39 measurements, as said in [32], are divided into the 7 types of current, which results in 6 samples per test series.

Figure 3 schematically shows the current curves for the further operational durability tests. As there are no scientific investigations in this field, these current curves have been chosen freely. They have been selected with the objective to provide a reduction of current in a cycle, to cover different types of ramps/platforms as well as to investigate dwell times as seen in curve b) and d) of 150ms within a cycle. While the current curve e) performs a complete sinus curve, the curve f) performs only half a sinus curve within the same time. All six current types are now compared with a rectangular activation reference.

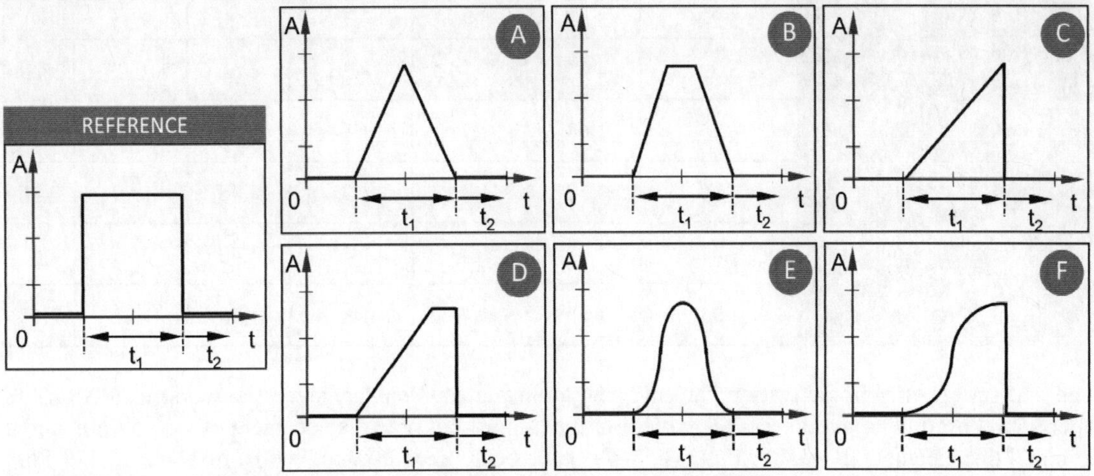

Figure 3. Schematic diagram of the current curves (A) rectangular, (B) trapezoid, (C) saw-tooth, (D) saw-tooth plateau, (E) sinus, (F) sinus-saw-tooth.

Table 1 shows both the investigated currents of the preliminary tests and the relevant parameters for the endurance tests such as the measurement frequency, the absolute time for one cycle as well as the mechanical tension. The electrical current which is used for the rectangular activation is comparable to the current of the electrical controlled activation.

Table 1. Summary of parameters for the tests

Type of power	*rectangular*	*triangle*	*saw-tooth*	*saw-tooth-plateau*	*trapezoid*	*sinus*	*sinus-saw-tooth*
current	2.35 A	3 A	3.2 A	3 A	3 A	3 A	3 A
measurement frequency	50Hz						
activation time	1s						
cooling time	9s						
absolute cycle time	10s						
mech. stress	400MPa						

2.5. Results and discussion

As illustrated in table 2, the criterion for the technological relevance of $\Delta\,\mu = 580$ cycles has been fulfilled. This means the deviation of all cycles is higher than 580 cycles to the reference. As a result of the proved technological relevance, there is a considerable influence of the selected activation current on the operational lifetime.

Table 2. Summary of all average values from the tests

Type of Power	*triangle*	*sinus*	*trapezoid*	*rectangular*	*sinus saw tooth*	*saw tooth*	*saw tooth platea u*
average cycles till crack	**15.349**	**11.590**	**8.483**	**7.487**	**6.695**	**5.949**	**1.299**
max. displacement (mm)	4.99	5.15	5.20	5.44	5.31	5.30	6.67
min. displacement(mm)	2.52	2.90	3.15	3.22	3.35	3.39	4.19
max. displacement (%)	5.02	5.16	5.25	5.48	5.31	5.35	5.38
min. displacement (%)	2.53	2.90	3.17	3.25	3.35	3.42	4.19

lost in displacement absolute (%)	49.60	43.80	39.62	40.69	36.91	36.07	22.12
elongation (%)	3.20	3.87	5.47	7.75	6.20	9.35	20.03
change of operation point (mm)	3.29	4.02	5.74	8.36	6.60	10.23	25.14
min. elec. resistant martensit (Ω)	1.75	1.67	1.71	1.74	1.68	1.72	1.70
max. elec. resistant martensit (Ω)	1.79	1.74	1.85	1.95	1.84	1.99	2.56
min. elec. resistant austenite (Ω)	1.47	1.45	1.44	1.46	1.40	1.45	1.41
max. elec. resistant austenite (Ω)	1.71	1.65	1.73	1.81	1.70	1.82	2.17

The test series with the activation current "saw tooth plateau" has achieved the worst results on the operational lifetime of average 1,299 cycles but the highest possible displacement of s= 6.67 mm and a minimal displacement of s= 4.19 mm. The wires have been predetermined with a displacement of 5% , which amounts to 5mm. The effect of a displacement of over 6.6% indicates a great elongation of the wire which compensates the displacement loss. In comparison, the current method with the triangle curve has achieved the highest cycle number on an average of 15,349 cycles but the least maximal and minimal displacement of s= 4.99 mm and s= 2.52 mm.
The measurement results show that a symmetrical current curve with an increasing and decreasing course in the activation cycle has a positive influence on the operational lifetime. Furthermore, the measurement results demonstrate that the conventional activation of wires with a rectangular curve has the least operational lifetime expectation of all symmetrical types of activation.

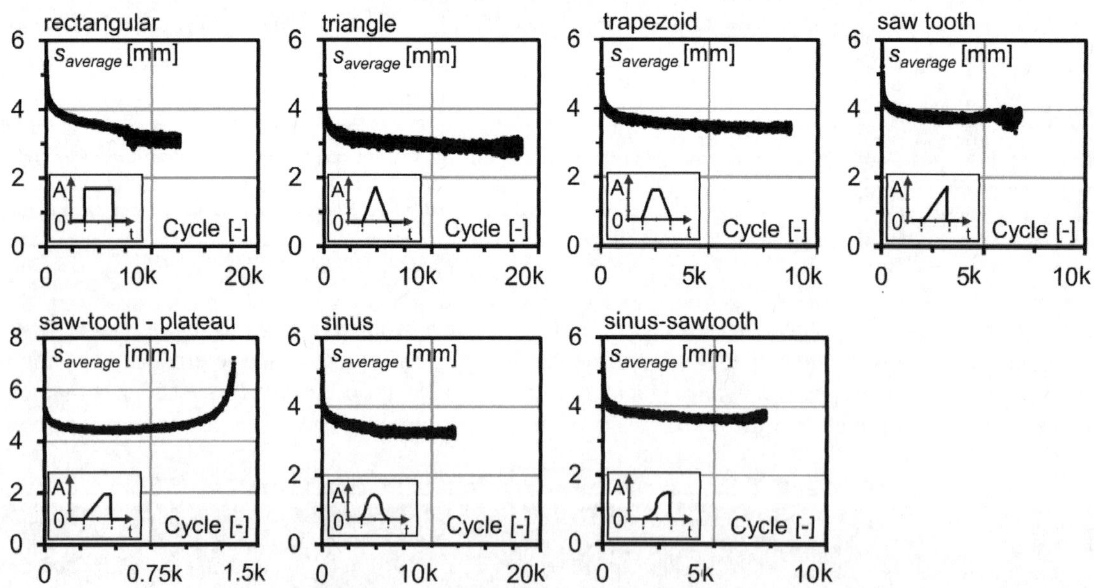

Figure 4. average loss of displacement over the operational lifetime.

Figure 4 shows the average loss of displacement. In every diagram, the six curves of the test procedure superpose each other. One exception is the larger number of cycles compared with table 2 observed in the rectangular, triangle and saw-tooth graphs.

2.6. Discussion

The investigations were able to determine significant statistical distinctions regarding the type of current within the time-based activation frequency that is relevant for technical actuators. [21] The limits of the sample are a contraction of 5%, an electrical activation time of 1 s as well as a mechanical tension of 400 MPA when using SM wires with a diameter of 0.3mm. This research has investigated seven different types of current (time-controlled) with six single tests in relation to their influence on the operational lifetime. The results have shown substantial distinctions in comparison to the standardized electrical activation [24, 28, 31] by means of rectangular current course. Symmetrical current curves and especially the triangle curve, which provides an increase and a decrease of current within a cycle, has shown an operational lifetime of 15,349 cycles, which corresponds to a double operational lifetime in comparison to the reference (rectangular current). In this context, they have nevertheless shown the greatest effect in the loss of displacement of 50%.

In comparison, the current curves with increasing current, which present their maximum at the end of the cycle, deliver the worst results. The number of cycles is significantly below 10,000. However, this type of current, especially the "saw tooth plateau", demonstrates a strong change of the elongation of about 20%. This elongation compensates the loss of displacement and leads to a higher displacement at the end of the operational lifetime. This investigation has also shown that the criterion for electrical controlled activations by reaching the Austenite finish temperature with a rectangular current is not sufficiently accurate. In comparison to other forms of activation especially the triangle form revealed that the fatigue of the SMA is accelerated by the electrical current. This leads to the assumption that the electrical current has an additional but not further specified effect on the regular fatigue based on the thermo-mechanical cyclization.

3. Conclusion and outlook

The investigations show that an adjusted time-controlled current with comparable parameters can crucially influence the operational lifetime of NiTi-SMA. With this investigation, it is possible to find operational lifetime-optimized currents for an activation of 1s for technical applications.

As a result of the investigations, it is seen that the electrical current / the electrical activation by using comparable parameters has an additional effect on fatigue and lifetime of SMA.

3.1. Outlook

In further investigation will need to carry out a higher number of tests, which goes beyond the statistical minimal requirement, in order to reduce the influences of material scatter and the scatter of prepared samples even further. In addition, further tests with different mechanical tensions and temperatures should be performed in order to newly evaluate the operational lifetime of alloys in relation to the current, especially in comparison to the existing manufacturer's specification. It is further recommended to investigate additional current curves and their influence on the operational lifetime.

4. Acknowledgment

The authors acknowledge the funding of the FLAAI project through the aviation research program LuFo V-1 by the Federal Ministry of Economic Affairs and Energy.

5. References

[1] E. Jochen, C. Jaeger, et. al., Investments in climate friendly Germany. Study carried out by the Ministry for the Environment, Agriculture and Reactor-protection, Ministry for the Environment, Agriculture and Reactor-protection, Potsdam, (2008).

[2] F. Butera, A. Coda, G. Vergani, Shape memory actuators for automotive applications, Nanotec IT newsletter, AIRI/nanotec IT, Roma. 12–6 (2007).

[3] P. Abrahamsson, E. Møster, Demands on shape memory alloys from the application designer's point of view, J Phys IV France. 07 (1997) C5-667-C5-72.

[4] BCC Research, Smart materials and their applications: technologies and global markets, BCC Research Advanced Materials Report (2011) p. 161.

[5] BCC Research, Smart materials and their applications: technologies and global markets, BCC Research Advanced Materials Report (2015) p. 161.

[6] M. Mertmann,G. Vergani, Design and application of shape memory actuators. The European Physical Journal-Special Topics 158(1) (2008) 221-230.

[7] J. Mohd Jani, M. Leary, A. Subic, M. Gibson, A review of shape memory alloy research, applications and opportunities, Materials & Design. 56 (2014) 1078–1113.

[8] M. A. McEvoy, N. Correll, Materials that couple sensing, actuation, computation, and communication, Materials science, Science 347 (6228) (2015) p. 1261689.

[9] M. Elahinia, Shape Memory Alloy Actuators: Design, Fabrication and Experimental Evaluation, John Wiley & Sons, USA, (2015).

[10] R. Velazquez, Contribution to the Design and Implementation of Portable Tactile Displays for the Visually Impaired, Paris VI, (2006).

[11] K. Otsuka, T. Kakeshita, Science and Technology of Shape-Memory Alloys. New Developments, MRS Bull 27 (02) (2002) 91–100.

[12] C. Rice, Shape Memory Alloys, Applications, in: M. Schwartz (Eds.), Encyclopedia of smart materials, J. Wiley, New York, (2002).

[13] M. Schwartz, Encyclopedia of smart materials, J. Wiley, New York, (2002).

[14] D. Lagoudas, Shape memory alloys. Modeling and engineering applications, Springer, New York, 2008.

[15] S.H. Liu, T.S. Huang, J.Y. Yen, Tracking control of shape-memory-alloy actuators based on self-sensing feedback and inverse hysteresis compensation, Sensors. 10 (1) (2010) p. 112–127.

[16] J.J. Zhang, Y.H. Yin, J.Y. Zhu, Electrical resistivity-based study of self-sensing properties for shape memory alloy-actuated artificial muscle, Sensors. 13 (10) (2013) p. 12958–12974.

[17] J. Pilch, L. Heller, P. Šittner, Final thermomechanical treatment of thin NiTi filaments for textile applications by electric current, ESOMAT 2009 - 8th European Symposium on Martensitic Transformations (2009).

[18] Z.G. Wang, X.T. Wang, X.D. Feng, S. Zhu, J. Deng, L.M. Wang, Effect of electrothermal annealing on the transformation behavior of TiNi shape memory alloy and two-way shape memory spring actuated by direct electrical current, Physica B: Condensed Matter. 349(1-4) (2004) 365–370.

[19] R.B. Gorbet, K.A. Morris, R.C.C. Chau, Mechanism of bandwidth improvement in passively cooled SMA position actuators, Smart Materials and Structures. 18:095013 (2009).

[20] H. Ashrafiuon, M. Eshraghi, M.H. Elahinia, Position Control of a Three-link Shape Memory Alloy Actuated Robot, Jnl of Intell Mat Systems and Struct. 17 (5) (2006) 381–392.

[21] Isermann, Rolf Mechatronic systems. Vol. 2. Berlin, Heidelberg, New York: Springer, ISBN: 3540325123. (2008)

[22] LEARY, M.; HUANG, S.; ATAALLA, T.; BAXTER, A.; SUBIC, A.: Design of shape memory al-loy actuators for direct power by an automotive battery. In: Materials & Design (2012) (43), S. 460–466

[23] R. Casati, F. Passaretti, A. Tuissi, Effect of electrical heating conditions on functional fatigue of thin NiTi wire for shape memory actuators, Procedia Engineering 10 (2011) 3423–3428.

[24] C. Zanotti; P. Giuliani; A. Tuissi; S. Arnaboldi; R. Casati; C. Zanotti, et al., Response of NiTi SMA wire electrically heated, ESOMAT 2009 (2009) 7-11.

[25] H.N. Bhargaw, M. Ahmed, P. Sinha, Thermo-electric behaviour of NiTi shape memory alloy, Transactions of Nonferrous Metals Society of China. 23 (2013) 2329– 2335.

[26] Fleczok B, Lubisch M and Kuhlenkötter B., Development of a pre-evaluation method to design an electrically activated NiTi-based shape-memory wire actuator. In: Higher Education Forum (ed.) Hong Kong International Conference on Engineering and Applied Sciences 2015. Hong Kong, VR China, (2015) pp. 1–8.

[27] Stephens, R. I.: Metal fatigue in engineering. Wiley, New York, NY, 2001.

[28] S. Miyazaki, Thermal and Stress Cycling Effects and Fatigue Properties of Ni-Ti Alloys, in: T.W. Duerig, et al. (Eds.), Engineering Aspects of Shape Memory Alloys, Butterworth-Heinemann Ltd, Essex, 1990, pp. 394–413.

[29] F. Schiedeck, S. Mojrzisch, Design of a robust control strategy for the heating power of shape memory alloy actuators at full contraction based on electric resistance feedback, Smart Mater, Struct. 20 (4) (2011) 45002.
[30] A.R. Pelton, G.H. Huang, P. Moine, R. Sinclair Effects of thermal cycling on microstructure and properties in Nitinol, Materials Science and Engineering. A 532 (2012) 130–138.
[31] T, Luchetti, A. Zanella, M. Biasiotto, A. Saccagno, Electrically Actuated Antiglare Rear-View Mirror Based on a Shape Memory Alloy Actuator, J. of Materials Engineering and Performance 18 (5-6) (2009) 717–724.
[32] W. Kleppmann, Pocketbook design of experiments: optimise products and processes, Hanser, München, 2013.
[33] S. Lunau, Six Sigma+Lean Toolset Mindset to the successful implementation of improvement projects, Springer, Berlin, 2012.

IOP Conf. Series: Materials Science and Engineering **216** (2017) 012009 doi:10.1088/1757-899X/216/1/012009

A Preliminary Investigation of Temperature Dependency of a Shape Memory Actuator with Time-Based Control in Aircraft Interiors

Dennis Otibar, Antonia Weirich, Marcus Kortenjann and Bernd Kuhlenkötter
Ruhr-Universität Bochum, Universitätsstraße 150, 44801 Bochum, Germany
Email: otibar@lps.rub.de; weirich@lps.rub.de; marcus.kortenjann@rub.de; kuhlenkoetter@lps.rub.de;

Abstract. Shape memory alloys (SMA) possess an array of unique functional properties which are influenced by a complex interaction of different factors. Due to thermal sensitivity, slight changes in the environmental temperature may cause the properties to change significantly. This poses a huge challenge especially for the use of SMAs as actuators. The most common and elementary activation strategy of SMA actuators is based on the duration of activation and cooling with constant activation parameters. However, changing environmental influences cause the necessity to modify these parameters. This circumstance needs to be especially considered in the design process of actuator controls. This paper focuses on investigating the influence of environmental temperature changes on time-based activated SMA actuators. The results of the described experiments form the base for designing reactive control strategies for SMA actuators used in alternating environments. An example for application fields with changing environments and particularly changing temperatures are aircraft related implementations. This area also stands to benefit from the actuators' advantages in ecological efficiency.

1. Introduction

Actuators which are based on functional materials might lead to significant progress in fields with mainly conservative technologies. These fields benefit highly from the advantages offered by unconventional actuators. Examples for these innovative technologies are actuators based on shape memory alloys (SMA). These SMA actuators provide some major benefits because of their high energy density and the resulting possibilities in lightweight construction and miniaturization. An application field which can strongly profit from these benefits is aviation. Especially in civil aviation, aircraft weight and ecological efficiency are becoming more and more important [1]. The area of aircraft interiors offers many possible applications suitable for SMA actuators. There are many conventional systems used in comfort and safety applications that can be substituted and improved by innovative SMA actuators because of their unique properties. Beyond that, there are numerous possibilities of implementing SMA actuators in new applications which are only feasible because of this functional material's advantages such as compact design, silent actuation or energy efficiency.
However, there are only few established applications of SMA actuators in aircraft industry. In order to use this untapped potential, the research project FLAAI ("Shape Memory Lightweight Actuators for Applications in Aircraft Interiors") of the Ruhr-University of Bochum is investigating the applicability of SMA actuators in the interiors of civil aircrafts. Systems in this application field require, amongst others, different ranges of environmental temperatures and temperature alterations which may not have any influence on the functionality of the actuators. As SMAs are a thermally sensitive material, their properties have an especially high interdependence on temperature variations. For this reason, the

authors have investigated the properties of SMA actuators regarding the environmental temperature requirements in cabins of civil aircrafts using the example of an antagonistic actuator with a time-based control.

2. Shape Memory Alloys and Actuators

2.1. Shape Memory Alloys (SMA)
SMAs are functional materials which can remember a previously defined shape by thermal activation and also show high elastic properties. These effects are based on a reversible, diffusionless transition between the low temperature phase (martensite) and the high temperature phase (austenite). The thermal effect can be described as the one-way effect that shows shape recovery during heating as well as the two-way effect that is characterized by an additional shape modification during cooling. The shape recovery can be triggered by the intrinsic effect or external loads. These effects make SMAs suitable for a variety of actuator applications [2].
Generally, SMA actuators are based on the extrinsic two-way effect due to its higher amount of technical parameters as force and displacement [3]. When these actuators are activated electrically, the change in temperature is caused by joule heating [4, 5]. The most important electrical parameters for SMA performance concerning force, displacement and fatigue are the current, voltage and activation time. [6]
The resetting of the actuator position can be realized by different technical principles (illustrated in fig. 1). The three common ones are a constant mass, a return spring or an antagonistic SMA element, which is subject to an opposed activation strategy [7, 2].

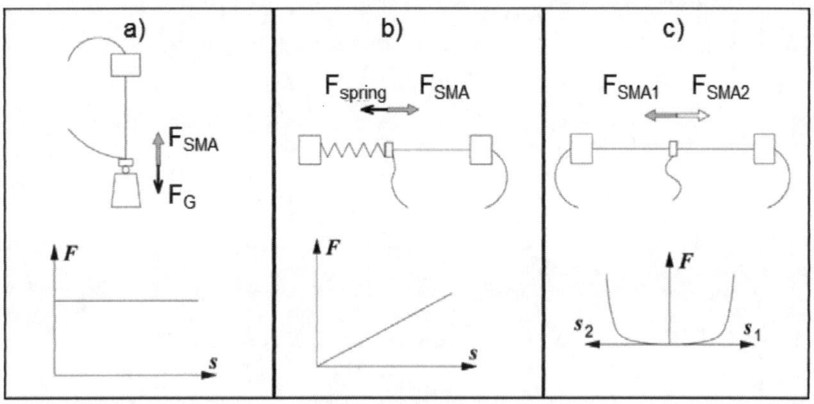

Figure 1. Three principles for resetting an SMA element, a) weight force, b) return spring, c) another SMA wire (agonist-antagonist-principle), based on Kohl [7].

2.2. SMA Actuators in Aircraft Interior Systems
The Fraunhofer institute for machine tools and forming technology developed an unlocking mechanism based on SMAs for small oxygen masks (see fig. 2) [8]. Nowadays, these masks are electrically released by a solenoid. For functionality or safety checks, a manual release mechanism is designated [9]. Replacing the solenoid by an SMA actuator, it has been possible to reduce the weight of the system by half [9]. Further constructive measures have lowered the weight down to 66 % compared to the conventional actuator. In addition, the power consumption has been reduced to 34 % and the number of system parts has fallen from 27 to 18, which is a reduction of 67 % [10].
Other products, for which SMA actuators can be used, are the backrest adjustment and the extension of seating areas in aircrafts (see fig. 3). In order to use less powerful SMA actuators instead of electric motors, the passenger weight is used to reduce the required actuator force.
For extending the seating or lying area, a pivotal expansion element (2) can be folded out. This is done by a tubular SMA actuator (3) and a surface element (4) which is loaded by the passenger's weight (1) and thus supports the SMA actuator. Furthermore, there is a locking mechanism (5) to hold the

expansion element in the desired position. It consists of a catch which is unlocked by an SMA wire actuator.

When a passenger is sitting on the seat, the backrest is pre-tensioned in vertical direction by a flexible joint (7) and an element (7) within the backrest. The flexible joint (7) has the possibility to store energy. If an angular adjustment of the backrest is desired, the backrest is pressed into a flat position by the weight of the passenger (1). This further increases the preload on the flexible, elastic joint (7). Only a small force is required to bring the back into the upright position. This movement is realized by an SMA wire [11].

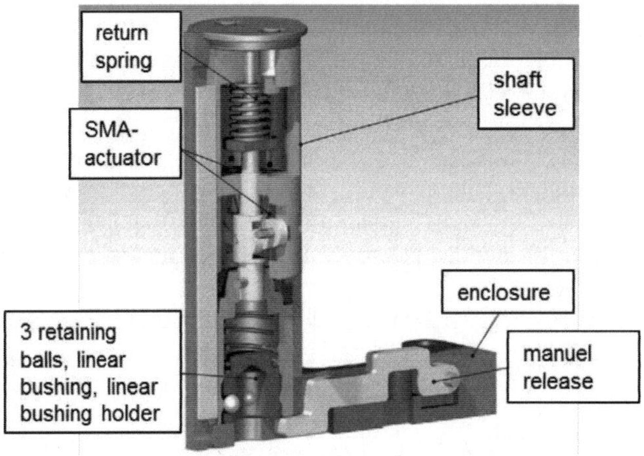

Figure 2. SMA based unlocking actuator for oxygen masks, developed by Fraunhofer IWU, based on Eppler [10].

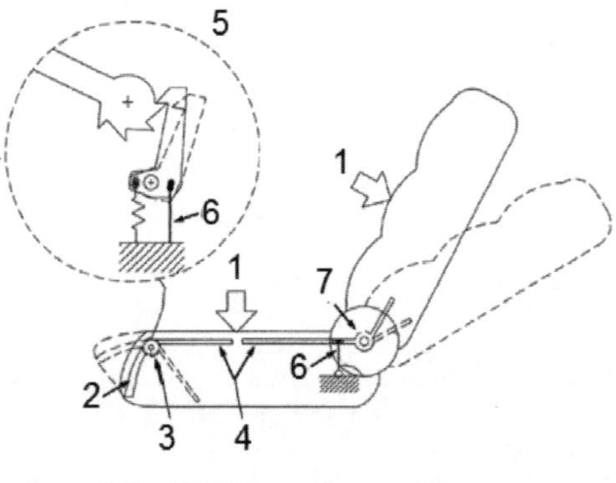

Figure 3. Schematic illustration of the SMA based seat adjustment systems, based on Lawall et al. [12].

3. Temperature requirements for electronic devices in aircraft interiors

To enhance the operation field of an actuator, it is necessary to consider the temperature requirements of category D1 of the RTCA (Radio Technical Commission for Aeronautics) directive DO-160D. The temperature range in operation is between -20°C to +55°C and the temperature maxima between -55°C to +70°C (see table 1). If an actuator fulfills these requirements, it can be used in an air-conditioned but not pressure-controlled system in the aircraft at a flight altitude of up to 15,200 m. In this case, actuators can be easily used in commercial aircrafts since the maximum flight altitude is 13,100 m according to this directive.

Furthermore, the temperature change is limited to 2°C per minute, since the actuator is located in the temperature-regulated aircraft interior. Therefore, only category C of the RTCA DO-160D has to be fulfilled instead of category A with 10°C per min [13, 14].

Table 1. Extract from the RTCA/DO-160D directive concerning temperature requirements [13].

temperature requirements for SMA actuators within aircraft interiors		
demand / wish	**requirement**	**specification**
demand	operating temperature	-20 [°C] up to +55 [°C]
demand	maximum temperature	-55 [°C] up to +70 [°C]

4. Material and methods

4.1. Experimental Procedure

The experimental procedure contains the following steps: First, it is necessary to experimentally determine the parameters for the time-based control for the current supply at room temperature. Preliminary studies as described below are needed to provide a basis for the main experiment. The main experiment investigates temperature influences on the activation behavior and thus on the reliable functionality of a time-based controlled SMA actuator.

4.2. Preliminary Studies

Thanks to its availability and simple programming language, the programmable logic control type LOGO! by Siemens AG is used as control unit. The first step is to determine the activation and cooling times. Due to their dependency on parameters like wire length, wire diameter, applied electrical voltage and load on the wire, the activation and cooling times have to be identified on an experimental setup as is illustrated on figure 4.

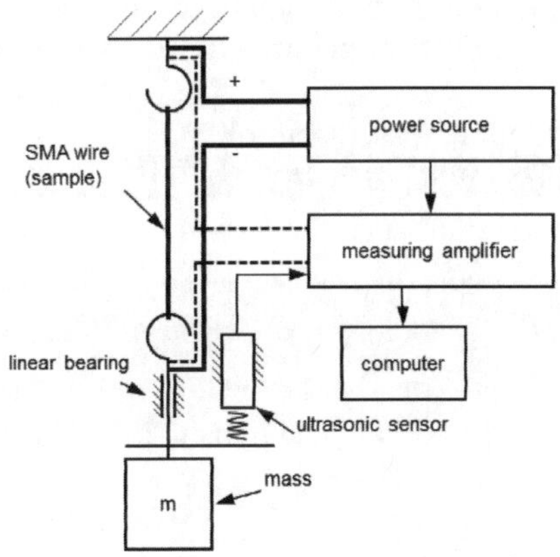

Figure 4. Schematic experimental setup for the determining of the activation and cooling time parameters.

For this experiment, a wire specimen of 100 mm in length and 0.3 mm in diameter is clamped and loaded with a constant mass of 1,080 g which corresponds to a wire tension of 150 N/mm². As specimen material, SmartFlex 90 wires by SAES Getters are used for all tests that are described in this paper. The mounted wire specimen is activated by means of a current source type EA-PS 3016-20 B, by Elektro-Automatik. The displacement of the SMA wire is detected by the ultrasonic position sensor, type CH-8501 DADM by Baumer Holding AG. The data of the displacement sensor as well as the applied voltage and current are transmitted via a MGC Plus measuring amplifier, by Hottinger Baldwin Messtechnik GmbH (HBM), to a measuring computer. The measurement software Catman Easy AP by HBM processes the data for the visualization and output. Five wire specimens with different activation and cooling times are tested in parallel. One activation cycle contains a heating and a cooling phase. The heating times are equidistant from 0.8 to 1.4 s. A voltage value of 2.4 V is set according to the SMA characteristic current course [15, 16]. Figure 5 shows the intervals at which current is applied to a specimen as well as the displacement curves of each specimen.

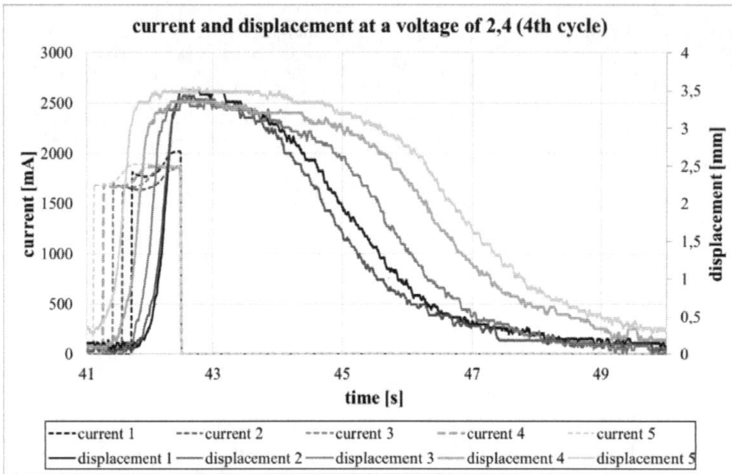

Figure 5. current and displacement of 5 SMA wire specimens as a function of time.

According to the applied voltage and the resistance change due to the phase transformation from martensite to austenite, the current value is within the range of 1,600 to 2,000 mA as can be seen in figure 5. Ideally, the current course should end at the zenith to avoid overheating of the SMA wire. Figure 5 shows also the displacement courses of the specimens, while the specimens 3 and 4 provide the best results regarding a displacement of 3.5 %, which complies with the manufacturer specifications [17]. For the control program, 1.3 s as the activation time and 7.5 s as the minimum cooling time are determined, which corresponds approximately to the values of specimen number 4. Table 2 shows the determined parameters for the time-based control.

Table 2. applied parameters for the time-based control.

specimen no.	voltage	activation time	cooling time
[-]	[V]	[s]	[s]
4	2.4	1.3	7.5

4.3. Experimental Setup and Material

The SMA actuator (see fig. 6) is based on an antagonistic principle with two SMA wires that can be activated alternatingly to switch between two actuator positions. Each of these end positions, retracted or extended, are held by a mechanical locking device. The actuating displacement is approx. 7 mm at room temperature.

The temperature influences on the functionality of the SMA actuator with time-based control are investigated on an experimental setup as shown schematically in figure 7. The SMA actuator is mounted inside a climate chamber, type SU-221, by ESPEC CORP. Two current sources, type TOE 8805, by TOELLNER Electronic Instrumente GmbH, are used to activate each of the SMA wires of the actuator. To measure the actuator position, a laser displacement sensor, type OADM 12U6460/S35A by Baumer Holding AG is used. An anchor plate, which reflects the laser beam, is attached to a linear guidance shaft. This shaft is mounted in an air-bearing to minimize the friction loss that would influence the measurement results. A ball joint connects the actuator with the shaft in order to avoid torsional moments on the linear shift. Measured values are the actual temperature inside the climate chamber and the displacement position of the SMA actuator. All these values are sent to the measuring computer for data processing via the measuring amplifier, type MX840A by HBM. The temperature of the climate chamber is controlled by the PChamber software.

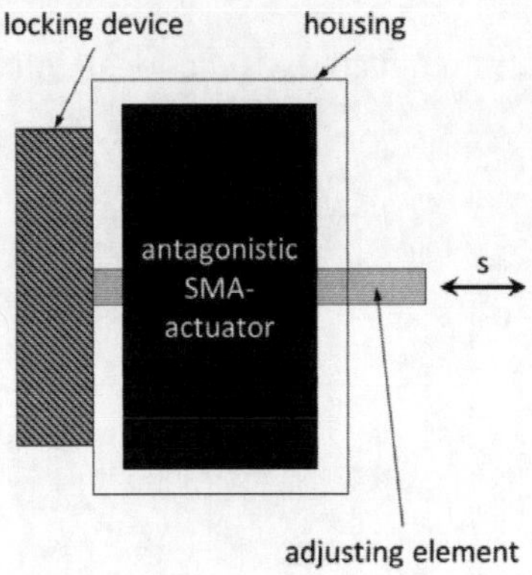

Figure 6. Schematic illustration of the antagonistic SMA actuator with a locking device to fix the retracted or extended actuator position.

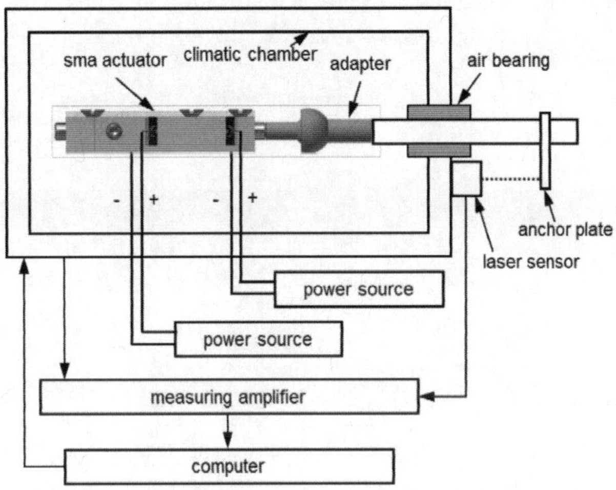

Figure 7. Schematic experimental setup for the temperature tests.

5. Results

5.1. Temperature Influence

This experiment investigates the temperature dependence of the functionality of an antagonistic SMA actuator which is activated by a time-based control. The applied temperature range from -20°C up to +70°C complies with the maximum operating temperatures for aircraft interior actuators further tests can be made if a reliable functionality of the actuator in an operating temperature range from -20°C to +55°C is fulfilled. The detailed test procedure is conducted according to the directive RTCA DO-160 (see above).

The two diagrams (see fig. 8 and fig. 9) show the displacement of the antagonistic SMA actuator exposed to defined environmental temperature variations as a function of time. Due to the time-based control, the time intervals between each activation cycle are the same.

The left diagram illustrates the influence of a decreasing temperature profile on the reliable functionality of the SMA actuator. At a temperature of about 3.5°C, the first malfunction can be detected. The actuator fails to switch to the displacement from about +7 mm to -1 mm. After one complete switching cycle of the actuator, it fails completely in the further course of the test at environmental temperatures of about 2.5°C.

The right diagram shows the influence of an increasing temperature profile up to 70°C on the functionality of the actuator. It can be seen that the absolute displacement is reduced from approx. 7 mm at room temperature to less than 4 mm at 70°C. Generally, the actuator is still switching its position, which leads to the conclusion that the functional reliability is provided at a maximum temperature limit.

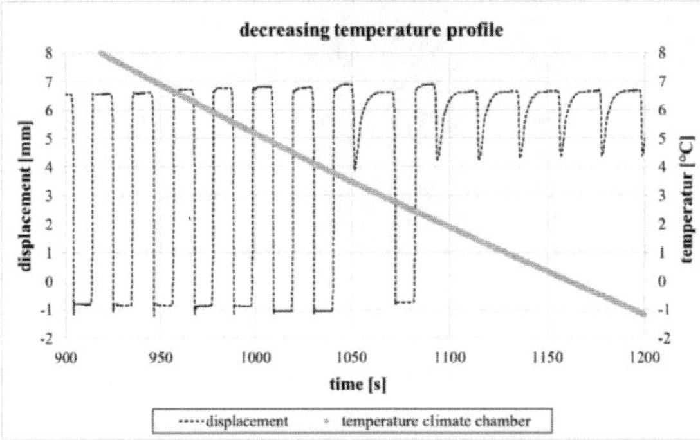

Figure 8. Displacement of the antagonistic SMA actuator influenced by a decreasing temperature profile (low temperature dependency).

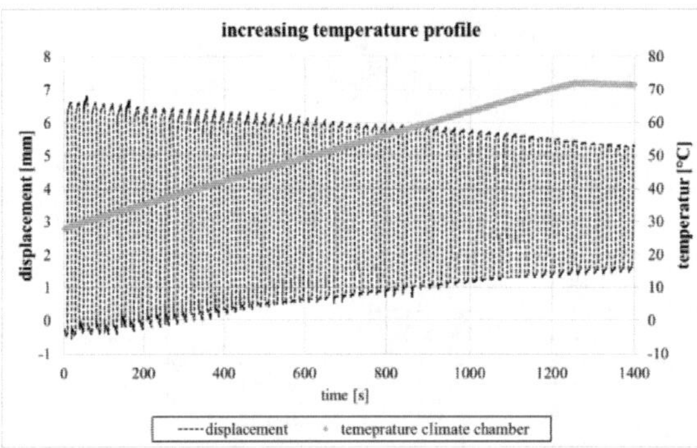

Figure 9. Displacement of the antagonistic SMA actuator influenced by an increasing temperature profile (high temperature dependency).

5.2. Discussion

There is a temperature dependency regarding the reliable functionality of the time-based controlled SMA actuator. Especially temperatures below 3.5°C can be considered as critical. According to the RTCA DO-160 directive, the actuator would not meet the requirements and may not be applied. The actuator fails to switch the displacement position which indicates that the SMA wire is not activated completely due to the low environmental temperature. The most probable reason for this is that the agonistic wire does not receive enough energy to be heated up to the austenite finish temperature and thus to transform completely into the austenite phase. For a phase transformation at lower temperatures, it is possible to either raise the current time of the control system or to raise the current value. Therefore it is necessary for the actuator control to tap the environmental temperature and adjust activation parameters.

Regarding the increasing temperature profile up to +70°C, it has been possible to prove the functional reliability of the actuator with time-based control. Although the absolute displacement of the actuator is reduced to less than 4 mm, the actuator position is still switching. The reduced displacement can be explained by a slight pre-activation of both SMA wires. In high environmental temperatures, the material is not able to transform completely into the low temperature phase. As a result of this, there are two main dissipation factors according to the antagonistic principle. On the one hand, there is a higher opposing force for the activated agonistic SMA wire to overcome because the opposed wire is slightly activated by the environmental temperature. On the other hand, the agonistic wire is also slightly pre-activated for the same reason and has thus already contracted before the controlled activation.

6. Conclusion and Outlook

Generally, a time-based control can be applied on SMA actuators under the following considerations: There may not be any extreme environmental temperature variations or – as most application fields, e.g. aircraft interior systems, involve such requirements – the control system has to be temperature dependent. In the latter case, it is necessary to apply an additional temperature sensor. It enables defined temperature ranges with related control parameters. For example, the activation time could be longer or the electrical current needs to be increased at low temperatures. The exact parameters are to be determined for each particular SMA actuator. In order to reduce the high effort needed to investigate this huge number of parameters, another approach is a resistance-based control system. The martensitic and austenitic phase have different values of electrical resistance. A change of the resistance value can be detected, which provides conclusions on the current material phase and thus on the actuator condition. In case of an incomplete phase transformation, the control keeps the SMA element activated or vice versa and it stops the activation in case of an earlier phase transformation. Furthermore, the resistance-based control can be considered as a condition monitoring device, which predicts a possible failure of the SMA actuator. Although there are still a few systematical limits, this investigation has served to prove that a time-based control can be applied as a rather simple activation solution for aircraft interiors.

Acknowledgment

The authors acknowledge the funding of the FLAAI project through the aviation research program LuFo V-1 by the Federal Ministry of Economic Affairs and Energy.

7. References

[1] M. Hinsch, J. Olthoff, Pulse Generator aviation. Industrial leadership through aviation specific setup and process concepts, Springer, Berlin Heidelberg, 2013.

[2] A. Czechowicz, S. Langbein, Design technology shape memory technology, Springer, Wiesbaden, 2013.

[3] P. Guempel; S. Glaeser, N. Jost, Shape memory alloys. Opportunities in mechanical engineering, medical technology and actuators with 119 pictures and 9 tables, expert verlag, 2004.

[4] A. Musolff, Shape memory alloys, experimental studies and development of adaptive structures, Technical University of Berlin, Berlin, 2005.
Information on https://opus4.kobv.de/opus4-tuberlin/files/1064/musolff_andre.pdf

[5] B. Fleczok, M. Lubisch, B. Kuhlenkoetter, Development of a pre-evaluation method to design an electrically activated NiTi-based shape-memory wire actuator. Higher Education Forum, Hong Kong International Conference on Engineering and Applied Sciences, Hong Kong, VR China, 2015, pp. 1-8.

[6] H. Janocha, Adaptronics and smart structures: Basics, materials, design, and applications, Springer, Berlin, New York, 2007.

[7] M. Kohl, Development of micro-actuators based on shape memory alloys, Dissertation, Institution for micro-structure technology, University Kalsruhe (TH), Kalsruhe, 2002.

[8] Fraunhofer IWU.
Information on https://www.iwu.fraunhofer.de/de/forschung/leistungsangebot/kompetenzen-von-a-bis-z/mechatronik-adaptronik/smart-materials/entriegelung-sauerstoffmasken.html

[9] Fraunhofer IWU: Smart-FLUOX: Unlockling donning masks with shape memory alloys, Fraunhofer IWU, Chemnitz, 2016.
Information on http://www.iwu.fraunhofer.de/de/forschung/leistungsangebot/kompetenzen-von-a-bis-z/mechatronik-adaptronik/smart-materials/entriegelung-sauerstoffmasken.html

[10] C. Eppler, Unlockling donning masks with shape memory alloys, Fraunhofer IWU, Chemnitz, 2013.

[11] J.P. Lawall, N.D. Mankame, R.J. Skurkis, Using resting load to augment active material actuator demand in power seats, DE Patent 102013211923 B4, Germany, 2013.

IOP Conf. Series: Materials Science and Engineering **216** (2017) 012009 doi:10.1088/1757-899X/216/1/012009

[12] J.P. Lawall, N.D. Mankame, R.J. Skurkis, Using resting load to augment active material actuator demand in power seats, US Patent 20120267928 A1, USA, 2012.
[13] Radio Technical Commission for Aeronautics Inc., RTCA DO 160D Enviromental Conditions and Test Procedures for Airborne Equipment, Washington D.C., 1997.
[14] A. Joern, Aircraft development through the example of Airbus A380. From the idea to the admission, Bachelor and Master publication, Hamburg, 2015.
[15] D.C. Lagoudas, Shape memory alloys. Modeling and engineering applications, Springer, Boston Massachusetts, 2008.
[16] M. Kristen, Investigation of the electrical control of shape memory drives in handling technology, Mechanics centre of the technical university of Braunschweig, 1994.
[17] M. Mertmann, H. Borgmann (ed.), Fatigue in Nitinol actuators, Proceedings of the actuator, Bremen, Germany, 2006.

IOP Conf. Series: Materials Science and Engineering **216** (2017) 012010 doi:10.1088/1757-899X/216/1/012010

Experimental Validation of Condition Monitoring for Electrically Activated Shape Memory Alloys for an Unlocking Device

Christian Rathmann, Benedict Theren, Benjamin Fleczok, Bernd Kuhlenkötter
Ruhr-University of Bochum, Universitätsstraße 150, 44801 Bochum, Germany

Email: rathmann@lps.rub.de; theren@lps.rub.de; fleczok@lps.rub.de; kuhlenkoetter@lps.rub.de;

Abstract. Shape memory alloys (SMA) belong to the group functional materials which can be activated thermally. Along with a phase transformation, they can remember a previously imprinted shape and have a special resistance behavior. Therefore, they can also be used as a sensor and may be capable of detecting various system states in technical systems. This paper makes a contribution by evaluating the measurability of measured variables by SMA elements. Furthermore, it investigates the technically relevant states of "blockade" and "activation" of electrically activated shape memory actuators. It develops and validates an algorithm which is able to detect a possible "blockade". Moreover, this work presents a hardware concept for a condition monitoring system of shape memory actuators.

1. Introduction

The electrical activation of SMA actuators has several advantages compared to conventional actuators, e.g. reduced space or simple methods to monitor the system [1]. Evaluating experimental data has revealed that there is a high dependency of the electrical resistance or current as well as specific actuator states such as the remaining lifetime [2], elongation [3], stress [4], temperature [3][5] or blockage [6]. By condition monitoring it is possible to provide recommendations for action in case an error is detected [7]-[9]. The following study develops a condition monitoring for the actuator state "blocked" as well as a proper "activation" in order to give specific recommendations regarding the SMA actuator system.

1.1. Structure

Preliminary different measurands are classified regarding their ability to be monitored by an SMA element. Based on this classification, specific actuator errors are assigned to certain measurands. A fault tree analysis is performed in order to derive certain recommendations for the representative actuator state "blocked". A series of experiments are performed in order to find a specific criterion to differ a blocked actuator from a fully activated one.
After evaluating the data, an algorithm is derived which is able to detect the blockage of an SMA actuator. Finally, this work develops a physical condition monitoring system with a Raspberry Pi.

2. Available Measurands and Status Detection with SMA

All measurands are grouped according to [8]-[13] and presented in Figure 1. Based on the results taken from the publications of [3][5][6][14] and experience, a qualitative assessment is performed if

SMA elements are capable to measure the specific measurands. Through measuring variables it is possible to formulate statements concerning the monitoring of system states. Consequently, this leads to a certain system intelligence.
Mechanical variables are generally measurable except for torque and rotational speed. Torque can only be applied for small ranges and rotational speed is not detectable. Process variables such as the number of produced goods or quality are measurable indirectly through SMA elements. Regarding the number of produced goods, the number of activations of the SMA element is an indicator. Information on the product quality can be gathered if the quality characteristic depends on the proper closing or opening of the SMA actuator. The thermal sensitivity of SMA elements makes them suitable for measuring thermal variables especially during the phase transformation. Electric measurands are easy to access during the electrical activation. However, it is not advantageous to exclusively use them as a sensor for measuring electrical variables. Fluidic variables can be measured by using an SMA sensor element, as shown on the basis of fluid velocity [14]. Therefore, it is possible to measure e.g. the flow rate and loss.

measurand		SMA	
		s	a/s
mech-anical	force and pressure	++	++
	stress	++	++
	velocity	o	++
	torque	o	o
	speed of rotation	n.d.	n.d.
acou-stical	vibration	o	n.d.
	sound	n.d.	n.d.
pro-cessual	process time	o	++
	production	o	+
	resource consumption	++	++
	energy consumption	o	++
	product quality	o	o
thermal	temperature	o	+
	pressure	+	+
	efficiency	n.d.	n.d.
electrical	current	n.u.	++
	voltage	n.u.	++
	resistance	n.u.	++
	power consumption	n.u.	++

measurand		SMA	
		s	a/s
fluidic	throughput	++	+
	fluid loss quantity	++	+
	pressure change	o	+
	fluid level	n.u.	n.u.
	humidity	n.d.	n.d.
geo-metric	length, displacement	+	++
	distance	+	+
	clearance	+	++
	thickness	n.u.	n.u.
	angle	o	o
material changes	change of shape	n.d.	++
	corrosion	n.d.	n.d.
	temperature	o	+
material stress	mechanical	+	++
	thermal	o	+
	environmental	n.d.	n.d.
	frequency	+	++
optical	color, texture, reflection, pattern	n.d.	n.d.
chemical	substance properties, impurities	n.d.	n.d.

o limited detectable; + detectable; ++ well detectable; n.d. not detectable; n.u. not useful; s sensor element; a/s actuator-sensor element

Figure 1. Overview of the detectable states of SMA actuators and SMA sensors.

3. Condition Monitoring with a Binary Moving SMA Actuator

The state of the actuator "blocked" is further analyzed by using a fault tree analysis. This is presented in figure 2. Different reasons for the error are summarized. The main triggers leading to a functional failure are blockage and the lack of performance of the SMA actuator system.
In the event of a loss of function due to a lack of performance of the SMA actuator, relevant aspects to be considered are SMA-unspecific events like damaged reset elements as well as material-specific events like material fatigue. The functional and structural fatigue can both be detected by the electrical resistance. Therefore, it is possible to predict the time until the required displacement cannot be reached. The other trigger blockage is particularly interesting because it directly indicates a faulty operation and thus should be prevented or at least easy to detect.

IOP Conf. Series: Materials Science and Engineering **216** (2017) 012010 doi:10.1088/1757-899X/216/1/012010

It is assumed that by changing the electrical resistance during the phase transformation it is possible to detect blockage if it occurs [6]. As it is possible to evaluate blockage, the functionality of the actuator system can be evaluated after each activation. In case of such an event, it is recommended to perform the next activation with reduced energy in order to validate the sustainability of the error. If the blockage still exists, it is possible to initiate countermeasures. However, there has not been any reliable investigation which directly focuses the blockage of an SMA actuator system. To validate the state and fault detection with SMAs, further experiments are required for its validation. For this purpose, the case of blockade is examined more closely in the following paragraph.

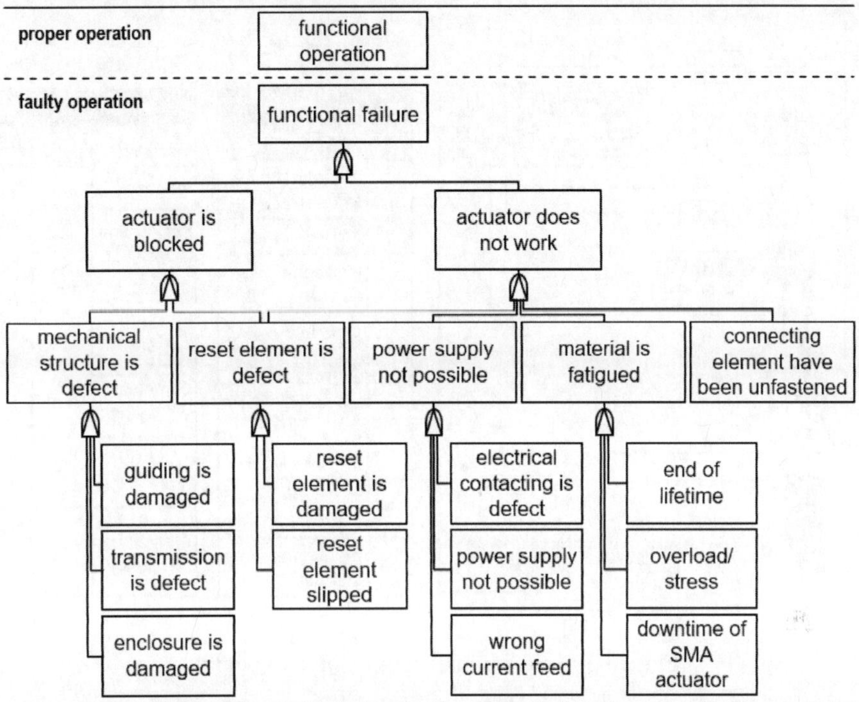

Figure 2. System states and faults of a binary SMA actuator system.

4. Material and Methods

4.1. Experimental Setup and Material

The series of experiments are performed in order to propose a model for the blockade detection of an SMA actuator. Smartflex90 from SAES Getters are used as samples. When the wires are activated, the electrical voltage is controlled. Therefore, the current arises out of voltage. For this reason, this analysis uses the current instead of the electrical resistance. The analyzed SMA actuator wire should show significant differences in its behavior of the electrical current if it is blocked compared to a regular activation. Figure 3 shows the experimental setup for this study.

The sample is mounted between an isolating layer and a copper block. This minimizes the electrical resistivity. The blocking of the wire is performed by a firm connection of the lower specimen mounting and the test frame. A blocking of the wire is defined as a reduction of the wire's movement by 90%. A complete activation is defined as a movement of 4 mm or more. All required experimental parameters are summed in figure 3. 16 wires in total are blocked five times and each wire is activated 70 times. The newly mounted samples are activated 30 times at the beginning of the measurement. Afterwards they are blocked once and activated ten times. This cycle is performed five times.

Figure 4 presents the current-time diagrams for an activated (left) and a blocked (right) SMA wire. For a better overview, the scale is fitted to the relevant area. The current is scattered on purpose. The higher current leads to a higher wire movement and allows an analysis of blocking at these movement levels. The range of the movement is from 4 mm to 6 mm. Significant differences in the current-time plots can be identified both by a qualitative analysis of the plot's trend and by a quantitative analysis of the current's level. The activated wire follows a rather cubic trend while the blocked trend can be described as rather squared. For the activated wire the current's level in the beginning of the activation process is higher than in its end. For the blocked wire, the levels vary inversely to the trend of the activated wire (See table 1).

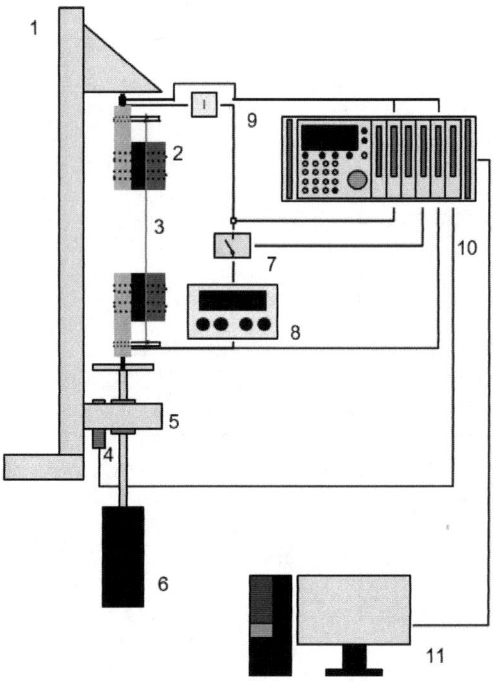

N#	component
1	test frame
2	mounting system
3	SMA sample
4	ultrasonic sensor
5	plain bearing
6	weight
7	electronic relay
8	laboratory power supply
9	measuring amplifier
10	measuring computer

parameter	unit	value
diameter	mm	0,5
length	mm	~ 120
connection	-	clipped
stress	mPa	400
current	A	~ 5
complete activation	mm	4
cooling time	s	20
measuring frequency	Hz	50

Figure 3. Experimental setup and parameters.

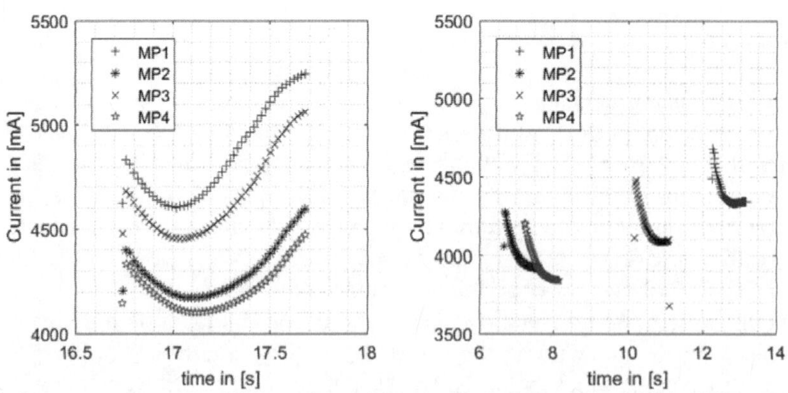

Figure 4. 2 current time plot of an activated SMA wire (left) and a blocked SMA wire (right).

5. Results

For a data analysis, a test algorithm is derived in order to distinguish blocking from activation. The process is shown in Figure 5.

The criterion "higher than 250 mA" is chosen to make a clear difference between the beginning of the activation and measurement noises. The value is based on experience and is specifically chosen for a

IOP Conf. Series: Materials Science and Engineering **216** (2017) 012010 doi:10.1088/1757-899X/216/1/012010

wire diameter of 0.5 mm. The power source delivers its electrical output after 1 ms. By a measure frequency of 50 Hz, this leads to five values until the required values are reached. In order to increase the reliability, this value was set to ten.

The algorithm is used for 80 data sets of a blocked and for 80 data sets of an activated wire. One activation and one blocking are wrongly detected. The false detected activation leads back to a lack of movement. The minimum movement is set to 4 mm but due to a deterioration, the sample movement decreased to 3.5 mm. In conclusion, the criterion "complete activation" is not met. The wrongly detected block is located in an insufficiently installed blocking tool. This causes a reduction of movement by just 80% instead of the required 90%. This makes this variation to an outliner and it will not be further regarded.

It has been possible to detect a total of 100% of the blockings and 98.6% of the activation. This leads to the conclusion that the algorithm works properly.

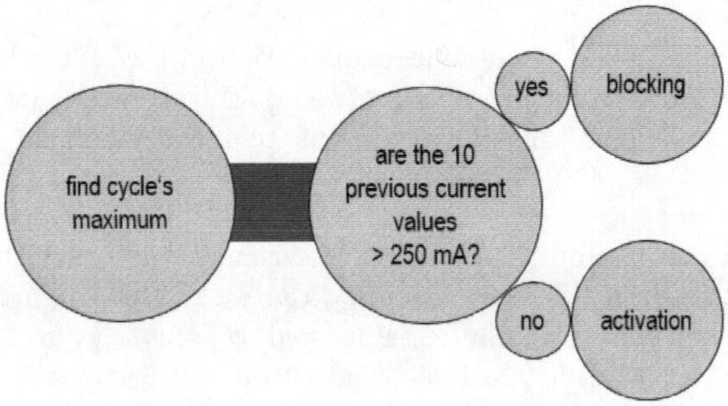

Figure 5. Test algorithm for condition monitoring of a blocked SMA wire.

Table 1. Differences depending on the actuator state.

actuator state	
activated wire	blocked wire
lower current at beginning of activation compared to end	**higher current** at beginning of activation compared to end
roughly cubic trend	roughly square trend

6. Concept of a Condition Monitoring System for Electrically Activated SMA Actuator Systems operating binary

Setting up a condition monitoring system requires further components such as a sensor to acquire current and voltage. The sensor sends its signal to a microcontroller with enough power for data processing and saving as well as an Ethernet or Wi-Fi connection to send out messages in case of detected errors. In this case a raspberry is chosen. For a prototype, the LMP92064 by Texas Instrument is used as a sensor. It delivers its data via an SPI Interface. The most fitting micro controller is a raspberry pi because it combines all requirements and is an open source system. Figure 6 shows the circuit diagram.

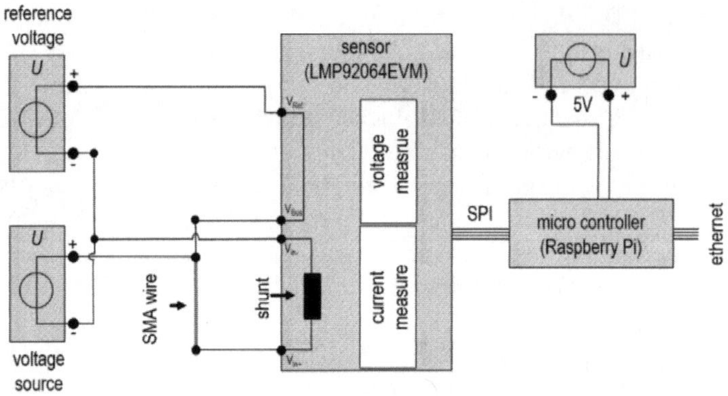

Figure 6. Overview of the hardware concept of an SMA actuator condition monitoring system.

This concept can be seen as a hardware concept of [15][16]. The reliability of SMA actuators is increased by the more precise surveillance of system states. This decreases failure and, in consequence, the availability of technical systems which rely on the usage of SMA actuators.

7. Conclusion and Outlook

The paper has evaluated the possibility of using SMA elements as a sensor for acquiring different measure variables. As a result, it has been possible to prove that SMAs are well suited to measure especially mechanical, processual or fluidic variables and to be used as an actuator sensor system to detect errors and states. This work has examined how to detect the blockade of an operating SMA actuator system. Moreover, an algorithm for this purpose has been developed and validated. In conclusion, the approach to develop a condition monitoring system for SMA actuators is a promising field of research. This also increases the possible applications, which, however, will also require new business models.

7.1. Outlook

To put sensoring and condition monitoring with SMAs into practice, more research is needed. Further studies should focus on other faults and states than those presented in this paper. Additionally, it is necessary to investigate the extent to which factors such as environmental issues or actuator-specific factors influence the possibility of error detection. Furthermore, investigations are required to develop proper electronics which are customized to the needs of SMAs. These electronics have to consider the requirements of SMAs and should realize low-cost condition monitoring solutions. Finally, it will be necessary to run a long-term test of SMA actuator systems under real conditions.

8. Acknowledgment

The authors acknowledge the funding of the FLAAI project through the aviation research program LuFo V-1 by the Federal Ministry of Economic Affairs and Energy.

9. References

[1] Butera, F.; Coda, A.; Vergani G. (2007): Shape memory actuators for automotive applications. In: Nanote IT newsletter 2007, S. 12–16, revised 07.03.2017.
[2] Rathmann C, Fleczok B and Kreimeier D (2015a) An investigation using self-sensing to estimate the lifetime of shape memory actuators. In: Proceedings of the SICASE 2015 Seoul International Conference on Applied Science.
[3] Zhang JJ, Yin YH and Zhu J-Y (2013) Electrical resistivity-based study of self-sensing properties for shape memory alloy-actuated artificial muscle. Sensors (Basel, Switzerland) 13(10): 12958–12974.
[4] Lan CC and Fan CH (2010) An accurate self-sensing method for the control of shape memory alloy actuated flexures. Sensors and Actuators A: Physical 163(1): 323–332.

[5] Novák V, Šittner P, Dayananda GN, Braz-Fernandes FM and Mahesh KK (2008) Electric resistance variation of NiTi shape memory alloy wires in thermomechanical tests: Experiments and simulation. Materials Science and Engineering: A 481-482: 127–133.

[6] Herrera GA, McKnight GP, Gao X, Johnson N and Browne AL (2011) Use of intrinsic electrical resistance changes in shape memory alloys as robust actuator state and fault detection sensors. In: Proceedings of the SMASIS 2011 Conference on Smart Materials, Adaptive Structures and Intelligent Systems. New York: ASME Digital Collection, pp. 133–138.

[7] Davies A (2012) Handbook of Condition Monitoring: Techniques and Methodology. Dordrecht: Springer Netherlands.

[8] Klein U (2008) evaluation regarding vibration diagnosis of machinery. Düsseldorf: Verl. Stahleisen.

[9] VDI-Society of Production and logistics(1999) Condition Monitoring Beuth 03.080.10(2888).

[10] Czichos, H. (2015): Mechatronik. Grundlagen und Anwendungen technischer Systeme. 3., überarb. u. erw. Auflage. Wiesbaden: Vieweg +Teubner.

[11] Hering E and Schönfelder G (2012) Sensors in Science and Technics: Functionality and field of use. Wiesbaden: Vieweg +Teubner.

[12] Hesse S and Schnell G (2014) Sensors for Process and Fabrication automation: Function, Execution and application. Wiesbaden: Springer Vieweg.

[13] Standards committee acoustics, noise canceling and vibration technics (2011) Condition Monitoring and Diagnosis of machinery – general usage: Beuth 25.020(17359).

[14] Seelecke S (2015) Sensing Properties of SMA Actuators and Sensorless Control. In: Czechowicz A and Langbein S (eds) Shape Memory Alloy Valves: Basics, Potentials, Design. Cham, s.l.: Springer International Publishing, pp. 73–87.

[15] Rathmann C, Remmetz T and Kreimeier D (2015b) Maintenance of Shape Memory Actuator Systems - Applications, Processes and Business Models. Procedia CIRP 30: 84–89.

[16] Rathmann C, Czechowicz A and Meier H (2013) An Investigation of Service-Oriented Shape Memory Actuator Systems for Resource Efficiency. In: Proceedings of the SMASIS 2013 Conference on Smart Materials, Adaptive Structures and Intelligent Systems, pp. V001T04A006.

IOP Conf. Series: Materials Science and Engineering **216** (2017) 012011 doi:10.1088/1757-899X/216/1/012011

Seismic Response Analysis of an Unanchored Steel Tank under Horizontal Excitation

Zhang Rulin, Cheng Xudong and Guan Youhai
College of Pipeline and Civil Engineering, China University of Petroleum, Qingdao 266580, China

Email: zhangrulin@upc.edu.cn;

Abstract. The seismic performance of liquid storage tank affects the safety of people's life and property. A 3-D finite element method (FEM) model of storage tank is established, which considers the liquid-solid coupling effect. Then, the displacement and stress distribution along the tank wall is studied under El Centro earthquake. Results show that, large amplitude sloshing with long period appears on liquid surface. The elephant-foot deformation occurs near the tank bottom, and at the elephant-foot deformation position maximum hoop stress and axial stress appear. The maximum axial compressive stress is very close to the allowable critical stress calculated by the design code, and may be local buckling failure occurs. The research can provide some reference for the seismic design of storage tanks.

1. Introduction

With the increasing demand for soil energy in the world, the large storage tanks have become important foundation for the energy reserves. Storage steel tanks often store flammable and explosive liquid media, once suffered earthquake damage, a large number of flammable liquid leakage and burning, and even lead to devastating secondary disasters. So the seismic performance of steel tanks has attracted the attention of scholars all over the world. The research of storage tank often uses the classic Haroun-Housner model to simulate the coupling of tank and liquid [1], and different simplified mechanical models of tank are studied by literature [2-3]. Based on the mechanical model with fluid masses and convective spring stiffness given by Newmark and Rosenblueth [4], Seleemah A.A. and El-Sharkawy M. [5] investigated the seismic response of elevated broad and slender liquid storage tanks isolated by elastomeric or sliding bearings, and studied the accuracy of predictions of SAP2000 vs. 3DBASIS-ME programs.

However, the simplified model is difficult to get tank wall stress and displacement distribution. Actually, the seismic problem of tank belongs to a three-dimensional space problem, and the performance of tank is very complicated. Especially, it is difficult for the unanchored tank to apply the simplified method due to the nonlinear lift-off effect at the tank bottom. Bayraktar [6] studied the issue of the tank uplift and showed that the displacement and stress response became more obvious after uplift. ZHANG [7] studied the horizontal seismic response of unanchored tanks based on the numerical simulation method. Hosseinzadeh [8] compared the seismic assessment of existing steel oil storage tanks by API650-2008 provisions with FEM analyses, the results demonstrate there are some imperfections in the code requirements that require further investigation in some cases.

In this paper, a 3-D FEM model of an unanchored tank is established based on the software ANSYS, in which it considers the large amplitude sloshing of liquid, liquid and tank coupling effect and contact characteristics between the bank bottom and concrete foundation, and the deformation and stress distribution along the height of tank under horizontal earthquake action has been studied.

2. FEM model of steel tank and input seismic wave

2.1. FEM Model and parameters

The studied tank is a steel tank, and the tank body part adopts Q235, the inside diameter is 18.9 m, high 14.5m, the tank wall is 12.30 m high, the vault height is 2.25m, the liquid level is 10.5m height. Tank wall thickness is divided into 4 sections, from the bottom are 12mm, 10mm, 8mm and 6mm. The vault thickness is 6mm. The material parameters of the tank are as flows: the elastic modulus is 206GPa, the poisson ratio is 0.3, the density is 7800kg/m^3, and the coefficient of linear expansion is 1.0*10^{-5}. The reinforcing ring uses angle steel 63*8*100. In this part of the research, steel storage tanks were analyzed using nonlinear FEM analysis using ANSYS software [9]. The FEM model of fluid-tank coupling system based on ANSYS is shown in figure 1.

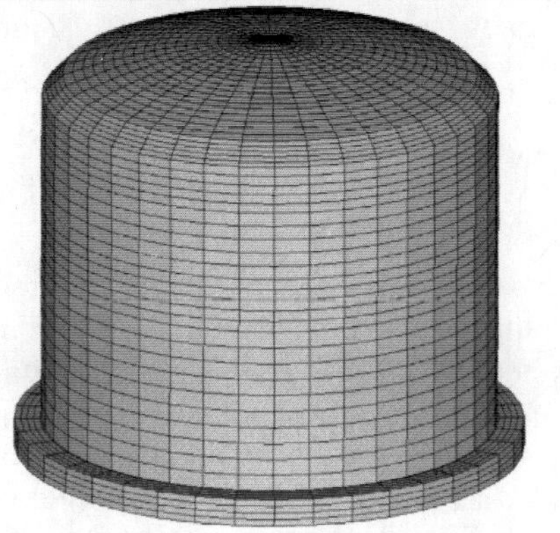

Figure 1. FEM model of the liquid storage tank.

In the FEM model, the tank wall, the bottom plate and the vault use the element of Shell181. Assuming that the fluid is incompressible and uses the element of Fluid80, which is well suitable for fluid-solid coupling analysis. The stiffening ring is made of Beam188 elements. The Beam188 and elements of Shell181 are used to share the node in the same joints, the FEM mesh is divided and the tank wall is coordinated.

This paper considers the coupling relationship between the liquid and the tank wall. The radial degrees of freedom are coupled between the liquid element and the tank wall element at the same location, and coupling the vertical degrees of freedom at the same locations at the bottom of the tank. In this case, along the height of the tank wall, the shell and fluid will keep the displacement coordination in the tangential direction, and no mutual penetration in the radial direction. At tank bottom the fluid is allowed the relative sliding along the horizontal direction, and also not allowed to penetrate in the vertical direction.

Under strong earthquake nonlinear lift-off may appear between the foundation and the tank bottom. Face-to-face contact is adopted between the tank bottom and the base of the platform, and the contact element and the target element which are matched with the shell element and the solid element are selected. Goal element is set at the tank bottom, and surface contact element is set on the cap.

2.2. Seismic wave input

When the seismic wave is input, the bottom of the concrete platform is fixed with three directions. In this paper, the tank is located in the seismic fortification area of 8 degree, which belongs to class III site. According to site condition and requirements of code for design of vertical cylindrical welded steel oil tanks, El Centro (North-South direction) earthquake is selected as the horizontal input exaction. The peak acceleration of seismic wave is adjusted to 0.4g according to Chinese seismic code, and the duration of seismic wave is 30s. The time history curve and the corresponding Fourier spectra of seismic wave are shown in figure 2 and figure 3.

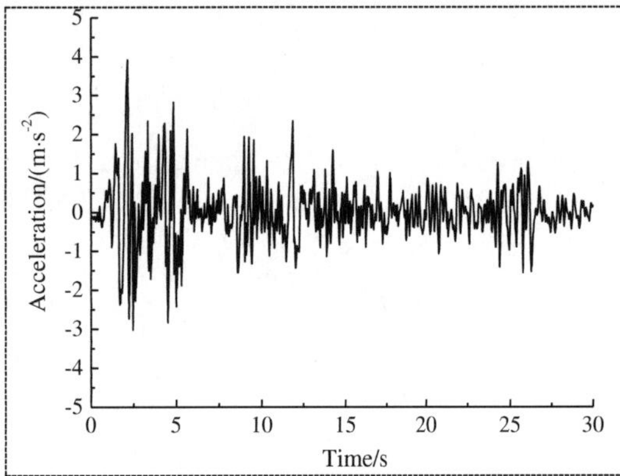

Figure 2. Acceleration of input seismic wave.

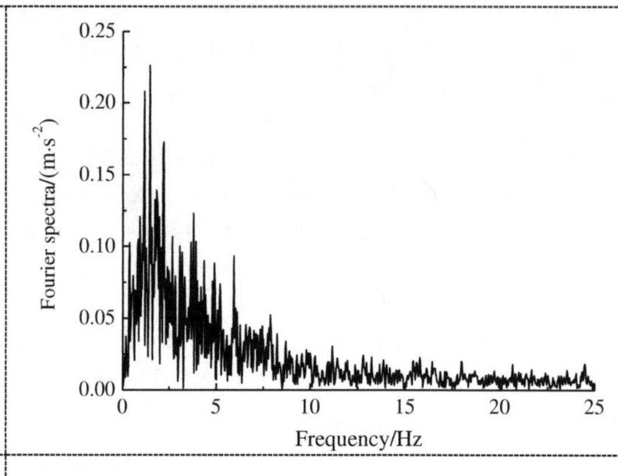

Figure 3. Fourier spectra of input seismic wave.

3. Analysis of seismic response results

3.1. Modal analysis results

Modal analysis is the basis of dynamic analysis, and it is also a means to check whether the finite element model is established and the parameters of model calculation are correct or not. According to the FEM model of tank liquid system established in this paper, and modal analysis is carried out. The liquid sloshing frequency and liquid tank coupled vibration fundamental frequency are 0.23Hz and 7.05Hz respectively, which is very close to the calculation results by the formula in the latest National Standards of the People's Republic of China, "Code for design of vertical cylindrical welded steel oil tanks"(GB50341-2014) [10] are respectively 0.22Hz and 7.19Hz. That is to say, the finite element analysis model of the tank-liquid system in this paper is reasonable and can be used for dynamic analysis.

3.2. Sloshing displacement of liquid

Choose the left and the right position on the free liquid surface near the tank wall, along the input direction of seismic wave, and denoted as “LEFT” and “RIGHT”, and the time history curve of vertical sloshing displacement is shown in Figure 4.

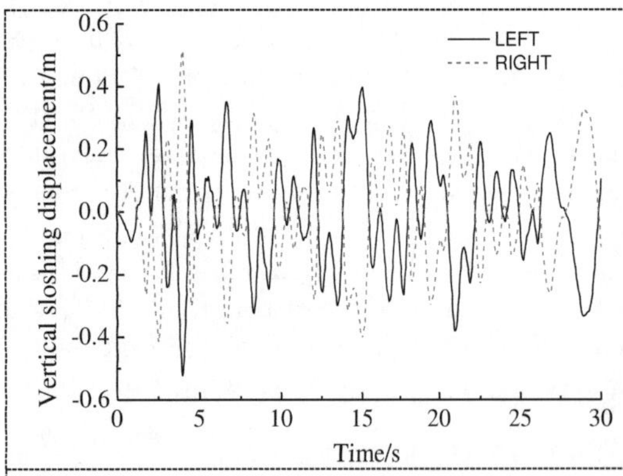

Figure 4. Vertical sloshing displacement of liquid surface.

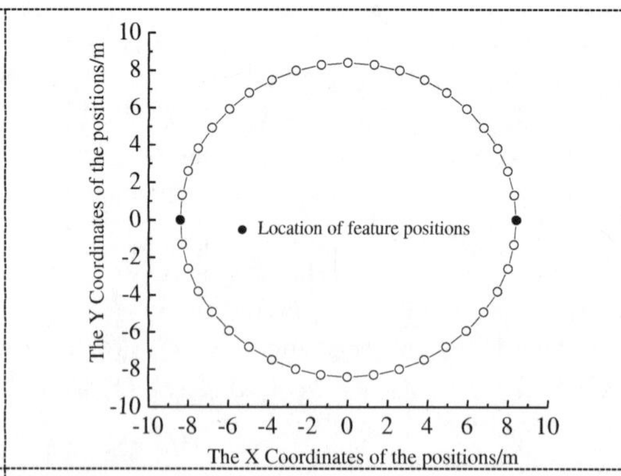

Figure 5. The selected feature points on the tank (plan)

It can be seen from the Figure 4, on both sides of the tank liquid surface, it occurs large amplitude sloshing with long period under the horizontal seismic wave excitation, and sloshing history time curves of the two ends are basically symmetrical. Among them, the maximum sloshing height at the

IOP Conf. Series: Materials Science and Engineering **216** (2017) 012011 doi:10.1088/1757-899X/216/1/012011

left end reaches 0.409m, the maximum sloshing height of the right end reaches 0.515m, but both do not exceed the maximum height of the tank wall. In addition, due to the difference of stiffness between the liquid and the tank body is great, and the maximum shaking height value appears at 3.98s(RIGHT), which is later than peak value time of the input seismic wave, which is at 2.14s.

3.3. Displacement analysis of tank

Select the feature point's location on the tank section along the input direction of seismic wave, and study the displacement distribution along the height of tank wall, which are shown in figure 5.

The horizontal displacement on the left side and right side appeared maximum response at 2.44s and 2.30s respectively, and it is later than seismic wave peak value time 2.14s. Figure 6 is the maximum horizontal displacement the distribution along the tank wall at 2.44s (LEFT) and 2.30s (RIGHT).

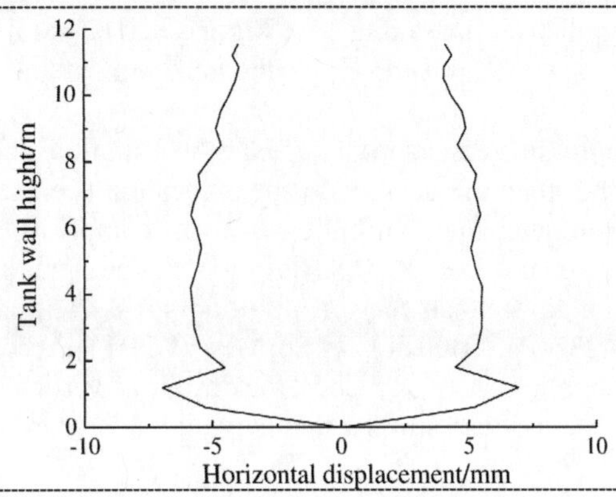

Figure 6. The horizontal displacement distribution along the tank wall height

From the Figure 6, the tank wall near the bottom appears to be a certain elephant-foot deformation, and the largest displacement is 6.97mm (LEFT) at the height of 1.2m from the bottom of the tank. At the height of 1.8m from the bottom of the tank, due to the interaction of both axial compressive stress and hoop tensile stress, the tank wall has a larger shrinkage deformation. At the same time, the horizontal displacement distribution at 8.5m from the bottom of the tank is reduced due to the constraint of the stiffening steel ring.

3.4. Stress results of tank

Select the same location points as shown in figure 5. As the radial stress distribution along the height of the tank wall is very small, it is not listed here, and the hoop stress and the axial stress distribution along the height direction of the tank wall are given in figure 7 and figure 8.

As seen from figure 7, the hoop tensile stress increases rapidly at the beginning and then decreases from the bottom of the tank. Due to the effect of stiffening ring, the hoop tensile stress at 8.5m premises decreases. The maximum tensile stress appears in the pot at 1.2m from the tank bottom, which is also in the neighborhood of the elephant-foot deformation. The maximum hoop tensile stress is 103.92MPa (RIGHT). The peak hoop stress is smaller than the yield stress of steel tank, which is 235 MPa. That is, the maximum deformation of the tank does not reach the yield limit state of the steel material.

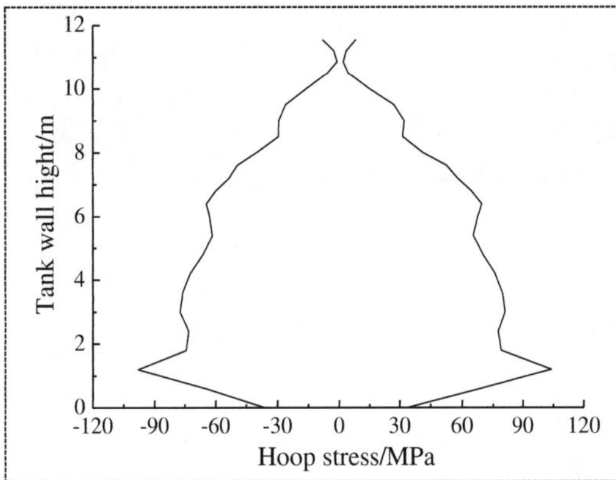

Figure 7. The hoop stress distribution along the tank wall height.

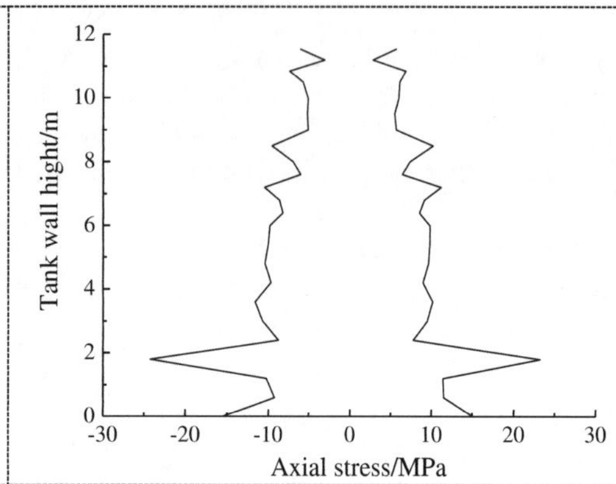

Figure 8. The axial stress distribution along the tank wall height.

From figure 8, the axial compressive distribution appears the mutation step shape, and the peak value occur at position slightly higher than the peak hoop stress location (i.e. elephant-foot position), that is at 1.8m from the bottom. Moreover, the distribution of axial compressive stress on the tank wall is larger than that of the base plate, and the upper part of the axial compressive stress decreases sharply, and the maximum value is far lower than that of the hoop stress. The maximum axial compressive stress of the tank wall is 23.23MPa (RIGHT), 24.26 MPa (LEFT). According to the latest National Standards of the People's Republic of China, "Code for design of vertical cylindrical welded steel oil tanks"(GB50341-2014)[10], the critical stress of tank wall under earthquake is calculated by the following equation,

$$[\sigma_{cr}]=0.22Et/D \tag{1}$$

where $[\sigma_{cr}]$ is the allowable critical stress for the tank wall, MPa; E is the elastic modulus of the bottom shell material at the design temperature, MPa; t is effective thickness of bottom shell, m; and D is the inner diameter of the tank, m.

According to the equation (1), the allowable critical stress of the tank is 28.77MPa. From the figure 8, the maximum axial pressure value along the whole tank is 24.26 MPa, and it is very close to the allowable compressive stress value 28.77MPa. Therefore, in this case, there may be local buckling failure occurs near the tank wall bottom under the input El Centro earthquake with PGA=0.4g.

4. Conclusions

By using the El Centro seismic wave as input, the results of liquid sloshing, the displacement and stress distribution along tank wall are analysed, and the following conclusions can be obtained,

(1)The liquid surface in the tank occurs large amplitude vertical sloshing with long period characteristic, and the sloshing time history curve at the two end points on the surface is basically symmetrical. The maximum sloshing height reaches 0.515m, and does not exceed the height of tank wall under earthquake.

(2)The elephant-foot deformation occurs at 1.2m height form the tank bottom. Due to the volume of the tank is relatively small, and the biggest elephant-foot deformation does not reach the material yield limit state. The peak axial compressive stress does is very close to the allowable compressive stress value in code, so there maybe local yield instability destruction near the bottom of the tank wall.

(3) It is suggested that the soil-structure dynamic interaction is considered in the future study because the foundation effect may have a great influence on the dynamic response of the tank.

5. Acknowledgments

The authors appreciate the support of National Natural Science Foundation of China (Grant No. 51408609) and Shandong Province Higher Educational Science and Technology Program (Grant No. J14LG51).

IOP Conf. Series: Materials Science and Engineering **216** (2017) 012011 doi:10.1088/1757-899X/216/1/012011

6. References

[1] Haroun, M A 1983 Vibration studies and tests of liquid storage tanks. *Earthq Eng Struct D.***11** 179-206.

[2] Goudarzi M A and Sabbagh-Yazdi S R 2009 Numerical investigation on accuracy of mass spring models for cylindrical tanks under seismic excitation. *Int J Civ Eng.* **7** 190-202.

[3] Sun J, Cui L, Hao J, and Wang, Z 2013 The simplified mechanical model and the seismic response for isolation tank with floating roof. *J Harbin Inst Tech.* **45** 118-22.

[4] Newmark N M and Rosenblueth E, 1971 Fundamentals of earthquake engineering, Prentice-Hall, Englewood Cliffs, N.J.

[5] Seleemah A A and Mohamed E S 2011 Seismic analysis and modelling of isolated elevated liquid storage tanks *Earthq. Struct.* **2** 397-412.

[6] Bayraktar A, Sevim B, Altunışık A C, and Türker T 2010 Effect of the model updating on the earthquake behavior of steel storage tanks. *J Constr Steel Res,* **66** 462-69

[7] Zhang X C, Cai Y Q, 2013 Horizontal seismic response numerical analysis of large unanchored storage tank, *J Wuhan Univ Tech.* **35** 94-7

[8] Hosseinzadeh N, Kazem H, Ghahremannejad M, Ahmadi E, and Kazem N 2013 Comparison of api650-2008 provisions with fem analyses for seismic assessment of existing steel oil storage tanks. *J Loss Prevent Proc.* **26** 666-75.

[9] ANSYS software. Engineering simulation software, 2007 version 14. Pennsylvania, United States: Canonsburg.

[10] National Standards of the People's Republic of China 2014 *Code for design of vertical cylindrical welded steel oil tanks*(Beijing: China Planning Press)

IOP Conf. Series: Materials Science and Engineering **216** (2017) 012012 doi:10.1088/1757-899X/216/1/012012

Simulation of Corrosion Process for Structure with the Cellular Automata Method

M C Chen[1,2], Q Q Wen[1]

[1]School of Civil Engineering and Architecture, East China Jiaotong University, Nanchang, 330013, China,
[2]Jiangxi Provincial Key Laboratory of Simulation and Control of Building Process

Email: mcchen@ecjtu.edu.cn; 550358136@qq.com;

Abstract. In this paper, from the mesoscopic point of view, under the assumption of metal corrosion damage evolution being a diffusive process, the cellular automata (CA) method was proposed to simulate numerically the uniform corrosion damage evolution of outer steel tube of concrete filled steel tubular columns subjected to corrosive environment, and the effects of corrosive agent concentration, dissolution probability and elapsed etching time on the corrosion damage evolution were also investigated. It was shown that corrosion damage increases nonlinearly with increasing elapsed etching time, and the longer the etching time, the more serious the corrosion damage; different concentration of corrosive agents had different impacts on the corrosion damage degree of the outer steel tube, but the difference between the impacts was very small; the heavier the concentration, the more serious the influence. The greater the dissolution probability, the more serious the corrosion damage of the outer steel tube, but with the increase of dissolution probability, the difference between its impacts on the corrosion damage became smaller and smaller.To validate present method, corrosion damage measurements for concrete filled square steel tubular columns (CFSSTCs) sealed at both their ends and immersed fully in a simulating acid rain solution were conducted, and Faraday's law was used to predict their theoretical values. Meanwhile, the proposed CA mode was applied for the simulation of corrosion damage evolution of the CFSSTCs. It was shown by the comparisons of results from the three methods aforementioned that they were in good agreement, implying that the proposed method used for the simulation of corrosion damage evolution of concrete filled steel tubular columns is feasible and effective. It will open a new approach to study and evaluate further the corrosion damage, loading capacity and lifetime prediction of concrete filled steel tubular structures.

1. Introduction

Concrete filled steel tube (CFST) has been widely used in high-rise building, bridge, high voltage transmission tower, oil drilling platform and offshore structure and other engineering structures owing to its excellent mechanical properties [1]. However, since long-term exposure to air or seawater, the outer steel tube is vulnerable to aggressive agents like acid rain, ocean climate, sea water and so on. As etching time goes, there will be large area corrosion and losing on the surface of outer steel tube, decreasing the rigidity of CFST structure, role of combined action and degradation of structural resistance, and thereby affecting its service life, that is durability and safety. On the study of the mechanical performances of CFST structures in corrosive environment, many experimental researches have been carried out [2, 3]. Nevertheless, none of these studies takes into account that the environmental corrosion damage process is a complex dynamic system and the degradation of structural resistance is time-dependent. Although many scholars have used the Monte Carlo [4] and Cellular Automata method [5,6] to study the damage evolution process of localized corrosion on steel structure, none of them focused on corresponding simulation researches on the uniform corrosion damage evolution process of steel structure. Therefore, study on the simulation of uniform corrosion damage evolution process for the CFST structure in the aggressive environment is of theoretical value and practical significance. Due to many influencing factors in environment, corrosion damage can not be accurately measured [7]. As an effective tool to research complex nonlinear system and non-uniform dynamic system, the cellular automata method has better universal property and stability in simulating the corrosion damage evolution process.

In this paper, a two-dimensional cellular automata (CA) model will be used to simulate uniform corrosion damage evolution process of CFST in corrosive environment. The influence of solution concentration and dissolution probability of corrosive agents is discussed. As a benchmark example, the model is used to simulate the process of corrosion damage of outer steel tube of CFST under acid rain environment. Present simulation is compared with theoretical solutions and experimental results.

2. The model of cellular automata

2.1 Composition

Cellular automata (CA) is a complex dynamic system which is discrete in time and space [8]. It consists of cellular, neighbor, space, status, local evolution rule, initial conditions, boundary conditions and so on [9]. Figure 1 shows the relationships among them.

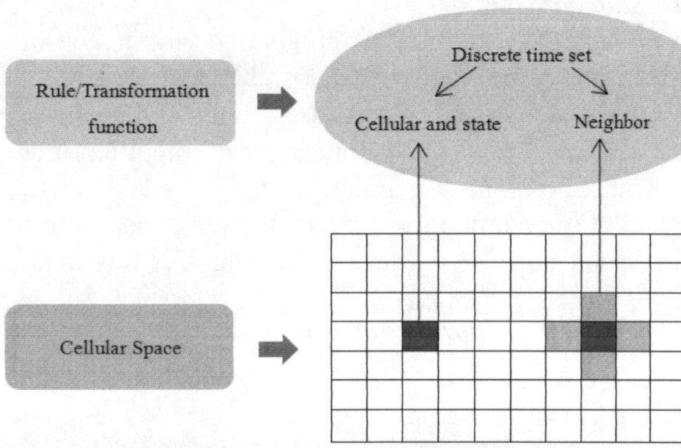

Figure 1. Schematic diagram of cellular automata composition

Cellular is the most basic component of cellular automata. Neighbors in two-dimension cellular automata have many forms, three common types of them are Von Neumann, Moore and Extended Moore. The local evolution rule is a dynamic function, which determines the state of central cell at next moment from the state of central cell and its neighbors at present moment. It is also called the state transition function expressed as in the following:

$$C_i^{t+1} = \varphi(C_i^t, C_{i,1}^t, C_{i,2}^t, \cdots, C_{i,j}^t, \cdots, C_{i,n}^t) \quad (1)$$

Where C_i^{t+1} is the state of central cell i at next moment t+1, C_i^t is the state of central cell i at current moment t; $C_{i,j}^t$ indicates the state of cellular neighbor j at current moment t; φ is local evolution coefficient, and n is the number of neighbors around central cell. Boundary conditions are used to determine the special physical characteristics of the system during simulation process, and four types of boundary conditions are usually periodic, fixed, adiabatic and reflection. In this paper, the periodic boundary condition was adopted.

2.2 Local evolution rule

The metal and etchant solution space are discretized in CA model, and the dynamic characteristics of metal corrosion are described through the evolution rules of cellular state at discrete time. In this paper, the system is composed of metal and etchant solution, and it is divided into a 1000 × 1000 ordered cellular grids, as shown in Figure 2. Among them, A and B represent corrosive and non-corrosive cell, respectively, which they have direction, also can randomly spread. They are used to simulate diffusion process in electrolyte system. Solution concentration *c* is determined by the ratio of A/B. M representing the metal cell, it can be corroded by A, and cannot move at will during simulation process. A and B can occupy the cell position of M at will.

A	B	A	B	A	B
B	A	B	A	B	A
A	B	A	B	A	B
M	M	M	M	M	M
M	M	M	M	M	M
M	M	M	M	M	M

Figure 2. Cellular grid

According to the above local reaction rules along with MATLAB software programming, the corrosion damage simulation under different conditions is carried out.

2.3 Characterization of corrosion damage

The shape of metal surface will change with time. After a period of time, uneven pattern on the surface has been formed. For uniform corrosion, the method of characterizing the degree of corrosion are usually based on corrosion weight, depth, and current density. From the engineering application point of view, the effective section size of corroded components is an important index to reflect the life and safety of structure or equipment. Therefore, it is more practical to characterize the extent of corrosion by the corrosion depth method. The etch depth can be expressed as the ratio of *N(t)* (total number of metal cells which are corroded during *t* (time)) over *M(t)* (which is the number of metal cells inside each layer of cells), that is,

$$H(t) = N(t) / M(t) \qquad (2)$$

Where *M(t)* in present paper is set to be 1000.

3. Comparison of simulation results with theoretical prediction and experimental results

In terms of the equivalent corrosion depth method, this paper defines the overall height of the metal cell M equal to 3.640mm in thickness of experimental steel tube, and thus the length of single metal cellular is 3.64/5000=0.00728mm. The simulation time step t=100 is equivalent to the actual corrosion time of one week during the simulation process. Due to consideration of the influence of gravity, when defined the direction of solution cellular, the probability that cellular moves to the left and the right is 1/4, respectively and the ratio of the probability that cellular moves to upward over downward is 1:9. The corrosion depth on the surface of concrete filled steel tube (CFST) under acid rain attack was predicted by using the cellular automata model based on the above local transition rules, with concentration c=0.1, 0.3, 0.5, 0.7 and dissolution probability p=0.1, 0.3, 0.5, 0.7, 0.9, respectively. The curves of time-dependent corrosion thickness under different dissolution probability when c=0.1 and the curves of time-dependent corrosion thickness under different solution concentration when p=0.1 were obtained, as shown in Figure 3 and Figure 4.

Figure 3 and Figure 4 indicate that the corrosion depth of outer steel tube wall with the variation of etching time is stable after second weeks either for different dissolution probabilities or different solution concentrations.

Comparing the data when solution concentration c=0.1 and dissolution probability p=0.1 with theoretical predictions by the Faraday's first law [10] and experimental results. The thicknesses of steel tube at different etching time are shown in Table 1, the curve of time-corrosion depth of the outer steel tube each week and the curve of time- thickness of the outer steel tube are shown in Figure 5 and Figure 6.

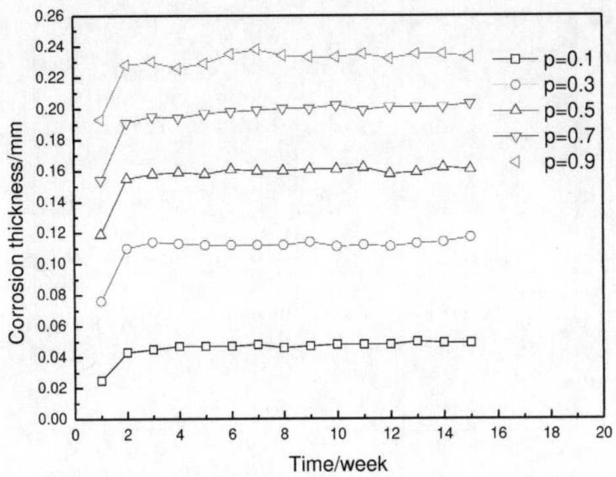

Figure 3. The curve of time-dependent corrosion thickness under different dissolution probability

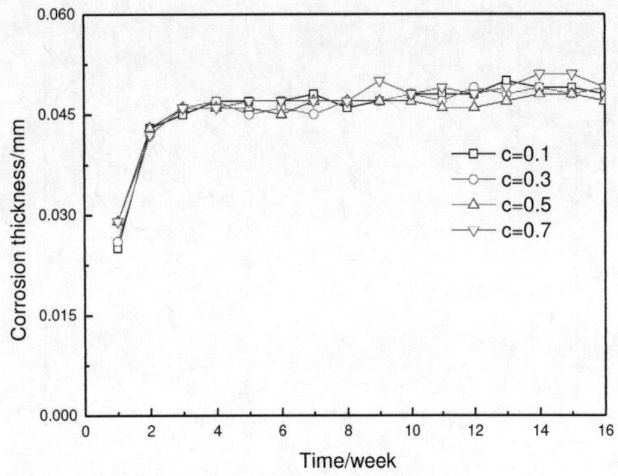

Figure4. The curve of time-dependent corrosion thickness under different solution concentration

Table 1. Chart of steel thickness date (unit: mm)

Time/week	0	1	2	3	4	5	6	7	8
theoretical wall thickness	3.640	3.595	3.550	3.505	3.460	3.415	3.370	3.325	3.280
measured wall thickness	3.640	3.599	3.556	3.512	3.467	3.421	3.374	3.327	3.280
simulated wall thickness	3.640	3.615	3.572	3.527	3.480	3.433	3.386	3.338	3.292
weekly theoretical corrosion depth	0.000	0.045	0.045	0.045	0.045	0.045	0.045	0.045	0.045
weekly measured corrosion depth	0.000	0.041	0.043	0.044	0.045	0.046	0.047	0.047	0.047
weekly simulated corrosion depth	0.000	0.025	0.043	0.045	0.047	0.047	0.047	0.048	0.046

Time/week	9	10	11	12	13	14	15	16	
theoretical wall thickness	3.235	3.190	3.145	3.100	3.055	3.010	2.965	2.920	-
measured wall thickness	3.232	3.184	3.136	3.087	3.036	2.985	2.935	2.882	-
simulated wall thickness	3.245	3.197	3.149	3.101	3.051	3.002	2.953	2.905	-
Weekly theoretical corrosion depth	0.045	0.045	0.045	0.045	0.045	0.045	0.045	0.045	-
weekly measured corrosion depth	0.048	0.048	0.048	0.049	0.051	0.051	0.050	0.053	-
weekly simulated corrosion depth	0.047	0.048	0.048	0.048	0.050	0.049	0.049	0.048	-

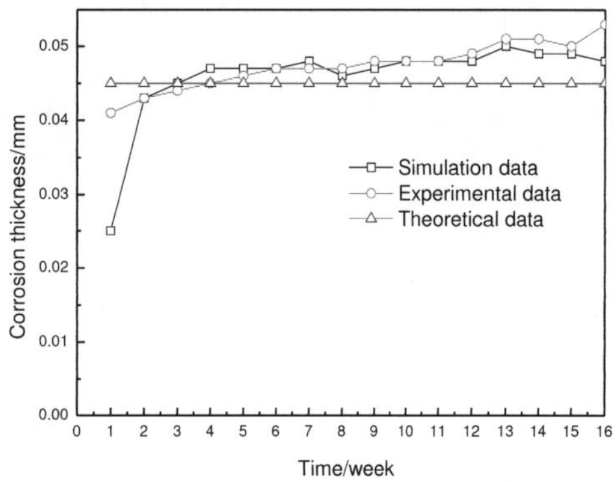

Figure 5. Time-dependent average corrosion thickness

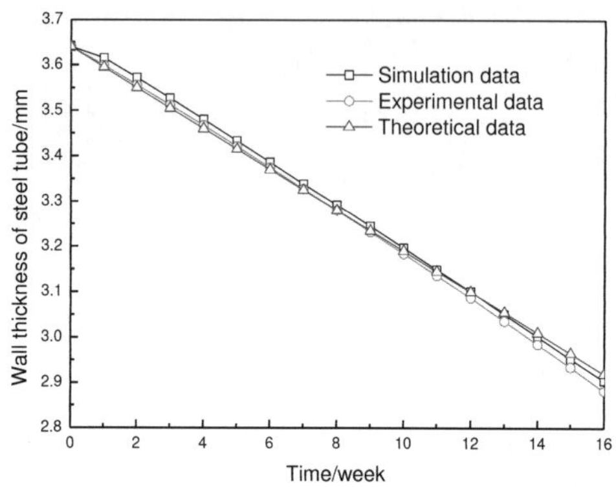

Figure 6. Time-dependent thickness of steel tube wall

From Table 1, Figure 5 and Figure 6, we can see that the curves of the variation of surface depth of CFST simulated with the CA model agree well with the theoretical curve and the experimental curve. The corrosion depth at previous weeks is slightly less than that of the theoretical corrosion depth, because there is a piece of passivation film on the surface. As etching time increases, the corrosion depth of the week is lightly higher than the theoretical corrosion depth. To a certain extent, it also confirms the Faraday first law.

4. Conclusion

This paper has established a model of uniform corrosion damage evolution on metal surface based on the cellular automata (CA) method, considering several factors such as the initial solution concentration, etching time, and dissolution probability et al, and examined the evolution of corrosion damage. In addition, the process for corrosion damage evolution on the surface of CFST under acid rain environment has been simulated. Theoretical predictions and experimental results are used to confirm the model. Based on the basis of above studies, the following conclusion can be drawn:
First of all, the corrosion depth increases with increasing etching time. Secondly, as the dissolution probability rises, the relationship between corrosion depth and etching time is increased. Thirdly, although the average corrosion depth was slightly changed with the increase of the solution concentration, on the whole, the difference among the effects of solution concentration on the corrosion depth is not large. Besides, the corrosion depth of the outer steel tube wall with the increase of etching time is stable after second weeks either for different dissolution probability or different solution concentration. What's more, it is feasible and effective to use the CA model to simulate the corrosion depth of CFST under acid rain attack. By defining different local evolution rules and changing the parameters in the simulation process, the CA model could be applied to the study of process for corrosion damage evolution of metal materials in aviation, petrochemical and nuclear power and other fields.

References

[1] Han LH 2007 *Encased structures-Theories and Practice* (Beijing: Academic press)
[2] Han LH, Hou C, Wang QL. 2013 Square Concrete Filled Steel Tubular Members Under Loading and Chloride Corrosion: Experiments. *Journal of Constructional Steel Research*.**71(1)** 11-25
[3] Hou C, Han LH, Zhao XL. 2013 Full-range analysis on square CFST stub columns and beams under loading and chloride corrosion. *Thin-Walled Structures*. **68(1)** 50-64
[4] Malki B, Baroux B. 2005 Computer simulation of the corrosion pit growth. *Corrosion Science*. **47(1)** 171-182
[5] Rusyn B P, Tors'Ka R V, Pokhmurs'Kyi A Y. 2015 Modeling of the Evolution of Corrosion

Pitting with the Use of Cellular Automata. *Materials Science.* **50(5)** 706-713
[6] Torska R and Rusyn B 2016 Application the cellular automata method for modeling the growth of surfaces corrosion defects. *Lviv-Slavske, Ukraine.* 23-26
[7] Zhang R X, Mahadevan S. 2001 Reliability-based reassessment of corrosion fatigue life, *Structure Safety.***23(1)** 77-91
[8] Burks A W. 1971 Essays on cellular automata. *The Quarterly Review of Biology.*
[9] Margolus N, Toffoli T 1987 *Cellular automata machines: a new environment for modeling* (Cambridge: MIT Press)
[10] Ary C, Newman N R B B. 1987 Assessment of simple methods of determining the free chloride ion content of cement paste. *Cement & Concrete Research.* **17(87)** 907-918

IOP Conf. Series: Materials Science and Engineering **216** (2017) 012013 doi:10.1088/1757-899X/216/1/012013

Numerical Study on Section Constitutive Relations of Members Reinforced by Steel-BFRP Composite Bars

Tongliang Xiao [1], Hongxing Qiu [2]

[1]College of Architectural Engineering, Nanjing Institute of Technology, Nanjing 211167, China
[2]School of Civil Engineering, Southeast University, Nanjing 210096, China

Email: axtl@njit.edu.cn; bqiuhx@seu.edu.cn

Abstract. Steel-Basalt FRP Composite Bar (S-BFCB) is a new kind of substitute material for longitudinal reinforcement, with high elastic modulus, stable post-yield stiffness and excellent corrosive resistance. Based on mechanical properties of S-BFCB and the plane cross-section assumption, the moment-curvature curves of beam and column members are simulated. Some parameters such as equivalent rebar ratio, postyeild stiffness, concrete strength and axial compression ratio of column were discussed. Results show that the constitutive relation of the cross section is similar with RC member in elastic and cracking stages, while different in post-yield stage. With the increase of postyeild stiffness ratio of composite bar, the ultimate bearing capacity of component improved observably, member may turn out over-reinforced phenomenon, concrete crushing may appear before the fibersarefractured. The effect of concrete strength increase in lower postyeild stiffness ratio is not obvious than in higher. The increase of axial compression ratio has actively influence on bearing capacity of column, but decreases on the ductility.

1. Introduction

Traditional reinforced concrete structure suffering an earthquake is characterized by rapid increase of deformation, essentially constant bearing capacity and secondary stiffness close to zero in yielding stage. The structure gets damaged rapidly with high residual plastic deformation in this stage, thus hard to be repaired after earthquake.

Forthe purpose of achieving the earthquake fortification goal that buildings do not fall down after a major earthquakeor are reparable after a medium earthquake, Chinese and overseas scholars have investigated a lot to ensure theconcrete structure to have a stable secondary stiffness and control the residual deformation after an earthquake.Such researches include tactfully mixing unbonded reinforcement with bonded reinforcement for use[1], applying aprestress on bridge piers [2], using CFRP bars for column bases[3] and using high strength reinforcement to strengthenconcrete frames[4]. In recent years, fiber reinforced polymer (FRP) bars are a competitive option in civil engineering. The use of FRP reinforcement is attractive for structures that operate in aggressive environments, such as in coastal regions, or for buildings that host magnetic resonance imaging (MRI) units or other equipment sensitive to electromagnetic fileds [5]. FRP has high strength, a low elastic modulus, poor ductility and good durability, while steel is the opposite. Combining the advantages of the two materials is expected to have outstanding comprehensive properties. A hybrid rod made with fiber-reinforced polymer (FRP) skin over a steel core was first proposed by Nanni et al. (1993) [6] to achieve good anti-corrosion performance for concrete beams. Based on the abovementioned research, a new kind of hybrid rebar named Steel-fiber Composite Bar was proposed by Wu et al.(2010) [7] (Figure 1). S-BFCB is a kind of longitudinal bar wrapped by fiber reinforced polymer (FRP) in longitudinal direction, which is a compound of elastic-plastic steel and linear elastic FRP with stable postyield stiffness [8-10].

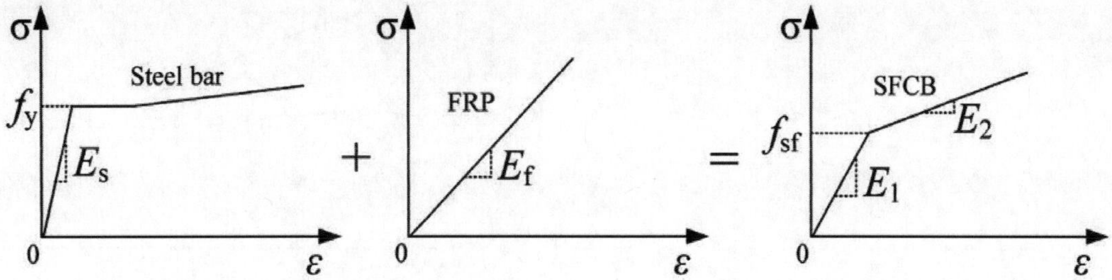

Figure 1. Stress-strain relation of Steel and FRP

2. Material constitutive relation model

Steel-Basalt FRP Composite Bar (S-BFCB) is a material of tension and compression asymmetry constitutive relation [11]. Relevant program is compiled by OpenSees software open platform, concluding the proposed S-BFCB tension and compression asymmetry material constitutive relation.The moment-curvature curve of beam and column is simulated, S-BFCB member is compared with RC memberby equal stiffness and strength principle and some parameters such as concrete strength are also discussed.

2.1. Constitutive relation of bars

This paper adopts the reinforcingsteel01material as the steel constitutive model, its tensile model is Chang and Mander model.It is completed by defining the steel yield point, harden point position, hardening stiffness andstrength of the limit point.For composite bar concrete members, composite bar is divided into steel bar and FRP to simulate respectively, which is used in FRP and linear elastic elements.

2.2. constitutive relation of concrete

For the constitutive relation of the cross section, concrete constitutive relation is divided into the core concrete bounded by stirrups andconcrete of protective layer.Concrete constitutive relation usethe concrete02 material, which seeks a balance between simplified and the precision.This paper adopts the correction of Kent - Park model [12].

3. Moment-curvature curveof beam

Material and geometrical parameters are as follows: the cross section size of beam is 300 mm *200 mm, compressive and extensive strength standard values of C30 concrete are $f_{ck} = 20.1MPa$ and $f_{tk} = 2.01MPa$ .Yield strength and ultimate strength of steel bar are 400 *MPa* and 540 *MPa*, and the corresponding strainsare 0.002 and 0.08 respectively.The ultimate strength of Basalt fiberis 1540MPa andnominal yield and ultimate strains are $\varepsilon_{sfy} = \varepsilon_y = 0.002$ and $\varepsilon_{sfu} = 0.022$.

In Figure 2, the bearing capacity of S10B20 and S10B138 is more than S10 beam due to the fiber, S10B20 beam is correspond to S12 beam by the equal-strength principle design while S10B138 beam is correspond to S12 beam by the equal-stiffness principle design. In the pre-yield stage, stiffness is relatively consistent, but the bearing capacity of S10B138 beam improved significantly while the section ductility is not significantly lower.

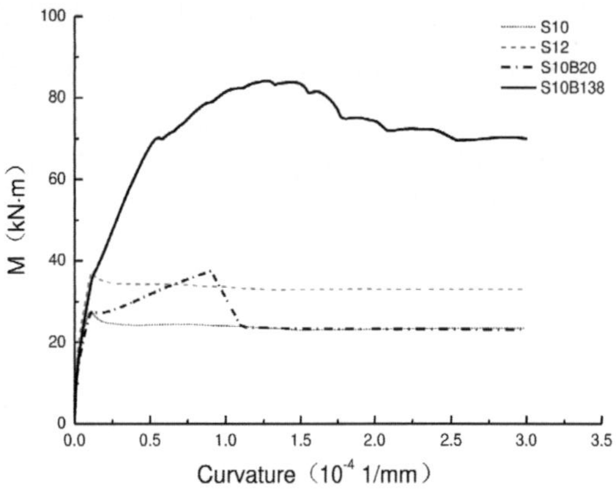

Figure 2. M-φ relationship of RC and S-BFCB concrete beam

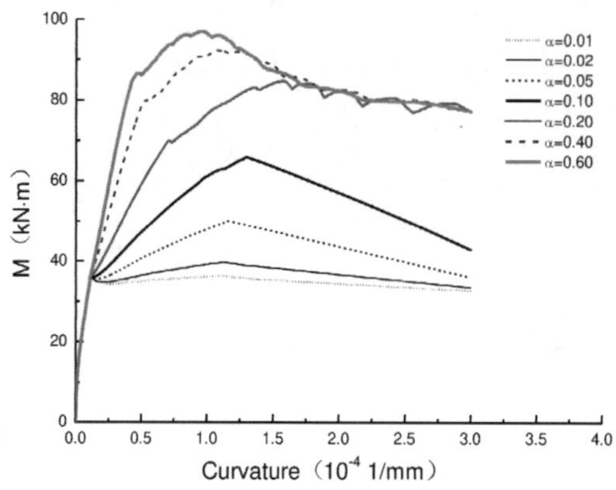

Figure 3. M-φ curve of beam in different stiffness ratio

Parameter α in Figure 3 is the reinforcement materials elastic modulus ratio of post yield to pre-yield, with the range from 0.01 to 0.60.The ultimate bearing capacity improvedobviously with the increment of parameter α .

4. Moment-curvature curve of column

Material and geometrical parameters of column are as follows: section size is 300 mm×300 mm, concrete compressive strength and tensile strength standard values of C30 are $f_{ck} = 20.1MPa$ and $f_{tk} = 2.01MPa$, axial compression ratio of column is 0.2. The material mechanics performance of steel, basalt fiber and composite baris equal tothe value of beam section.

Contrast curve in Figure 4, S10B24column is correspond to S14column by the equal-strength principle design while S10B163column is correspond to S14column by the equal-stiffness principle design. In the pre-yield stage, stiffness is relatively consistent, but the bearing capacity of S10B163column improved significantly while the section ductility is not significantly lower. In Figure 5, it shows that with the increasesof axial compression ratio from 0 to 0.6, the bearing capacity of S12B24 column improves obviously but the ductility of column members is reduced.

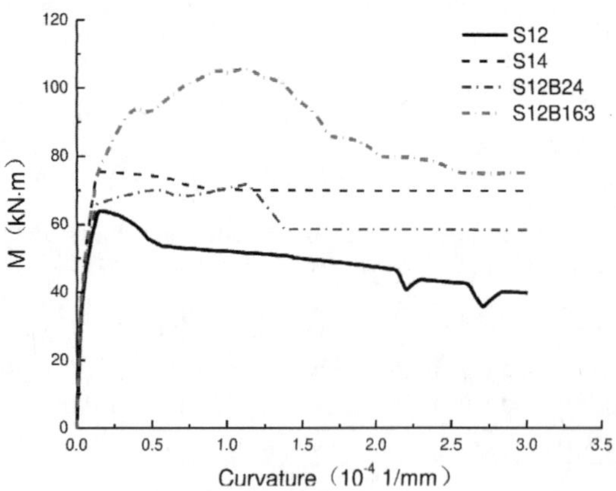

Figure 4. M-φ curve of RC and S-BFCB column

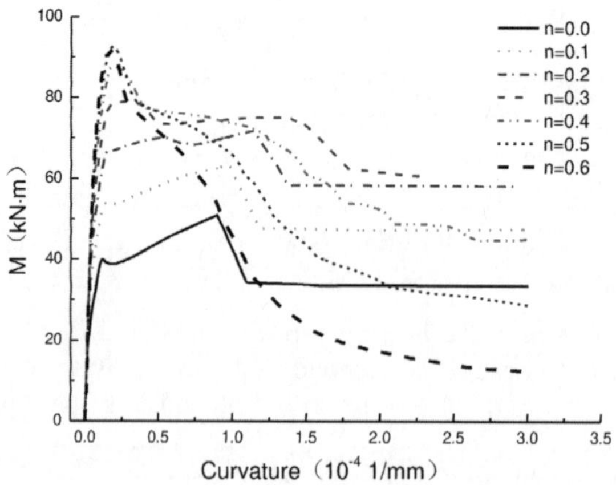

Figure 5. M-φ curve of column in different axial compression ratio

In Figure 6, with the increase of parameter α from 0 to 0.60, M-φ curve of column is similar with the curve of beam, but when reach the ultimate compressive strain of concrete, the effect of bearing capacity improvement is limited and the decline period is coincide. With the increase dosage of fibers, column may appear the phenomenon of overreinforced, concrete crushing may occur before fiber being pulled break. Figure 7 shows that with the improvement of concrete strength, the ultimate bearing capacity of S10B24 column improves not obvious.

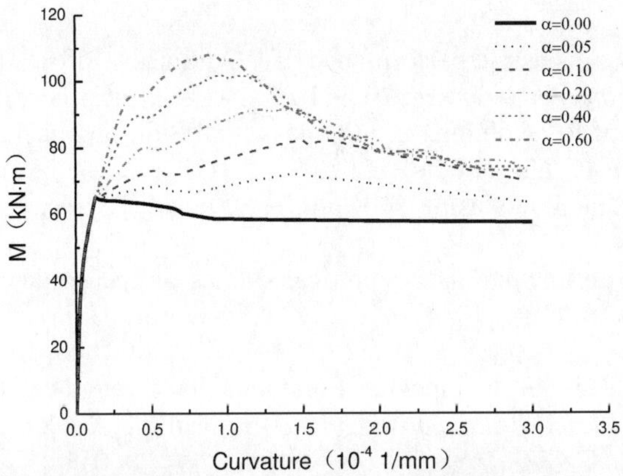

Figure 6. M-φ curve of column in different postyeild stiffness

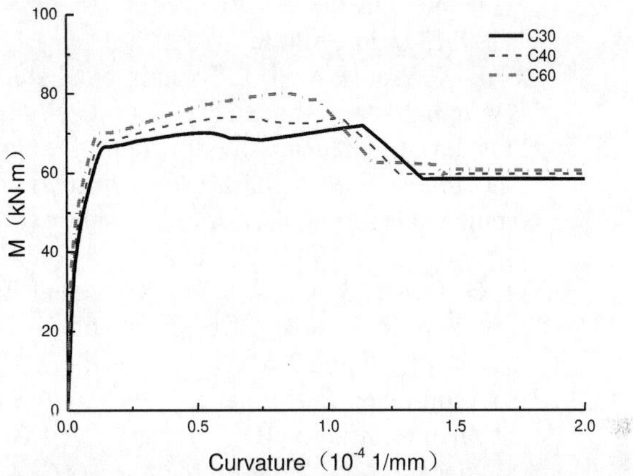

Figure 7. M-φ curve of column in different concrete strength

By the contrast of numerical simulation in this paper, secondary stiffness of composite reinforcement material improve the bearing capacity of composite reinforced concrete members, but the performance of the composite reinforced concrete member not only depends on the composite reinforcement material, but also related to the strength of the concrete. Blindly increase the secondary stiffness reinforced materials need more basalt fiber, the member design should be considered not only the ratio of reinforcement and fiber, but also concrete strength index, to ensure the best comprehensive performance.

5. Summary

From the comparison of data, the conclusionin this paper is confirmed as follows:
Based on mechanical properties of S-BFCB and the plane cross-section assumption, the moment-curvature relationship of beam and column members have been analyzed by numerical simulation. The relationship relevant to some parameters such as equivalent rebar ratio, postyeild stiffness, concrete strength and axial compression ratio of column were discussed. Results show that the constitutive relation of the cross section is similar with RC member in elastic and cracking stages, while different in post-yield stage. With the increase of postyeild stiffness ratio of composite bar, the ultimate bearing capacity of component improved observably, member may turn out overreinforced phenomenon, concrete crushing may appear before the fiber is snap. The effect of concrete strength increase in lower postyeild stiffness ratio is not obvious than in higher. The increase of axial compression ratio has actively influence on bearing capacity of column, but decreases on the ductility.S-BFCB has good prospects for engineering applications.

6. Acknowledgements

The authors would like to acknowledge financial support from the National Natural Science Foundation of China (No. 51078077), the National Science and Technology Pillar Program during the 12th Five-Year Plan Period (2012BAJ14B02) and Nanjing Institute of Technology Foundation of China (No. YKJ201511).

References

[1] Iemura H, Takahashi Y, Sogabe N. Two-level seismic design method using post-yield stiffness and itsapplication to unbounded bar reinforced concrete piers [J]. *Structural Engineering/Earthquake Engineering, JSCE*. 2008, **23(1)**: 109s-116s.

[2] Ou, Y C, Chiewanichakorn M, Aref A J, et al. Seismic performance of segmental precast unboundedposttensioned concrete bridge columns [J]. *Journal of Structural Engineering, ASCE,* 2007, **133**: pp 1636-1647.

[3] Xie X, Bu Z Y. Elementary study on usefulness of carbon fiber for controlling bridge pier seismic damage [J]. *Journal of Zhejiang University (Engineering Science)*. 2005, **(10)**: pp 131-137. (in Chinese)

[4] Zhang X, Wei H, Ye L P. Experimental study on seismic performance of RC frames reinforced withhigh-strength steel [J].*China Civil Engineering Journal*. 2009, (05): 74-78. (in Chinese)

[5] Hany JawaheriZadeh, Antonio Nann. Design of RC Columns Using Glass FRP Reinforcement. [J]. *Journal of Composites for Construction,* **17**, No. 3, June 1, 2013.

[6] Nanni A. Flexural behavior and design of RC members using FRP reinforcement. [J] *Structure Engineering* ,1993; **119(11)**: pp 3344–3359.

[7] Wu G, Luo Y B, Wu Z S, Hu X Q, et al. Experimental and theoretical studies on mechanical properties of steel-FRP composite bars (SFCB) [J]. *China Cvil Engineering Journal*. 2010,**43(3)**: pp 53-61. (in Chinese)

[8] Luo Yun-biao. Performance Study on Steel Fiber Composite Bars and Its Strengthened Concrete Structure[D]. *Master's Degree Thesis of Southeast University*, Nanjing, 2008.(in Chinese)

[9] Sun ZY, Wu G, Wu ZS, et al. Seismic behavior of concrete columns reinforced by steel-FRP (fiber-reinforced polymer) *composite bar. J Compos Constr* 2011; **15(5)**: pp 696–706.

[10] SUN Zeyang, XIAO Wenchao, WU Gang, et al. Preliminary Study on Moment-curvature Behavior of Concrete Beams Reinforced by Steel-Fiber Reinforced Polymer Composite Bars[C]. *The 9th national construction engineering application of FRP academic communication*: 2015:5. (in Chinese)

[11] Xiao Tongliang, QiuHongxing, Sun Lanxiang. Experimental and Theory Study on the Basic Tensile and Compression mechanical properties of Steel-Basalt Fiber Composite Bars, *Journal of southeast university(Natural science edition)*, 2014, **44(4)** pp 805-810. (in Chinese)

[12] Scott B D, Park R, Priestley M J N. Stress-strain behavior of concrete confined by overlapping hoops at low and high strain rates[J]. *ACI Journal*, 1982,**79(1)**: pp 13-27

IOP Conf. Series: Materials Science and Engineering **216** (2017) 012014 doi:10.1088/1757-899X/216/1/012014

Ultimate Capacity of Uniaxially Compressed Steel Plates Strengthened by CFRP

Xin Tao[1], S Y Cao[1]
[1] Key Laboratory of Concrete and Prestressed Concrete Structures of Ministry of Education, Southeast University, Nanjing, China

Email: seutaoxin@gmail.com;101000873@seu.edu.cn;

Abstract. This paper presents experimental studies on the ultimate capacity of steel plates strengthened by Carbon Fiber-reinforced polymer (CFRP). Some 40 plates are tested under uniaxial compression along its longitudinal direction. The effect of CFRP pasting, plate slenderness ratio and the boundary conditions are illustrated on the studies. The study shows that the steel plates strengthened by CFRP exhibited higher capacity in comparison to similar plates without CFRP. It also can be seen that multi-layers CFRP pasting were better than the single layer, pasting multi-direction is better than pasting uni-direction, pasting the transverse layer outside is better than pasting the longitudinal layer outside, and the effect of longitudinal layer was better than the transverse layer.

1. Introduction

In the past several decades, Fiber-reinforced polymer (FRP) has been widely used in the strengthening of steel structures because of its high strength to weight ratios, excellent resistance to corrosion, environmental degradation, very flexible to form all kinds of shapes, and easy to handle during construction. At the same time, thin-walled plates and tubes made from metals have been used extensively in many different applications [1], including the construction, infrastructure, aeronautical, aerospace, automotive, marine and sporting industries. A relatively recent advent to compression members is the combination of metal and FRP tubes produced by external bonding fibers to the metal section.

In axial compression applications, the motivation for strengthening thin-walled steel structures with FRP is to control buckling deformations and/or provide increased compression capacity [2]. Shaat and Fam (2006) [3] experimented short and long HSS columns with high modulus CFRP sheets and increase the column axial strength by up to 23%. A nonlinear model of this experimental study was developed [4] to predict the axial load capacity, the results show that CFRP effectiveness increases for columns with larger out-of-straightness deficiencies and columns of higher slenderness. Then a simplified analytical model was proposed [5] to predict the ultimate axial load of FRP-strengthened slender steel columns. Bambach (2006,2010a,2010b) [6-8] experimented short hollow structural section(HSS) with different fiber layouts, and observed that the capacity and mean crush load of the composite metal-CFRP HSS exceed the sum of those for the individual metal HSS and CFRP HSS, by up to 1.8 times. A three dimensional finite element model has been developed by Devi [2], and the parametric studies show that the capacity increases with the increase in the number of CFRP layers.

It is clear from the above literature review that FRP is good at strengthening the compression member, especially to the HSS columns. However, more research is still required in this field. This paper focused on the ultimate capacity of uniaxially compressed steel plates strengthened by CFRP with different pasting method of FRP, different plate slenderness ratio and different boundary conditions.

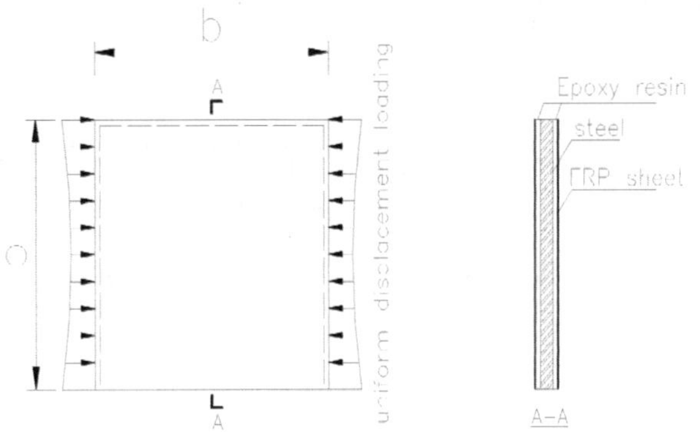

Figure 1. Geometry conditions of the strengthened plate

Figure 2. Test rig of the plates

Table 1. Details and the results of the plate specimens

Group	Thickness (mm)	b/t	CFRP pasting	boundary conditions	Specimen no.	P_{ult} (kN)	Specimen no.	P_{ult} (kN)	Specimen no.	P_{ult} (kN)
1	2.61	92	LT	4s	CP92-1	49.6	SFSCP92-1	64.0	FSCP92-1	67.9
				2c2s	CP92-2	61.4	SFSCP92-2	88.8	FSCP92-2	90.1
				2c1s1f	CP92-3	40.5	SFSCP92-3	41.8	FSCP92-3	53.5
				3s1f	CP92-4	22.2	SFSCP92-4	23.1	FSCP92-4	36.6
2	3.64	66	LT	4s	CP66-1	112.3	SFSCP66-1	113.6	FSCP66-1	142.3
				2c2s	CP66-2	122.8	SFSCP66-2	125.4	FSCP66-2	176.3
				2c1s1f	CP66-3	73.1	SFSCP66-3	88.8	FSCP66-3	113.6
				3s1f	CP66-4	58.8	SFSCP66-4	73.1	FSCP66-4	86.1
3	5.58	43	LT	4s	CP43-1	240.3	SFSCP43-1	266.4	FSCP43-1	286.0
				2c2s	CP43-2	279.5	SFSCP43-2	287.3	FSCP43-2	367.0
				2c1s1f	CP43-3	246.8	SFSCP43-3	244.2	FSCP43-3	272.9
				3s1f	CP43-4	163.2	SFSCP43-4	197.2	FSCP43-4	206.3
4	3.64	66	T	2c2s	–	–	–	–	FSCP66-5	137.1
			L	2c2s	–	–	–	–	FSCP66-6	152.8
			TL	2c2s	–	–	–	–	FSCP66-7	158.0
			LT2	2c2s	–	–	–	–	FSCP66-8	210.3

Table 2. Material properties

Material Properties	t(mm)	Young's modulus(MPa)	Yield stress(MPa)	Tensile strength(MPa)
Steel plate	2.61	206	287.9	434.1
	3.64	206	253.2	369.3
	5.58	206	228.7	327.8
CFRP	t_f(mm)	Young's modulus(MPa)	Fiber mass per unit area(g/m^2)	Tensile strength(MPa)
	0.167	260	300	3492.5

Epoxy resin	----	Tensile shear strength(MPa)	Compressive strength(MPa)	Tensile strength(MPa)
	----	19.6	87.9	48.4

2. Scope of the investigation

In the investigations reported in this paper, the ultimate capital of steel plates strengthened by CFRP is considered. As shown in Figure 1, the length and width of the plate is b, the thickness of steel and CFRP sheets are given as t and t_f, respectively. All the plates tested were: (i) square, since they can be assumed to represent a single wave of a buckled plate; (ii) subjected to uniform displacement loading along its longitudinal direction. Some parametric variations have been considered as follows:

1. The situation of the CFRP pasting. Some of them were plates without CFRP, some were plates with single side CFRP pasting and some were plates with double sides CFRP pasting, the direction and the number of layers were also studied in this paper.
2. Plate slenderness ratio b/t, which is an important parameter that governs the plate strength. In this study, the Plate slenderness ratio b/t was 92, 66 and 43, respectively, corresponding to the steel plate's width t of 2.61mm, 3.64mm and 5.58mm.
3. The boundary conditions of the plates. Four kinds of boundary conditions were studied in this paper, those are: (i)all four are simply supported (4s); (ii) two opposite loaded edges are clamped while the remaining two edges are simply supported (2c2s) ;(iii) two opposite loaded edges are clamped, one edge is simply supported, and the fourth edge is free (2c1s1f); and (iv)three edges simply supported while the fourth edge which is unloaded is free (3s1f).

3. Test setup

The ultimate capacity of uniaxially compressed steel plates strengthened by CFRP was studied with experimental method, and a test rig has been fabricated in the laboratory (see Figure 2), the test rig was used in combination with existed universal hydraulic jack of 2000 kN maximum capacity. The applied loads were measured by the static strain testing systems, which were composed by Force sensor, DH3816 and computer. The displacement and deformation of the plates were achieved by displacement meters

A total of 40 specimens were manufactured, of which 12 were steel plates without reinforcement (CP), 12 were steel plates with single side FRP pasting (SFSCP), and the rest 16 specimens were plates with double sides FRP pasting (FSCP). The specimens were divided into four groups (see Table1); the tests in Group 1, 2 and 3 were to study the influence of the boundary condition and the CFRP pasting. By comparing the tests in those three groups, the influence of the plate slenderness ratio b/t also can be studied. The details of the specimens were in Table 1. In the table, L and T means one longitudinal layer of CFRP sheet and one transverse layer of CFRP sheet, respectively. LT means one longitudinal layer pasting inside and then one transverse layer pasting outside, TL is exactly the opposite. LT2 means two longitudinal layers pasting inside and then two transverse layers pasting outside. The Material properties were in Table 2.

4. Results and discussion

4.1 Failure patterns

Figure 3 shows the typical failure pattern of these plates. The reason why it isn't so obvious to see the final Out-of-plane deflection is that the experiments were ended soon after the bearing capacity declines. From Figure 3, we can see that the failure mechanism consists four yield lines which start from the four corners and end near the hole edges, also the steel plates and CFRP have a good coordinated working performance throughout the experiment.

4.2 Load-deflection curves

Some typical relationships between compressive load and in-plane displacement of the plates have been plotted in Figure 4. Traditional three-phase curves occurred to those specimens. In these cases, deflection increased linearly with enhancing load at the beginning of the test. As the load continued to increase, non-linearity between load and displacement became evident due to the yielding of local materials. Great increases in in-plane deflection caused by small increments of load were unavoidable in the final phase. From Figure 4 we can see that the pasting of CFRP can increase the stiffness before buckling, also it can improve the mechanical property afterwards.

Figure 3. Typical failure patterns of specimen

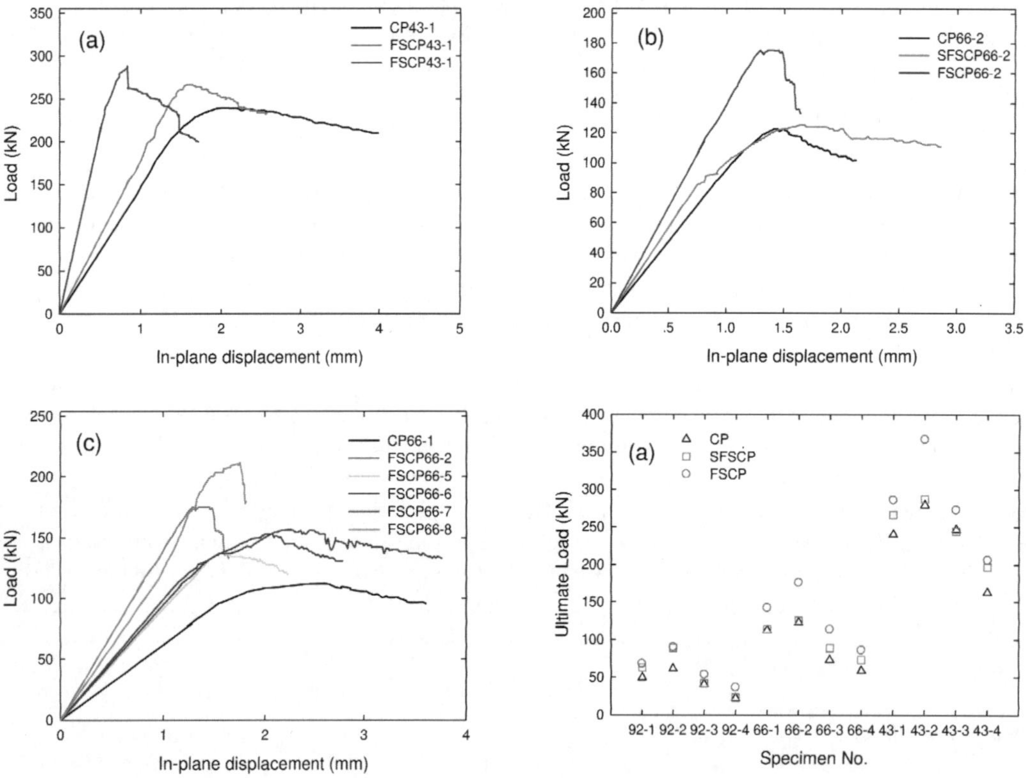

Figure 4. Load-deflection curves of specimens

4.3 Ultimate load

4.3.1 The situation of the CFRP pasting. Figure 5 shows the influence of the CFRP pasting. Some conclusions can be obtained as following:

1. As shown in Figure 5-a, the ultimate load of FSCP specimens is bigger than CP specimens, but the ultimate load of SFSCP specimens was not stable. Considering of the final buckling mode of the plate, it comes out that if CFRP was pasted on the convex surface of the final buckling mode of the plate, it

will improve the capacity of steel plates very well; otherwise, it will be almost of no effect. So in the practical, if we can be sure that the FRP sheet is pasting in the drum surface, the influence of FRP can be considered, otherwise, it should be carefully considered. Take the compression of hollow steel tube for example, it has be foreseen that one of the opposite sides will be outward buckling and the other couple will be inward buckling, so only the FRP sheets pasted in the outward buckling sides will be considered in design.

2. It also can be seen from Figure 5-b that multi-layers were better than the single layer (LT2 has the highest ultimate capacity). Pasting multi-direction is better than pasting uni-direction (LT and TL have better ultimate capacity than T and L). Pasting the transverse layer outside is better than pasting the longitudinal layer outside (LT has better ultimate capacity than TL). The effect of longitudinal layer was better than the transverse layer (L has better ultimate capacity than T).

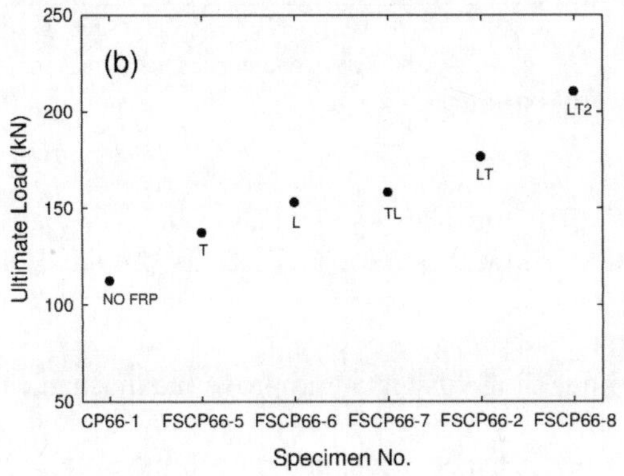

Figure 5. The situation of the CFRP pasting

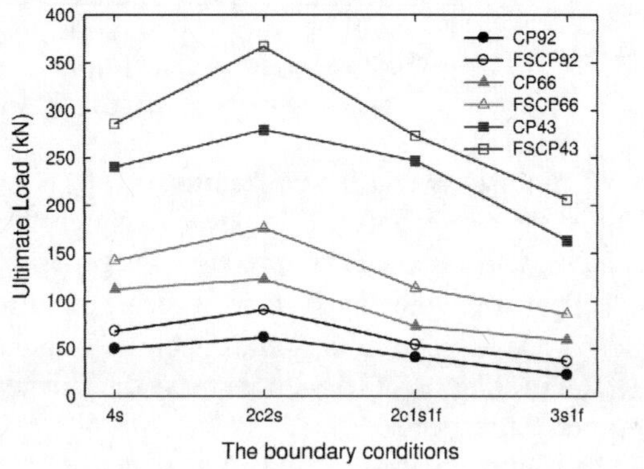

Figure 6. The boundary conditions of the plates

4.3.2 The boundary conditions of the plates. The influence of the boundary conditions of the plate was in Figure 6. It can be seen that the boundary conditions affect the ultimate load of the plates great. Results the ultimate load of plates with the different boundary conditions were as follows: 2c2s, 4s, 2c131f, 3s1f. We also find that the smaller the plate slenderness ratio, the plates is more sensitive to the boundary conditions.

4.3.3 Plate slenderness ratio b/t. Figure 7 shows the influences of plate slenderness ratio b/t on ultimate capacity of plates. It can be seen that the ultimate capacity variations of plates strengthened by CFRP are very similar to the pure plates, that is, the ultimate capacity first degrades quickly as b/t is gradually enlarged, then degrades slowly. So in the design, a modified thickness method can be used to consider the role of CFRP sheets.

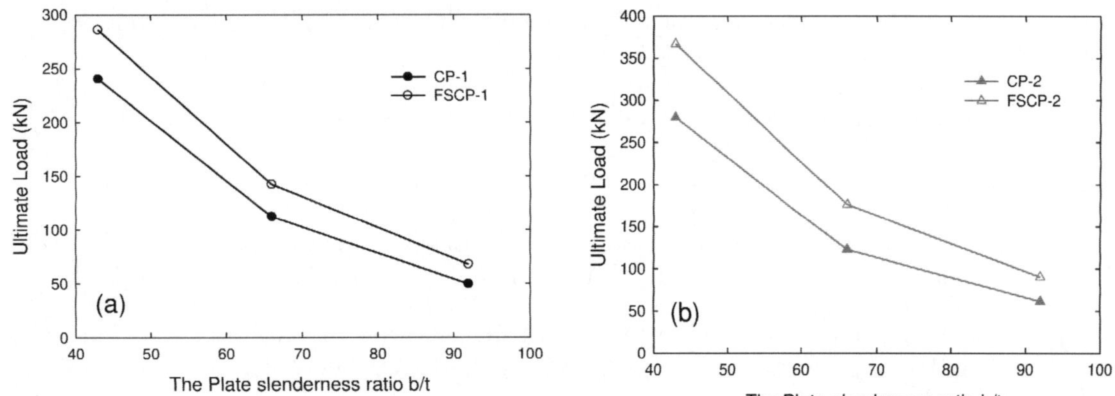

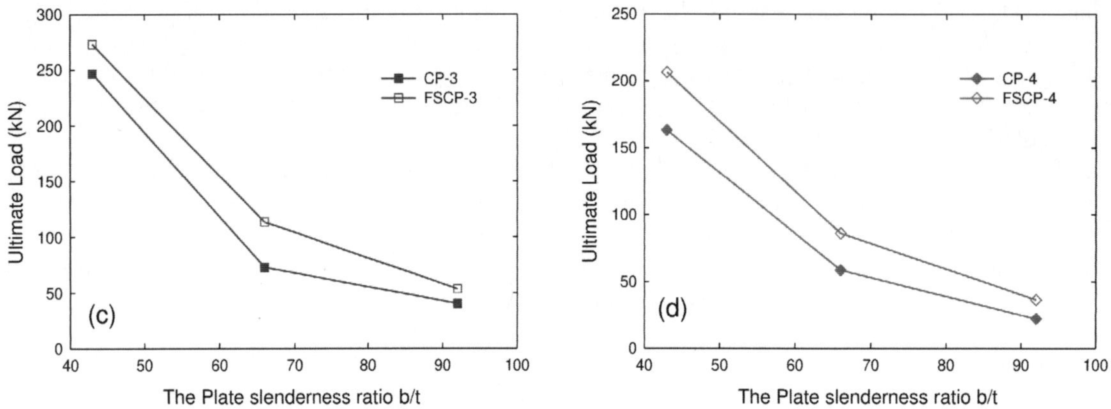

Figure 7. Plate slenderness ratio b/t

5. Conclusions and future trends

The study presented in this paper explores the use of CFRP sheet for strengthening of steel plates subjected to uniaxial compression. In general, test results show that the CFRP sheets provided an excellent strengthening system for steel structures.

Based on the test results, some conclusions and the further studies can be drawn:

1. The pasting of CFRP can increase the stiffness before buckling, also it can improve the mechanical property afterwards.
2. The ultimate load of FSCP specimens is bigger than the CP specimens, and the ultimate load of SFSCP specimens was not stable.
3. Multi-layers of CFRP were better than the single layer, pasting multi-direction is better than pasting uni-direction, pasting the transverse layer outside is better than pasting the longitudinal layer outside, the effect of longitudinal layer was better than the transverse.
4. The smaller the plate slenderness ratio, the plates is more sensitive to the boundary conditions.
5. The results and insights developed from the present study will be very useful for the propose of a modified thickness method, which the role of CFRP sheets can be considered.

In the further research, a FE model will be established, and parameter analysis are also needed to propose a modified thickness method to calculate the buckling load, and then the ultimate load can be calculated according to the specification for structural steel buildings. The elastic buckling of perforated plates strengthened with FRP under uniaxial compression also will be studied through the experimental and numerical simulation methods.

References

[1] Bambach, M. Strengthening of Thin-Walled (Hollow) *Steel Sections Using Fibre-Reinforced Polymer (FRP) Composites-6*. (2014).

[2] Devi, U. and Amanat, K.M. Non-Linear Finite Element Investigation on the Behavior of CFRP Strengthened Steel Square HSS Columns under Compression. *International Journal of Steel Structures* 15, 3 (2015), pp 671–680.

[3] Shaat, A. and Fam, A. Axial Loading Tests on Short and Long Hollow Structural Steel Columns ... *Canadian Journal of Civil Engineering* 33, 4 (2006), pp 458–470.

[4] Shaat, A. and Fam, A. Fiber-Element Model for Slender HSS Columns Retrofitted with Bonded High-Modulus Composites. *Journal of Structural Engineering* 133, 1 (2007), pp 85–95.

[5] Shaat, A. and Fam, A.Z. Slender Steel Columns Strengthened Using High-Modulus CFRP Plates for Buckling Control (Vol 13, Pg. 2, 2009). *Journal of Composites for Construction*18, 1 (2009), pp 2–12.

[6] Bambach, M.R. Local Buckling and Post-Local Buckling Redistribution of Stress in Slender Plates and Sections. *Thin-Walled Structures* 44, 10 (2006), pp 1118–1128.

[7] Bambach, M.R. Axial Capacity and Crushing Behavior of Metal–fiber Square Tubes – Steel, Stainless Steel and Aluminum with CFRP. *Composites Part B: Engineering* 41, 7 (2010), pp 550–559.

[8] Bambach, M.R. Experiments of Edge-Stiffened Plates in Uniform Compression. *Thin-Walled Structures* 49, 2 (2011), pp 343–350.

IOP Conf. Series: Materials Science and Engineering **216** (2017) 012015 doi:10.1088/1757-899X/216/1/012015

Analysis on Main Influence Factors for Ultrasonic Flaw Detection of Steel Weld

Tao Du[1], Jiandong Sun[1], Dong Liu [2], Yujun Guo[3], Qing Gao[4]
[1]Shandong Academy of Building Research Institute of Structure Jinan; 250031
[2]Feicheng Construction Engineering Quality Supervision Station Tai'an; 271000
[3]Shandong Provincial Department of housing and urban rural development Jinan; 250000
[4]Shandong Hi-speed Company Limited Jinan; 250000

dt7777@126.com

Abstract. The paper determined several main factors influencing ultrasonic detection results through research and analysis on the influence results and accuracy during the ultrasonic flaw detection of steel structure weld, and made a summary how to select and operate instruments correctly, utilize favorable factors of detection results and avoid unfavorable factors, as well as improve detection accuracy during the practical detection.

1. Overview

The implementation of new Code for Construction Quality Acceptance of Steel Structure Work GB50205-2001 strengthens the site inspection of steel structure, especially the approaches adopted by site random inspection, which properly ensures safety and durability of steel structure work. Quality inspection of welds is vital to quality control of steel structure work. The new acceptance code puts more emphasis on the ultrasonic flaw detection of steel weld, and the research of this part plays an important role as well in out subject "analysis on NDT technology of steel structure".

When an ultrasonic unit is used, the defect position and size are mainly determined by the position and height of defect echo on the fluorescent screen. The position and the height of the factors influencing the echo appear are manifold, we summarized several main factors affecting the accuracy of ultrasonic detection results from many years of practical experience and subject research in engineering detection (excluding the human-made factor), and it is quite important to learn about these factor for correctly selecting instruments and accessories and detection methods, making full use of favorable factors and avoiding unfavorable factors in ultrasonic detection results, and enhancing the accuracy of detection results.

2. Selection and Use of Detector

Union Pxut-3300 six channels full digital ultrasonic detector is selected. It is equipped with a 0.8Kg battery, and the light weight, high luminance and long continuous working hours are suitable for site detection. In particular, its performance indexes represent the highest level of current ultrasonic detection of steel structure.

3. Several Main Influence Factors

3.1 Influence of couplant materials (acoustic impedance)

To improve the detection results, a layer of acoustically transparent media between probe and workpiece surface, i.e. couplant. The couplant is used to exhaust air between probe and workpiece surface and further to effectively transmit the ultrasonic wave into workpiece. Ultrasonic coupling refers to sound intensity transmissivity of ultrasonic wave on the detection surface. Better ultrasonic coupling effect causes higher sound intensity transmissivity with high echo amplitude, which is easy for analyzing defects. Couplant should meet the following requirements:

(1) Be able to wet workpiece and probe surface, with proper fluidity, viscosity and adhesion, easy to be cleaned;

(2) With high acoustic impedance and good sound transparent property;

(3) With wide source, low cost, no corrosion to workpiece and no harmful to human body;

(4) Stable property, not easy to go bad, long storage life.

Engine oil, water and chemical paste are selected to be the study objects considering the above requirements based on the previous detection practices.

Acoustic impedance Z is the ratio between sound pressure P at a certain point and particle vibration velocity V at this point, i.e. $Z=P/V=\rho C$ (where, ρ is media density, c is ultrasonic velocity in media). The acoustic impedances of the above three couplants are determined as follows by reference to related material standards: (See Table 1)

Table 1. The acoustic impedances of the above.

Media	Z (kg/m2.s)
Engine oil	1.28
Water	1.50
Carboxymethyl cellulose adhesive	2.20

The influence degree of coupling material on defected wave height is tested, and the experiment is carried out on CSK-IIIA standard test block. The horizontal holes with depth of 10mm, 20mm, 30mm and 40mm are taken as standard defects to test its maximum wave height. The data obtained is shown as follows:

Engine oil as couplant: (See Table 2)

Table 2. Engine oil as couplant.

Maximum wave height of attenuation values	Cover depth of holes (mm)			
	10	20	30	40
	50	46	42	37

(dB)	45	48	41	37
	49	45	41	37
Averagevalue (m)	49	45	41	37

Notes: The attenuation value which reduces the maximum echo height to 80% of full screen is taken as a comparative value.

Water as couplant: (See Table 3)

Table 3. Water as couplant.

Maximum wave height of attenuation values (dB)	Cover depth of holes (mm)			
	10	20	30	40
	50	46	41	38
	50	46	41	37
	49	46	41	37
Average value (m)	50	46	41	37

(3) Chemical paste as couplant: (See Table 4)

Table 4. Chemical paste as couplant.

Maximum wave height of attenuation values (dB)	Cover depth of holes (mm)			
	10	20	30	40
	51	47	41	38
	51	47	42	39
	51	46	42	39
Average value (m)	51	47	42	39

See Figure 1 for curve graph.

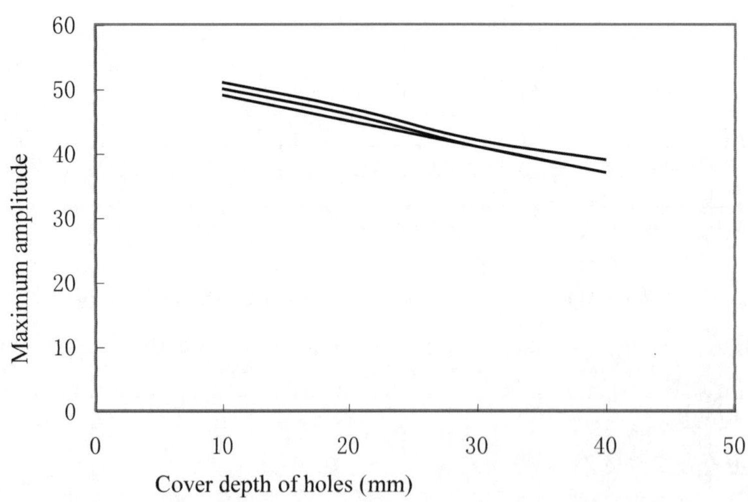

Figure 1. The curves from top to bottom represent chemical paste, water and engine oil, respectively.

It can be observed from the standard defect and wave height that the average height of water as couplant is about 0.5dB higher than engine oil as couplant. The reading of digital ultrasonic amplitude attenuator is generally accurate to 1dB, and that of analog type ultrasonic amplitude attenuator is accurate to 2dB. The influence of water and engine oil on defect amplitude can be ignored. On the other hand, the average echo height of chemical paste is 2dB higher than engine oil as couplant, which shows that the acoustical coupling effect of chemical paste is the best. However, if the steel workpiece surface is not processed timely, water and chemical paste (its main component is water) will cause corrosion. In contrast, engine oil is featured with proper viscosity, fluidity and cohesion, and the practical test shows that the engine oil can effectively reduce the friction between probe and the workpiece surface to be tested for more convenient detection. To sum up, the engine oil is recommended to be adopted as couplant.

3.2 Influence on external force applied to test probe
It can be deferred from detection experience that when ultrasonic detection is carried out by manual with the same couplant, the larger the external force applied to the probe, the higher the defect reflection echo obtained is. The experiment is carried out on an electronic scale, and the horizontal holes with depth of 10mm, 20mm, 30mm and 40mm are also taken as standard defects to test its maximum wave height. The data is obtained as Table 5.

Table 5. Maximum wave height.

	Classification	Cover depth of holes (mm)			
		10	20	30	40
Maximum wave height of attenuation value (dB)	①	46	41	37	34
	②	49	45	40	37
	③	50	46	42	38
	④	51	46	42	39
	⑤	51	46	42	39

Notes: The ①②③④⑤ in the above table are the comparative values to the average attenuation values when the maximum wave height is reduced to 80% of the full screen under the external forces of 0.1kgf, 0.25kgf, 0.5kgf, 0.75kgf and 1kgf.

See Figure 2 for curve graph.

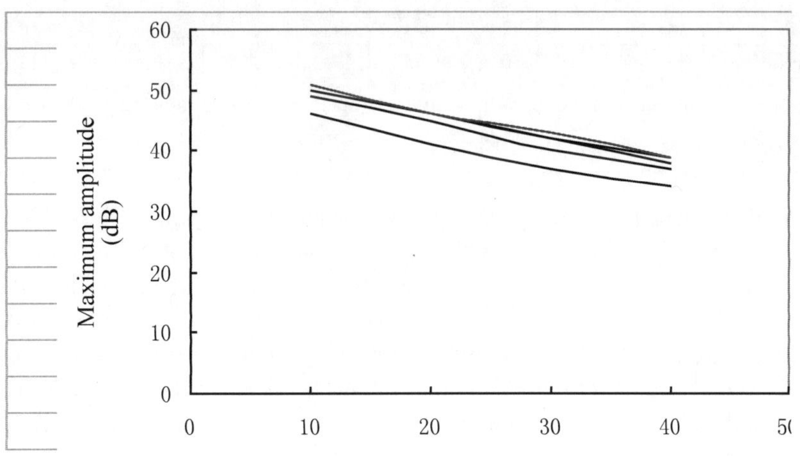

Figure 2. The external forces from top to bottom are 1kgf, 0.75kgf, 0.5kgf, 0.25kgf and 0.1kgf.

It can be seen that 0.5kgf is the demarcation point of external forces applied to the test probe. When the external force is less than 0.5kgf, the reflection wave height drops fast, and when it is larger than 0.5kgf, the reflection wave height increases slowly, which can be ignored considering the detector attenuation accuracy of 1dB or 2dB.

Therefore, it is suggested that the detection personnel should apply even force of approximately 0.5kgf on the probe during the practical detection.

3.3 Influence of workpiece surface roughness

The practical detection proves that the roughness of workpiece surface has a significant influence on acoustical coupling. Two blocks of steel plate in size of 300mm×300mm×10 mm were selected to be jointed by fusion-through welding. Simulate a cavity defect at the weld, use the probe to detect at the same side, position and direction, and then quantify the defects. The roughness of steel plate surface was handled by four methods:surface non-treatment, sanding (surface finishing), oxidized surface layer polishing by steel brush and mechanical polishing, as shown in Figure 3:

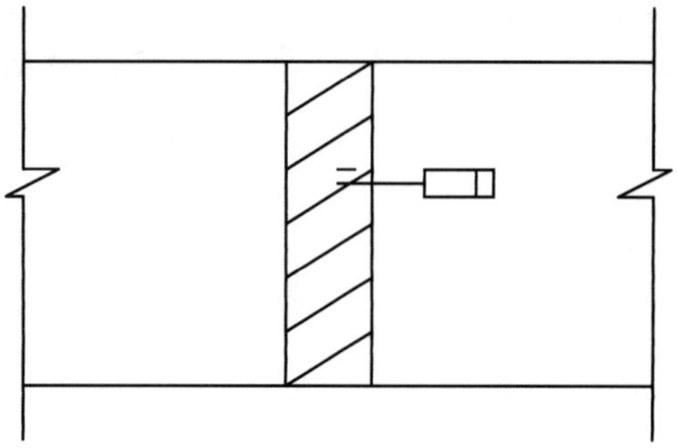

Figure 3. Detection of schematic diagram.

Attenuation reading of defect wave height: (See Table 6)

Table 6. Attenuation reading of defect wave height.

Media /instrument reading (dB)	①	②	③	④
Engine oil	32	36	37	37
Water	34	37	37	38
Carboxymethyl cellulose adhesive	34	36	37	37

Notes:①②③④ represents the four surface treatment methods: namely surface non-treatment, sanding, oxidized surface layer polishing by steel brush and mechanical polishing.

The table infers that the higher surface roughness generally generates poorer coupling effect with lower reflection echo when the same couplant is used. With regard to the couplant with lower acoustic impedance, the higher surface roughness generally have greater influence on reflection amplitude. It also can be observed that after the workpiece surface is polished by abrasive paper, the reflection amplitude is seldom effected if further surface treatment is carried out to improve its degree of finish. Thus the workpiece surface is only required to be polished by manual to ensure there is no welding slag or other foreign matters and the probe can move smoothly on workpiece surface during the practical detection.

3.4 Influence of workpiece shape

Workpieces to be tested are not only flat and straight steel plate, but plates or pipes with different curvature radius to be connected by welding. The probe surface is flat when it is manufactured, so it may have point contact or line contact when contacting with a hook surface, which will cause a loss of effective acoustic energy. The incoming sound pressure of ultrasonic wave against reflector is in direct proportion to the effective chip area. If the coupling area decreases, the incoming sound pressure will reduce accordingly, which will result in a reduction of reflection wave height and detection sensitivity. It certainly will affect the defect quantitative and leak detection. To increase the effective contact area between probe and surface to be tested, it is necessary to polish the curvature to be tested of the probe. The polished probe is shown in Figure 4:

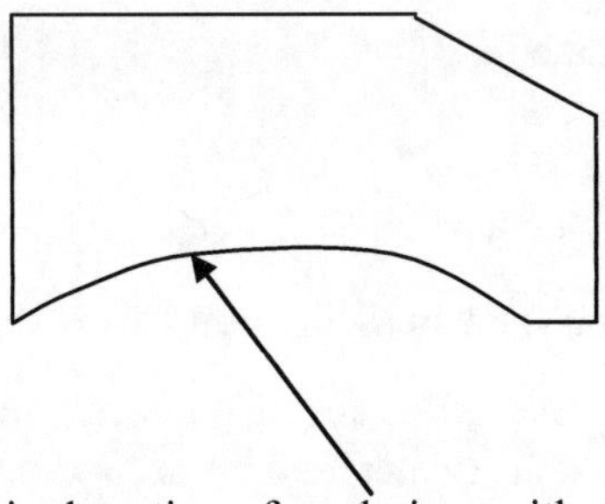

Figure 4. Polished probe in detection of workpiece with small curvature radius.

The research on butt-welded seam of steel pipe with different curvature showed that the defect reflection echo is relatively stable when the curvature radius is larger than 20mm, and similar echo amplitude can be received at the same position in repeated detections. But the reflection echo is not stable when the curvature radium is less than 20mm, and reflection echo may not be received at the same position in repeated detections. The research results indicated that the defects is possible to be undetected when the curvature radius of a workpiece is less than 20mm.When the probe surface is polished to a similar status of workpiece curvature, the defect reflection echo will recover to stable. For the purpose of conservation, it is suggested that when the curvature radius of a workpiece is less than 40mm during the practical operation, the possibility of leak detection can be considerably decreased if the probe surface is polished to as similar status with workpiece curvature.

3.5 Influence of working temperature

We studied the influence of working temperature on ultrasonic detection results. The common wave conversion media such as water and chemical paste have a considerable change in sound velocity under different temperatures. For example, when the media is water, the sound velocity of ultrasonic wave under different temperatures: (See Table 7)

Table 7 The sound velocity of ultrasonic wave under different temperatures

Temperature t (℃)	10	20	25	30	40	50	60	70	80
Sound velocity (m/s)	1448	1483	1497	1510	1530	1544	1552	1555	1554

However, when the incidence angle of ultrasonic wave is slant, the reflection wave and the propagation direction of refracted wave are determined by reflection-refraction law (i.e. Snell's law). Namely:

Sinα l/CL1= Sinα ls/CL1= Sinα s/Cs1= Sinβ l/C12= Sinβ s/Cs2

Wherein:α l---incident angle of longitudinal wave

α ls----reflex angle of longitudinal wave

α s---reflex angle of horizontal wave

β l----refraction angle of longitudinal wave

β s----refraction angle of horizontal wave

CL1----longitudinal wave velocity of the first media
Cs1----transverse wave velocity of the first media
C12----longitudinal wave velocity of the second media
Cs2----transverse wave velocity of the second media

The longitudinal wave detection of angle probe is taken as an example, i.e. :Sinα l/CL1= Sinβ l/C12,it is known that when the temperature changes, the incidence angle of probe remain unchanged, but the sound velocity of ultrasonic wave in the first media (water) and the second media (steel) change, which resulting in a change of Sinβ l, i.e. a change in refraction angle tgβ l (K value).In the experiment, a CSK-IA test block is put in water with a temperature of 60℃.Water is adopted as couplant (media) to test K value of 2.5PK2 angle probe. When the temperature of test block and water temperature reach 50℃ after 5 minutes, K value of the probe is measured to be 2.32, but the K value is measured to be 1.95 when the probe is in an ambient temperature of 22℃. We can see that the K value reduces 0.57, which is significantly different from the nominal value (K=2.0).

In the practical detection, if the working temperature is not significantly different from the ambient temperature, measured K value under the ambient temperature can be used directly without modification. But when the working temperature is considerable different from the ambient temperature, a test must be conducted to the probe performance by simulating the working temperature to correct the detection results.

4. Conclusion

It is of great help for our practical operation to understand the above five factors influencing ultrasonic detection results. During the practical detection, the test accuracy can be improved by selecting correct instruments and operating methods, making the most of favorable detection results and avoiding the unfavorable ones.

References

[1] National Standard of the People's Republic of China, "Code for Construction Quality Acceptance of Steel Structure Work GB50205-2001", Beijing, China Pressing Press, 2001
[2] National Standard of the People's Republic of China, "Classification for Manual Ultrasonic Flaw Detection Methods and Results of Steel Weld GB/T11345-89", Beijing, 1990
[3] Ultrasonic Flaw Detection Group "Ultrasonic Flaw Detection", Labor and Personnel Press, 1989
[4] Japan Nondestructive Inspection Association, Ma Yukuan et al ., "Translation of Ultrasonic inspection", Jilin Science and Technology Press, 1984

IOP Conf. Series: Materials Science and Engineering **216** (2017) 012015 doi:10.1088/1757-899X/216/1/012015

[5] Hu Tianming “Ultrasonic Flaw Detection”, Wuhan University of Science and Technology of Surveying and Mapping Press, 1994
[6] Ultrasonic Flaw Detection Group "Ultrasonic Flaw Detection", Electric Power Industry Press Press, 1980
[7] Yun Qinghua et al., “Nondestructive Testing”, labor press, 1983
[8] Yun Qinghua, "Nondestructive Testing Technology for Boiler Pressure Vessel", Tianjin science and Technology Press, 1985
[9] Jiang Weiping, “Ultrasonic Testing”, Wuhan University of Surveying and Mapping Press, 1991
[10] Zhang Junzhe et al. “Nondestructive Testing Technology and Its Application”, Science Press, 1993
[11] Nondestructive Testing Society of Japan, Ma Ming et al., "Ultrasonic Flaw Detection A", Jiangsu Science and Technology Press, 1982

IOP Conf. Series: Materials Science and Engineering **216** (2017) 012016 doi:10.1088/1757-899X/216/1/012016

Research on Ultrasonic Flaw Detection of Steel Weld in Spatial Grid Structure

Tao Du[1], Jiandong Sun[1], Shengguang Fu[2], Changquan Zhang[3], Qing Gao[4]
[1]Shandong Academy of Building Research Institute of Structure Jinan; 250031
[2]Feicheng Construction Engineering Quality Supervision Station Tai'an; 271000
[3]Shandong Provincial Department of housing and urban rural development Jinan; 250000
[4]Shandong Hi-speed Company Limited Jinan; 250000

dt7777@126.com

Abstract. The welding quality of spatial grid member is an important link in quality control of steel structure. The paper analyzed the reasons that the welding seam of small-bore pipe with thin wall grid structure is difficult to be detected by ultrasonic wave from the theoretical and practical aspects. A series of feasible detection methods was also proposed by improving probe and operation approaches in this paper, and the detection methods were verified by project cases. Over the years, the spatial grid structure is widely used the engineering by virtue of its several outstanding characteristics such as reasonable structure type, standard member, excellent space integrity and quick installation.
The wide application of spatial grid structure brings higher requirements on nondestructive test of grid structure. The implementation of new Code for Construction Quality Acceptance of Steel Structure Work GB50205-2001 strengthens the site inspection of steel structure, especially the site inspection of ultrasonic flaw detection in steel weld. The detection for spatial grid member structured by small-bore and thin-walled pipes is difficult due to the irregular influence of sound pressure in near-field region of sound field, sound beam diffusion generated by small bore pipe and reduction of sensitivity. Therefore, it is quite significant to select correct detecting conditions. The spatial grid structure of welding ball and bolt ball is statically determinate structure with high-order axial force which is connected by member bars and joints. It is welded by shrouding or conehead of member bars and of member bar and bolt-node sphere. It is obvious that to ensure the quality of these welding positions is critical to the quality of overall grid structure. However, the complexity of weld structure and limitation of ultrasonic detection method cause many difficulties in detection. No satisfactory results will be obtained by the conventional detection technology, so some special approaches must be used.

1. Analysis of Welding Process

Spatial grid structure is generally made from low-carbon steel Q235 or low alloy steel 16Mn, and the member bar is small-bore steel pipe with thin wall in size of Φ48×3.5mm- Φ15×98mm. When the shrouding or conehead of member bars and of member bar and bolt-node sphere, if the wall thickness of a member bar is less than or equal to 6mm, it can be welded through even if not setting a groove; When the wall thickness of a member bar is larger than 6mm, an approx. 300mm grove is required to be set. The common defects in welding seam include incomplete penetration or fusion of root, slag inclusion and air hole, but cracks seldom appear. Manual arc welding generally uses Φ2.5mm or Φ3.2mm welding rod, otherwise the root is not easy to be welded through. If downgrade welding is adopted during the gas shielded welding, the deposited metal is easy to flow down, resulting in incomplete fusion. If the welding spot is too high to realize smooth transition, incomplete penetration and slag inclusion will appear on two sides of the welding spot, and the root at arc striking position and arc suppression position will have incomplete penetration and slag inclusion.

2. Analysis on the Reasons for Ultrasonic Wave Which Is not Easy to Detect

2.1. Influence of near-field region

Based on the fundamental principle of ultrasonic detection, the thin-walled weld must be detected by special probe. Its main reason is that the large probe chip used in general weld detection causes a length of near-field region, which is larger than the distance from the detected thin-walled weld to sound source. The severe inference in near-field region and uneven sound pressure distribution plate make it impossible to quantify defects and give accurate and reliable judgment on defect echo. Additionally, the front distance of probe is above 12mm. The main sound beam with a perpendicular incidence of the probe cannot reach the weld even if the large retracted angle (e.g. 700) is adopted. The scanning on the whole beam can realized only by the second-trace echo and third-trace echo. But nearly three times of manufacturing error of wall thickness will appear in the third-trace echo reflection, and the sound velocity of multi-trace wave detection is too wide against the plate thickness, which will significantly affect the detection and positioning of defects. In the detection of thin-walled weld, the following requirements should be met to finish the detection by the first-trace echo and the second-trace echo:①

small-sized chip;② large refracted angle (large K value);③ Short front distance L.

2.2. Interference of head wave

If the defect echo of thin-walled workpiece is too close to head wave, the shape of head wave will have a significant influence on defect echo, and the head wave width and non-detection zone will be directly affected.

Head wave width: It refers to the time of duration emitting pulse, and is indicated by the ultrasonic propagation distance which is equivalent to the horizontal distance between scale "0" to trailing edge of head wave and the intersection of vertical scale 20% line under a certain upper sensitivity. The wider the head wave is, the poorer capacity for detecting defects on sub-surface will be. For example, the actually measured head wave width of commonly used CTS-22 ultrasonic detector (manufactured by Shantou Ultrasonic Instrument Plant) and 2.5PKz angle probe is 5mm.When the butt-welded seam of 3mm thickness steel plate is detected, the scanning ratio is adjusted deeply as 1:1 as a result, the defect echo cannot be determined by the primary wave (horizontal distance ≤3mm) and secondary wave (horizontal distance ≤6mm), but by the tertiary wave. The sound pressure and sound intensity of ultrasonic wave

have been increased considerably after tertiary wave attenuation, so the defect echo is totally impossible to be identified.
Non-detection area is an area near the detection surface, where no defects can be detected, and indicated by the minimum distance from the detection surface to the surface where defects can be detected. If the non-detection area is small, the defects near the detection surface can be detected; on the contrary, only the defects away from the detection surface can be detected. The non-detection area is not only related to probe performance, also inseparably interconnected with head wave width. For example, the measured non-detection area for CTS-22 detector and 2.5PK2 angle probe on CSK-IA test block is within 0～5mm.
It can be inferred that this non-detection area is also near to the defect echo of the secondary wave from the butt weld of 3mm plate, and its defects only can be determined by the tertiary wave as well.

2.3. Influence of curvature

Spatial grid member is small-bore and thin-walled steel pipe with a diameter of Φ48～Φ159mm and large surface curvature, which cause the contacting area between probe and workpiece reduces from surface contact to line contact and even point contact. The poor coupling condition will result in severe loss of acoustic energy, which will significantly reduce wave signal. In addition, the curvature has a large influence on positioning and length measurement of defects. It is obvious that the defect length measured from excircle will exceed its actual length, so its indicating length should be calculated by a formula, which is extremely inconvenient on site.
Other influencing factors
Weld detection of thin-walled workpiece is also influenced by signal-to-noise ratio, surplus sensitivity, resolving ability and other factors. These factors are composite performance indexes of detector and probe, and the composite performance of general detector and probe does not meet the demand of weld detection of thin-walled workpiece.
The above influence factors make it extremely difficult to weld detection of thin-walled workpiece by general instrument and detection methods. In practice, we have made repeated tests and researches on the particularity of grid structure joint weld and other factors influencing ultrasonic detection, and prepared a set of simple and practical technique with better effect.

3. Test method

The detection method is described by an example of steel pipe (diameter 50mm, wall thickness 4mm) butt weld in welded spherical joints.

3.1. Detector

Union Pxut-3300 six channels full digital ultrasonic detector is selected. It is equipped with a 0.8Kg battery, and the light weight, high luminance and long continuous working hours are suitable for site detection. In particular, its performance indexes represent the highest level of current ultrasonic detection of steel structure.
Its main performance:

① Horizontal linearity. Its horizontal linearity is tested to be less than and equal to 1%, see test method in Ultrasonic Detection P89.

② vertical linearity. Its vertical linearity is measured as $\Delta d \leq 4\%$;

IOP Conf. Series: Materials Science and Engineering **216** (2017) 012016 doi:10.1088/1757-899X/216/1/012016

③ Dynamic range. Its dynamic range is measured as D≥30dB;

④ Attenuator precision. The attenuator precision has an impact on reliability of defect detection and the accuracy of defect quantitative. Its attenuation range is 0-120dB, which can be stepped by 0.1dB.

3.2. Probe

Selection of probe parameters:

①Probe frequency: The attenuation coefficient of the primary wave can be ignored because its sonic path distance is too short for the thin-walled member bars. To improve the directivity of ultrasonic beam, reduce noise wave and improve the resolution, it is better to use higher detection frequency. We use a probe with a detection frequency of 5MHz.

② Probe K value; Considering the particularity of grid structure joint weld and the distribution rule of defects, the detection can only be carried out from one side of member bar, and the primary wave should be used as far as possible. To make the primary wave step over the root of weld, sonic path distance and span should be long enough, which mean K value should be large enough. We use the probe with a K value of 2.5.

③Frontal L:The front distance L of probe should be short enough. We use the probe with a front distance of L=6mm.

④ Composite performance of detector and probe: signal-to-noise ratio N=50dB, surplus sensitivity S=50dB, head wave width W0<2mm.

⑤ Probe curvature:

The probe surface is flat when it is manufactured, so it may have point contact or line contact when contacting with a hook surface, which will cause a loss of effective acoustic energy. The calculation formula of ultrasonic sound pressure:

$$\rho = \rho_0 \frac{A}{\lambda S}$$

ρ ---sound pressure incident on a certain reflector

ρ_0 --- radiated sound pressure

A----effective chip area

S-----beam path distance

λ -----Wavelength

Known that the incoming sound pressure of ultrasonic wave against reflector is in direct proportion to the effective chip area. If the coupling area decreases, the incoming sound pressure will reduce accordingly, which will result in a reduction of reflection wave height and detection sensitivity. It certainly will affect the defect quantitative and leak detection. The research on butt-welded seam of steel pipe with different curvature showed that the defect reflection echo is relatively stable when the curvature radius is larger than 20mm, and similar echo amplitude can be received at the same position in repeated detections. But the reflection echo is not stable when the curvature radium is less than 20mm, and reflection echo may not be received at the same position in repeated detections. The research results indicated that the defects is possible to be undetected when the curvature radius of a workpiece is less than 20mm.When the probe surface is polished to a similar status of workpiece curvature, the defect reflection echo will recover to stable. For the purpose of conservation, it is suggested that when the curvature radius of a workpiece is less than 40mm during the practical operation, the possibility of leak detection can be considerably decreased if the probe surface is polished to as similar status with workpiece curvature. We made curvature finishing to the probe by polishing the polymethyl methacrylate of the probe with steel file and fine abrasive paper to match the detection surface to the curvature radius of test block. The curvature radius of the finished probe is 27mm, which is approximate to the curvature radius 25mm of pipes to be tested. The polished probe is shown as follows(see figure 1):

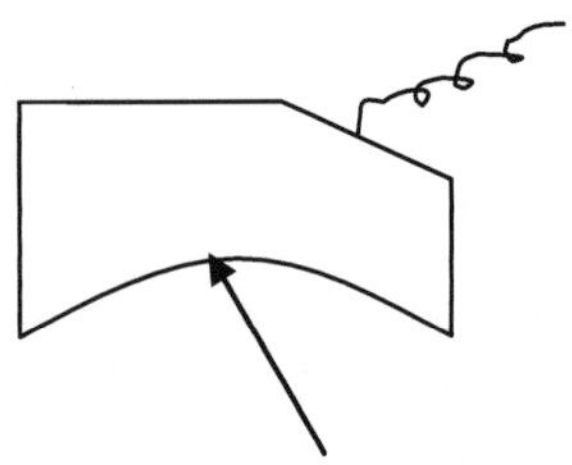

Figure 1. Polished probe.

3.3. Test block
CSK-1C test block specially for profile steel grid structure is adopted. Every three test blocks are divided into one group, and the curvature radius of the three test block are 27mm, 40mm and 60mm respectively, which can meet the requirements in JG/T3034.1-1996 that the curvature radius of detection surface should be 0.9-1.5 times of test block.

3.4. Couplant
Engine oil is selected.

3.5. Detection surface
Considering the particularity of grid structure joint weld, the detection can only be carried out from one side of member bar. The primary wave should be used as far as possible to detect the root defects; the secondary wave is mainly used to detect the defects at the upper part; sometimes, the tertiary wave is used to detect the defects at one side of shrouding or conehead. Single-face and bilateral inspection can be adopted for butt weld between steel pipes; for single-face and single-side inspection can be used for the butt weld of pipe sphere limited by operating conditions. The demand of primary wave and secondary wave detection can be met only if the width of detection surface is determined to be 30mm.

3.6. Selection of detection direction
The direction specified in JG/T3034.1-1996 can be used, i.e. the moving mode of a probe should be in a shape of "W" with swing in angle of 100-150°.To determine the position and shape of defects, observe dynamic waveform of defects and distinguish echo signal, various scanning methods such as forward and backward scanning, left-to-right scanning, turning scanning and circle scanning should be added.

3.7. Adjustment of scanning speed (scanning rate)
The horizontal distance adjustment method is adopted and the scanning rate is determined as 3:1, the results proved that defect echo can effectively avoid the interference from noise wave of head wave, and the waveform is clear, which is beneficial to identifying the defects.

IOP Conf. Series: Materials Science and Engineering **216** (2017) 012016 doi:10.1088/1757-899X/216/1/012016

3.8. Positioning and quantitative of defects

During the practical operation, the sensitivity can be detected according to JG/T3034.1-1996.Adjust the detector to the assessment curve to detect the sensitivity, and then increase to 10dB to measure it roughly. Fix the probe after finding defect echo, and draw a line along the direction of main sound beam of the probe. Mark the position where the line crosses through the weld, which is the position of weld defect. Adjust the detection sensitivity to the specified sensitivity and quantify the defects according to attenuator reading at this time.

4. Application example

The top floor of complex office building of a certain mechanical manufacturing enterprise in High-tech Development Zone of Jinan is a spatial grid structure covering 1000m2. It is connected by welded spherical joints and 6 kinds of member bars are used. Ten joints and 30 pipe-sphere butt welds and pipe-pipe butt welds are detected by ultrasonic wave, the flaw detector, probe and the couplant is engine oil. The surface coupling loss compensation is 4dB, and Φ3 horizontal holes are taken as standard defects, as well as a distance-amplitude curve is made.

There are 25 defects in total, and the defects are planed with a gouge immediately. It is proved that the actual defects is roughly consistent with the defects inferred by ultrasonic detection, and the coincidence rate reach above 90%.See Table 1:

Table 1. Comparison of detected and actual defects on test piece.

No.	Pipe specification	Nature of defects	Defect length	Wave height
1	Φ48×3.25mm	Slag inclusion	5	RL+2dB
2	Φ48×3.25mm	Incomplete penetration	5	RL+2dB
3	Φ48×3.25mm	Incomplete penetration	7	RL+8dB
4	Φ48×3.25mm	Incomplete penetration	9	RL+6
5	Φ60×3mm	Incomplete penetration	10	RL+6
6	Φ60×3mm	Slag inclusion	5	RL+6
7	Φ60×3mm	Crack	8	RL+6
8	Φ60×3mm	Slag inclusion	5	RL+4
9	Φ60×3mm	Incomplete penetration	8	RL+6
10	Φ75.5×3mm	Incomplete penetration	6	RL+8
11	Φ75.5×3mm	Incomplete penetration	7	RL+6
12	Φ75.5×3mm	Incomplete penetration	5	RL
13	Φ75.5×3mm	Incomplete penetration	5	RL+4
14	Φ75.5×3mm	Incomplete penetration	10	RL+6
15	Φ75.5×3mm	Incomplete penetration	8	RL+6
16	Φ88.5×3.25mm	Incomplete penetration	6	RL+6
17	Φ88.5×3.25mm	Incomplete penetration	7	RL+6
18	Φ88.5×3.25mm	Incomplete penetration	5	RL+6
19	Φ114×3.5mm	Incomplete penetration	5	RL+2
20	Φ114×3.5mm	Incomplete penetration	5	RL+2
21	Φ114×3.5mm	Incomplete penetration	5	RL+2
22	Φ114×3.5mm	Incomplete penetration	5	RL+2
23	Φ114×3.5mm	Incomplete penetration	5	RL+2
24	Φ140×3.5mm	Incomplete penetration	7	RL+2
25	Φ140×3.5mm	Incomplete penetration	7	RL+6

IOP Conf. Series: Materials Science and Engineering **216** (2017) 012016 doi:10.1088/1757-899X/216/1/012016

Table 2. Defect types.

Defect class	Incomplete penetration	Crack	Slag inclusion or air hole
Quantity	21	1	3

Table 3. Reflection wave height of each defect type

Defect class	Incomplete penetration	Crack	Slag inclusion or air hole
Reflection wave height	RL+0～+8	RL+6	RL+2～+6

The following can be deferred from Table 2 and Table 3:

(1) Incomplete penetration accounts for 84% in the 25 defects, which can be considered that the incomplete penetration is more frequent than other defects and has strong reflection under the normal welding condition. For all incomplete penetrations exceeding 5mm, the defect echoes are at or above the defect line. Too narrow clearance at groove root is the main reason for incomplete penetration.

(2) Cracks and slag infusion of air holes seldom appear under the normal welding condition, and the reflection wave heights of cracks and slag infusion of air holes in this detection are above the defect line.

(3) By a comparative analysis, the wave height of defect echo is not completely direct proportional to the indicating length of defects.

(4) The coincidence rate between detection results and actual results reaches above 90%, which indicates that instrument, probe, test block and detection methods used in this study are feasible and the detection results can meet the requirements of quality control.

References

[1] HU Tianming et.al, Ultrasonic Flaw Detection, China Boiler and Pressure Vessel Safety Periodical Office

[2] The national standard Code for Quality Acceptance of Steel Structure Work Construction GB50205-2001

[3] MOC Standard, Classification for Manual Ultrasonic Flaw Detection Methods and Results of Welded Spherical Joints JG/T3034.1-1996

[4] Ultrasonic Flaw Detection Group "Ultrasonic Flaw Detection", Labor and Personnel Press, 1989

[5] Japan Nondestructive Inspection Association, Ma Yukuan et al ., "Translation of Ultrasonic inspection", Jilin Science and Technology Press, 1984

[6] Hu Tianming "Ultrasonic Flaw Detection", Wuhan University of Science and Technology of Surveying and Mapping Press, 1994

[7] Ultrasonic Flaw Detection Group "Ultrasonic Flaw Detection", Electric Power Industry Press Press, 1980

[8] Yun Qinghua et al., "Nondestructive Testing", labor press, 1983

[9] Yun Qinghua, "Nondestructive Testing Technology for Boiler Pressure Vessel", Tianjin science and Technology Press, 1985

[10] Jiang Weiping, "Ultrasonic Testing", Wuhan University of Surveying and Mapping Press, 1991

[11] Zhang Junzhe et al. "Nondestructive Testing Technology and Its Application", Science Press, 1993

[12] Nondestructive Testing Society of Japan, Ma Ming et al., "Ultrasonic Flaw Detection A", Jiangsu Science and Technology Press, 1982

IOP Conf. Series: Materials Science and Engineering **216** (2017) 012017 doi:10.1088/1757-899X/216/1/012017

Performance Using Bamboo Fiber Ash Concrete as Admixture Adding Superplasticizer

Gunalaan Vasudevan

Faculty of Engineering and Built Environment, Lecturer, Tunku Abdul Rahman University College Malaysia

Email: gunalaanv@acd.tarc.edu.my

Abstract. The increasing demand on natural resources for housing provisions in developing countries have called for sourcing and use of sustainable local materials for building and housing delivery. Natural materials to be considered sustainable for building construction should be 'green' and obtained from local sources, including rapidly renewable plant materials like palm fronds and bamboo, recycled materials and other products that are reusable and renewable. Each year, tens of millions of tons of bamboo are utilized commercially, generating a vast amount of waste. Besides that, bamboo fiber is easy availability, low density, low production cost and satisfactory mechanical properties. One solution is to activate this waste by using it as an additive admixture in concrete to keep it out of landfills and save money on waste disposal. The research investigates the mechanical and physical properties of bamboo fiber powder in a blended Portland cement. The structural value of the bamboo fiber powder in a blended Portland cement was evaluated with consideration for its suitability in concrete. Varied percentage of bamboo fiber powder (BFP) at 0%, 5%, 10%, 15%, and 20% as an admixture in 1:2:4 concrete mixes. The workability of the mix was determined through slump; standard consistency test was carried on the cement. Compressive strength of hardened cured (150 x 150 x 150) mm concrete cubes at 7days, 14days and 28days were tested.

1. Introduction

Nowadays, construction development in Malaysia increase very fast. At the same time, the cost of properties is become expensive. Not very numerous white collar class Malaysians can bear the cost of owning a rooftop over their heads. Property costs are presently high as can be and the centre pay wage specialists are constrained to go for leased homes or pick a spot at the edges of the city. Malaysia as a developing nation additionally got no special case to get new innovation in the development business. Since the large demand has been put on building material industry particularly in the most recent decade inferable from the expanded populace which causes an interminable deficiency to building materials, the structural designer has been tested to change over the modern waste to helpful building and development material. [1]

2. Literature Review

Concrete is most common and widely used in construction industry because of its strength and durability. Concrete is made with the mixture of cement, aggregates and water. The aggregates can be either fine aggregates or coarse aggregates. In this research, bamboo is selected as an additive material in concrete design. It also can be one of the recycle materials to cut down the using quantity of cement and aggregates in concrete design. Bamboo is easy to break into pieces by using saw. It is good to use bamboo because bamboo would not abandon in jungles. By using bamboo will be more sustainable to our environment. If over use the quantity of cement and aggregates will facing the deplete of resources. Besides that, bamboo material can benefit to construction industry. Bamboo is easy to handle and cut without using any specialist equipment. Bamboo is a non-polluting material. Bamboo can considered as saving material and cost because it does not waste any parts. In addition, bamboo's shape is circular form and light in weight [1].

2.1 Properties of Bamboo

In bamboo, specific gravity can be measure the density. The specific gravity of bamboo is a comparison of its density to that of water. The range is from 0.4 to 0.8 due to anatomical structure. The dampness substance moves vertically from the base to the top parts and horizontally from the outside layer to the inward layers. The dampness content in bamboo can has high. Green bamboo may have 100% percent dampness (oven-dry weight basis). It can be up to 155% for the deepest layers and 70% for the fringe layers. The vertical variety from the top (82%) to the bottom (110%) is comparatively less. Next, for the bamboo's fiber saturation point is about 20-22%. [2] Bamboo do not has secondary growth which unlike wood. All gains after all gains reach their own full height are depends on additional of material to cells.[4]

2.2 Mechanical Properties of Bamboo

Bamboo has strong mechanics and great flexibility, simple to be prepared which cause it for extensive variety of utilization in construction industries. With the increase in bamboo's age, the compressive stress also increases. One year old bamboo has the lowest compressive stress, which is 16.1 MPa. For five years old bamboo has the highest compressive stress. The compression properties are equal to the longitudinal direction. Bamboo known as light material and can strengthen product. Meanwhile, it is stable because of its cavities are light and elastic. Original texture of wood is similar to the bamboo texture and the technical conditions. Wood has harder center and make it weaker in the outer and inner parts.

2.3 Chemical Composition of Bamboo

The chemical composition of bamboo is entirely comparative with wood. Their major compositions are cellulose, hemi-cellulose and lignin. The amount of composition is over 90% of the mass. The minor compositions are resins, tannins, waxes and inorganic salts. Thus, bamboo has higher alkaline extracts, ash and silica contents compared to wood. Other than that, bamboo contains some organic composition as well. Bamboo contains 2-6% starch, 2% deoxidized saccharide, 2-4% fat and 0.8-6% protein. The durability and life is depending on the carbohydrate content bamboo. For the natural durability, bamboo is depending on the climatic condition. Next, the composition of starch can prevent attack from fungi. Moreover, the ash content of bamboo has inorganic minerals, primarily silica, calcium and potassium.

3. Experimental Process

In this research is to determine the performance of the concrete that contain the bamboo fiber powder by preparing concrete cube sample and tested to obtain some of the basic engineering properties. The concrete mix design is done by systematic analysis and chooses the proportion of the ingredient to use the concrete mix to produce an economical concrete and also with strength that desired when the cube is hardened. The variables which can be controlled are water cement ratio, maximum aggregate size, aggregate grading and use of admixture.

4. Analysis and Discussion

The Figure 1 gives the results of test conducted on hardened concrete with 0-20% bamboo fiber powder for 7, 14 and 28 days. From figure 2, the results show that the compressive strength increases with increasing curing time. It seems the compressive strength obtained for concrete with 15%

replacement by bamboo fiber powder showed average value compared to control concrete for 7 days, 14 days and 28 days respectively.This increment of the positive value shows that with the mixture of bamboo fiber powder into the concrete will achieve the early strength ratio's for the 14 days.

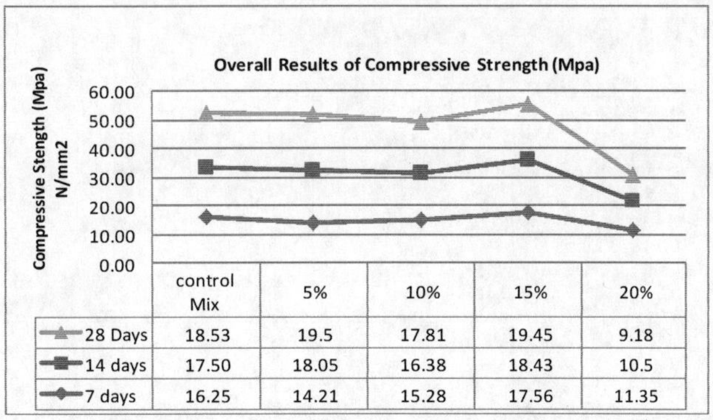

Figure 1. Overall results of compressive strength (Mpa) between control mix and different percentage bamboo fiber powder (BFP) of concrete mix

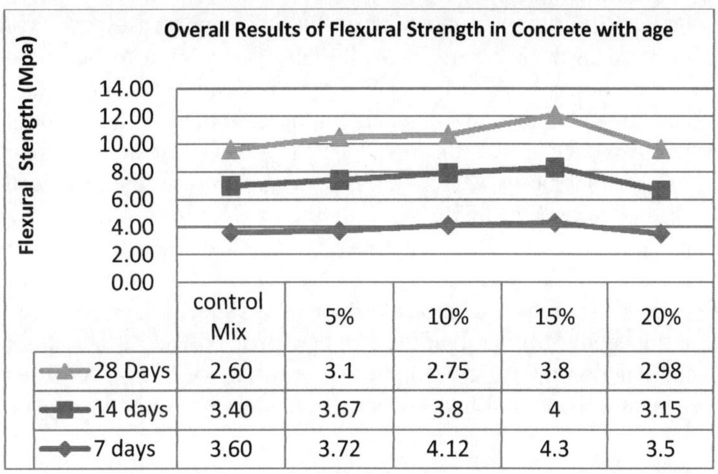

Figure 2. Overall results of flexural Strength (Mpa) between control mix and different percentage bamboo fiber powder (BFP) of concrete mix

The Figure 2 shows the result of variation of flexural strength of concrete with cement replacement by bamboo fiber powder for 7, 14 and 28 days. It seems flexural strength of concrete with 15% cement replacement by bamboo fiber powder showed average higher value compared to control concrete for 7 days, 14 days and 28 days respectively. As the percentage of replacement of cement with bamboo fiber powder increases strength increases up to 10% and beyond that it decreases with 20%. The increase in strength up to 15% replacement of cement by bamboo fiber powder may be due to the pozzolanic reaction of bamboo fiber powder due to high silica content. Also it effectively fills the voids and gives a dense concrete microstructure.

4.1 Scanning Electron Microscopy Test (SEM)

After the concrete samples were prepared for SEM studies, images were extracted from each sample. A total of two samples taken from the center and edge of the concrete cylinders in axial direction was studied. The images was acquired by the image analyzer at a magnification of ×60 and digitized into an array of 512 × 512 pixels with 256 gray levels (1 pixel = 3.3 μm). Based on figure 3, SEM test can show the location and present of bamboo fiber powder in concrete. The use of SEM to extract images from the concrete specimens for different testing conditions is explained [3]. The images rendered were analyzed using an image analyzer. The image analyzer identifies Wood's metal, which represents the crack network in concrete specimens. The SEM identifies the geometric aspects of features in the microstructure, such as micro cracks in concrete represented by Wood's metal.

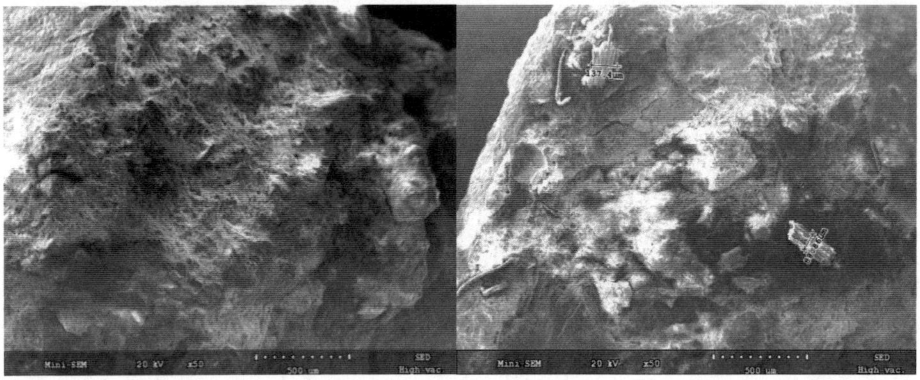

Fig 3. SEM test for control mix concrete and bamboo fiber powder concrete

5. Conclusion and Recommendation

From the results obtained from the laboratory test effect of bamboo ash powder in concrete mix does not give a positive optimism from the result of the laboratory experiment although slightly weak in strength while carry out compression test but the strength is still consider good just it is too weak in workability. This may due to bamboo ash powder can increase the water absorption therefore make the workability of concrete reduce. From the result of the slump test and compaction factor test, the conclusion can be make is the more the bamboo ash powder substitute to the fresh concrete, the lower the workability of the concrete. This might cause by the assumption for the proportion of bamboo ash powder and also the water content. But, superplasticizer (Master Rheobuild 1100 RM)[4] is a good product to use in concrete. The compressive strengths of SP concrete are usually higher than the corresponding strengths of the reference mixes. When SP is used, a water reducer up to 32-33% can be achieved [5]. From this study, I found that the bamboo fiber powder as an admixture in concrete is a be done in actual construction industries. Due to the performance of bamboo fiber powder, it can mix with concrete. The reason is bamboo fiber powder will absorb the water during concrete mix

References

[1] Kishen, J., D.P. Ghosh, and M.A. Rehman. 1956. Studies on moisture content, shrinkage, swelling and intersection point of mature (Dendrocalamus strictus) male bamboo. Indian For. Rec... **1**: 1-30. (Viewed on 4 Dec 2015)

[2] Lee, A.W.C, X.S. Bai, and P.N. Peralta. 1994. Selected physical and mechanical properties of giant timber bamboo grown in South Carolina. Forest Prod. J.. **44(9)**:40- 46 (Viewed on 3 Dec 2015)

[3] Microanalysis Consultants Ltd, Concrete Examination by SEM. (Viewed on 2 August 2016)

[4] Master Rheobuild 1100, High Range Water Reducing Superplasticising Admixture. (Viewed on 1 August 2016)

[5] Nadav Malin and Jessica Boehland, 2006. Bamboo in Construction, **15** (Viewed on 19 Nov 2015)

IOP Conf. Series: Materials Science and Engineering **216** (2017) 012018 doi:10.1088/1757-899X/216/1/012018

Research on Durability of Recycled Ceramic Powder Concrete

M C Chen[1,2,3], W Fang[1,2], K C Xu[1,2] and L Xie[1,2]

[1]Jiangxi Provincial Key Laboratory of Simulation and Control of Building Process,
[2]School of Civil Engineering and Architecture, East China Jiaotong University, Nanchang 330013, China

Email: 919367042@qq.com; fangweiwei23@163.com; 505958373@qq.com; boshini@163.com;

Abstract. Ceramic was ground into powder with 325 mesh and used to prepare for concrete. Basic mechanical properties, carbonation and chloride ion penetration of the concrete tests were conducted. In addition, 6-hour electric fluxes of recycled ceramic powder concrete were measured under loading. The results showed that the age strength of ceramics powder concrete is higher than that of the ordinary concrete and the fly ash concrete. The ceramic powder used as admixture would reduce the strength of concrete under no consideration of its impact factor; under consideration of the impact factor for ceramic powder as admixture, the carbonation resistance of ceramic powder concrete was significantly improved, and the 28 day carbonation depth of the ceramic powder concrete was only 31.5% of ordinary concrete. The anti-chloride-permeability of recycled ceramic powder concrete was excellent.

1. Introduction

Waste ceramic is a new renewable material. It is famous for its light weight, low price, high pozzolanic reactivity and rich source of SiO_2. Waste ceramic as construction materials has been wildly applied for various practical engineering and got remarkable economic and social effects. In the past few decades, many domestic and foreign scholars did researches on the utilization of ceramics in cement and concrete industry, and obtained a large number of reliable data and useful conclusions with waste ceramic as coarse aggregates, fine aggregates and admixtures in concrete. The various performance of that recycled ceramic concrete was analysed [1-6]. The existing researches has shown that crushed ceramic as coarse aggregates has a certain effects on the performance of the recycled concrete, but the treatment of ceramic coarse aggregates can meet the requirements of concrete strength[7-8]. Previous studies have shown that waste ceramic used as regenerated ceramic sand in concrete can improve effectively mechanical properties of the concrete [9-10]. As Wang [6] indicated polished ceramic brick dust as admixtures has better pozzolanic reactivity; the SiO_2 in the brick dust has strong activity to participate in chemical reactions than fly ash does. In summary, waste ceramic used as parts of recycled concrete were feasible and practicable.

In the experimental study of ceramic recycled concrete, most scholars have devoted to using waste ceramic as coarse and fine aggregates. There are very few researches on employing waste ceramic powder as admixtures, especially on the analysis of its recycled concrete durability [11-13].

IOP Conf. Series: Materials Science and Engineering **216** (2017) 012018 doi:10.1088/1757-899X/216/1/012018

The author's previous researches showed that the optimal amount of ceramic powder as admixture is 25%, and the age strength growth curve of its recycled ceramic cement mortar is similar to that of fly ash. Therefore, in present paper, ceramic powder of 25% will be used to investigate the basic mechanical properties, carbonization and anti-chloride-permeability performances, which provides a theoretical basis for developing and popularizing ceramic powder concrete.

2. Materials and methods

2.1. Materials

The waste ceramic of apparent density 2.375 kg/m^3 was collected from the Jiangxi provincial Jingdezhen. Table 1 shows the chemical compositions of ceramic. The specific surface area of ceramic powder is 458m^2/kg. The grain size distribution of ceramic powder is shown in Table 2.

Table 1. The chemical compositions of ceramic (unit: %).

SiO_2	Al_2O_3	Fe_2O_3	CaO	MgO	K_2O	Na_2O	SO_3
62.72	23.31	1.24	6.67	0.65	1.55	0.96	0.07

Table 2. Particle size distribution of ceramic powder.

grain composition (μm)	<10	10~20	20~30	30~60	>60
percentage (%)	51.0	20.4	11.1	9.5	8.0

2.2. Concrete production

C40 concrete test specimens consisted of 5 groups. There are 12 specimens in each group. Tests of their compression, carbonation resistance and chloride permeability were performed. Mix proportion of the concrete was designed according to "Specification for mix proportion design of ordinary concrete"(China standard specification: JGJ 55-2011). OC is ordinary concrete, CC-1 and FC-1 indicate concrete made from ordinary concrete whose 25% cement is replaced by recycled ceramic and fly ash, respectively. CC-2 and FC-2 are respectively concrete based on CC-1 and FC-1 whose influence coefficients of admixture were chosen to be 0.8 by virtue of the specification. The mixture ratio of concrete shown in Table 3.

Table 3. Mix proportions of concrete (unit: kg/m^3).

No.	Strength grade (F_{cu})	Cement	Fly ash	Ceramic powder	Sand	Coarse aggregate	Water
OC	C40	381.11	0	0	591.57	1201.07	186.30
CC-1	C40	285.836	0	95.278	567.48	1152.15	186.30
FC-1	C40	285.836	95.278	0	567.48	1152.15	186.30
CC-2	C40	350.30	0	116.77	567.48	1152.15	186.30
FC -2	C40	350.30	116.77	0	567.48	1152.15	186.30

2.3. Test procedure

The compressive strength of concrete specimens was respectively measured at 3d, 7d, 14d, 28d and 56d.

The depth of carbonation of specimens were measured after carbonized for 3d, 7d, 14d and 28d. The carbonation tests were conducted with TH-074 concrete carbonation machine based on 'standards for test method of long-term performance and durability of ordinary concrete '(China standard: GB/T 50082-2009). Variations of the quality and relative dynamic elastic modulus for post-carbonated test specimens were analyzed to prove present recycled ceramic powder concrete to be sufficient in resisting carbonation.

The chloride diffusion coefficient of concrete was measured by electric flux method through RCM-F concrete durability machine. The chloride ion penetration tests were carried out by standard ASTM C2012 and China standard GB/T 50082-2009.

3. Text results and analysis

3.1. Compressive strength

It can be seen from Figure 1 that growth trends of age strengths of all concrete specimens are similar. The 28d compressive strength of CC-1 is the lowest, 31.80Mpa; that of CC-2 is the highest, 47.45Mpa, which increase by 6.9%, 49.2%, 32.1% and 40.4% by comparison of those of OC, CC-1, FC-1 and FC-2, respectively. It means that the decrease of the influence coefficients of admixture for CC-2 is helpful to increase its compressive strength, and its growth rate is greater than those of the other cases. Therefore, in the following CC-2 material is used to study the durability of recycled ceramic powder concrete.

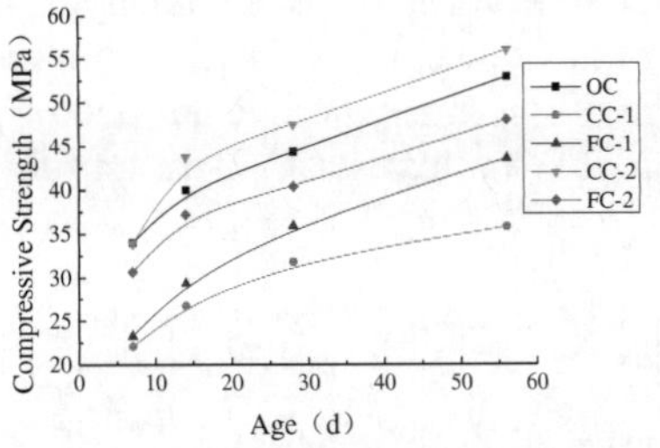

Figure 1. The age strength of five groups concrete.

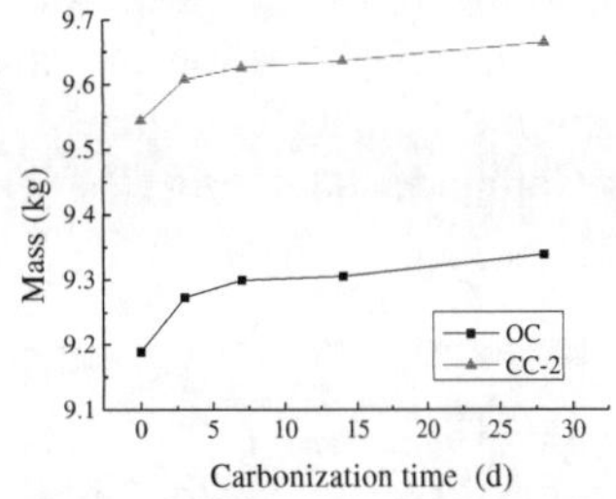

Figure 2. Change in mass.

3.2. Carbonation resistance

Carbonation test of CC-2 material was conducted to investigate the impact of carbonation on its mass and relative dynamic elastic modulus. All experimental data are graphed in Figures 2-4 For the sake of comparison, relevant experiment data for OC material is also given in the figures.

It is shown from Figure 2 that all curves rise quickly in the first 3 days of carbonization, and then become stable with increasing carbonization time; apparent mass of CC-2 is greater than that of OC, implying that the recycled ceramic powder concrete is more denser.

At the early stage of carbonization, moisture absorption by pore structure in concrete is the mainly reason of quality variation. Absorption saturated and calcium carbonate filler in capillary resulting in the absorption of moisture difficult, the quality in later period increases slightly.

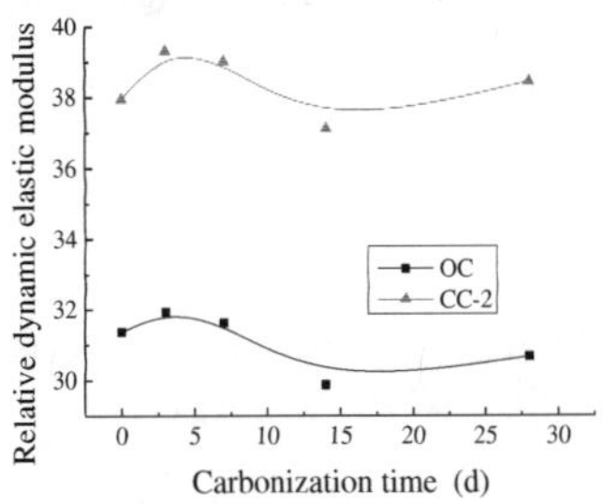

Figure 3. Change in relative dynamic elastic modulus.

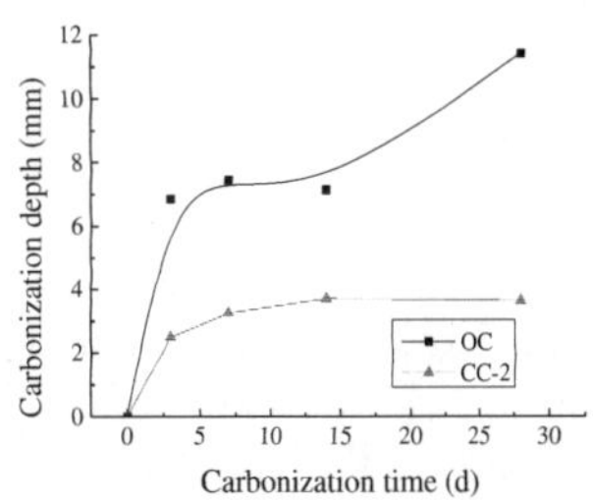

Figure 4. Carbonation depth changes with time.

Figure 3 shows relationships between relative dynamic elastic modulus and carbonization time. From this Figure, we can draw similar conclusions to those from Figure 2. In general, 2 curves in Figure 3 comply with a trend of up-down-up.

The age strength of the concrete is still developing in the early phase of carbonation, and the concrete structures are becoming more dense for hydration and carbonation reactions. Therefore, its relative dynamic elastic modulus have a certain increase. Thereafter, since the alkalinity of concrete pore solution and concrete matrix decrease in the carbonation reaction process, the concrete generates shrinkage, and thus leading to the decrease of its relative dynamic elastic modulus. During the later phase of carbonation, compounds resulting from the carbonation are filled in microcracks and pores in concrete structure, which gives rise to denser concrete microstructures, and therefore the concrete relative dynamic elastic modulus increases

Figure 4 shows the carbonization depth of OC and CC-2 against carbonation time. It can be seen that the carbonation depth of CC-2 is only 3.6 mm while the OC already among 11.4mm at 28 day carbonation. Pozzolanicity and micro-aggregate effects produce when ceramic powder is added into concrete. The microstructure of concrete filled with ceramic powder forbids the invasion and diffusion of carbon dioxide in the concrete. Furthermore, pores in concrete structure is filled with calcium carbonate from the carbonation reaction to prevent carbon dioxide from diffusing further in the concrete. As carbonation time increase, the carbonation progresses gradually. Due to reasons aforementioned, we can draw a conclusion that ceramic powder used as admixtures is helpful to improve the carbonation resistance of concrete. Figure 5 presents experimental photos of carbonation for OC and CC-2. Comparison of Figure 5(a) with 5(b) indicates CC-2 has stronger carbonation resistance than OC does.

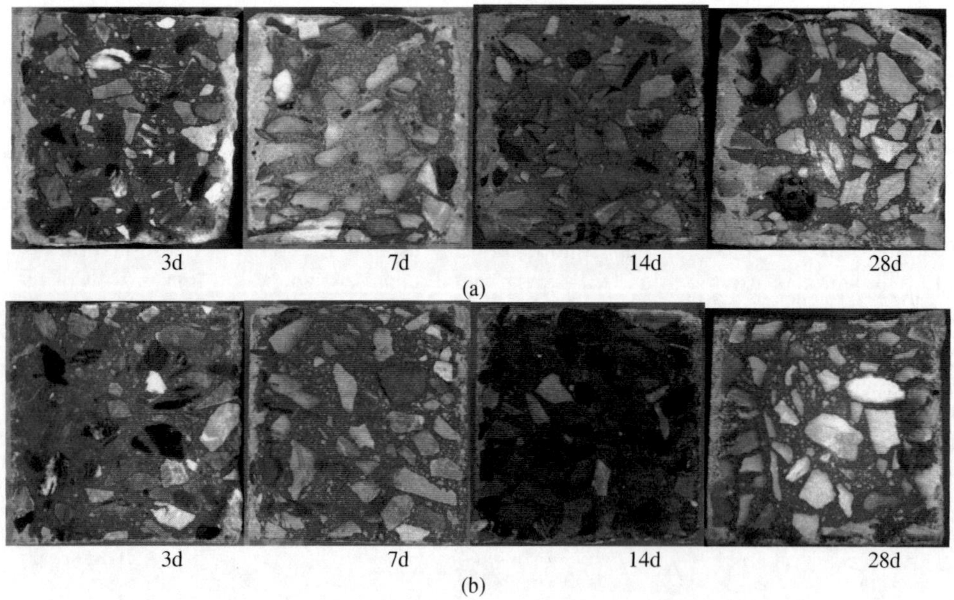

Figure 5. Photos of carbonation (a) OC; (b) CC-2.

3.3. Anti-chloride-permeability

As the corrosion of reinforcements in concrete is mainly induced by the attack of chloride ions, the effects of ceramic powder on chloride ion penetration were studied. 6-hour electric fluxes of ceramic powder concrete were measured. The standard of chloride ion permeability in concrete shown in Table 4.

It can see from Figure 6(a) that 6-hours electric flux of OC increased 58.99% than its of CC-2, implying that the resistance of chloride ion permeability of CC-2 is higher than that of OC. From

Table 5 and Figure 6(a) we can conclude that the classification of anti-chloride-permeability of OC belongs to C, but that of CC-2 E, which is excellent in anti-chloride-permeability.

Table 4. Relations of 6-hours electric flux and chloride ion permeability in concrete.

6-hour electric flux (C)	Levels of durability of concrete	Chloride ion permeability	Classify
>4000	poor	high	A
2000~4000	weak	medium	B
1000~2000	average	low	C
500~1000	good	very low	D
<500	excellent	negligible	E

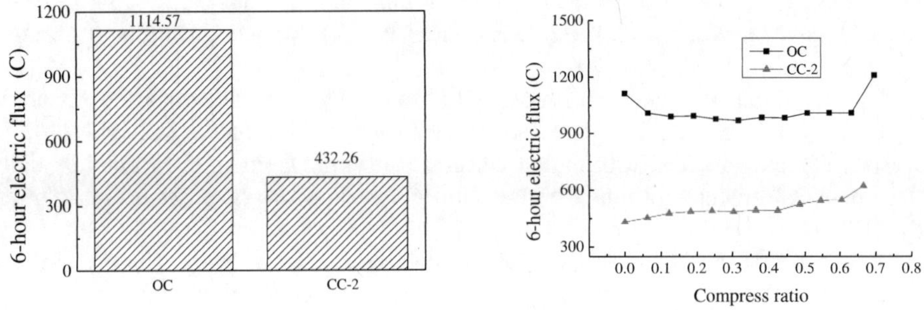

(a) (b)

Figure 6. 6-hours electric flux of OC and CC-2 (a) no stress; (b) loading.

It is shown from Figure 6(b) that, for OC, 6-hour electric flux of concrete decreases when compress ratio is small, which may be the compactness of concrete microstructures by the compress at initial stage, and then keep stable growth until compress ratio reaches 0.63. Afterward, it increase quickly; for CC-2, 6-hour electric flux grows stably until compress ratio reaches 0.61 and then increase quickly but relatively slower than OC does. When compress ratio is higher, the reason why the growth of 6-hour electric flux is rapid is that the higher compress ratio accelerate concrete damages. Comparison of OC curve with CC-2 curve in Figure 6(b) shows it is stronger in electric flux for CC-2 than for OC, representing that ceramic powder used as admixtures can effectively decrease the chloride ion permeability of concrete. Summation of section 3.2 and 3.3 tells us a fact that ceramic powder used as admixtures can improve durability of recycled ceramic powder concrete.

4. Conclusion

1. Ceramic powder as admixtures can effectively improve the compressive strength of concrete. Under the same conditions, the basic mechanical properties of ceramic powder concrete are better than that of fly ash concrete.

2. Rational addition of ceramic powder can effectively improve the carbonization resistance and anti-chloride-permeability of concrete.

3. In actual mix proportion design of recycled ceramic powder concrete can refer to that of fly ash concrete whose influence factor is 0.8 given in the Specification.

5. Reference

IOP Conf. Series: Materials Science and Engineering **216** (2017) 012018 doi:10.1088/1757-899X/216/1/012018

[1] Liu F L, Liu J H, Ma B G, Huang J and Li H N. 2015 Basic Properties of Concrete Incorporating Recycled Ceramic Aggregate and Ultra-fine Sand *Journal of Wuhan University of Technology-Mater*. **30(2)** 352–60

[2] Guerra I, Vivar I, Llamas B, Juan A and Moran J. 2009 Eco-efficient Concretes: the Effects of Using Recycled Ceramic Material from Sanitary Installations on the Mechanical Properties of Concrete *Waste Management*. **29(2)** 643–6

[3] Halicka A, Ogrodnik P and Zegardlo B. 2013 Using Ceramic Sanitary Ware Waste as Concrete Aggregate *Construction and Building Materials*. **48(11)** 295–305

[4] Medina C, Sánchez de Rojas M I and Frías M. 2013 Properties of Recycled Ceramic Aggregate Concretes: Water Resistance *Cement and Concrete Composites*. **40(40)** 21–9

[5] Jiménez J R, Ayuso J, López M, Fernández J M and Brito J D. 2013 Use of fine recycled aggregates from ceramic waste in masonry mortar manufacturing *Construction and Building Materials*. **40** 679–90

[6] Wang G X. 2010Pozzolanic Activity of Ceramic Polishing Powder as Supplementary Cementitious Material *Journal of the Chinese Ceramic Society*. **38(7)** 1229–34

[7] Chen M C, Ding X M, Wang K and Xie L. 2012 Research on Recycled Building Ceramic Concrete *Materials Review*. **26** 303–6

[8] Xu K C, Fang W, Chen M C and Ding X M. 2014 On the Mechanical Properties of Ceramic Recycled Coarse Aggregate Concrete *Journal of Experimental Mechanics*. **29(4)** 478–80

[9] Silva J, Brito J D and Veiga R. 2008 Fine Ceramics Replacing Cement in Mortars Partial Replacement of Cement with Fine Ceramics in Rendering Mortars *Materials and Structures*. **41(8)** 1333–44

[10] Zhang C Z, Liu F L and Liu J H. 2011 Research on the Impact of Waste Ceramic Reclaimed Sand Mortar Performance *China Concrete and Cement Products*. **7** 54–7

[11] Medina C, Sánchez de Rojas M I and Frías M. 2013 Freeze-thaw Durability of Recycled Concrete Containing Ceramic Aggregate *Journal of Cleaner Production*. **40(2)** 151–60

[12] Cheng Y H, Huang F, Li G L, Xu L S and Hou JL. 2014 Test Research on Effects of Ceramic Polishing Powder on Carbonation and Sulphate-corrosion Resistance of Concrete *Construction and Building Materials*. **55(2)** 440–6

[13] Kulovaná T, Vejmelková E, Keppert M, Rovnaníková P, Keršneret Z and Cerný Robert. 2016 Mechanical, durability and hygrothermal properties of concrete produced using Portland cement-ceramic powder blends *Structural Concrete*. **17(1)** 105–15

IOP Conf. Series: Materials Science and Engineering **216** (2017) 012019 doi:10.1088/1757-899X/216/1/012019

Development of an Improved Crack Propagation Model for Corrosion-Induced Cover Cracking in RC Structures

Hilyati S.[1,a], Nizam Z. M. [1,2,b] *, Zurisman M. A. A.[1,c] and Azhar A. T. S.[1,2,d]

[1]Faculty of Civil and Environmental Engineering, Universiti Tun Hussein Onn, Malaysia (UTHM), 86400 Batu Pahat, Johor, Malaysia
[2]Research Centre for Soft Soil (RECESS), Universiti Tun Hussein Onn Malaysia (UTHM), 86400 Batu Pahat, Johor, Malaysia

Email:[a]hf150093@siswa.uthm.edu.my; [b]mnizam@uthm.edu.my; [c]zurisman@uthm.edu.my; [d]saifulaz@uthm.edu.my;

Abstract. During the last two decades, reinforced concrete (RC) has been extensively used in most of the world as one of the common construction material due to its advantages and durability. However, RC structures exposed to marine environments are subjected to chloride attack. Chlorides from seawater penetrate into RC structures are not only causing severe corrosion problems but also affect the durability and serviceability of such structures. This paper investigates the influence of transverse reinforcement and spacing of reinforcing bars on concrete cover cracking of two-way RC slab specimens using accelerated corrosion tests. The experimental program involved the testing of four RC slab specimens and was generally designed to observe the crack width and the time of crack to propagate. An improved model for predicting the timing of crack propagation based on the experimental data was then developed.

1. Introduction

Reinforced concrete structures are normally exposed to environmental variability during their intended service life. An important phenomenon that should be considered for RC structures located adjacent to seawater is chloride-induced reinforcement corrosion [1]. Penetration of chloride is the main cause of corrosion in reinforced concrete structures located adjacent to seawater. Steel bars are naturally protected against corrosion by passivation of steel surface due to the high alkalinity of the concrete. The corrosion begins when a sufficient amount of chlorides penetrate into a concrete cover and destroy the inhibitive property by permeating the passivating layer of steel surface and this will increase the risk of corrosion.

Corrosion-induced cover cracking is an important measure for evaluating the service life of RC structures. The deterioration due to chloride-induced cover cracking leads to structural failure and reduces the service life of RC structure [2-5]. For these reasons chloride-induced cover cracking becomes a significant concern to asset owners. Structures suffering from chloride-induced cover cracking require a high frequency of site investigation, repair, maintenance and rehabilitation work as it demands large expenditure from asset owners.

Numerous studies have been conducted using non-destructive methods to study the corrosion process; however, it is impossible to accurately evaluate the corrosion process along the reinforcing bars in-situ with the concrete cover. Hence, an experimental work based on natural corrosion processes was carried out to study the empirical link between the occurrence and widths of cracks and the amount of corrosion [2]. The finding from the study supported that generally, the corrosion pattern accelerated by imposing direct electrical current is closer to a general corrosion like the natural corrosion during the second stage of propagation. Therefore, the use of accelerated corrosion test in the laboratory has become the most common technique to speed up the corrosion-induced damage in reinforced concrete structures [3-5].

2. Research Significance

Several attempts have been made to study corrosion-induced cracking by using accelerated corrosion tests [3, 4]. However, results from these studies were not very encouraging due to limitations which have been highlighted in the study by Mullard and Stewart [3]. These limitations include the unrealistic geometry of specimens as well as limited investigation of variables such as concrete cover, bar diameter and concrete strength. The occurrence of the confinement effect on crack propagation was observed when cracks over the reinforcing bars located at the edge of specimens have a higher propagation rate than the cracks located at the internal positions and this is particularly significant for columns and other structural elements which have many edges. The Mullard and Stewart [3] study was based on experimental data for one-way RC slab specimens and did not include any transverse reinforcement in the RC slab specimens. The limitations present in these previous works provide an opportunity for pursuing research in this area by developing more realistic models for corrosion-induced damage. The first series of accelerated corrosion test, ACT (1) studied by Sabtu and Stewart [5] was successfully demonstrated that the presence of transverse reinforcement and spacing of reinforcing bars have a significant effect on the rate of crack propagation. Therefore, the second series of the accelerated corrosion test, ACT (2) was conducted to further investigate the cracking behaviour of chloride-induced cracking included crack patterns, crack width, time to first cracking, T_{1st} (the time taken by the crack to achieve 0.05 mm) and T_{ser} (the time taken by the crack to achieve a certain crack width) on the two-way RC slab specimens. Therefore, ACT (2) proposes an improved empirical model for predicting the time of crack propagation in the presence of transverse of reinforcement and spacing of reinforcing bars.

3. Experimental design for Accelerated corrosion tests

3.1. RC Slab Specimens.

The accelerated corrosion test (2) was conducted on four RC slab specimens made with the concrete compressive strength of 30 MPa and 50 MPa. The specimens used a ready mix concrete and the mixtures were prepared by the nominated company. The four specimens had an identical size of 850 mm × 850 mm × 250 mm thick. The specimens were moist-cured for 28 days to achieve the desired compressive strength. The process of inducing the corrosion along the length of the reinforcing bars is by adding the calcium chloride ($CaCl_2$) by weight of cement to the concrete mixture to simulate the corrosion process along the length of the reinforcing bars. There are ten reinforcing bars in each RC slab specimen. Five of the reinforcing bars are the main reinforcing bars and the remaining five reinforcing bars are transverse reinforcing bars. Figure 1 and Figure 2 show the arrangement of the reinforcing bars for two-way slab specimens and the experimental set-up for two-way RC slab specimens respectively. In this study, measurement of crack width was taken on the main reinforcing bars only to measure the crack width in the presence of transverse reinforcement. The crack initiation and propagation process, which is from zero crack width to crack width of 1.00 mm, was monitored using a microscope and crack width measurement card.

3.2. Crack observations of ACT (2).

The duration of the ACT (2) is 2784 hours which is approximately four months. The crack began randomly over the reinforcing bar and measurements of crack were taken at the specific location when the cracks achieved 0.05 mm.

3.3. Measured corrosion rate ACT (2).

The Gravimetric weight loss method as specified in ASTM G1-90 (1990) was performed after the completion of the accelerated corrosion tests to accurately determine the amount of corrosion at the end of the testing. This is due to the actual corrosion rate in the accelerated corrosion test can vary from the theoretical applied current density. Previous experimental works based on the accelerated corrosion tests found that the measured corrosion rates were differed from the applied corrosion rate [3-6]. The time to cracking was corrected by multiplying the experimental time by the ratio of experimental corrosion rate/100, where 100 $\mu A/cm^2$ is the applied corrosion rate. The applied corrosion rate, 100 $\mu A/cm^2$ was chosen in order to be consistent with previous research works [3-5].

Figure 1. Arrangement of reinforcing bars for two-way RC slab specimens.

Figure 2. Experimental set-up for two-way RC slab specimens.

3.4. Crack initiation phase (T_{1st}).

The ACT (2) specimens took a longer time to achieve the first crack when compared to other studies eg., Andrade et al. (1993), Mangat and Elgarf (1999), Cabrera and Ghoddussi (1992), El-Maaddway et al. (2007) and Mullard and Stewart (2011). The ACT (2) specimens consider more variables, such as spacing of the main bars and spacing of transverse bars; therefore more factors contribute to the slower cracking process. All of the ACT (2) specimens are two-way RC slab specimens and the presence of the transverse bars will prolong the time for the crack to reach the concrete surface.

3.5. Crack propagation phase (T_{ser}).

The chosen magnitude of the current density allowed the specimens to achieve the desired crack widths within the specific time frame of approximately four months. By applying an electrical current to reinforcing bars through a current regulator, a constant value of current was supplied to each reinforcing bar. Cracks were observed through a microscope with a graduated scale of 0.02 mm increments. Based on the first observation, immediately after the visible cracks appeared, the cracks were found to begin at the external bar and were followed by the internal bars. The cracks were also found to be less in the intersection area between the main and transverse bars. With more intersection of reinforcing bars, this can reduce the size of crack and can even slower or prevent crack propagation.

3.6. Data observation.

Data points were plotted and appear to follow a straight line. Most of the data points are taken manually using a microscope with an accuracy of 0.02 mm and a crack width measurement card. The complexity of the two-way RC slab specimens with additional variables included in this test, such as spacing of the main bar, the spacing of the transverse bar, concrete cover, diameter bar and tensile strength, were found to contribute to the variation of the data. However, it would be reasonable to conclude that there is a linear relationship between crack width and time; as time increases, the crack width increases.

3.7. Calculating the rate of crack propagation, r_{crack} for internal bars.
In this study, r_{crack} was calculated based on the slope of the line of best fit for each individual reinforcing bar. r_{crack} was calculated by using the Least square method.

3.8. Effect of transverse reinforcement on concrete cover cracking.
By conducting the ACT tests on two-way RC slab specimens, the effect of confinement on RC structures is able to be studied in detail. It was found that the confinement effect on two-way RC slab specimens is due to two factors. As the confinement effect can reduce the rate of crack, the first factor that contributes to the slower cracking is the existence of reinforcing bars on either side. This condition happened to the internal bars. The second factor that can reduce the rate of crack, is the presence of transverse bars on top of the main bars that act as a confinement to the main bars. Figure 3 supports the confinement effect by comparing the one-way and the two-way RC slab specimens and clearly, the two-way RC slab specimens appears to have significantly reduced rates of crack propagation when compared to one-way RC slab specimens.

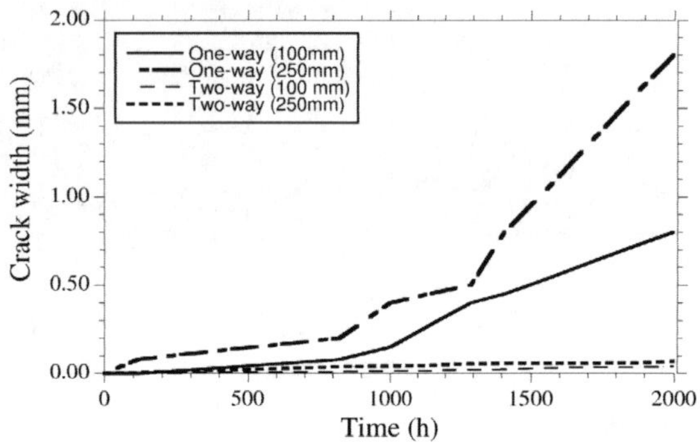

Figure 3. Comparison of one-way (ACT (1)) and two-way (ACT (2)) RC slab specimens [5].

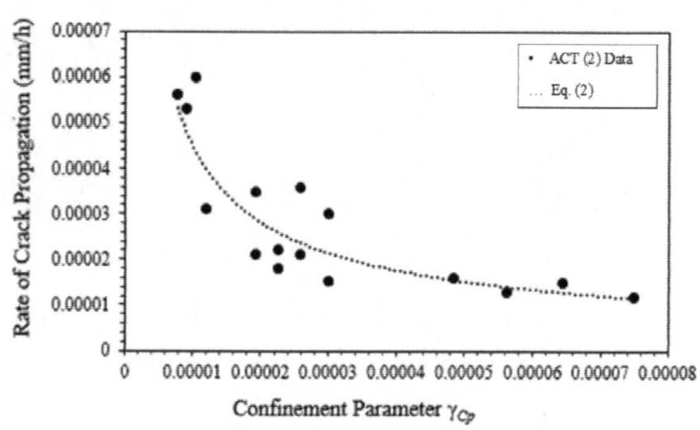

Figure 4. Crack propagation as a function of concrete confinement parameter γ_{cp}.

3.9. Effect of reinforcement spacing on concrete cover cracking.
Reinforcement spacing or distance between bars was observed to have an important effect on the cracking process. From observation made throughout the experimental work, the crack propagation is higher when the spacing is larger and the crack propagation is lower when the spacing is smaller. By increasing the spacing or by using the larger spacing, in this case 250 mm, the bond between concrete and steel bar is decreasing and the rate of crack propagation will increase. However, by decreasing the spacing or by using smaller spacing, in this case, 100 mm spacing, the bond between concrete and the steel bar is adequate and stronger hence, reducing the rate of crack propagation.

4. Crack propagation model based on two-way RC slab specimens
A predictive model for crack propagation developed by Mullard and Stewart [3] incorporates cover, diameter bar and concrete tensile strength. The model was found to be robust, having reasonable agreement between experimental and predicted time to crack propagation, however the study was based on the one-way RC slab specimens which excludes the presence of transverse and spacing of reinforcing bars. Therefore, the current study improved the Mullard and Stewart [3] model by proposing a new normalised parameter, "concrete confinement parameter" which incorporates spacing of main and transverse reinforcing bar. Concrete confinement parameter (γ_{cp}) defined as

$$\lambda_{cp} = \frac{C}{Df_t} \times \frac{1}{S_M S_{TR}} \tag{1}$$

where C is the concrete cover in mm, D is the reinforcing bar diameter in mm, and f_t is the concrete tensile strength in MPa, S_M is the spacing of the reinforcing bar in mm and S_{TR} is the spacing of transverse reinforcing bar in mm. Figure 4 shows the influence of the concrete confinement parameter on crack propagation rate in mm/h. Data from each reinforcing bar were plotted independently, therefore, all the data points represent a single bar. The line of best fit the for all the data points from internal bars is

$$r_{crack} = 2.0e^{-8\gamma_{cp}^{-0.674}} \qquad 1 \times 10^{-5} \leq \gamma_{cp} \leq 8 \times 10^{-5} \qquad (2)$$

where r_{crack} is the crack propagation rate in mm/h and γ_{cp} is the concrete confinement parameter. The crack propagation rate was found to decrease as the concrete confinement parameter is increased. The findings of the current study are consistent with other research which found that the concrete confinement such as transverse reinforcement, stirrups or other type of orthogonal reinforcement could reduce the rate of crack propagation.

5. Conclusions

This paper presented data obtained from the accelerated corrosion test conducted on two-way RC slab specimens. Clearly, the presence of transverse reinforcing bar reduces corrosion-induced cover cracking by a significant extent. The two-way RC slab specimens appear to have significantly reduced rates of crack propagation when compared to one-way RC slab. The data of the two-way RC slab specimens was used to develop an improved empirical crack propagation model is not only considered the effect of concrete cover, concrete compressive strength and bar diameter, the new model also included the influence of transverse reinforcement and spacing of reinforcing bars which then refers as the 'concrete confinement parameter'.

6. Acknowledgements

The authors acknowledged the assistances from Prof Mark G. Stewart from the University of Newcastle, Australia and the financial support from the Australian Research Council.

7. References

[1] X. Zhang, J. Wang, Y. Zhao, L. Tang, and F. Xing, Time-dependent probability assessment for chloride induced corrosion of RC structures using the third-moment method, Construction and Building Materials, Elsevier Ltd, 76 (2015) 232–244.

[2] R. Zhang, A. Castel and R. François, Concrete cover cracking with reinforcement corrosion of RC beam during chloride-induced corrosion process, Cement and Concrete Research 40 (2010) 415–425.

[3] J.A. Mullard and M.G. Stewart, Corrosion-Induced Cover Cracking: New Test Data and Predictive Models, ACI Structural Journal, 108 (2011) 1.

[4] K.A.T Vu, M.G. Stewart and J.A. Mullard, Corrosion-Induced Cracking: Experimental Data and Predictive Models, ACI Structural Journal, 102 (2005) 5.

[5] H.B. Sabtu and M.G. Stewart, The Effect of Reinforcement Confinement on Concrete Cover Cracking in Reinforced Concrete Structures, From Materials to Structures: Advancement through Innovation, Taylor & Francis Group (2013)

[6] C. Alonso, C. Andrade, J. Rodriguez and J.M. Diez, Factors Controlling Cracking of Concrete Affected by Reinforcement Corrosion, Material and Structures, 31 (1998) 435-441.

IOP Conf. Series: Materials Science and Engineering **216** (2017) 012020 doi:10.1088/1757-899X/216/1/012020

Dynamic Stiffness of Non-Loaded and Loaded Beams

M Musiał

Faculty of Civil Engineering, Wroclaw University of Science and Technology, Wybrzeże Wyspiańskiego, 50-370 Wroclaw

E-mail : michal.musial@pwr.edu.pl

Abstract. The problem of the bending stiffness of reinforced concrete beams is considered. The research so far has shown that static stiffness (determined on the basis of deflections) is lower than dynamic stiffness (determined on the basis of eigenfrequencies). The analyses focus on eigenfrequencies determined in different ways. The dynamic rigidities of cracked RC beams were found to differ between non-loaded and loaded beams.

1. Introduction

By measuring the parameters of a structure one can assess its condition. One of the basic parameters of reinforced concrete beams, measured for diagnostic purposes, are their deflections. Knowing the deflections, the static scheme and the load one can calculate a beam's bending stiffness EI_{eff} which provides valuable information about the condition of the beam. In the case of RC beams under bending, stiffness EI_{eff} is not constant and depends on the degree of load advancement – a ratio of the moment acting on the beam to the latter's resistance to bending. The stiffness of an element is not constant (decreases) mainly due to cracking (resulting in stiffness degradation).

There are several theories on the basis of which one can calculate beam stiffness EI_{eff}, taking into account cracking. Most of the theories (e.g. [1, 2]) take into account a combination of the beam's stiffness in stage I (before cracking) and in stage II (after cracking). According to [1], the relation for stiffness is as follows:

$$EI_{eff} = \frac{E_{cm} I_{II}}{1 - \beta \left(\frac{M_{cr}}{M} \right)^2 \left(1 - \frac{I_{II}}{I_I} \right)}, \tag{1}$$

where: E_{cm} – the mean Young's modulus of the concrete, M_{cr} – the cracking moment, M – the maximum bending moment acting on the beam, I_I, I_{II} – moments of inertia in respectively stage I and II, β – a coefficient depending on the type of load (1.0 for single short-term loading, 0.5 for sustained loads or many cycles of repeated loading).

According to [2], the stiffness of a beam can be calculated from the following relation:

$$EI_{eff} = E_{cm} \left\{ \left(\frac{M_{cr}}{M} \right)^3 I_I + \left[1 - \left(\frac{\alpha M_{cr}}{M} \right)^3 \right] I_{II} \right\}. \tag{2}$$

Another parameter measured when evaluating the condition of a structure (e.g. by means of operational modal analysis [3]) is eigenfrequency. Knowing the eigenfrequency for the given loading diagram and the vibrating mass one can calculate dynamic stiffness EI_d. The latter is different from static stiffness EI_{eff}. Numerous papers (e.g. [4-7]) have been devoted to determining the dynamic stiffness of RC beams on the basis of eigenfrequencies. Usually it is proposed (on the basis of a series of test results) to modify relation (1) or (2) so that they can be used to calculate dynamic stiffness. However, it should be noted that because of the methodology used the analyses made by different researchers yield, in most cases, different results.

Two most often used methods of determining the eigenfrequencies of RC beams are discussed in this paper. In one of the methods the measurements are performed on elements loaded with solely their dead weight. In the other method, ballast masses are used besides the dead load.

2. Dynamic stiffness acc. to [6]

As part of this research experiments on suspended elements (free-free beams) were conducted in order to determine their dynamic stiffness. The eigenfrequencies of the beams, in which only their dead load was the vibrating mass, were recorded. The eigenfrequencies would decrease as the beams were loaded at an increasing rate in each step of the three-point bending test (figure 1a). Dynamic tests were conducted after the beams had been completely unloaded and suspended on elastic ropes (figure 1b). The advantages of testing beams in this way are presented in [8]. One of the advantages is that the influences of the decreasing stiffness and the increasing mass are separated. Thus the influence of the increasing forces of inertia was eliminated in the experiments.

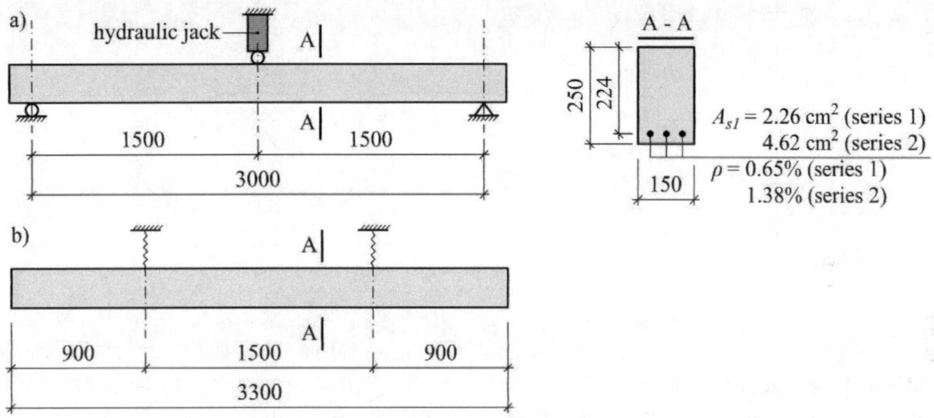

Figure 1. Test stands acc. to [6] (dimensions in mm).

On the basis of the experiments the following modified form of relation (1) was proposed:

$$EI_d = \frac{\alpha_B E_{cm} I_{II}}{1-\beta\left(\frac{M_{cr}}{M}\right)^{2\beta_B}\left(1-\frac{I_{II}}{I_I}\right)}, \tag{3}$$

where: α_B – a coefficient of conversion from the static Young's modulus of concrete to the dynamic Young's modulus, β_B – a coefficient modifying the character of the change in stiffness after the beam cracks (in stage II).

Studies (e.g. [9, 10]) have shown that coefficient α_B for ordinary concrete ranges from 1.10 to 1.30 (hence $E_{cm} < E_d$). This is mainly due to the fact that static Young's modulus E_{cm} is determined through the classic axial compression test on a cylindrical specimen subjected to a load of $0.1f_c$-$0.3f_c$ (where f_c is the compressive strength of the concrete). In the present experiments the dynamic Young's modulus of the concrete was determined using specimens loaded with only their dead weight. As it is known, the Young's modulus of concrete depends on the degree of strain and it is inversely proportional to the latter.

Parameter β_B pertains to the rate of dynamic stiffness decrease after cracking. Studies carried out by the author showed that the decrease was the more intensive, the lower the reinforcement ratio. The economic tension reinforcement ratio of a typical RC beam is in a range of 0.7%-1.5%. As part of the present research two series of beams with tension reinforcement ratio ρ = 0.65% (9 elements) and 1.38% (3 elements) were tested. One can assume that the above values come from the extremes of the recommended economic reinforcement ratio interval. If parameter β_B is assumed to vary linearly within the interval, the following relation can be proposed:

$$\beta_B = -2.22\rho + 0.0669 \ . \tag{4}$$

Figure 2 shows static and dynamic rigidities as a function of the bending moment, calculated from relations (1) and (3) for the two extreme reinforcement ratios. The values of breaking moment M_R and cracking moment M_{cr} have been marked on the diagram.

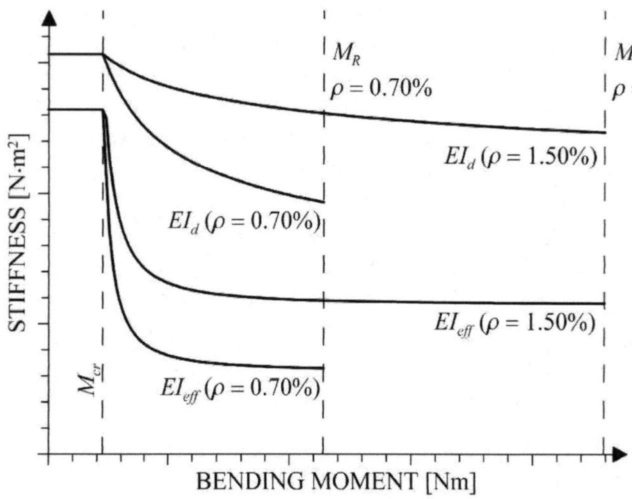

Figure 2. Static and dynamic stiffness versus bending moment.

3. Dynamic stiffness acc. to [4]

Tests of the eigenfrequencies of RC beams conducted differently than in [6] are reported in [4]. In the latter case, four-point bending tests with ballast masses were carried out, which means that the eigenfrequencies of the loaded beams were measured. The load would be increased stepwise after each recording of eigenfrequencies and deflections. The test stand and one of the three tested series of beams are schematically shown in fig. 3. The beams in the other two series were not considered since their reinforcement ratios were much higher (1.76% and 2.64%). Otherwise the comparisons presented further in this paper could not have been made due to the fact that the relations given in section 2 do not take into account so high reinforcement ratios.

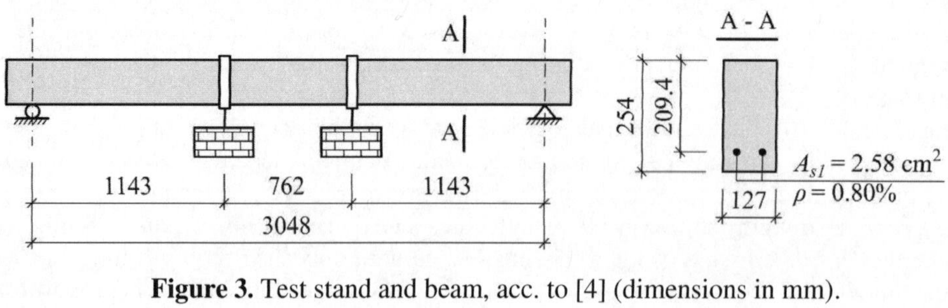

Figure 3. Test stand and beam, acc. to [4] (dimensions in mm).

The tested beams were made of concrete with mean compressive strength f_{cm} = 41.37 MPa and mean static Young's modulus E_{cm} = 30.48 GPa. The above values were determined using cylindrical specimens made of the same batch of concrete mixture and cured in the same conditions as the beams. Ribbed reinforcing bars made of steel with mean yield strength f_{ym} = 276 MPa and mean Young's modulus E_{sm} = 200 GPa were used as the reinforcement.

On the basis of the test results relation (2) was modified to the form:

$$EI_d = E_{cm}\left[\left(\frac{\alpha M_{cr}}{M}\right)I_I + \left(1 - \frac{\alpha M_{cr}}{M}\right)I_{II}\right]. \tag{5}$$

The tests showed that empirical coefficient α was in an interval of 0.6-0.8. Moreover, eigenfrequencies would decrease during the tests due to the degradation of stiffness and to the simultaneous increase in the vibrating mass (as the load on the hangers was increased). The authors of [4] took this into account by giving a relation for dynamic stiffness (calculated on the basis of experimental eigenfrequencies), covering the beam's dead weight and the suspended ballast masses.

4. Comparative analyses

Comparative analyses concerning the relations presented in [6] were carried out using the experimental results and the relations given in [4]. Considering the fairly representative test elements, the similar dimensions of the elements and the similar properties of the concrete and the steel used in [4] and [6], such comparative analyses are justified. Figure 4 shows the dynamic stiffness of the beams, calculated on the basis of the measured eigenfrequencies of two tested elements [4]. The curves in the diagram are described by relation (5) for the extreme parameter α values (0.6 and 0.8). Using linear regression it was found that the experimental results are best described by relation (5) with parameter α = 0.785, which is also shown in the diagram.

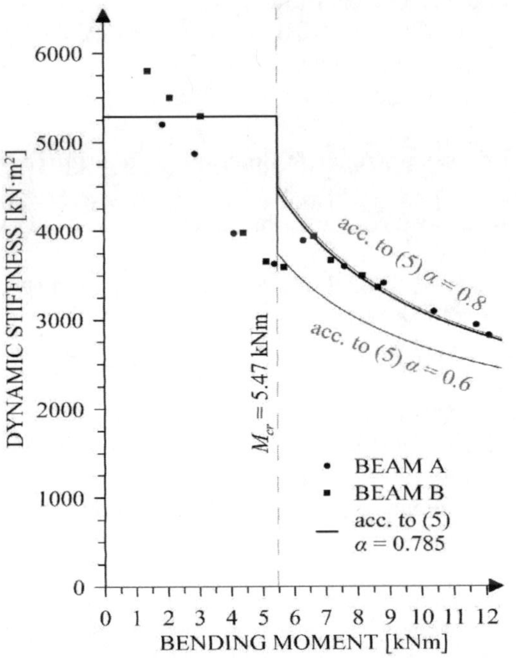

Figure 4. Dynamic stiffness values from tests [4] and calculations (5).

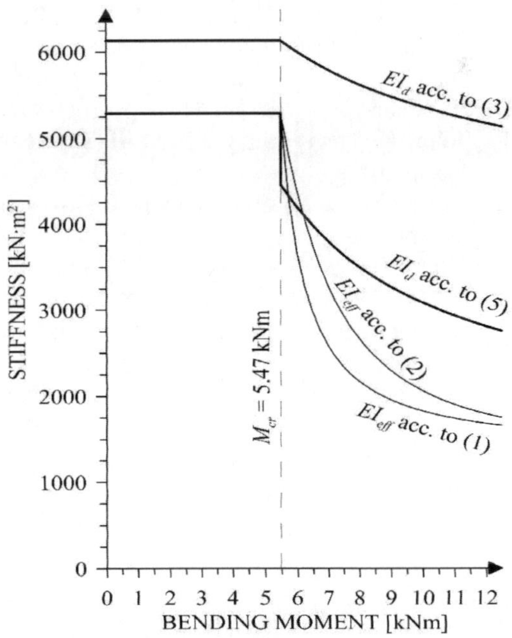

Figure 5. Static and dynamic stiffness calculated from the different relations.

For the beam shown in figure 3 dynamic stiffness EI_d values were calculated from relations (3) and (5), while static stiffness EI_{eff} values were calculated from relations (1) and (2). The calculation results are compared in the diagram in figure 5.

The dynamic stiffness calculated from formula (3) has the highest value and the smallest gradient of decrease after cracking. The dynamic stiffness value calculated from relation (5) is lower. The static rigidities have the lowest values. The static stiffness values calculated from respectively relation (1) and (2) are comparable. However, this fact is not elaborated since static stiffness is not the main subject of this paper.

5. Conclusions

The following conclusions can be drawn from the analyses:

- The dynamic stiffness values determined on the basis of the eigenfrequencies of the RC beam are higher than the values of its static stiffness determined on the basis of its deflections.
- The dynamic stiffness value determined on the basis of the eigenfrequency of the non-loaded RC beam with cracks is higher than the dynamic stiffness value of the RC beam with cracks and ballast masses.
- The observed behaviour of the beams indicates that the influences of stiffness and inertia on eigenfrequencies interact, which is reflected in the dynamic stiffness values. *The author does not mean here the natural decrease in eigenfrequencies as the vibrating mass is increased (observed for any type of structure), but the effect of this mass on the stiffness of the structure.*
- The presented results can be exploited in, e.g., the diagnosis of structures, based on eigenfrequency measurements.

The analyses should be regarded as preliminary, nevertheless their results indicate that such studies are worthwhile. Currently experiments on a larger number of beams differing in their reinforcement ratios are being planned.

References

[1] EN 1992-1-1: Design of concrete structures - Part 1-1: General rules and rules for buildings

[2] ACI 380: Building Code Requirements for Reinforced Concrete Buildings

[3] Grosel J, Sawicki W and Pakos W 2014 Application of classical and operational modal analysis for examination of engineering structures, *Procedia Engineering* **91** pp 136–41

[4] Jerath S and Shibani M M 1985 Dynamic Stiffness and Vibration of Reinforced Concrete Beams, *ACI Journal* **82–18** pp 196–202

[5] Johns K C and Belanger M D 1981 Dynamic Stiffness of Concrete Beams, *ACI Journal* **78–18** pp 201–5

[6] Musiał M 2012 Static and dynamic stiffness of reinforced concrete beams, *Archives of Civil and Mechanical Engineering* **12** pp 186–91

[7] Numayr K S, Al-Jallamdeh S A and Al-Akhras N M 2002 Dynamic stiffness of reinforced concrete beams, *Structures & Buildings* **156** pp 373–9

[8] Musiał M 2016 Laboratory tests of reinforced concrete beams with the use of operational modal analysis, *Materials Science Forum* **866** pp 124–8

[9] Musiał M and Grosel J 2016 Determining the Young's modulus of concrete by measuring the eigenfrequencies of concrete and reinforced concrete beams, *Construction and Building Materials* **121** pp 44–52

[10] Musiał M 2015 Young's modulus test in two directions on the basis of eigenfrequencies in concrete beams, *Technical Transactions* **3–B** pp 133–42

IOP Conf. Series: Materials Science and Engineering **216** (2017) 012021 doi:10.1088/1757-899X/216/1/012021

Challenges in Strengthening of Concrete Elements with PBO-FRCM Materials

T Trapko[1], D Urbańska[1] and M Musiał[1]

[1] Faculty of Civil Engineering, Wroclaw University of Science and Technology, Wybrzeże Wyspiańskiego, 50-370 Wroclaw, Poland

E-mail: tomasz.trapko@pwr.edu.pl

Abstract. In the paper a technology of reinforced concrete elements strengthening with PBO mesh bonded with a mineral mortar was presented. The possibilities of application of this system in shear strengthening of reinforced concrete beams and compressed elements are in the range of interests of the authors. In the paper authors experiences in strengthening of such type elements with PBO mesh bonded with mineral mortar were described. Research has shown that destruction of elements strengthened with PBO-FRCM system is caused by a debonding of the fibres from the matrix layer or delamination at the final overlap of PBO mesh. In order to obtain the most efficient use of PBO fibres and prevent premature loosening of composite appropriate anchorage should be used.

1. Introduction

The subject of reinforcing concrete structures using FRP (*Fiber Reinforced Polymers*) composite materials has been of interest for over two decades to many research centers in the world. For more than a decade, the subject has also been present in several Polish universities. It is evidenced by the growing number of publications, appearing in monographs, reputable scientific journals and at thematic conferences. The first laboratory work, engineering applications and computational models related mainly to bent beam elements. Materials used in them were based on *Carbon Fibre Reinforced Polymers* (CFRP), *Glass Fibre Reinforced Polymers* (GFRP) and *Aramid Fibre Reinforced Polymers* (AFRP).

Reliability and effectiveness of the FRP system depends on the epoxy resin, which connects composite and concrete. As has been shown in several studies parameter determining the efficiency of the strengthening is the temperature at which strengthened element is used [1]. Limit temperature for the FRP systems is glass transition temperature of epoxy resin, after which it loses its adhesive properties. The elimination of this problem is possible in two ways, through the use of thermal protection on strengthened elements or by eliminating resin as a binder combining composite with concrete.

In FRCM system (*Fiber Reinforced Cementitious Matrix*) resin was replaced with mineral mortar, which covers the reinforced element and which connects to the composite fiber grid, providing its adhesion to concrete. One of the reinforcement material used in this system is a mesh made from PBO fibers (*p-Phenylene BenzobisOxazole*), commonly called PBO mesh. This system is referred in the literature as PBO-FRCM. Table 1 compares the technical characteristics of the PBO mesh and the most popular composite carbon fiber sheet CFRP [2].

IOP Conf. Series: Materials Science and Engineering **216** (2017) 012021 doi:10.1088/1757-899X/216/1/012021

Table 1. Technical parameters of PBO mesh and CFRP sheet [2]

No.	Type of fibres	Tensile strength [MPa]	Modulus of elasticity [GPa]	Ultimate strain [%]	Density [g/cm^3]	Thickness of fibres [mm]
1	PBO	5270 $^{(a)}$/5800 $^{(b)}$	270	2,15	1,56	0,0455
2	CFRP	2670 $^{(a, c)}$/4300 $^{(b, d)}$	238	1,80	1,76	0,1310

(a) according to own tests; (b) according to the producers; (c) data for the laminate; (d) data for the fibres

2. Shear strengthening of reinforced concrete beams

The PBO-FRCM can be effectively used as shear strengthening of reinforced concrete beams as shown inter alia in own studies performed at the Faculty of Civil Engineering, at Wroclaw University of Science and Technology [3, 4].

Beams strengthened with PBO-FRCM system show more ductile nature of the work, than elements strengthened with FRP system. It is connected with the phenomenon of slip that occurs between mortar and fibres. This phenomenon does not appear in FRP systems. In most studies of shear strengthened beams with PBO-FRCM system the open U-shape grid [5-7], without an additional anchoring of the ends of the composite has been used. The mechanism of destruction was due to premature debonding of composite and the loss of the shear capacity, and thus incomplete and not very effective use of PBO mesh. For example, strain in the composite in studies [6, 7] only amount to approx. 2.5-3.5‰. Better use of the PBO grid is possible with proper anchorage of composite stirrups, which was the subject of research performed by the authors [3, 4].

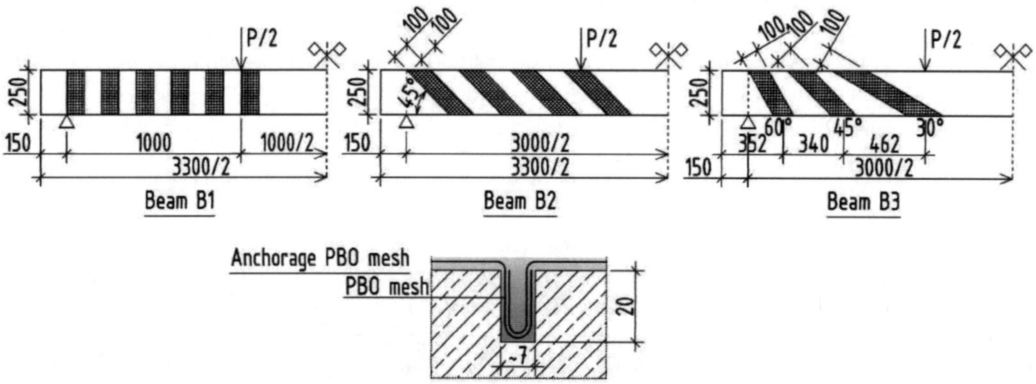

Figure 1. Scheme of strengthened beams – series 1

Reinforced concrete beams strengthened on a shear with external stirrup of PBO mesh with mortar mineral were constructed and tested. Research was conducted in two series. The first series consisted of three beams of a rectangular cross-section 150x250 mm (Fig. 1). The B1 beam was strengthened by wrapping the entire cross-section with PBO stirrups at an angle 90° to the longitudinal axis, with overlap at the compressed surface of the cross-section. Beam B2 was strengthened with composite strips inclined to the longitudinal axis of the beam at 45°, and for beam B3 strips were inclined at three angles 30, 45 and 60°. The beams B2 and B3 were longitudinal incised at the bottom and at the top, in the middle of the width of the cross-section to a depth of 20 mm. The strips for beam B2 were anchored by adhering the ends of the mesh into incision at both sides with epoxy resin. A method of anchoring for beam B3 was the same as in the beam B2, but the ends of strips were attached into the incision at the bottom with epoxy resin, and at the top with mineral mortar. All beams were strengthened with strips of 100 mm width.

Destruction of the beams was a result of the weakening of external overlap/anchorage and rapid development of diagonal cracks. The failure mechanism of the beam B1 consisted of delamination of the overlap on the upper surface of the element (Fig. 2). Beams B2 and B3 failed after pulling out PBO mesh from the anchoring (Fig. 3). Strain of the anchored PBO stirrup measured at half height of the beam reached the value of 3.7-5.0‰. This represents significant improvement of the effectiveness of PBO-FRCM strengthening compared to research conducted on beams strengthened without end-anchoring (U-shaped strengthening).

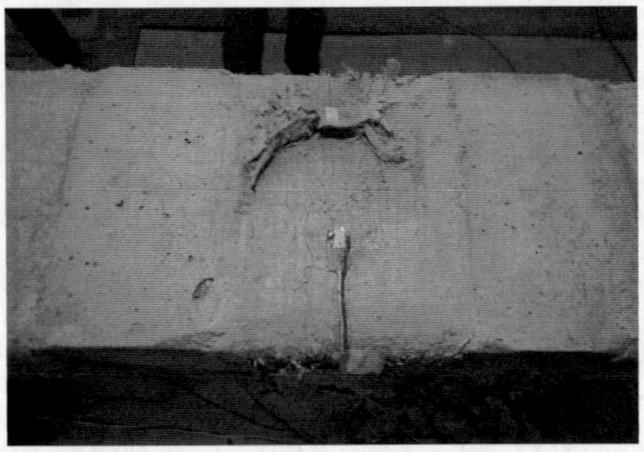

Figure 2. Failure of beam B1 – delamination of the lap on the top surface

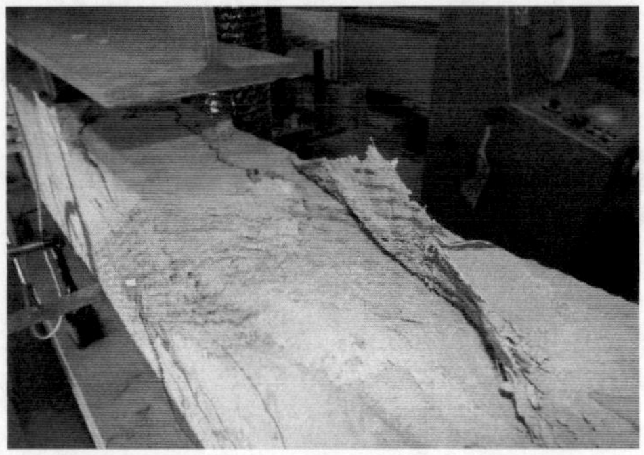

Figure 3. Failure of beam B3 – pulling the endings of PBO stirrups out of the lap

A second series of tests was carried out on 3 T-shaped beams strengthened in a shear with PBO mesh with anchorage in the flange of the beam. Beams were tested as a simply supported in three-point bending All beams were strengthened with strips of 150 mm width, and the inclination of fibres in the main direction toward the longitudinal axis of the beam was 90°. Each of the beams had the same type of anchorage. A gap of 20x20 mm and a length equal to the length of the beam was made under the flange, and glass fibre bar with PBO strip wrapped around was glued into a gap. The bar was glued using a mineral mortar which is also used for bonding PBO mesh to the surface of element (Fig. 4).

All specimens failed in shear (Fig. 5). Before the damage, a large diagonal crack from the force to the bar and then along the bar was created in the flange (Fig. 6). Anchoring was not pulled out. For each of the beams debonding of the composite in a thin layer of mortar-fibre contact, at the location of diagonal cracks, were observed. After debonding, composite further took part in the transfer of shearing forces, but separately from the concrete, therefore the diagonal crack could increase to the destruction. Deformation of anchored PBO stirrups in the second series of tests, measured at mid-height of the composite, amounted to 7.5-8.0‰. This value is higher than that obtained in the previous test series and higher than for composites without anchorage. This may indicate that the proposed anchoring allows for better use of composite materials than in the beams without anchors. It should be noted that in all presented tests no rupture of PBO fibers was observed.

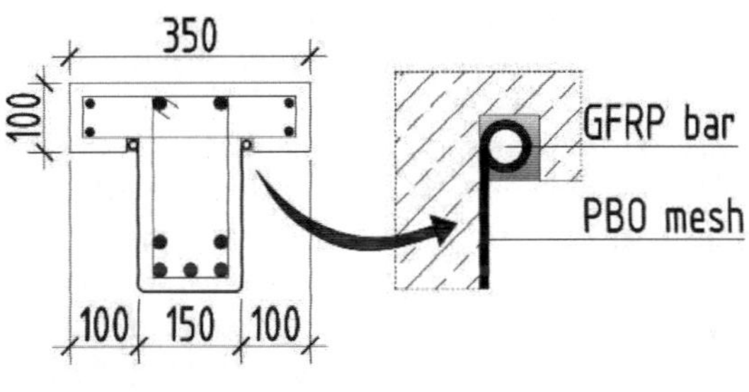

Figure. 4. Cross-section of the beams – series 2

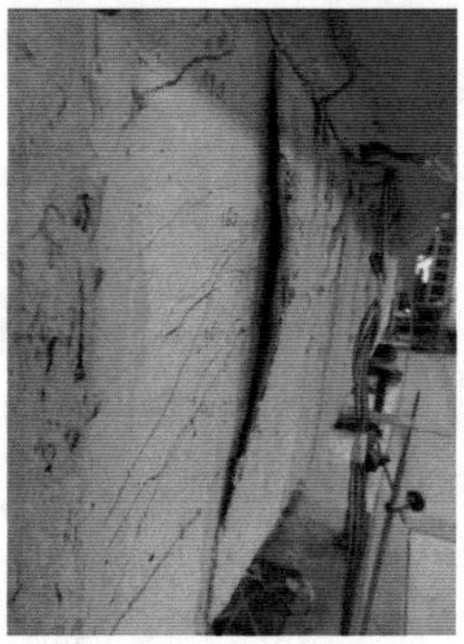

Figure 5. Failure of beam– debonding of the PBO mesh

Figure 6. Failure of beam – crack along the bar and PBO debonding

3. Strengthening of compressed elements

The advantages of wrapping the concrete with the CFRP outer sheet are well-known. However, when introducing a new reinforcement system, it was necessary to assess the effectiveness of the reinforcement and compare the behaviour of the elements confined in the CFRP sheet and the PBO mesh [1, 2].

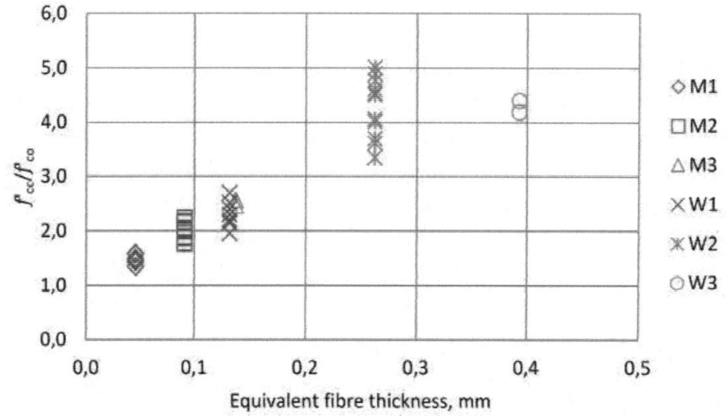

Figure. 7. Compressive strength of confined concrete f'_{cc} relative to fibre thickness [1, 2] designations: W – CFRP sheet, M – PBO mesh, *i* – number of layers, f'_{cc} – *compressive strength of confined concrete*, f'_{co} – *compressive strength of unconfined concrete.*

As expected, the CFRP system produced noticeable effect of number of the sheet layers circumscribing the section on the increase in the load-bearing capacity. The thickness of the composite coat and the type of fibres applied affect the measured limit load-bearing capacities. When applying one layer of the CFRP sheet, the similar value of the failure load to the one with the application of the three layers of the PBO mesh is obtained (Fig. 7).

The comparison of tensile strength of the fibers in the PBO mesh and CFRP sheet (Table 1) suggests that the PBO-FRCM system should give even better reinforcement results. However, greater tensile strength of the PBO mesh does not translate directly into greater increase in load-bearing capacity and the value of the compressive strain. It is caused by graininess of the mortar and filling the composite with it as well as adhesion to concrete and between its layers.

From the destruction mechanism (Fig. 8), to the nature of the paths of strain, to the limit values of load-bearing capacity and strain, research has shown that these two systems cannot be compared to each other at the same intensity of reinforcement (number of layers). All elements reinforced with the CFRP sheet were destroyed in a violent manner, by breaking the shell. The mechanism of destruction of the elements reinforced with the PBO mesh was quite different. None of the attempts resulted in breaking the fibre and the initiation of destruction always occurred on the external overlapping zone.

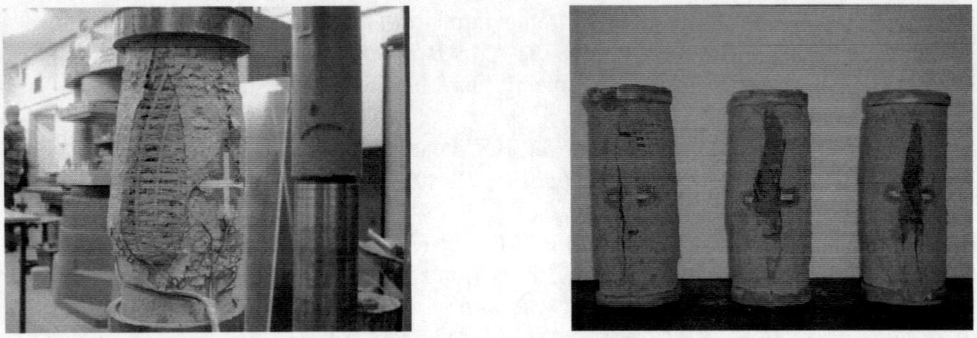

a) overlapping zone z=70 mm, where z/u= 0.2 (u – perimeter of the sample, 113 mm)

b) elements without overlapping zone

Figure 8. Typical failure of PBO confined concrete cylinders in compression test

The value of the limit peripheral strains depends essentially on the length of the overlapping zone and the factors that determine delamination are the smallest strains at the length of the overlapping zone. At the same time, it should be noted that the distribution of the strains along the length of the composite overlapping zone is non-linear. Along with the increase of the overlapping zone, delamination strains increase. The coefficient of utilization of the PBO mesh in the end zone of the overlapping zone is within the range of 0.28-0.44, depending on its length. At the same time, by analysing the strains of the transverse outside the overlapping zone, it can be seen that the coefficient of utilization of the PBO mesh is 0.55-0.80.

Undoubtedly, the PBO-FRCM system is competitive with the CFRP system. The nature of functioning of the elements reinforced with the PBO mesh is more predictable to use in more complex structural systems rather than separate compressed elements. The indicated and slow increase in strains in the area of the destructive load allows the redistribution of the internal forces and joining of other, less stressed elements.

4. Conclusion

Therefore, the PBO-FRCM system is not intended to replace or eliminate the FRP resin composites. However, the presented papers demonstrate great potential of the composites based on mineral mortar as the reinforcement of concrete structures. Both systems - FRP and PBO-FRCM differ a lot, even though both of them derive from the same idea involving the restriction of freedom of stretched and compressed concrete element strains.

The conscious decision to choose the reinforcement system must be dictated by the fact that the selected system can ensure the achievement of those objectives both temporarily and in a long term. This selection should be accompanied by full knowledge of the reinforced element, including one which will help predict potential threats to the reinforced structure and help choose the most suitable system. On the basis of the acquired knowledge it is important to take into account what benefits and risks are offered by both technologies – FRP and PBO-FRCM, and the fact that there is no perfect system, and for each one you can specify both the number of advantages, as well and defects.

The most important issue for the future is to develop an effective method of anchoring the PBO mesh on the external overlapping zone, in order to fully utilize its strength properties. Only then will it be possible to analyse, among others, rheological characteristics, the impact of variable loads, etc.

Based on the described research it can be seen that the use of appropriate anchorage effectively improves the use of the composite. In the case of PBO-FRCM shear reinforcements problem is the occurrence of premature debonding of composite from the surface of the element in a thin layer of mineral mortar. This is related to the use of mineral mortar which loses its properties after exceeding its resistance to tensile stresses. In order to better use of characteristics of PBO fibres is recommended to prevent this phenomenon.

IOP Conf. Series: Materials Science and Engineering **216** (2017) 012021 doi:10.1088/1757-899X/216/1/012021

5. References

[1] Trapko T 2013 The effect of high temperature on the performance of CFRP and FRCM confined concrete elements, *Composites Part B-Engineering* **54** pp 138-145

[2] Trapko T 2013 Fibre Reinforced Cementitious Matrix confined concrete elements, *Materials and Design* **44** pp 382-391

[3] Trapko T, Urbańska D and Kamiński M 2015 Shear strengthening of reinforced concrete beams with PBO-FRCM composites, *Composites Part B-Engineering* **80** pp 63-72

[4] Urbańska D, Trapko T and Kowalik T 2015 Nośność na ścinanie belek żelbetowych wzmocnionych materiałami PBO-FRCM, *Materiały Budowlane* **6** pp 66-67 (in Polish).

[5] Ombres L 2012 Shear capacity of concrete beams strengthened with cement based composite materials, *Proceedings of the 6th International Conference on FRP Composites in Civil Engineering CICE 2012* (Roma, Italy) Paper No. 01–281.

[6] Ombres L 2015 Structural performances of reinforced concrete beams strengthened in shear with a cement based fiber composite material, *Composites Structures* **122** pp 316-329.

[7] Ambrisi A D and Focacci F, Flexural strengthening of RC beams with cement based composites, *Journal of Composites for Construction ASCE* **15** pp 707–720.

IOP Conf. Series: Materials Science and Engineering **216** (2017) 012022 doi:10.1088/1757-899X/216/1/012022

Effect of Metakaolin and Slag blended Cement on Corrosion Behaviour of Concrete

Anita N Borade[1, 3] and B Kondraivendhan[2*, 3]
[1] Research Scholar, [2*] Assistant Professor,
[3] Applied Mechanics Department, Sardar Vallabhbhai National Institute of Technology, Surat-395007, Gujarat, India

[1]Email: boradeanita@yahoo.com; [2*] Email: kondraivendhan78@gmail.com;

Abstract. The present paper is aimed to investigate the influence of Metakaolin (MK) and Portland slag Cement (PSC) on corrosion behaviour of concrete. For this purpose, Ordinary Portland Cement (OPC) was replaced by 15% MK by weight and readymade available PSC were used. The standard concrete specimens were prepared for both compressive strength and half- cell potential measurement. For the aforesaid experiments, the specimens were cast with varying water to binder ratios (w/b) such as 0.45, 0.5 and 0.55 and exposed to 0%, 3%, 5% and 7.5% of sodium chloride (NaCl) solution. The specimens were tested at wide range of curing ages namely 7, 28, 56, 90 and 180 days. The effects of MK, w/b ratio, age, and NaCl exposure upon concrete were demonstrated in this investigation along with the comparison of results of both MK and PSC concrete were done. It was also observed that concrete with MK shows improved performance as compared to concrete with PSC.

1. Introduction

Concrete is the most widely used man-made construction material in the world today due to availability of raw materials, low cost of manufacturing and maintenance. It is also excellent in compression, durability and corrosion aspects. However, over the last few decades the durability of concrete becomes a serious problem which consumes major part of financial budget for inspections, repairs and maintenance of deteriorated structures [1]. Among the different factors affecting corrosion, chloride induced corrosion attract the more attention of the industry as well as researchers due to its severity. The sources of harmful chloride ions into the concrete are either internal (added through the ingredients of the concrete mix through chloride contaminated aggregate, contaminated water and through admixtures in fresh concrete) or externally in hardened concrete through de-icing salts, penetration of chloride ions from seawater in marine structures and from the soil and ground water contains chloride salts [2, 3]. The use of Supplementary Cementing Material (SCM) proves the influence on the hydration of cement compound either by physical or chemical effects [4, 5].
In addition to improving the durability, the use of these materials in construction reduces the waste deposits and the demand of cement. Hence helps in reduction in environmental pollution. The utilization of calcined clay in the form of metakaolin (MK) as a pozzolan for concrete has received considerable interest in recent years [6]. Thorough research has been carried out over the effect of different percentage replacement of cement with MK and many of them concluded that concrete incorporated with 15% MK exhibits premium level engineering properties comparable to silica fume concrete [7, 8]. From the literature it is also depicted that Portland slag cement is intensively studied and accepted as the reliable binder in corrosion resistance.

The present paper is aimed to investigate the influence of replacement of 15% of MK & PSC in concrete through compressive strength and half- cell potential measurements. For this purpose, six different concrete mixes with three water to binder ratio (w/b) of 0.45, 0.5 and 0.55 were designed and exposed to 0%, 3%, 5% and 7.5% of sodium chloride (NaCl) solution over varying curing ages such as 7, 28, 56, 90 and 180 days. The effects of MK replacement, w/b ratio, age, and curing procedure upon concrete properties were discussed and the results are compared with the concrete prepared with PSC.

2. Experimental Procedure

For the present investigation, Ordinary Portland cement (OPC) conforming to IS 12269-1999 was replaced with 15% MK confirming to IS 1489- 1991 and Portland slag Cement (PSC) confirming to IS 455-1995 were used. The chemical compositions of cements were evaluated from XRF (X-ray fluorescence) and the physical properties are shown in Table 1. The crushed gravels with maximum size 10mm and 20mm in 30:70 proportion were used as coarse aggregate (C.A). River sand confirming to zone-II of IS-383-2002 was used as fine aggregate (F.A). The specific gravity and water absorption of C.A were 2.88, and 1.10% respectively and the corresponding values for F. A. were 2.61, and 1.40%. Both the aggregates were confirming to (IS) 2386-2002 (Part I and III).

Table 1. Chemical and Physical Properties of Different Binders.

Bin-ders	SiO_2 (%)	Al_2O_3 (%)	Fe_2O_3 (%)	CaO (%)	MgO (%)	SO_3 (%)	Na_2O_3 (%)	K_2O (%)	LOI[a] (%)	Sp. Gra[b].	Sp. Sur.[c] (m^2/kg)
OPC	19.60	5.69	4.37	66.07	--	3.52	0.38	0.37	0.9	3.16	316.6
PSC	25.53	12.89	1.53	48.17	5.29	5.44	0.42	0.79	1.8	3.00	361.8
MK	56.73	41.66	0.89	0.19	0.14	0.17	--	0.23	3.3	2.64	1495.9
[a] Loss on Ignition,				[b] Specific Gravity,				[c] Specific Surface			

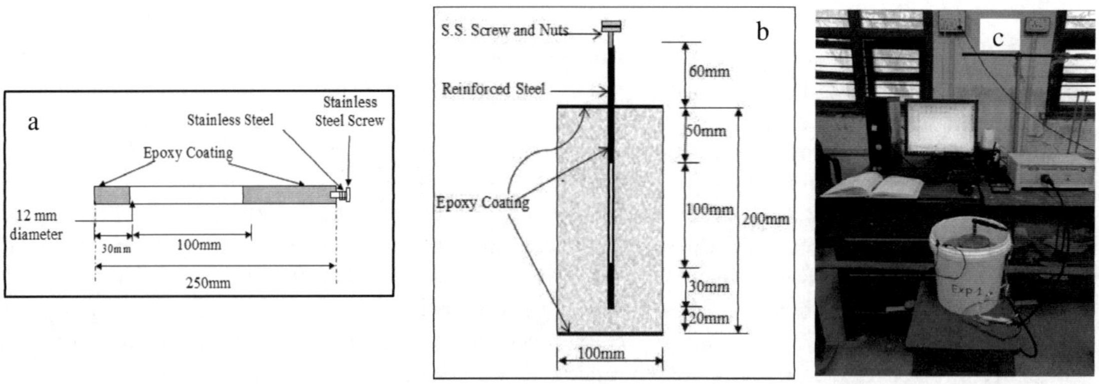

Figure 1. Line Diagram of a) Steel Preparation, b) Cylinder - G109- 2007, c) Lab Set-up for Potential Monitoring.

Concrete Cubes of 150mm size were prepared with PSC and OPC replaced with 15% MK Concrete mixture as per IS 10269-2009 with varying cement contents of 369 kg/m^3, 332 kg/m^3 and 302 kg/m^3 for 0.45, 0.5 and 0.55 w/b ratio respectively. The specimens were prepared for the aforesaid experiments were exposed to four different levels of Sodium Chloride (NaCl) solutions namely 0%, 3%, 5% and 7.5%. The compression strength testing was carried out at 7, 28, 56, 90 and 180 days of curing ages. The cylindrical specimens for half-cell potential measurement were prepared for the same concrete mix design with Thermo-Mechanically Treated (TMT) steel. The contents of carbon (C), phosphorus (P), Sulphur (S), Chromium (Cr) and Copper (Cu) were 0.23%, 0.018%, 0.016%, 0.006%

and 0%. The preparation of the steel was carried out as per the guidelines [9] as show in figure 1(a). Cylindrical samples of 100mm diameter and 200mm length with centrally embedded TMT steel bar of 12mm diameter were cast as show in figure 1(b).

3. Results and Discussions

The compression strength testing was carried out through Compression Testing Machine (CTM) at the rate of 140 kg/cm^2/min as per IS 519-2004 and results were plotted for compressive strength versus curing ages as shown in figure 2. It was observed from figure 2 that the early age strength was higher at higher w/b ratio compared to lower w/b ratio.

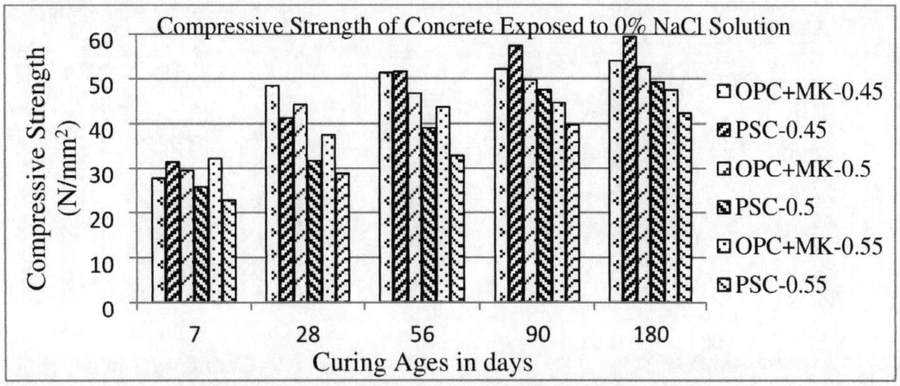

Figure 2. Compressive Strength for control concrete.

This may be due to large surface area of cementitious material demands more water for hydration process. MK is accelerating the early hydration of calcium silicate to produce extra polymeric silicate gel [5]. Strength of the concrete with MK was increased with increase in curing age and decrease with increase in w/b ratio. In the case PSC concrete, the hydration reaction is very slow at initial age, hence the rate of gain of compressive strength of PSC was lower at an early age and it was even less than concrete with MK but after 90 days, there was not much difference in strength of both concrete mixes. For all w/b ratios, compressive strengths were increased with increase in curing ages as expected.

3.1. Effect of NaCl Exposure on Compressive Strength

The results of effect of exposure condition on compressive strength over varied curing ages for PSC and MK are presented in figure 3 (a, b, c).

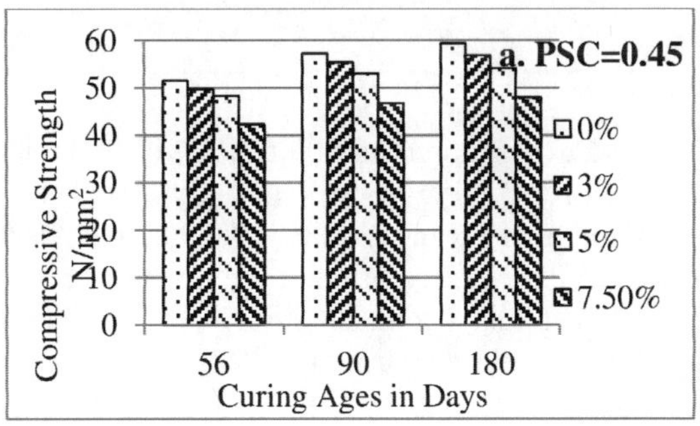

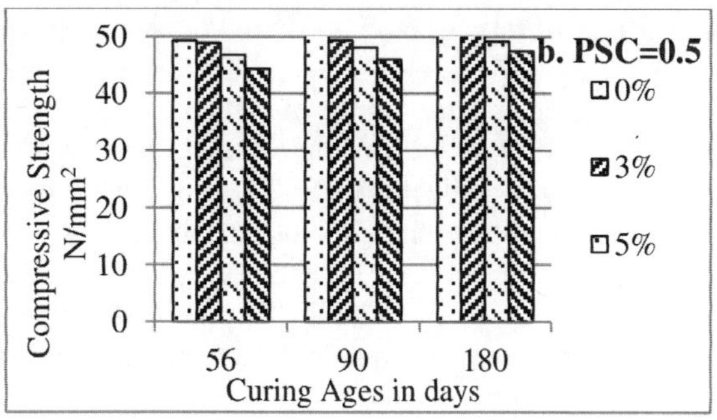

IOP Conf. Series: Materials Science and Engineering **216** (2017) 012022 doi:10.1088/1757-899X/216/1/012022

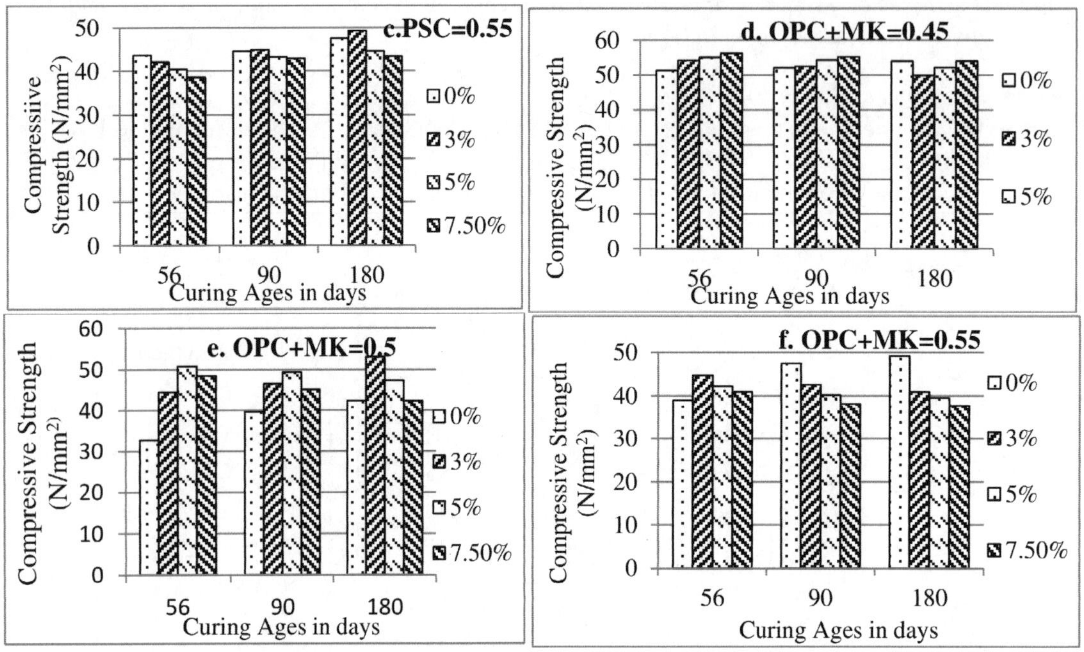

Figure 3. Compressive Strength versus Curing Ages.
(**Note:** The legend in the graph i e. 0%, 3%, 5% and 7.5% are percentage exposure of NaCl solution).

In the concrete with PSC, the compressive strengths were decreased with increased w/b ratio at all curing ages, whereas in case of concrete with MK showed reverse results at only lower w/b ratio. In addition, it was also observed that the compressive strength of MK mostly for various w/b ratios, were increased with increase in chloride exposure as compared to control concrete (0%) which was found reverse in case of PSC concrete. With increased chloride concentration, compressive strengths were decreased as shown in figure 3 (a, b, c) and figure 3 (d, e, f) depicted that compressive strength increase with age but decrease with chloride exposure for both the concrete as expected [10]. This may be because presence of higher percentage of alumina reacted with chlorides, forms friedel's salts and refines micro pores structure improves the strength [11]. In general the compressive strengths were increased for chloride exposure compared to control concrete for w/b ratio 0.45 and 0.5, which is lower for w/b ratio of 0.55.

3.2. Half Cell Potential Measurement:

The probability of corrosion of the reinforcing bars were monitored periodically with saturated calomel reference electrode (SCE) as shown in figure 1(c) over wetting (7 days) and drying (14 days) cycles in all types of the concrete specimens by the half-cell potential measurement [12] in and results were plotted for half-cell potential versus the curing ages in the NaCl solution environment as shown in figure 4 (a, b, c). The half- potential values of almost all specimens immersed in chloride environment were more negative at initial age which became stable over the time. This may be due to slow pozzolanic activity at initial age gives more negative potential. With the increase in the pozzolanic activity over the time refines the pore structure of concrete which may restrict the ingression of moisture and chloride up to rebar level improves the potential to positive. Concrete with MK gives more negative potential at lower w/b (0.45) ratio compared to higher w/b ratio (0.5 and 0.55). This phenomenon is attributed to the high fineness of metakaolin demands more water for hydration [13]. It was observed from the figures that the increase in w/b ratio increases the negative potential for PSC.

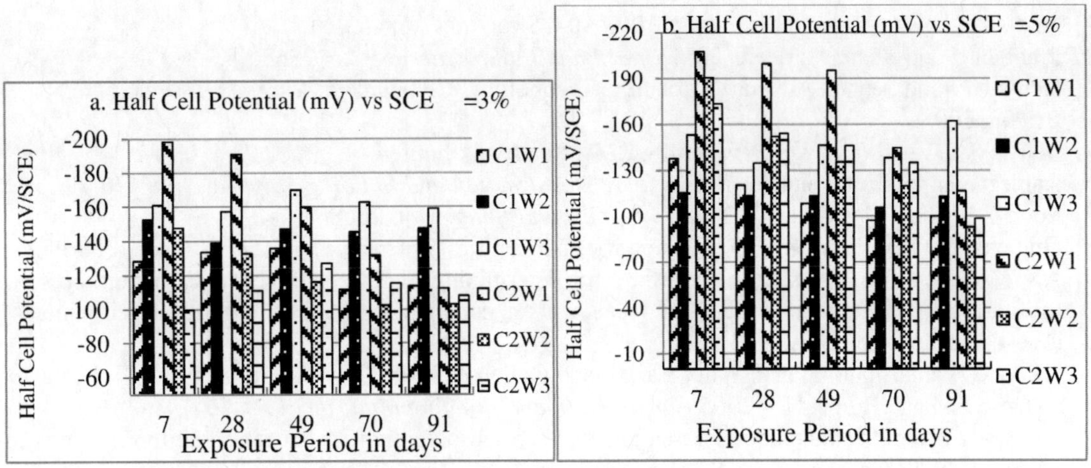

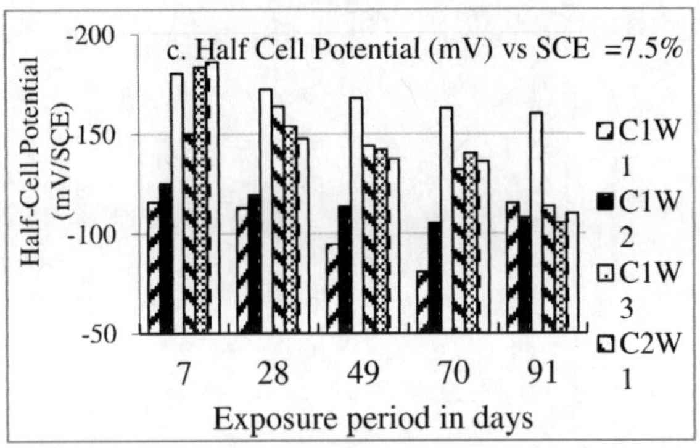

Figure 4. Potentials of Reinforced Concrete cylinder immersed in a.3%, b. 5%, c. 7.5% of NaCl.
Note: C1= PSC, C2=OPC+MK, W1 (w/b) =0.45, W2 (w/b) =0. 5 and W3 (w/b) =0.55.

Potential values were below (-126mV) up to 60 days which indicates corrosion risk is very less while the potential values were increased to the risk of intermediate corrosion (-126 to -256) [12].
From the observation, the concrete prepared with MK performs better than concrete with PSC in corrosion resistance. Potential values were exhibiting even at the end of 90 days less risk of corrosion. In general, it is also observed that the potential values of MK were negative at lower w/b ratio and positive at higher w/b ratio.

4. Conclusion

Based on the experimental studies on concrete with MK and PSC the following conclusions were drawn:

1. Concrete with MK shows high early-age compressive strength due to the higher rate of hydration contributed by MK.
2. At initial age the pozzolanic activity in case of concrete with PSC is sluggish and gives lower early age strength as compared to concrete with MK.
3. The Compressive strength of concrete with PSC reduced with the increase in sodium chloride exposure for all w/b ratios and all curing ages. Whereas, in concrete with MK, reverse nature was observed.
4. From half-cell potential measurements, it was observed that the concrete with MK at higher w/b ratio exhibited low negative potential values as compared to concrete with PSC.
5. Improved performance is exhibited in the case of concrete with MK specimens both the tests including compressive strength and half-cell potential. It is clear that for the concrete with MK performing better in corrosion resistance than that of the concrete with PSC.

5. References

[1] Trianaa V, Marriagaa J and Florezb J 2013 Mat. Research **16**, 1457-64.

[2] Andrade C, Alonso C and Molina F J 1993 Mat. Structures **26,** 453-64.
[3] Pradhan B and Bhattacharjee B 2011 *Const. and Build Materials*. **25**, 2565-75.
[4] Newman J and Seng B Advanced Concrete Technology, *Constituent Materials*, Elsevier, Great Britain, 2003.
[5] Martin S 1998, *Laboratory Report* **135**, RMC Readymix limited.
[6] Ramezanianpour A A and Jovein H B 2012 *Construction and Building Materials*, **30**, 470-79.
[7] Poon C S Kou S C and Lam L 2006 *J of Const and Build. Mat.* **20,** 858-65.
[8] Guneyisi E, Gesoglu M, Karaboga F and Mermerdaş K 2013 *Comp.Engg. Part B:,* **45**, 1288-95.
[9] ASTM G109 2005, Standard test method for determining the effects of chemical admixtures on the corrosion of embedded steel reinforcement in concrete exposed to chloride environments, (West Conshohocken, PA, ASTM).
[10] Batis G, Pantazopoulou P, Tsivilis S and Badogiannis E, 2005, *Cemt. & Con. Comp*. **27,** 125-30.
[11] Khatib J M and Hibbert J J, 2005 *Construction and Building Materials* **19**, 460-72
[12] ASTM C876 1999, Standard test method for half-cell potentials of uncoated reinforcing steel in concrete, (West Conshohocken PA, ASTM).
[13] Bai J, Wild S and Sabir BB. 2003, *Cement and Concrete Research* **33,** 353-62.

IOP Conf. Series: Materials Science and Engineering **216** (2017) 012023 doi:10.1088/1757-899X/216/1/012023

Numerical Modeling of Sliding Stability of RCC dam

O Mughieda[1], K Hazirbaba[2], K Bani-Hani[3] and W Daoud[4]
[1]Assoc. Prof., Dept. of Civil Eng., Abu Dhabi University, Abu Dhabi, UAE
[2]Asst Prof., Dept. of Civil Infrastructure and Environ. Eng., Khalifa University, Abu Dhabi, UAE
[3]Associate Prof. Dept. of Civil Eng. Jordan Univ. of Science and Technology, Irbid, Jordan
[4]Former graduate student at Jordan Univ. of Science and Technology, Irbid, Jordan

Email: omer.mughieda@adu.ac.ae; kenan.hazirbaba@kustar.ac.ae; khaldoon@just.edu.jo;omughieda@gmail.com;

Abstract. Stability and stress analyses are the most important elements that require rigorous consideration in design of a dam structure. Stability of dams against sliding is crucial due to the substantial horizontal load that requires sufficient and safe resistance to develop by mobilization of adequate shearing forces along the base of the dam foundation. In the current research, the static sliding stability of a roller-compacted-concrete (RCC) dam was modelled using finite element method to investigate the stability against sliding. A commercially available finite element software (SAP 2000) was used to analyze stresses in the body of the dam and foundation. A linear finite element static analysis was performed in which a linear plane strain isoperimetric four node elements was used for modelling the dam-foundation system. The analysis was carried out assuming that no slip will occur at the interface between the dam and the foundation. Usual static loading condition was applied for the static analysis. The greatest tension was found to develop in the rock adjacent to the toe of the upstream slope. The factor of safety against sliding along the entire base of the dam was found to be greater than 1 (FS>1), for static loading conditions.

1. Introduction

In the Mujib Canyon, some 60 km south of Amman, Jordan, a hybrid dam was constructed. The dam (Mujib Dam) is owned by the Jordan Valley Authority and was designed as a central roller compacted-concrete (RCC) gravity dam with adjacent earth fill dams at the valley flanks. Its maximum height reaches approximately 47 m and the total volume of the RCC structure is 720,000 m^3 [1].

The dam is located in south of Jordan in the Wadi Al Mujib, south of Madaba and north of Al Karak as shown in Figure 1. The Al Mujib drains toward the Dead Sea. The dam is a composite dam, consisting of a RCC middle section and clay core rock fill (CCR) sections at both abutments. The water intake consists of draw-off works with intakes at three levels. The reservoir will be mainly used for irrigation purposes. The catchment area is 4380 km^2 and the reservoir capacity is $35x10^6$ m^3. The maximum height above the foundation is 67 m and the spillway is free overflow with stepped chute [1]. The geometric configuration of the dam is shown in Figure 2.

IOP Conf. Series: Materials Science and Engineering **216** (2017) 012023 doi:10.1088/1757-899X/216/1/012023

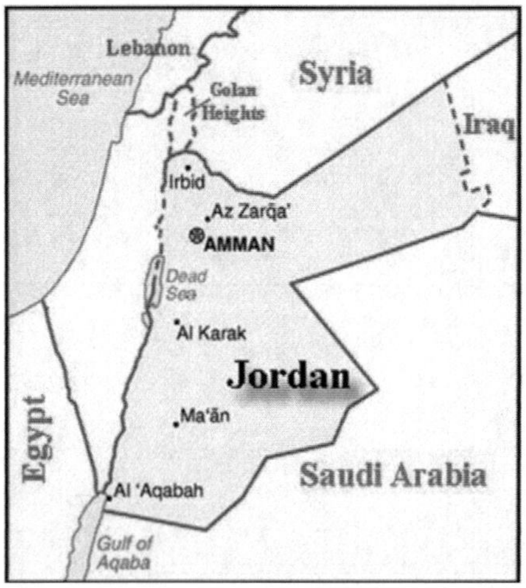

Figure 1. Geographic location of Mujib Dam.

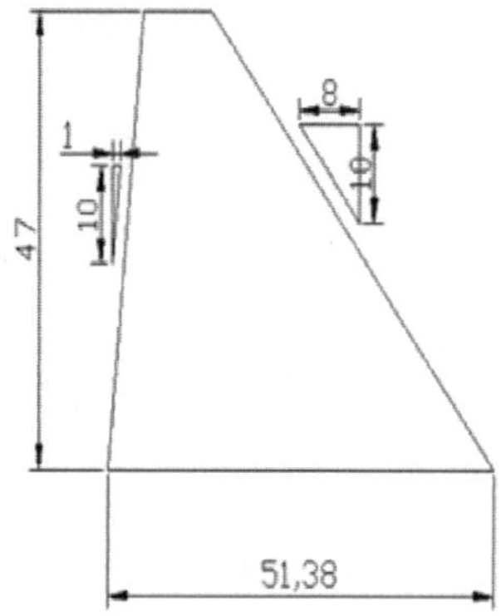

Figure 2. Geometric configuration of Mujib Dam.

2. Material properties

The properties of the dam material as well as those of the foundation material are shown in Table 1. [2].

Also, based on a comprehensive and extensive geotechnical exploration, the dam/foundation interface was characterized by the strength parameters presented in Table 2. [2]. The analyses carried out for this study used the properties and parameters listed in Tables 1 and 2.

Table 1. Properties of the dam and foundation material.[2].

Condition	Material	Modulus of elasticity, E(GPa)	Poisson's ratio, ν	Unit weight (kN/m^3)	Tensile strength kN/m^2	Compressive strength (MPa)
Static	Dam material	15	0.2	24	1.05	14.6
	Foundation	12	0.2	28		22

Table 2. Parameters of foundation.[2].

Rock location	Rack formation	Friction angle (ø)	Cohesion (c)	Compression strength
Dam/foundation interface	Naure limestone	47°	425 kpa	22MPa

3. Finite element modelling

The stress analysis was performed to obtain accurately characterize the distribution of stresses within the dam body under static loading conditions. The commercially available FEM program (SAP200) was used in the analysis [3].

4. Structural idealizations

The dam was modelled using 2-D plain strain isoperimetric elements. Two types of elements were used; the 4-nodded isoperimetric plain strain elements with 2 DOF per node, and the 3-nodded constant strain triangular elements (CSTE) with 2 DOF per node. The finite element mesh of the dam cross section was developed through mesh generation code, which was written by the authors in Fortran-Lahey77 Programming Language, in order to control the dimensions of the dam cross section and to carry out a mesh refinement. The FE analysis was performed based on the following assumptions: (i) plain strain linear elastic behavior, (ii) simplified soil-structure interaction entailing massless elastic foundation, and (iii) a uniform and homogeneous foundation. The boundaries of the foundation were fixed against translation and rotation movement. In addition, all out of plane DOFs were restrained for all nodes. A typical mesh refinement and boundaries of the dam cross section are shown in Figure 3.

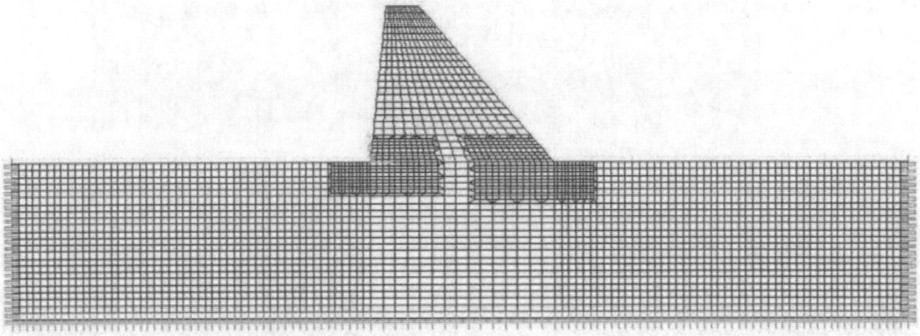

Figure 3. Mesh and boundaries of the foundation.

5. Static loading components

Static loading components comprises of: a) self-weight of the dam, b) hydrostatic uplift pressure: uplift at the concrete/rock interface assumed to vary as a straight line from full headwater pressure at the heel to 0 water pressure at the toe, over 100% of the base area, and put these values at the node in the finite element model to get the nodal uplift forces, c) hydrostatic pressure: a linear distribution of the static water pressure acting normal to the surface of the dam was applied varies from 0 at the water face to (W * h) at the dam base, then the values of hydrostatic nodal force were calculated and used in the analysis, and d) silt pressure: weight of accumulated silt resting up to 10 m on the upstream face

6. Results of analysis and discussion

The definition of dam failure usually refers to partial collapse or movement of a dam or its foundation that hinders the water-retaining ability of the dam [4]. Sliding or shear failure is the most common failure for dams constructed on rock [5]. Sliding occurs when the horizontal forces exceed the frictional resistance. Sliding can be divided in to three different kinds of failures; failure in the interface between concrete and foundation, failure in weak planes of the foundation, such as cracks, and failure in the solid foundation [6]. In this present research the potential failure along the interface plane was investigated for the Mujib Dam. The results of stress analysis of static loading for Maximum Stress (S11), normal stresses in x and y direction and shear stress along the base of the dam are shown in Figures 4 to 6, respectively. It was found that the maximum tensile stress (0.9 MPa) is developed in the rock adjacent to the toe of the slope. All calculated stresses were found to be within the compressive strength (22 MPa) and tensile strength (2 MPa) of the foundation rock and the compressive capacity of the RCC. (14.6 MPa).

Factor of safety against sliding was calculated as the ratio of the shear strength of the rock determined from Mohr-Coulomb criterion and the applied shear stress at the along the base of the dam at each nodes. The shear strength of rock is given by equation (1).

$$\tau_f = c' + \sigma_n' \tan\phi' \quad (1)$$

Where:

τ_f is the shear strength of foundation

c' and ϕ' are the shear strength parameters of the foundation

σ_n' is the effective normal stress along the base of the dam.

The factor of safety against sliding is given by equation (2).

$$FS(Sliding) = \frac{\tau_f}{S_{12}} \quad (2)$$

Where:
S_{12} is the applied shear stress along the base of the dam.

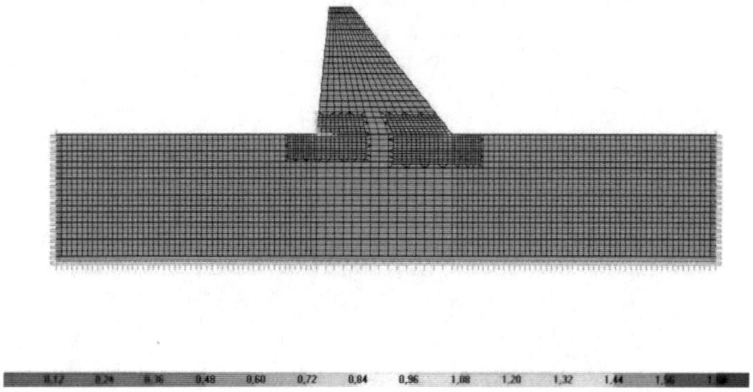

Figure 4. Envelope maximum stresses (S11) (Static Loads Condition)

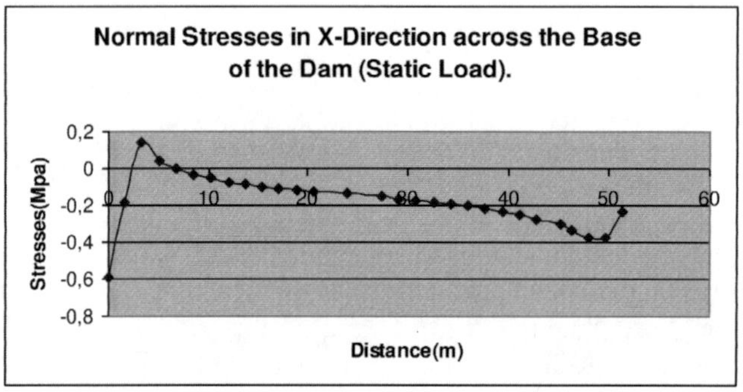

Figure 5. Normal stresses in X-Direction across the base of the dam (Static Load).

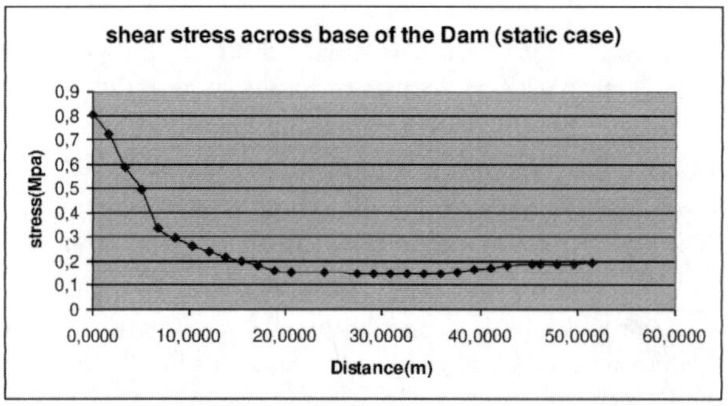

Figure 6. Shear stresses across base of the dam (Static Loads).

Using the cohesion of rock as 0.425 MPa and angle of internal friction as 47° the FS against sliding can be calculated using equation (2). The results of FS are shown in Figure 7. As shown in the figure the FS is above 1 at all nodes along the base of the dam. Most design criteria require a minimum FS against sliding of 2 to 4, based on normal high headwater and low tailwater conditions. This can drop to 1.5 to 2 under flood conditions, and typically is defined as greater than 1 for seismic loads. Although it is not considered by most codes and authorities, a true "fail safe" criterion for stability of an RCC dam is that the FS against sliding is greater than 1 for all load conditions [7].

IOP Conf. Series: Materials Science and Engineering **216** (2017) 012023 doi:10.1088/1757-899X/216/1/012023

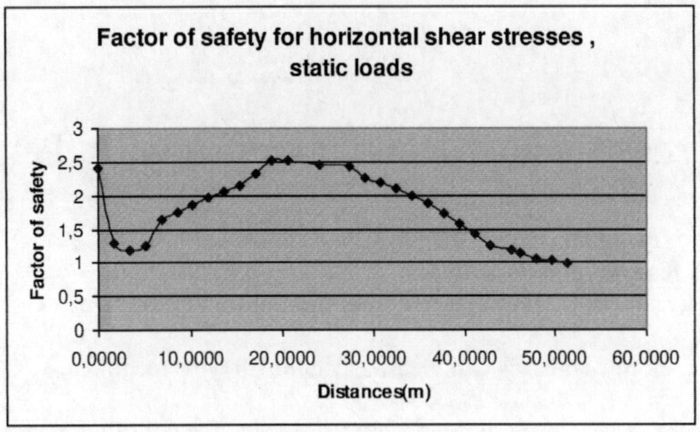

Figure 7. Factor of safety for horizontal shear stresses, Static Loads.

7. Conclusion

The stability of Mujib dam against sliding was carried out using 2D-FEM. The calculated stress levels were compared with the strength of RCC and rock foundation. Also, the factor of safety against sliding was analysed and it was found that under static loading conditions, all values of FS along the base of the dam exceeded 1.

References

[1] Information on http://www.lahmeyer.de/en/projects/hydropower-water-resources/dams.html

[2] HAMZA associates, "Mujib Dam project (Updated design report)", Ministry of Water and Irrigation Jordan Valley Authority, Amman. Jordan (2000).

[3] E. L. Wilson, and A. Habibullah, SAP90, A Series of Computer Programs for the Static and Dynamic Finite Element Analysis of Structures, Users Manual, Berkeley, California, USA, (1989).

[4] ICOLD, Dam failures statistical analyses. Bulletin 99. Commission Internationaldes Grands Barrages (1995).

[5] L. Broberge and M. Thorwid, Evalaution of Failure Modes of Concrete Dams, KTH Royal Institue of Technology. Stokholm, Sweden (2015).

[6] G. A., Johansson, F., Rytters, K., Stille, H. k., Betongdammars Glidstabilitet - Förslag på nya riktlinjer, Elforsk report 8:59, Energiforsk (Elforsk).(2008)

[7] E. K. Schrader, Roller-Compacted-Concrete Dams on Difficult Foundations: Practical Examples, HRW, Volume 14, No. 2, May (2006) 20-31.

IOP Conf. Series: Materials Science and Engineering **216** (2017) 012024 doi:10.1088/1757-899X/216/1/012024

Evaluating Freeze-Thaw Deterioration with Tensile Strength

A J K Komar[1] and A J Boyd[1]
[1] Department of Civil Engineering, McGill University, Montreal, QC, H3A 0C3, Canada

Email: andrew.komar@mail.mcgill.ca;andrew.boyd@mcgill.ca;

Abstract. Freeze-thaw damage is one of the leading contributors to infrastructure deterioration in temperate northern climates. Deterioration caused by freeze-thaw cycling is primarily induced by hydraulic pressures within the hydrated cement paste matrix that cause tensile cracking. Such damage should, therefore, be more effectively detected with tensile testing. This work presents the detection and evaluation of ongoing freeze-thaw (F/T) damage in plain concrete cylinders using the pressure tensile strength test, as it compares to compressive strength evaluation. Pressure tension test results exhibited significantly higher levels of deterioration compared to compression testing, with the samples losing up to 90% of their undamaged tensile capacity. Moreover, it was shown that tensile strength testing is far more sensitive to freeze-thaw deterioration, evidenced by a significant drop in the tensile to compressive strength ratio to below 5%.

1. Background

Portland cement concrete is among the most commonly used construction materials in the world, and is incorporated into many different kinds of structures and pavements. It is subject to inclement weather conditions that may adversely affect material strength. In particular, temperate northern climates go through months where the mean temperature hovers around the freezing point of the pore solution, thus resulting in repetitive cycles of freezing and thawing. Movement of water through the concrete microstructure under pressure, due to differential freezing, can induce very large hydrostatic pressures. These stresses can cause internal damage on the microstructural level such as cracking, spalling, delamination, aggregate pop outs and other such flaws that can lead to dramatically increased porosity [1].

By far the most common method for evaluating the strength concrete subject to deterioration is compressive strength testing. Compressive strength typically overestimates the true material resistance to crack propagation since the direction of loading will act to close active cracks. Other destructive test methods are more sensitive to the expansive damage associated with F/T cycles, such as the Pressure Tension (PT) test. This method applies compressed air over the curved surface of a cylinder, while the flat ends are left exposed to regular atmospheric pressure [2]. The pore water contained within the sample comes to a hydrostatic equilibrium with the applied gas pressure [3]. Since the applied gas pressure is a applied only along the curved surface a net tensile stress field develops parallel to the axis of the cylinder [4]. This indirect tensile effect will act to apply tension at every point within the cylinder at a magnitude equal to the applied stress, resulting in a tension failure crack that propagates across the specimen as the specimen is pushed apart from the inside.

The PT method, because its mode of action applies stress from within the sample as opposed to an external application (Fig 1), will be more sensitive to expansive damage such as the kind associated with F/T damage [4-8]. Any microstructural cracks that are present in a test specimen will be 'pushed apart' at the stress that is required to propagate them. Thus, this test will indicate deterioration as a loss of tensile strength which would not otherwise be detectable using more common methods [9].

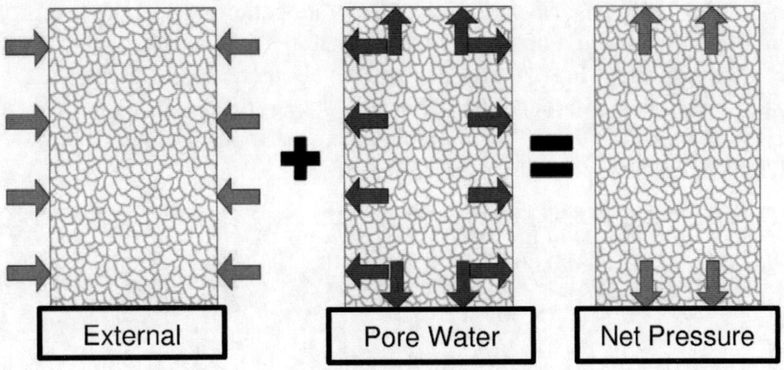

Figure 1. Pressure tension mechanism.

2. Materials and Methods

Two OPC mixtures were prepared with different water to cement ratios (W/C), 0.45 and 0.65, in order to ascertain the effect of different porosities. The purpose of this study was to demonstrate the effectiveness of the pressure tension test method in detecting ongoing F/T damage, so air entrainment was not used in order to ensure adequate damage within a reasonable time frame. Each mix produced one batch of thirty 100 mm diameter x 200 mm cylinders.

Each concrete cylinder was placed into an individual container conforming to the dimensional specifications laid out in ASTM C 666, specifically the maximum clearance between the edge of the container and the sample being less than 3 mm. The containers were filled with tap water to a depth of 3 mm above the specimen. The water was replaced during the thawing portion of each cycle.

For the freezing phase of the cycle the containers with the samples within were placed in commercially available freezers with a minimum operating temperature of -25 °C ± 3 °C. Thermocouples were embedded in two samples and indicated that a uniform temperature of -25°C was reached at the centre of the specimens after approximately 18 hours of exposure in the freezer. For the thawing portion of the cycle, the samples were transferred from the freezers into a large water bath equipped with a heating element and a circulating pump, ensuring that no thermal gradient could develop within the heat transfer medium of the water bath. From their frozen condition, it took on average four hours for the temperature of the water bath and the samples to reach a minimum temperature of 15 °C. An entire freeze-thaw cycle with this large temperature range could thus be performed in 24 hours. The extremes of the temperature range were beyond the standards as prescribed in ASTM C 666 but this was selected as a more severe environment to be more representative of extreme conditions which may occur in the field.

A baseline for strength comparison was established by testing samples at the conclusion of the curing period, prior to the initiation of freeze-thaw cycling. Subsequent testing intervals were established in order to track ongoing changes to the material strength as detected by the destructive testing methods. Destructive tests were performed after 1, 5, 10, 20 and 30 F/T cycles. Standard compressive strength tests as prescribed in ASTM C 39 were performed on all specimens selected for compressive strength testing. At every testing interval, two samples were tested in this manner. Pressure tension tests were also conducted at every testing interval on two samples, with the pressurizing rate set at 2.00 MPa/min.

3. Results and Discussion

At the conclusion of the testing, visual physical damage was observed on both the 0.45 and 0.65 samples. The extent of the deterioration was much more severe for the samples that had undergone more F/T cycles. As expected, the extent of surface scaling and deterioration was much more severe on the higher w/c specimens. The paste was particularly weakened, with some mass loss being observed through simple handling of the specimens, especially around the aggregates. In the case of the 0.65 w/c mixture, the extent of the surface deterioration led to some difficulties producing a solid seal during pressure tension testing, since 100% of the specimen exhibited surface scaling.

Compression tests failed in the usual manner (i.e. conical failure planes with paste crumbling). Pressure tension failure specimens showed a more disparate pattern across the levels of deterioration. Samples tested few cycles showed a break perpendicular to the tensile stress, with the fracture surface passing through the aggregates (Fig 2a). As damage levels increased, a greater area of the fracture surface was found to have passed around the surface of the aggregates (Fig 2b). This apparent failure through the interfacial transition zone (ITZ) was found to be more prominent in the higher porosity w/c mix, as well as specimens exposed to more F/T cycles.

Figure 2a (left) and **2b** (right): Failure planes from pressure tension testing.

The results of the compression testing can be found in Figure 3. The compressive strength decreased as a function of cyclic deterioration for both of the mixtures, with the 0.45 w/c dropping from 48 MPa to 35 MPa over 30 cycles, and the 0.65 w/c dropping from 25 MPa down to 15 MPa over the same period. This represents a compressive strength loss of 27% and 40%, respectively. Statistically, the decrease in compressive strength results reached significance after the 20th cycle for the 0.45 w/c, and the 10th cycle for the 0.65 w/c.

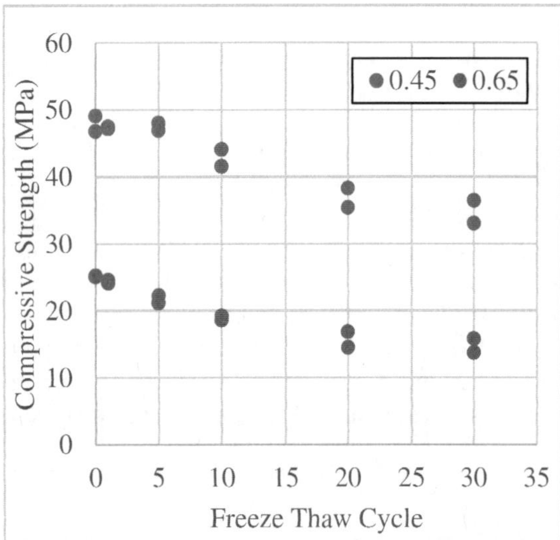

Figure 3. Compressive strength deterioration.

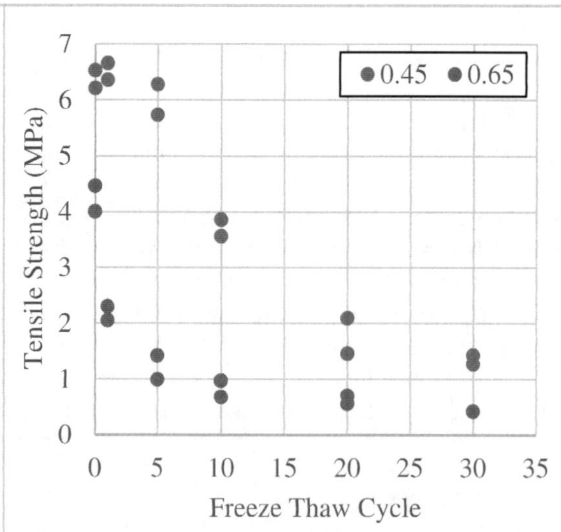

Figure 4. Pressure tensile strength deterioration.

The results of the pressure tension testing can be found in Figure 4. For both mixes, deterioration in the tensile capacity was observed, with the 0.45 w/c dropping from 6.4 MPa to 1.4 MPa, and the 0.65 w/c dropping from 4.2 MPa to 0.4 MPa. This represents a tensile strength loss of 78% and 90%, respectively. Of particular note was the dramatic reduction in tensile capacity for both mixes. After a single F/T cycle, the 0.65 w/c concrete lost 49% of its original strength. This falloff was statistically significant. The decrease reached significance for the 0.45 w/c by the 10th cycle, by which it had lost 42% of its original tensile capacity.

In both cases, the resulting tensile capacity resembled a negative exponential curve, with the damage converging on some minimal number for both mixtures (Fig. 5). It is important to note that the assumption of tensile capacity of concrete being roughly 10-15% of the compressive strength does not hold true for this study. As Figure 5 shows, the undamaged concrete at cycle zero reflected this assumption, though if the assumption remained valid then the figure would be a constant value for both mixes. However, by Cycle 20 for both mixes the ratio had dropped below 5%, where it bottomed out at 4% for the 0.45 w/c and 3% for the 0.65 w/c mix.

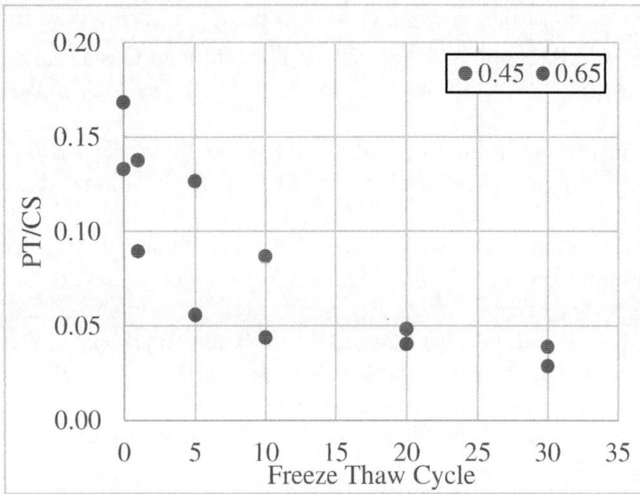

Figure 5. Tensile to compressive strength ratio.

This data demonstrates that the tensile capacity of concrete is more severely affected by F/T deterioration than the compressive strength, which is understandable given that the mode of damage tends to be cracks in the concrete paste. Only a tensile mode of loading will open these existing cracks, thus resulting in failure at a lower applied stress when compared to compression. A further observation was that the PT test detected a significant deterioration signal much earlier (i.e. at much fewer F/T cycles) than the compressive strength test.

4. Conclusions

It is clear that pressure tension testing detected F/T damage much earlier than compression testing. This result was particularly true for the higher porosity mix, where a significant loss in material strength was detected after a single F/T cycle. Considering that cracking has such a strong effect on tensile capacity, more research into this significant drop-off in tensile strength as a function of deterioration is clearly in order. The basic engineering assumption of a constant ratio between tension and compression was shown to be invalid in this case due to the greater effect of cracking on tensile capacity, which could lead to an overestimation in the residual material capacity with other modes of loading such as compression.

In addition, the fracture surfaces associated with the pressure tension failure specimens showed a systematic change in the material behavior due to the ongoing F/T cycles. The fracture surfaces first showed a cohesive bond between the aggregates and the surrounding paste in the undamaged state resulting in aggregate failure. Samples from increased F/T cycles, however, exhibited more intact aggregate particles while the paste appeared to fail around them within the ITZ. Since the ITZ is a zone of higher porosity, it follows that more severe damage from expansive processes would occur there, yet 'aggregate pullout' phenomenon was not observed on the compressive strength failure

IOP Conf. Series: Materials Science and Engineering **216** (2017) 012024 doi:10.1088/1757-899X/216/1/012024

specimens which also displayed a less significant deterioration trend. This aggregate pullout was only detected using the pressure tension method. This is expected, since that is the only method that will act to open existing cracks as opposed to compressing the existing fracture surfaces together. Considering that many structural loading modalities such as flexure incorporate tension, it is of integral importance to understand the true material behavior.

5. References

[1] Kosmatka S H, Kerkhoff B, Panarese W C, MacLeod N F and McGrath, R J 2002 *Design and Control of Concrete Mixtures 7th Edition*, Cement Association of Canada, Ottawa, Ontario, Canada

[2] Clayton N and Grimer F 1979 The Diphase Concept, with Particular Reference to Concrete *Developments in Concrete Technology*, Elsevier, Waterford, UK, 283-317

[3] Terzaghi K and Peck R 1967 *Soil Mechanics in Engineering Practice* Wiley, New York

[4] Uno T, Fujikake K, Mindess S and Xu H 2010 The Nitrogen Gas Tension Test of Concrete, Part 1: Effect of Boundary Conditions and Axial Strain Response *Materials and Structures.* **44** p 857-864

[5] Boyd A J, Bremner T W, Holm T A and Boyd S R 1998 *Tensile Testing to Evaluate the Effect of Alkali-Aggregate Reaction in Concrete* T192-2, IRC National Research Council, Ottawa, Canada

[6] Boyd A and Mindess S 2001 The Effect of Sulfate Attack on the Tensile to Compressive Strength Ratio of Concrete *3rd Int. Conf. on Concrete under Severe Conditions*, Vancouver, p 789-796

[7] Hartell J, Boyd A J and Ferraro C C 2011 Sulfate Attack on Concrete: Effect of Partial Immersion *Journal of Materials in Civil Engineering* **23** p 572-579

[8] Komar A J, Hartell J and Boyd A J 2013 Pressure Tension Test: Reliability for Assessing Concrete Deterioration *7th Int. Conference on Concrete under Severe Conditions*, Nanjing, China, p 337-344

[9] Komar A J K and Boyd A J 2014 Pressure-tension testing in the evaluation of freeze-thaw deterioration *10th Fib International PhD Symposium in Civil Engineering*, Québec, Canada, p 143-148

IOP Conf. Series: Materials Science and Engineering **216** (2017) 012025 doi:10.1088/1757-899X/216/1/012025

Review of Research and Application of Reinforced Concrete Structures Strengthened by Braces

Jing Gong, Zezhong Zhu, Cong Zeng

School of Civil Engineering and Architecture, Northeast Electrical Power University, Jilin City, Jilin, China

Email: gj6625@sina.com; 609698729@qq.com;

Abstract. Many of RC frame structures are urgently needed to be strengthened and maintained due to the increase of service life, the change of use function, and the impact of natural disasters. This paper reviews the research status of strengthening RC structures with braces; introduces the features and connection forms of joints connecting with braces; summarizes the engineering application of buckling-restrained braces, pointing out that buckling-restrained brace is an effective means to strengthen the RC frame structures with more reliable performance and broad application prospect.

1. Introduction

Many of RC structures had such serious problems as cracking and slumping due to the impact of aging and earthquake, causing great harm to people's lives and properties. Some existing structures cannot meet the requirement of seismic fortification intensity need for seismic reinforcement in accordance with *Code for Seismic Design of Buildings* [1]. China has issued *Design Code for Strengthening Concrete Structure* [2], summarizing many strengthening techniques, such as increasing section method, encasing steel, prestressed strengthening, bonding steel and carbon fiber polymer, introducing the requirements of design and construction and calculation of sections. Particularly, strengthening structures by adding braces has prominent advantages. Common steel brace not only can improve stiffness of structures, but weaken ductility of structures less because of its good ductility performance. Buckling-restrained brace can yield instead of buckling in tension and compression with better seismic performance, stronger energy-dissipated capacity in rare earthquake and more stable mechanical performance comparing with common brace.

2. Characteristics of strengthening structures by adding braces

Strengthening by adding braces is a common method to improve the bearing capacity of structures and reduce deformation of structures. Concentric braces, eccentric braces, buckling-restrained braces are classified. Reference [3-5] introduced the features of common steel braces in detail. Based on stiffness and bearing capacity, concentric braces were adopted with high lateral stiffness, however, buckling occurred after compression in the earthquake, resulting in a sharp decline in stiffness and bearing capacity. Based on energy dissipation, eccentric braces were produced with an important segment called the link beam. Energy from earthquake can be dissipated by inelastic deformation of the link beam. Reference [6] and [7] summarized and introduced detailedly different sections and composition of BRBs. Based on bearing capacity and energy dissipation, buckling-restrained braces were produced with better energy-dissipated capacity, composing of inner core, outer constraints and unbonded material. Under the strong earthquake, the inner core bore axial loads to dissipate seismic energy; the

IOP Conf. Series: Materials Science and Engineering **216** (2017) 012025 doi:10.1088/1757-899X/216/1/012025

outer constraints brought lateral restraint to the inner core, preventing from buckling; the unbonded material guaranteed the inner core to bear axial force.

3. Connection forms of joints connecting with additional braces

3.1 For new reinforced concrete frame structures

Firstly, the bolt sleeve is embedded in the beam-column, the anchor plate is fixed on the beam-column by bolts, and the adjacent anchor plate weld together, the gusset plate of brace is welded with the anchor plate as Figure 1.

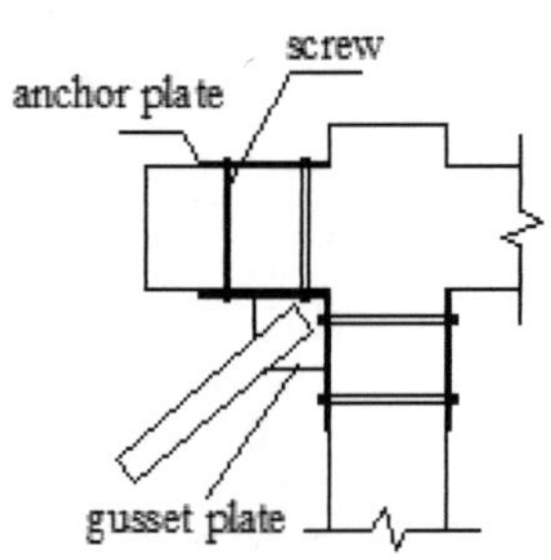

Figure 1. Screw connection

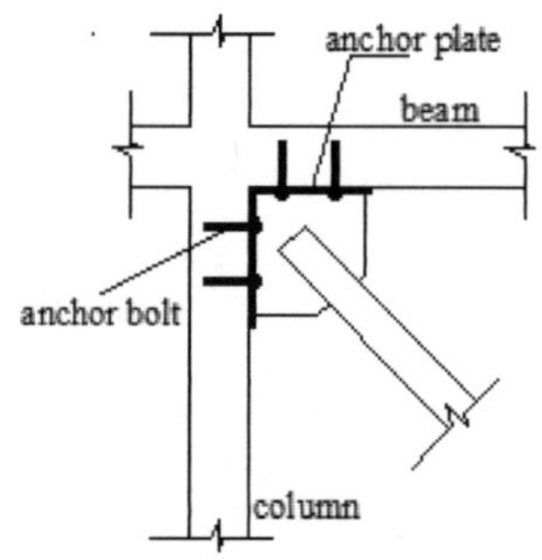

Figure 2. Anchor bolt connection

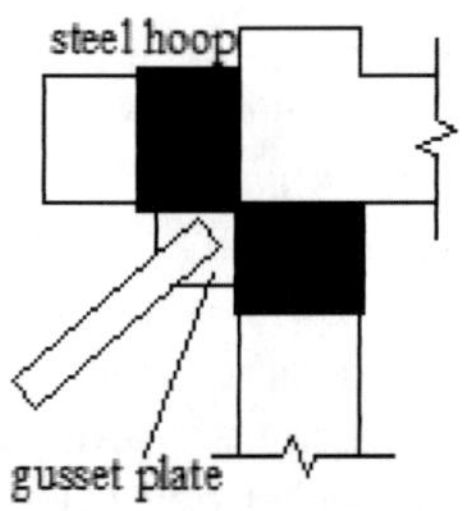

Figure 3. Encase steel hoop outside

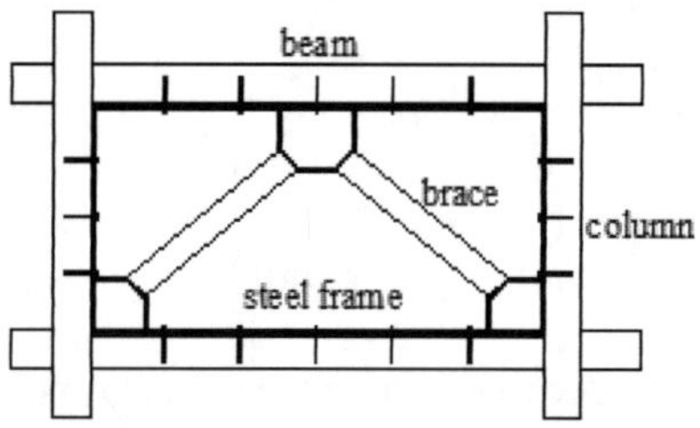

Figure 4. Embed steel frame inside

3.2 For existing reinforced concrete structures

The anchor plate is fixed on the surface of the beam-column by anchor bolts, the anchor plate is welded vertically and horizontally, and the gusset plate is welded with the anchor plate as Figure 2.

Surface of beam-column is polished and cleaned, then steel plate is bonded on the surface in epoxy resin, the gusset plate is welded with the steel plate as Figure3.

The steel frame is embedded in the reinforced concrete frame, and the gusset plate is welded with the steel frame. How to fix the steel frame? first reserving stud holes in the beam-column, then arranging studs on the steel frame and positioning accurately, finally pouring cement mortar between the steel frame and concrete frame as Figure 4.

4. Study on the performance of strengthening structures with braces

4.1 Common steel braces

Gonzalez [8], through the comparative study on the seismic performance of RC frame strengthened with Y-shaped steel braces and shear walls, showed that strengthening RC frame with Y-shaped steel braces was more economical and effective to improve the stiffness and ductility. Ghobarah [9], through the comparative study on the seismic performance of three-layer RC frames strengthened with Y-shaped eccentric steel brace and concentric steel brace, showed that eccentric bracing frame exhibited lower deformation and damage, the main damage was in the part of the link beam, dissipating energy by inelastic deformation; The frame inverted concentric steel brace damaged more. Man Xu [10], through elastic-plastic dynamic response analysis of failed middle and side column of five commonly concentrically braced steel frames, showed that the stiffness of structures was greatly improved, but the ultimate bearing capacity of structures was not improved obviously; V-typed, inverted V-typed and X-typed braces effectively reduced the displacement of the failure point; strengthening effect of the forward-brace for middle column was better than that of side column; the influence of backward-brace was approximate. Baocheng Zhao [11] made low cyclic loading test of RC frame strengthened with Y-shaped eccentrical steel brace with different connection forms between the link beam and RC-frame beam. U-steel hoop was encased on the beam to protect the connection point. Another form was just to anchor a 20mm-thick steel plate under the beam, showing that steel brace contributed to improve the lateral stiffness of structures and dissipate energy; the connection joint with U-steel hoop was more effective without damage, failure of structure was led by the failure of the link beam; the joint anchoring a steel plate damaged, leading to the failure of structure. Surong Fan [12] made low cyclic loading test of RC frames strengthened by encasing steel brace and embedding steel frame brace, showing that steel brace contributed to improve the stiffness and aseismic shear capacity of structures and reduce the story drifts, the beam-column joints were not damaged. In contrast, steel frame had better strengthening effect, working performance and stronger energy dissipation capacity.

4.2 Buckling-restrained brace

Mahrenholtz [13] made quasi-static test of RC frame strengthened with BRB, and introduced the design method of anchor connection, showing that the inter-story displacement angle of the frame was within 3%, and the mechanical properties of the brace and the joints were stable. Corte [14] made a of field test of a two-storey RC frame strengthened with buckling-restrained brace, showing that BRB occurred overall buckling, and the inter-story displacement angle of the frame was within 3%. Yooprasertchai [15], carried on quasi-static tests on RC columns connected with buckling restrained braces, showing that although BRB can greatly improve the lateral stiffness and strength of columns, but the overlap of steel bars has been seriously damaged. Hui Wu [16] carried on the experiment on seismic performance of three-storey RC frame strengthened with buckling-restrained brace, showing that the seismic performance of structure was improved remarkably, and the earthquake energy was mainly dissipated by BRB. Hui Wu [17] made a study on the seismic performance of replacing buckling-restrained brace. First a three-storey RC frame was strengthened by BRB, imposing loads until BRB yielded; then replaced the failed BRB, showing that BRB greatly contributed to improve the bearing capacity of structure and energy dissipation capacity. After the replacement of BRB, the structure regained good seismic performance. Bingzheng Cao [18], carried on the low cyclic test of two-layer RC frame strengthened with BRB, showing that the bearing capacity of structure was improved, but the adjacent beam-column connecting with BRB damaged, leading to the failure of main frame and decline of deformation capacity of frame. It was suggested that adjacent beam-column should be strengthened first.

5. The advantages of buckling-restrained brace

The lateral stiffness and bearing capacity of RC frames can be improved by common steel braces. However, it is easy to produce instability and buckle under the rare earthquake, so the seismic performance of steel braces is much poorer as Figure 5. BRB can not only effectively solve such problems as buckling in compression to ensure the stable seismic performance, but as a energy dissipation damper, it can effectively dissipate seismic energy to avoid the occurrence of serious damage of the main body of structures as Figure 6 .

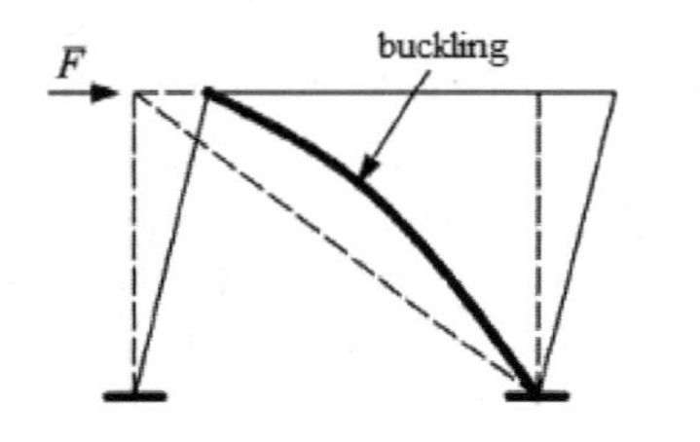

Figure 5. Steel brace buckling

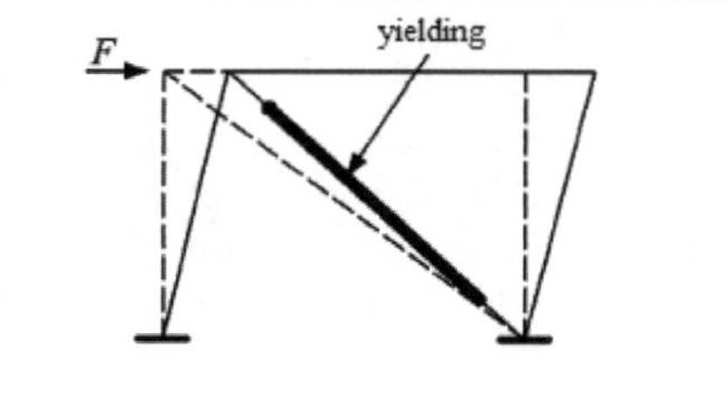

Figure 6. BRB yielding no buckling

6. Engineering application of buckling restrained braces

The building of crop and environment science in California University was the first use of buckling-restrained braces; 14-storey dormitory building in California University was strengthened by BRB [19]. The federal building creating in 1960s in Salt Lake City was retrofitted by 344 BRBs [20]. The project won the Utah annual outstanding engineering award in 2002.

A lot of researches on buckling restrained braces had been done in Taiwan. Cathay International Building in Taichung, a steel structure building, suffered the "9.21" earthquake in the construction process. After the earthquake, the seismic grade of the area was modified to improve. So 80 pure steel were adopted to strengthen the case as [21]. Yangming Culture University Gymnasium in Taipei was a elliptic structure. In order to improve the seismic performance of the structure, the diagonal part was provided with 96 buckling-restrained braces [21].

Hiroshima branch building of Takenake Corporation was built in 1971 as the RC frame structure. In order to meet the seismic fortification criterion of the new seismic code in Japan, the buckling restrained brace made of low yield point steel was used to retrofit. A total of 32 BRBs were designed to the four corners of each layer from the 2nd to 9th layer without effects on the appearance and lighting of the building [6].

Shanghai Earthquake Prevention and Disaster Reduction Center. The seismic fortification intensity was 8 degrees. If shear walls were adopted to strengthen, shear walls cracked and even failed and stiffness degradation was serious. After the comprehensive consideration, buckling-restrained braces system was used to provide the lateral stiffness and dissipate seismic energy under the rare earthquake [22]. Shanghai Expo Center used TJ-typed buckling-restrained braces developed by Professor Guoqiang Li from Tongji University and his team. In addition, Shanxi library, Shanghai Sanlin Sports Center Gymnasium, Shanghai Shenhong Office Building, Oriental Sports Center and Shanghai Hengfeng Middle School adopted TJ-typed BRB [23].

7. Conclusion

Strengthening RC frames by braces has better properties than other strengthening techniques. Particularly, buckling-restrained brace has such outstanding features and superiority as yielding without buckling in tension and compression, more stable mechanical performance, stronger energy dissipation capacity. However, this technology has been patented in many foreign countries without opening to the public, limiting the development and promotion of the technology. So we should enhance the research to enact technical specification for design and construction in accordance with the reality of our country. It should be noted that braces will have an impact on the internal force distribution of the beam-column joints connected with braces, which need to be strengthened to meet bearing capacity. The joints connected braces with the main structure are the weak link in the whole system. How to ensure the connecting joints are not damaged is needed solving and noticing in current design and construction.

8. Acknowledgments

The research work was supported by National Natural Science Foundation of China under Grant No. 51278091, Education Department of Jilin Province under Grant No. 2015[256], and Department of Science and Technology of Jilin Province under Grant No. 20150623024TC-16.

IOP Conf. Series: Materials Science and Engineering **216** (2017) 012025 doi:10.1088/1757-899X/216/1/012025

9. References

[1] GB50011 2010 *Code for Seismic Design of Buildings* (Beijing: China Building Industry Press)

[2] GB50367 2013 *Code for Design of Reinforced Concrete Structures* (Beijing: China Building Industry Press)

[3] Tremblay R 2002 Inelastic seismic response of steel bracing members *J Constr. Steel Res.* **58** pp 665-701

[4] Edoardo M and Nakashima M 2006 Seismic performance and new design procedure for chevron-braced frames *J Earthq. Eng. & Struct. Dyn.* **35** pp 433-52

[5] Jianbin L 2005 Study on design theory of buckling restrained brace and buckling restrained braced steel frame D Tsinghua Univ.

[6] Yun Z 2007 *Designing and application for buckling-restrained brace frame* (Beijing: China Building Industry Press) p 212

[7] Yanlin G Jingzhong T and Peng Z 2016 Research progress of buckling restrained braces: types, design methods and applications *J Eng. Mech.* **33** pp 1-14

[8] Gonzalez S 2007 Retrofitting of RC structures on gravity columns using Y-inverted steel bracing *D Puerto Rico: University of Puerto Rico, Mayaguez*

[9] Ghobarah A and Elfath H 2001 Rehabilitation of reinforced concrete frame using eccentric steel bracing *J Eng. Struct.* **23** pp 745-55

[10] Man X Shan G and Libin Z 2013 Progressive collapse dynamic analysis of concentrically braced steel frames *J Chin. Civ. Eng. Journal* **46** pp 329-38

[11] Baocheng Z Anlin Y and Sicheng J 2013 Experimental investigation of Y-eccentrically steel braced RC frames energy-dissipation *J Earthq. Eng. & Eng. Vibr.* **33** pp 47-53

[12] Surong F 2002 Experimental researchon the aseismic retrofitting of reinforced concrete frame *D Nanjing Univ. Techn.*

[13] Mahrenholtz C Lin P and Wu A 2015 Retrofit of reinforced concrete frames with buckling-restrained braces *J Earthq. Eng. & Struct. Dynam.* **44** pp 59-78

[14] Corte G D D'Aniello M and Landolfo R 2014 Field testing of all-steel buckling- restrained braces applied to a damaged reinforced concrete building *J Struct. Eng.* **141** D4014004-11

[15] Yooprasertchai E and Warnitchai P 2008 Seismic retrofitting of low-rise nonductile reinforced concrete building by buckling-restrained braces *Proc. the 14th World Conf. on Earthq. Eng.* (Beijing: China)

[16] Hui W Guowei Z and Jian Z 2013 Seismic performance of existing RC frame structures reinforced with buckling-restrained braces *J Chin. Civ. Eng. Journ.* **46** pp 37-46

[17] Hui W Yanxia Z and Guowei Z 2013 Experimental study on seismic performance of replaceable buckling-restrained braces in reinforced concrete frame *J Chin. Civ. Eng. Journ.* **46** pp 29-36

[18] Bingzheng C Chunming Z and Hao Zhen H 2012 Experiment study on seismic behavior of RC frames strengthened by buckling-restrained braces *J Earthq. Resist. Eng. and Retr.* **34** pp 1-7

[19] Parry B Aiken D and Jafarzadeh F 2001 Seismic retrofit of Wallance F. Bennett Federal Building *J Mod. Steel Constr.* 1-6

[20] Sabelli R Mahin S and Chang C 2003 Seismic demands on steel braced frame building with buckling-restrained braces *J Eng. Struct.* **25** pp 655-66

[21] Keh-Chyuan T and Junwei L 2004 Principle and application of buckling-rstrained braces *The 1st Nat. Symp. on Earthq. Prev. and Dis. Red. Eng.* (Beijing: China) p 31

[22] Lu Z Weihua Z and Rui H 2011 Application of buckling restrained brace in Shanghai Seismological Bureau Building *J Build. Struct.* **41** pp 125-30

[23] Guoqiang L Feifei S and Hai G 2009 Engineering application of TJ-type buckling restrained brace *J Build. Struct.* **39** pp 607-10

IOP Conf. Series: Materials Science and Engineering **216** (2017) 012026 doi:10.1088/1757-899X/216/1/012026

Effect of mineral admixtures on kinetic property and compressive strength of self Compacting Concrete

Srishaila Jagalur Mahalingasharma[1], Parasivamurthy Prakash[2], Vishwanath K N[3] and Veena Jawali[4]

[1]Department of Civil Engineering, RYMEC, Ballari, Karnataka, India
[2] Department of Construction Technology and Management, DSCE, Bangalore, India
[3]Department of Civil Engineering, DSATM, Bangalore, India
[4]Mathematics Department, BMS College of Engineering, Bangalore 560019, India

srishaila.sharma@gmail.com

Abstract: This paper presents experimental investigations made on the influence of chemical, physical, morphological and mineralogical properties of mineral admixtures such as fly ash, ground granulate blast furnace slag, metakaoline and micro silica used as a replacement of cement in self compacting concrete on workability and compressive strength. Nineteen concrete mixes were cast by replacing with cement by fly ash or ground granulated blast furnace slag as binary blend at 30%, 40%, 50% and with addition of micro silica and metakaoline at 10% as a ternary blend with fly ash, ground granulated blast furnace slag and obtained results were compare with control mix. Water powder ratio 0.3 and super plasticizer dosage 1% of cementitious material was kept constant for all the mixes. The self compacting concrete tested for slump flow, V-funnel, L-Box, J-Ring, T50, and compressive strength on concrete cube were determined at age of 3, 7, 28, 56, 90 days.

1. Introduction

Concrete is a construction material extensively used around the globe. Properties of concrete have changed tremendously to meet the field requirements. Adopting new technological advancements in construction industry few types of concrete have been developed to improve the various properties of concrete in fresh and harden state. Increase in the population growth necessities the need for housing and infrastructure demands leading usage of more concrete for construction industry. Recent developments in civil engineering constructions like high-rise buildings and long-span bridges necessitates the need concrete of higher compressive strength [1].
An advancement of concrete construction technology and its production has led to higher grades of concrete strength. In the pursuit of advancement of concrete construction an innovative type of concrete was used in construction such as self compacting concrete (SCC). The usage of self compacting concrete in high rise building and long span bridges has resulted in an increase in the homogeneity and better performance of the concrete structures [2]. Self compacting concrete can be use to mould the structure without the use of vibrators and consolidates by its own weight and can flow through densely and congested reinforced and complex structural elements [3]. Many researchers have conducted experimental studies made on various mineral admixtures like fly

ash, GGBS, micro silica and metakaoline incorporating in self compacting concrete. Thus incorporating high volumes of mineral admixtures can make it cost effective, increases the workability and strength at later ages. The preset research aimed to investigate the effect of chemical, physical, mineralogical, morphological properties of fly ash, ground granulated blast furnace slag, micro silica and metakaoline on the workability and compressive strength of self compacting concrete.

2. Experimental Investigation

2.1. Materials

In present study self compacting concrete was prepared with 53 grade Ordinary Portland cement confirming IS 12269-1987, Class F fly ash (PFA) in accordance to IS 3812-1981, Ground granulated blast furnace slag (GGBS) in accordance to IS 12089-1987, Micro silica (SF) in accordance to IS 15388-2003 and Metakaoline (MK).The chemical and physical properties of cement, and mineral admixtures are as shown in Table 1.
The illustrated SEM micrographs reveal that the cement particles (Figure1) are angular and non-spherical in shape, fly ash particles (Figure2) are spherical and smooth, hallow spheres called as cenospheres (microspheres) and plerospheres. GGBS particles (Figure3) are long, elongated and flaky in shape. Micro silica particles (Figure4) appear the spherical in shape and as denser particles. Metakaoline particles (Figure 5) smooth spherical shape powdered particle.

Table 1. Chemical and Physical Properties of Cement, flyash, GGBS, SF, MK.

	Cement	Fly ash	GGBS	SF	MK
SiO_2	16.34	62.63	33.77	91.9	52.0
Al_2O_3	6.95	23.35	13.24	0.4	42.2
Fe_2O_3	5.38	3.93	0.65	0.3	0.7
CaO	60.84	2.04	33.77	0.3	0.08
MgO	2.32	0.46	10.13	0.0	1.76
SO_3	1.99	0.53	0.23	---	---
Na_2O	1.50	1.35	--	---	0.07
K_2O	2.73	---	--	---	---
LOI	---	0.39	0.19	1.7	0.3
Specific Gravity	3.12	2.1	2.9	2.45	2.12
Specific surface (m^2/Kg)	297	311	436	22000	12000

X-ray diffract gram of cement shows the presence of crystalline phase of calcium, silica and alumina compounds. Fly ash is composed of 30 % crystalline compounds such as quartz, mullite, hematite, and 70% amorphous glassy material such as silica glass and glass of other oxides. GGBS shows amorphous silica, alumina and calcium. Micro silica is amorphous phase, significant presence of silica. Metakaoline is an amorphous phase of silica and alumina.

2.2. Aggregates

Manufactured sand confirming Zone II and the crushed coarse aggregate obtained from the local crushing plant are used in the present study. The properties of Coarse and Fine Aggregate are depicted in the Table 2. Super plasticizer (SP) poly carboxylic ether base Auromix 400 with specific gravity 1.08 and chloride ion content less than 0.2% and pH 6 are used. Fresh and potable water of pH 7.72 free from organic waste is used for mixing and curing of the SCC.

IOP Conf. Series: Materials Science and Engineering **216** (2017) 012026 doi:10.1088/1757-899X/216/1/012026

Table 2. Properties of fine and coarse aggregates

Properties	Coarse Aggregate	Fine Aggregate
Specific Gravity	2.7	2.62
Water Absorption	0.8%	3.8%
Fineness Modulus	6.86	2.68

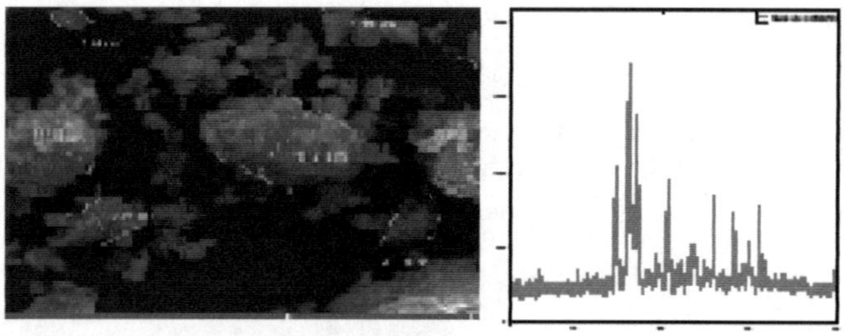

Figure 1. SEM and XRD images of Cement.

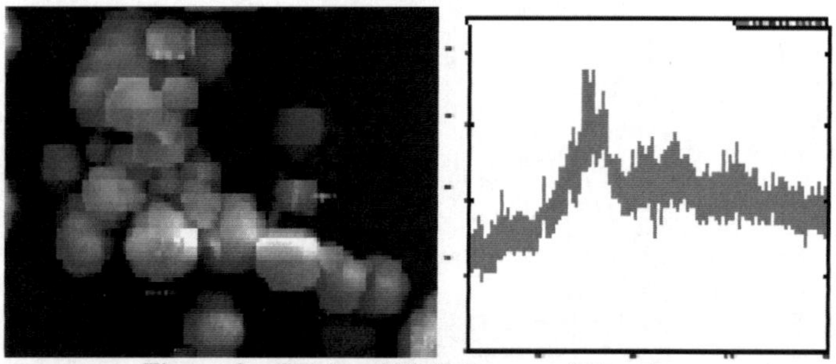

Figure 2. SEM and XRD images of fly ash.

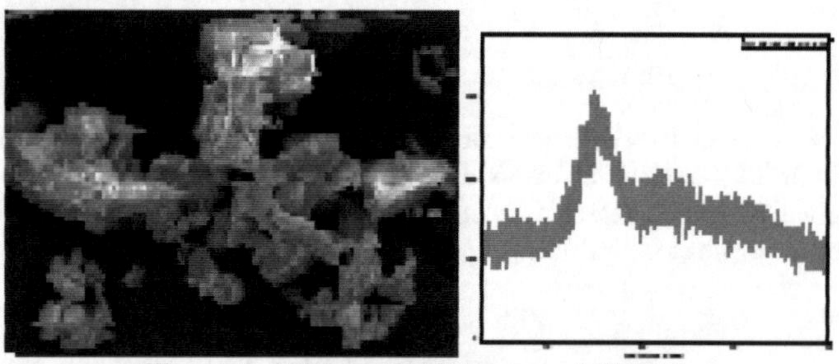

Figure 3. SEM and XRD images of GGBS.

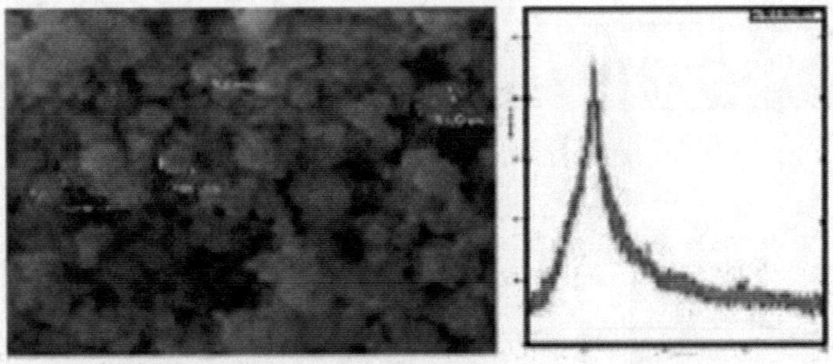

Figure 4. SEM and XRD images of Micro silica.

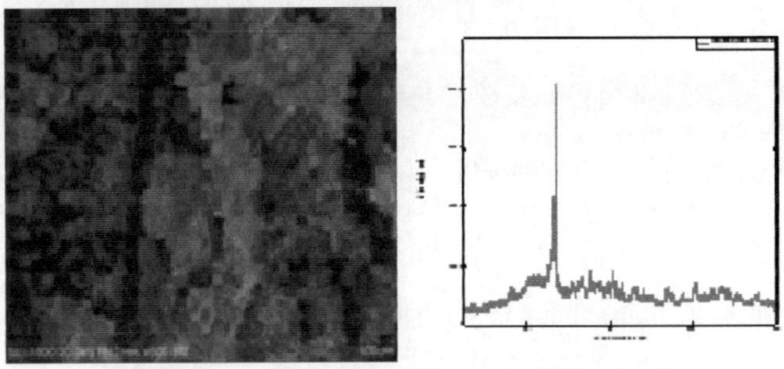

Figure 5. SEM and XRD images of Metakaoline.

2.3. Mix proportions

Total 19 mixes with 30%, 40%, 50% replacements of fly ash or ground granulated blast furnace slag and combined with micro silica or metakaoline at 10% as a replacement of cement were cast and examined on the kinetic properties and compressive strength of self compacting concrete and compared with control mix. Table 3 represents composition of self compacting concrete of various mineral admixtures.

Table 3: Mix Proportions (Kg/m3) of self compaction concrete with different replacement of Mineral admixtures

Mix No	OPC	Fly ash	GGBS	SF	MK	Fine aggregate	Coarse aggregate
Control Mix	576	---	---	---	---	879	777
PFA30%	455	121	---	---	---	879	777
PFA40%	415	161	---	---	---	879	777
PFA50%	374	202	---	---	---	879	777
PFA30% + SF10%	410	121	---	45	---	879	777
PFA40% + SF10%	370	161	---	45	---	879	777
PFA50% +SF 10%	329	202	---	45	---	879	777
PFA30% + MK10%	416	121	---	---	39	879	777
PFA40% + MK 10%	376	161	---	---	39	879	777
PFA50% +MK 10%	335	202	---	---	39	879	777
GGBS 30%	416	---	160	---	---	879	777
GGBS 40%	363	---	213	---	---	879	777
GGBS 50%	310	---	266	---	---	879	777
GGBS 30%+ SF 10%	371	---	160	45	---	879	777
GGBS 40%+ SF 10%	318	---	213	45	---	879	777

IOP Conf. Series: Materials Science and Engineering **216** (2017) 012026 doi:10.1088/1757-899X/216/1/012026

GGBS 50%+ SF 10%	265	---	266	45	---	879	777
GGBS 30%+ MK 10%	377	---	160	---	39	879	777
GGBS 40%+ MK 10%	363	---	213	---	39	879	777
GGBS 50%+ MK 10%	271	---	266	---	39	879	777

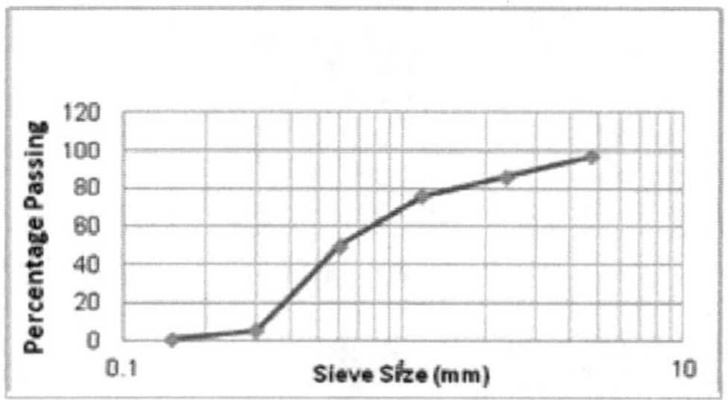

Figure 6. Particle Size distributions for manufactured sand.

3. Results and Discusions

In the present study physical, chemical, mineralogical, morphological properties of mineral admixtures their effect on fresh properties and compressive strength of self compacting concrete with different replacement are studied.

3.1. Fresh properties

The workability test on self compacting concrete such as slump flow, T50, J-Ring, L-Box and V-funnel with various replacements of cement by mineral admixtures are shown in the Table 4. The slump flows of the concrete mixtures are between 630 and 750 mm. The values of slump flow improve with the increase in flyash replacement and decreases with an increase in GGBS. An addition of micro silica in flyash and GGBS blended self compacting concrete decreases the flow. But with metakaoline, flow rate increases compared to micro silica as shown in Figure 7. The V-funnel time between 12 to 6 seconds indicating increase in the flow ability of concrete and flow time decreases with an increase in fly ash compared with other type of mineral admixtures as in the Figure 8.

The value L-box is between 0.8 to 1, it increases the passing ability of self compacting concrete with different replacement of mineral admixtures as shown in Figure 9. T_{50} slump flow test indicates high filling ability of concrete mixes. From Figure 10 it was observed that decrease J-ring value of SCC by increase in fly ash. It has been observed that increase in volume of fly ash increases in self compatibility properties; this is due to shape of flyash particles. Most of the particles are spherical in shape called as cenospheres or plerospheres, and amorphous glassy material such as silica glass and glass of other oxides as shown in the Figure 2. The spherical particles produce ball bearing action and take less water to wet the surface of fly ash, this result increases the workability of concrete. GGBS in concrete decreases the flow properties, as GGBS particles are flaky and elongated in shape and have larger surface area as shown in Figure 3. This results in higher water demand, and hence as a result concrete workability decreases. Introducing micro silica in concrete decreases the flow properties, due to micro silica are densified form of micro pellets and very larger surface area compared to other type of mineral admixture as shown in Figure 4. It consumes more water for wetting; this may decrease in workability of concrete.It has been observed that additions of metakaoline in the concrete decreases the flow properties due to high specific surface area, irregular particle shapes as shown in Figure 5, as increases in higher demands decreases the workability, it is comparatively higher workable than micro silica. It was also observed that increase in specific gravity of mineral admixtures decreases the flow property of self compacting concrete.

IOP Conf. Series: Materials Science and Engineering **216** (2017) 012026 doi:10.1088/1757-899X/216/1/012026

Table 4. Workability and compressive strength of self compacting concrete

Mix No	Slump Flow (mm)	L-Box (H_2/H1)	V Funnel (sec)	J-Ring (mm)	T 50 (Sec)	Compressive strength (MPa)				
						3 Day	7 Days	28 Days	56 Days	90 Days
Control Mix	675	0.9	7	5	4	14.35	40.62	49.13	50.16	51.18
PFA30%	725	0.97	5	3	3	9.14	29.42	37.27	60.04	70.79
PFA40%	735	1.10	3	2	2	7.05	28.00	35.13	53.71	63.04
PFA50%	750	1.20	3	1	2	4.35	24.53	31.12	45.15	58.24
PFA30% + SF10%	685	0.8	7.9	4	5	11.24	33.56	41.13	63.15	73.12
PFA40% + SF10%	690	0.85	6.3	3	6.5	8.36	31.36	39.13	58.15	66.51
PFA50% +SF 10%	710	0.85	5.8	2	6.3	6.35	28.47	35.23	50.16	61.16
PFA30% + MK10%	710	0.9	6	3	6	9.42	30.12	38.17	62.17	70.82
PFA40% + MK 10%	725	0.92	5	2	5.5	7.36	29.15	36.23	54.35	63.21
PFA50% +MK 10%	730	0.93	5	2	5	5.10	26.13	32.47	47.13	59.21
GGBS 30%	710	0.83	7	4	4	11.31	34.56	45.32	59.23	65.75
GGBS 40%	690	0.80	7	5	5	10.81	33.88	42.12	57.56	56.80
GGBS 50%	670	0.7	9	6	5.2	9.86	30.51	40.17	50.56	53.79
GGBS 30%+ SF 10%	660	0.8	10	7	7	12.56	36.15	49.12	61.13	66.38
GGBS 40%+ SF 10%	645	0.7	11.2	8	6.6	11.12	35.12	44.16	55.16	57.91
GGBS 50%+ SF 10%	630	0.6	12	10	7.2	10.12	31.12	41.28	50.16	54.12
GGBS 30%+ MK 10%	670	0.87	8	4	5.9	11.92	35.18	47.35	60.16	65.98
GGBS 40%+ MK 10%	655	0.84	9	4	6.3	11.15	34.13	43.15	53.12	57.18
GGBS 50%+ MK 10%	645	0.7	11	5	6.6	10.21	31.16	41.12	50.81	51.21

3.2 Compressive strength

Compressive strength is evaluated for all the mixes of self compacting concrete as in Table IV. GGBS has achieved maximum compressive strength at early ages, this is due more percentage of Cao (33.77%) in GGBS that react with water resulting in formation of tobermorite, leading to higher strength at earlier ages when compare to other type of mineral admixtures. The higher compressive strength at 56 and 90 days is observed in fly ash, this is due to glassy surface of alumina-silicate glass around the flash particles which may breaks when pore water turns to alkalinity at later ages, giving later age strength. Introducing of micro silica in both fly ash and GGBS series increases the strength due to higher percentage of ultra fine particles, it increases the packing density of self compacting concrete at initial stage later increase in Sio_2 reacting with CH obtained during hydration of cement.

The compressive strength after 28 days is increases, this was observed in the metakaoline series and this is due to content of Al_2o_3 which is more compared to other types of mineral admixtures, which react with portlandite which is formed during primary hydrates with cement and converts into tobermorite. This phenomenon increases the compressive strength of self compacting concrete.

The compressive strength after 28 days is increases, this was observed in the metakaoline series and this is due to content of Al_2o_3 which is more compared to other types of mineral admixtures, which react with portlandite which is formed during primary hydrates with cement and converts into tobermorite. This phenomenon increases the compressive strength of self compacting concrete.

IOP Conf. Series: Materials Science and Engineering **216** (2017) 012026 doi:10.1088/1757-899X/216/1/012026

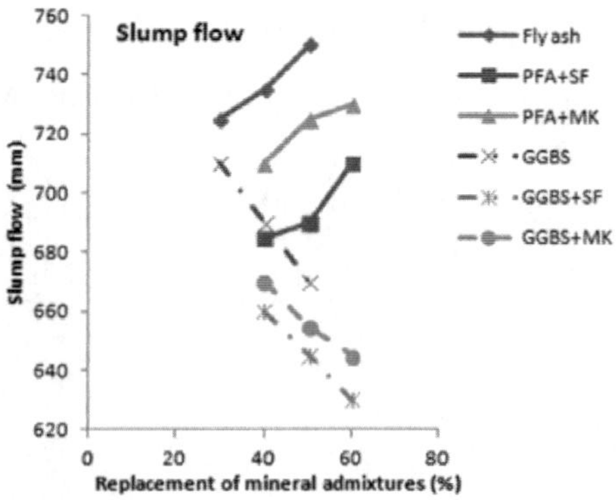

Figure 7. Slump flow test Vs Replacement of Mineral admixtures.

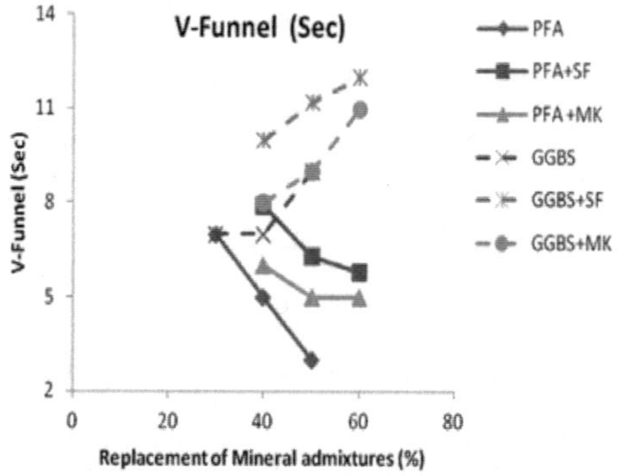

Figure 8. V-Funnel Test Vs Replacement of Mineral admixtures.

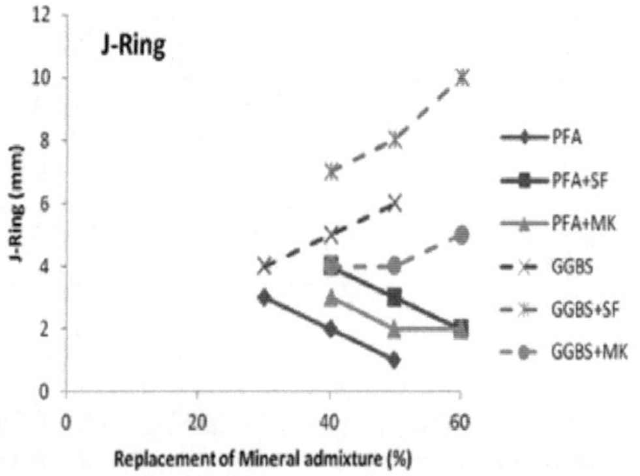

Figure 9. :L-Box Ratio Vs Replacement of Mineral admixture

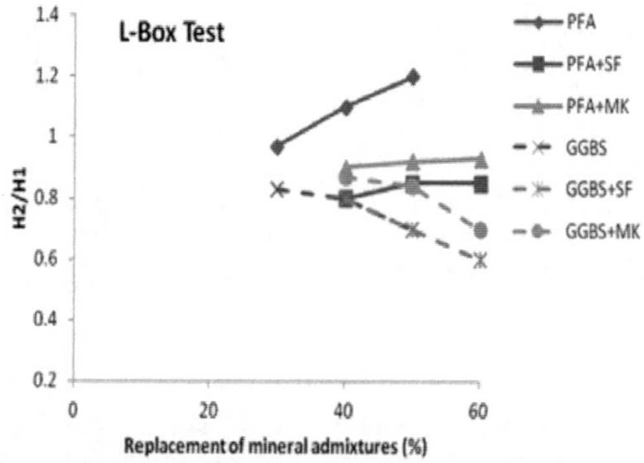

Figure 10. J-Ring Test Vs Replacement of Mineral admixtures

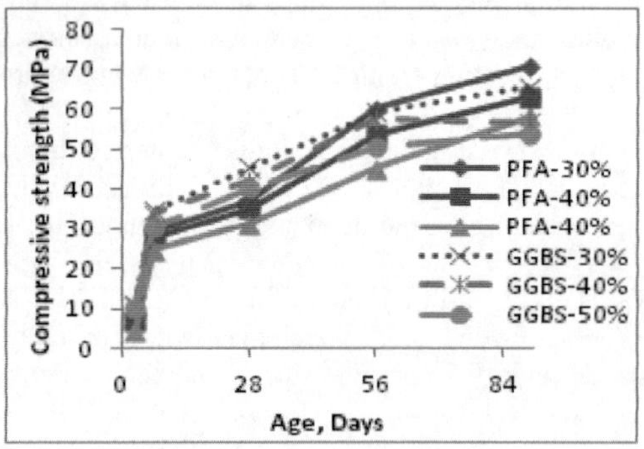

Figure 11. Compressive strength Vs Ages of with fly ash and GGBS.

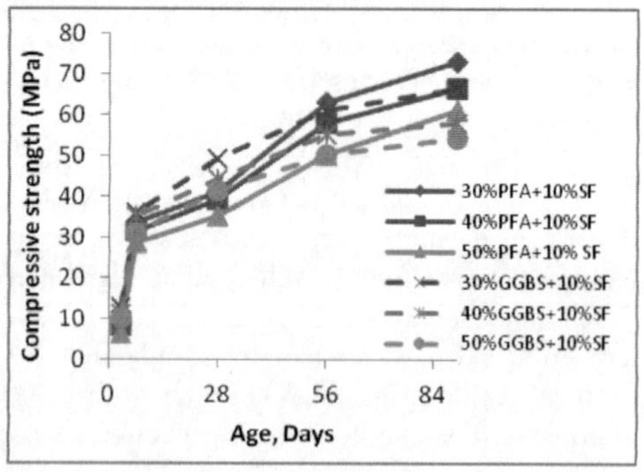

Figure 12. Compressive strength Vs Ages of SCC with fly ash and GGBS with SF.

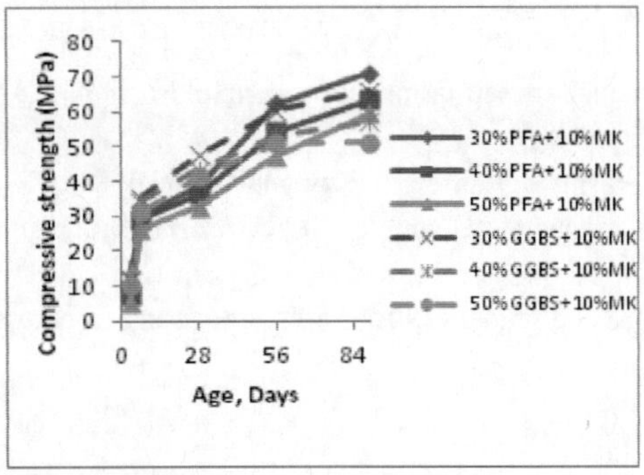

Figure 13. Compressive strength Vs Ages of SCC with fly ash and GGBS with MK.

4. Conclusion based on the results.

The following conclusions are drawn based on the above investigations

- The performance of self compatibility of fresh concrete can be improved by addition or replacement of cement by mineral admixtures.
- Use of fly ash in self compacting concrete increases in slump flow compared other series replacement of mineral admixtures.
- The workability of concrete depends on other factors such as specific gravity, shape, size, specific surface of mineral admixtures.
- Higher compressive strength has been obtained by 30% replacement of cement by ground granulated blast furnace slag and 10% of Silica fume series is 49.12 MPa at 28 day and this is equal to control mix. Also the increase in replacement levels has resulted in decrease in strength.

References

[1] P.Ramanathan, L.Baskar, P.Muthpriya and R.Venkatasubramani, "Performance of Self-Compacting Concrete Containing Different Mineral Admixtures", KSCE Journal of Civil Engineering 2013, pp 465-472.

[2] Seshadri Sekhar. T P. Sravana and Srinivasa Rao P " Some Studies on the Permeability Behaviour of Self Compacting Concrete" AKG Journal of Technology, Vol 1, No.2 pp 54-57

[3] Hamzah. A. F, Wan Ibrahim M. H, Jamaluddin N., Putra Jaya R., Zainal Abidin N. E, "Cementitious Materials usage in Self-Compacting Concrete: A Review", Proceeding of International Conferenceon Global Sustainability and Chemical Engineering (ICGSCE), Kuala Lumpur, 20th –22nd August 2014.

[4] Amrutha, Gopinatha Nayak, Mattur C. Narasimhan and S.V.Rajeeva "chloride-ion impermeability of self-compacting high-volume fly ash concrete mixes" International Journal of Civil & Environmental Engineering, Vol.11, No:04 2011, pp 29-35.

[5] Mustafa Sahmaran, Heru Ari Christianto and Ismail Ozgur Yaman, "The effect of chemical admixtures and mineral additives on the properties of self-compacting mortars", Cement & Concrete Composites,2006,pp 432–440.

IOP Conf. Series: Materials Science and Engineering **216** (2017) 012026 doi:10.1088/1757-899X/216/1/012026

[6] Paratibha Aggarwal, Rafat Siddique, Yogesh Aggarwal and Surinder M Gupta, "Self-Compacting Concrete - Procedure for Mix Design", Leonardo Electronic Journal of Practices and Technologies, Issue 12, 2008, pp 15-24.

[7] J. M. Khatib, "Performance of self-compacting concrete containing fly ash", Construction and Building Materials, 22, 2008, pp 1963–1971.

[8] Krishna Murthy.N, Narasimha Rao A.V, RamanaReddyI.V and Vijaya sekhar Reddy.M, "Mix Design Procedure for Self Compacting Concrete", *IOSR Journal of Engineering*, Volume 2, Issue 9 (September 2012), PP 33-41.

[9] EFNARC, Specifications and guidelines for Self Compacting Concrete (2005)

IOP Conf. Series: Materials Science and Engineering **216** (2017) 012027 doi:10.1088/1757-899X/216/1/012027

Hydrothermal Synthesis of Dicalcium Silicate Based Cement

N Dutta[1,2] and A Chatterjee[1]

[1]Aditya Birla Science and Technology Company Pvt. Ltd., Plot No. 1 & 1- A/1, MIDC Taloja, Taluka Panvel, Navi Mumbai, Maharashtra, India 410208

Email: nibedita.dutta@adityabirla.com

Abstract. It is imperative to develop low energy alternative binders considering the large amounts of energy consumed as well as carbon dioxide emissions involved in the manufacturing of ordinary Portland cement. This study is on the synthesis of a dicalcium silicate based binder using a low temperature hydrothermal route.The process consists of synthesizing an intermediate product consisting of a calcium silicate hydrate phase with a Ca:Si ratio of 2:1 and further thermal treatment to produce the β-Ca_2SiO_4 (C_2S) phase.Effect of various synthesis parameters like water to solid ratio, dwell time and temperature on the formation of the desired calcium silicate hydrate phase is reported along with effect of heating conditions for formation of the β-C_2S phase. Around 77.45% of β-C_2S phase was synthesized by thermal treatment of the intermediate phase at 820°C.

1. Introduction

Ordinary Portland cement clinker is composed of 50-70% of alite or Ca_3SiO_5 (C_3S) and 15-30% of belite or Ca_2SiO_4 (C_2S).The manufacturing process to produce these phases in clinker requires very high temperatures (~ 1450°C) leading to significant consumption of fuel as well as carbon dioxide emissions. As per the International Energy Agency (IEA) [1], such emissions amount to about 5% of all anthropogenic CO_2 emissions. The overall energy consumption could be reduced if a clinker rich in the belite phase, Ca_2SiO_4 (C_2S) rather than the alite phase, Ca_3SiO_5 (C_3S) is produced as C_2S forms at a lower temperature than that of C_3S [2]. Additionally, a clinker rich in belite can be produced using a lower lime saturation factor (LSF) than that of alite and hence would lead to lower carbon dioxide emissions due to decarbonation of lower amount of calcium carbonate used.

However while the alite phase is responsible for the early age strength of cement, the belite phase contributes mainly to the late age strength of cement i.e. after 28 days.Dicalcium silicate is known to exist in five polymorphic forms which are α, α'_H, α'_L, β and γ. The β polymorph of dicalcium silicate is usually present in ordinary Portland cement. γ-C_2S polymorphis regularly coordinated and non-hydraulic. The reactivity of C_2S with water is known to be dependent on a number of factors which include crystal imperfections, crystal size as well as polymorph in which it exists.Hence it is imperative to develop forms of belite phase with higher early age reactivity.

Reactive form of β-C_2S has been synthesized in the past [3,4,5] using low temperature hydrothermal synthesis route.The hydration of the synthesized β-$C_2S$was found to be complete at 28 days with curing at room temperature while β-C_2S synthesized by reaction of lime and quartz at 1400°C with stabilizing agents, at 50 days has been reported to be less than 40 % [5]. Ishida et al. [4] synthesized β-C_2S using hydrothermal route with a specific surface area of 7 m^2/g and found its hydration to be complete at 28 days with a water to solid ratio of 0.5. On the other hand, Shibata et al. [6] synthesized

pure β-C_2S by solid state reaction with a specific surface area of 1.59 to 2.10 m^2/g and found the degree of hydration to be 80 % after168 days with a water to solid ratio of 0.5 at 20°C.The higher reactivity of β-C_2S prepared using hydrothermal synthesishas been attributed to the high surface area.Unlike conventional kiln process, hydrothermal synthesis can be carried out at comparatively lowertemperatures. The intermediate products from hydrothermal reaction consist of crystalline calcium silicate hydrateswhich are further heated to produce the dicalcium silicate phase. α-C_2S hydrate $Ca_2(HSiO_4)(OH)$, Hillebrandite $Ca_2(SiO_3)(OH)_2$ and Dellaite $Ca_6(Si_2O_7)(SiO_4)(OH)_2$ are the three known forms of calcium silicate hydrates having a Ca/Si ratio of 2/1 [7].
Since researches in the past have observed a high reactivity of hydrothermally synthesized β-C_2S phase, in this work we have studied the parameters of the hydrothermal process controlling the synthesisof calcium silicate hydrate phase with a Ca:Si ratio of 2:1. Further, we have studied the effect of the heating conditions on formation of the β-C_2S phase from the intermediate hydrate phase.

2. Material and methods

The starting materials used for hydrothermal synthesis were $Ca(OH)_2$ (Sigma Aldrich), SiO_2 (Evonik) and water. The molar ratio of $Ca(OH)_2$ to SiO_2 was maintained at 2:1. The synthesis was carried out in a 500 ml autoclave (Amar Equipments, India) under saturated steam pressure at different temperatures, water to solid ratios and dwell times.After autoclaving, the intermediate products from hydrothermal synthesis were vacuum filtered and oven dried at 105°C.
The phase composition of the hydrothermally synthesized as well as the final calcined samples were investigated using powder X-ray diffraction (XRD). The data were collected using Bruker's D8advance diffractometer in a θ-θ configuration employing CuK_α radiation (λ = 1.54) with a fixed divergence slit 0.298° and a rotating sample stage at 40kV and 40 mA. The samples were scanned between 5 and 80° with a counting time of 0.1 second/step. Selected samples were analyzed by the Rietveld method using Bruker's TOPASversion 3software.
Surface areas of the thermally treated sample at 820°C were determined by Brunauer–Emmett–Teller (BET) analysis using Quantchrome Nova 1000eanalyzer with Nitrogen gas as the adsorbate. Powder sample of 0.5 gm was employed and degassed for 2 hours at 80 ± 50°C.
The hydrothermally synthesized sample with the desired intermediate phase was heated at different temperaturesin a high temperature Carbolite BLF furnace in platinum crucibles to produce the desired C_2S phases. The samples wereplaced in platinum crucibles and put in the preheated furnace at different temperatures (520°C, 570°C, 670°C, 720°C, 770°C, 820°C) and held there for 1 hour and taken out at that temperature itself and air quenched.

3. Results and Discussion

3.1 Effect of parameters of hydrothermal synthesis

Different parameters like dwell time, water to solid ratio and temperature of the hydrothermal synthesis process would have an effect on the product formed. Hence, we have studied the effect of each of these parameters on the formation of the desired calcium silicate hydrate phase with a Ca:Si ratio of 2:1.
3.1.1Effect of dwell time. Hydrothermal reactions were initially started with a water: solid ratio of 5:1 at 200°C and 400 rpm for 1 hour and 5 hours respectively to observe the effect of dwell time on the formation of the intermediates. As shown in Figure 1, longer duration was observed to increase the formation of C-S-H type intermediates. The intensities of the peaks of calcium hydroxide and calcium carbonate were observed to be considerably reduced on the increase of dwell time to 5 hours. At a dwell time of 1 hour, a hump at around 2Θ of 30°was observed which could be attributed to the unreacted amorphous silica present in the product. Also it was observed that calcium silicate hydrate phases with Ca:Si ratio of 2:1 were formed which included hillebrandite (ICSD 80127)and dellaite (ICSD 20148).Reinhardbraunsite,$Ca_5(SiO_4)_2(OH)_2$,(ICSD 31275) and jaffeite, $Ca_6(Si_2O_7)(OH)_6$(ICSD 39725) were formed along with the other dicalcium silicate hydrates and had a Ca:Si ratio of 2.5:1 and 3:1 respectively.

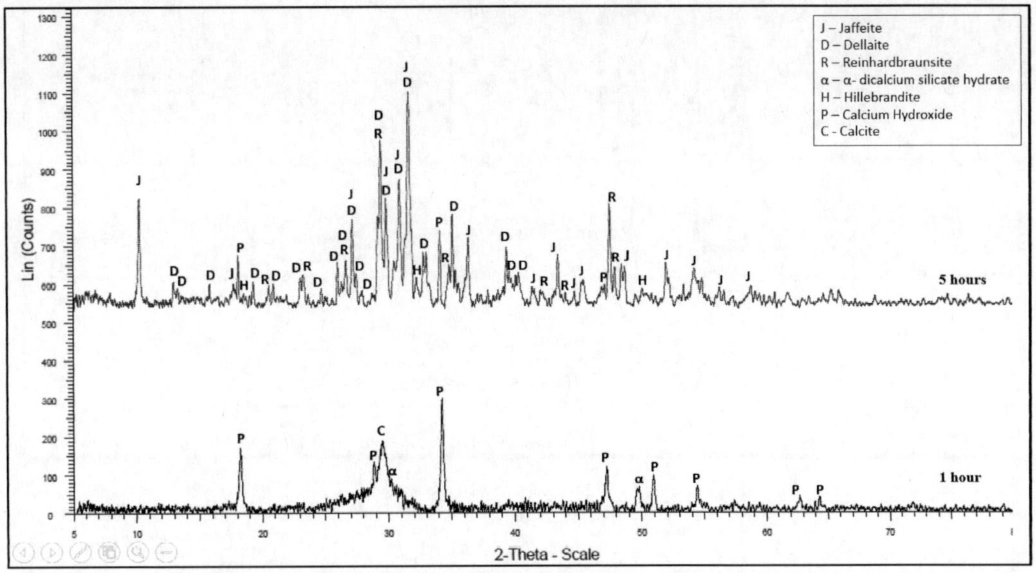

Figure 1. XRD patterns of hydrothermal synthesis product formed with dwell times of 1 hour and 5 hours

3.1.2 Effect of water to solid ratio. Hydrothermal reactions were carried out at water to solid ratios of 2:1, 3:1 and 5:1 at a temperature of 250°C for 5 hours at 400 RPM. Figure 2A shows the XRD patterns of the products formed from the hydrothermal synthesis carried out with these three water to solid ratios. At water to solid ratios of 2:1 and 3:1, peaks of unreacted calcium hydroxide along with calcium carbonate were observed. At a water to solid ratio of 2:1, calcium silicate hydrates with Ca:Si ratio less than and equal to 1 were observed to be formed, which included rosenhahnite, $Ca_3(Si_3O_8(OH)_2)$ (ICSD 100074), xonotlite, $Ca_6Si_6O_{17}(OH)_2$ (ICSD 20702) and truscottite, $Ca_{14}Si_{24}O_{58}(OH)_8.2H_2O$ (ICSD 5126). With an increase in the water to solid ratio to 3:1, peaks of calcium silicate hydrates with Ca:Si ratio above 1 were also observed to be formed, which included killalaite, $Ca_{3.2}(H_{0.6}Si_2O_7)(OH)$, (ICSD 200124). With further increase in the water to solid ratio to 5:1, peaks of calcium silicate hydrates with Ca:Si ratio of and above 2 were observed, which included jaffeite, dellaite and reinhardbraunsite. Hence, increasing water to solid ratio was observed to favor the formation of calcium silicate hydrates with a Ca:Si ratio of 2:1.

3.1.3 Effect of temperature. In order to observe whether the desired calcium silicate hydrate phases formed at lower temperatures or not, hydrothermal synthesis was carried out at 150°C with a water to solid ratio of 5:1 at 400 RPM for 5 hours. Figure 2B shows the XRD patterns of the products obtained from the hydrothermal synthesis carried out at 150°C and 250°C. Peaks of unreacted calcium hydroxide and calcite were mainly observed in the product synthesized at 150°C, indicating that the intial reactants did not take part in the reaction at that temperature.

3.2 Effect of thermal treatment on β-C_2S formation

Since the hydrothermal synthesis at 250°C for 5 hours with a stirring rate of 400 RPM was observed to favor the formation of a product with desired calcium silicate hydrate phases with Ca:Si ratio of 2:1, this product was used for further thermal treatment to synthesize C_2S. The product was heated at different temperatures ranging from 520°C to 820°C to study the effect of thermal treatment on the formation of the desired β-C_2S phase. The phases observed in the thermally treated samples have been presented in Table 1. The XRD patterns of the thermally treated product at 670°C, 720°C, 770°C and 820°C have been shown in Figure 3A.

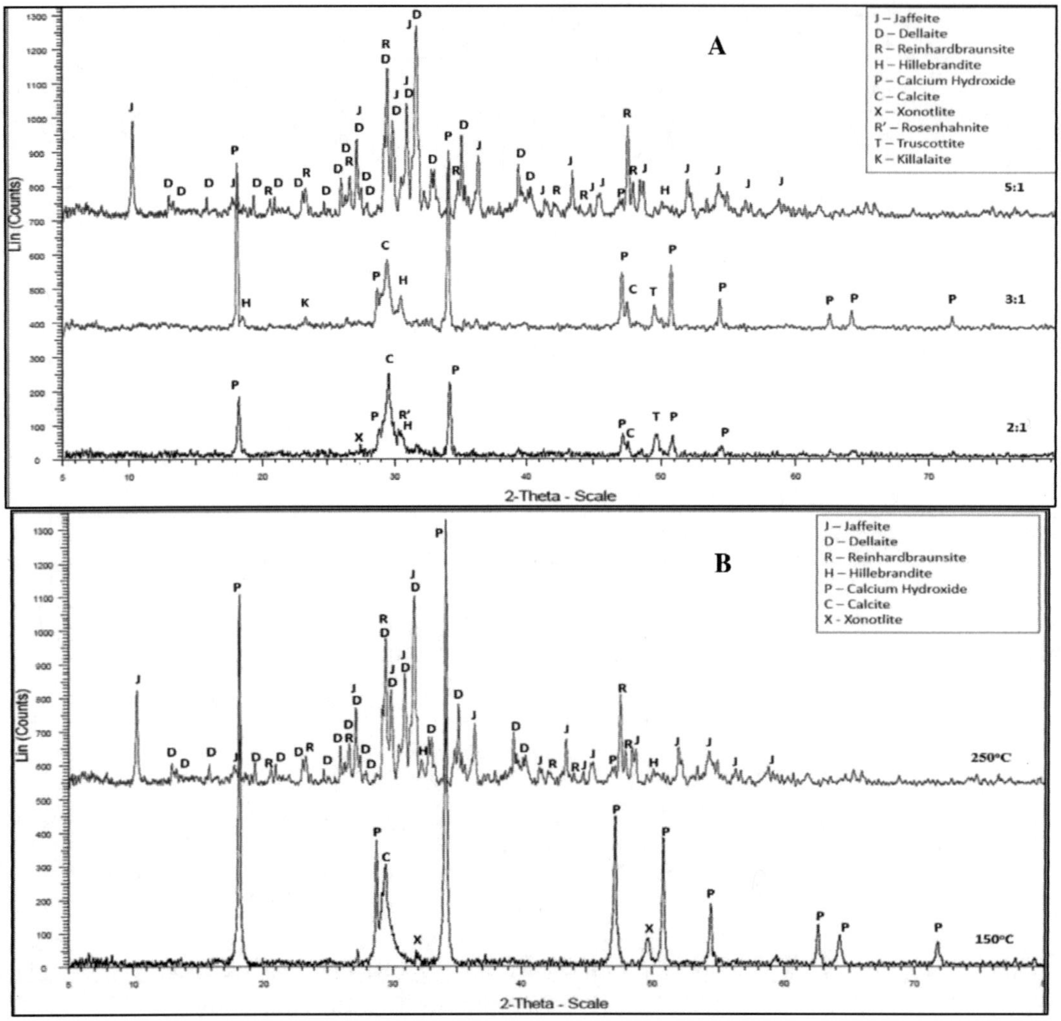

Figure 2. XRD patterns of hydrothermal synthesis product formed at– A) water to solid ratios of 2:1, 3:1and 5:1; B) temperatures of 150°C and 250°C

Table 1. Phases present in thermally treated samples at different temperatures.

Phases	570°C	620°C	670°C	720°C	770°C	820°C
Dellaite	✓	✓	✓	x	x	x
Reinhardbraunsite	✓	✓	x	x	x	x
$Ca_8Si_5O_{18}$	✓	✓	✓	✓	x	x
Calcite	✓	✓	x	x	x	x
Calcium Hydroxide	x	x	✓	✓	✓	✓
β -C_2S	✓	✓	✓	✓	✓	✓
α_H – C_2S	✓	✓	✓	✓	✓	x
γ-C_2S	✓	✓	✓	✓	✓	✓

As shown in Table 1, jaffeite and hillebrandite present in the hydrothermal synthesized sample are seen to decompose below a temperature of 570°C leading to the formation of β-C_2S, α_H'– C_2S, γ-C_2S and $Ca_8Si_5O_{18}$. On heating the sample to a temperature of 670°C, peaks of dellaite were observed which indicated that the phase has not been completely decomposed at that temperature. Peaks of calcite were not observed while peaks of calcium hydroxide appeared, indicating the decarbonation of calcium carbonate at a temperature of 670°C. Peaks of β-C_2S (ICSD 280995) and γ-C_2S (ICSD 81095) were detected along with peaks of calcium silicate, $Ca_8Si_5O_{18}$. Peaks of this $Ca_8Si_5O_{18}$ were observed to have disappeared on heating the sample at a temperature of 770°C. This possibly could have been due to the reaction between this phase and the excess calcium oxide present to form C_2S at a temperature of 770°C. The calcium silicate hydrate phases present in the hydrothermal synthesis product with a Ca: Si ratio above 2:1, jaffeite and reinhardbraunsite, were observed to be decomposed to the C_2S phases. Reaction of the decomposed product with any excess silica present in the system could have led to this formation of the C_2S phase at the heated temperature. Peaks of $\alpha_H' - C_2S$ (ICSD 81097) known to be a reactive form of belite, were detected on heating the sample at 670°C, 720°C and 770°C. Intensities of peaks of β-C_2S were observed to be maximum for the product obtained by heating the sample at 820°C. Phases present in the thermally treated products at 770°C and 820°C were quantified using Rietveld analysis as presented in Table 2. Figure 3B shows the phase quantification done for the thermally treated sample at 820°C. As evident from the phase quantification, $\alpha_H' - C_2S$ phase is observed to transform to the β-C_2S at 820°C.The BET surface area of the thermally treated sample at 820°C was found to be 8 m^2/g. Due to the large surface area obtained using the hydrothermal synthesis route, it has been observed in the past that the reactivity of β-C_2S synthesized is much higher as compared to that obtained by solid state reaction between CaO and SiO_2 [4].

Table 2. Rietveld phase quantification of thermally treated samples at 770°C and 820°C

Temperature (°C)	Phase percentage (wt %)			
	β-C_2S	γ-C_2S	$\alpha_H' - C_2S$	Calcium Hydroxide
770	49.97	24.25	22.55	3.24
820	77.45	19.67	-	2.87

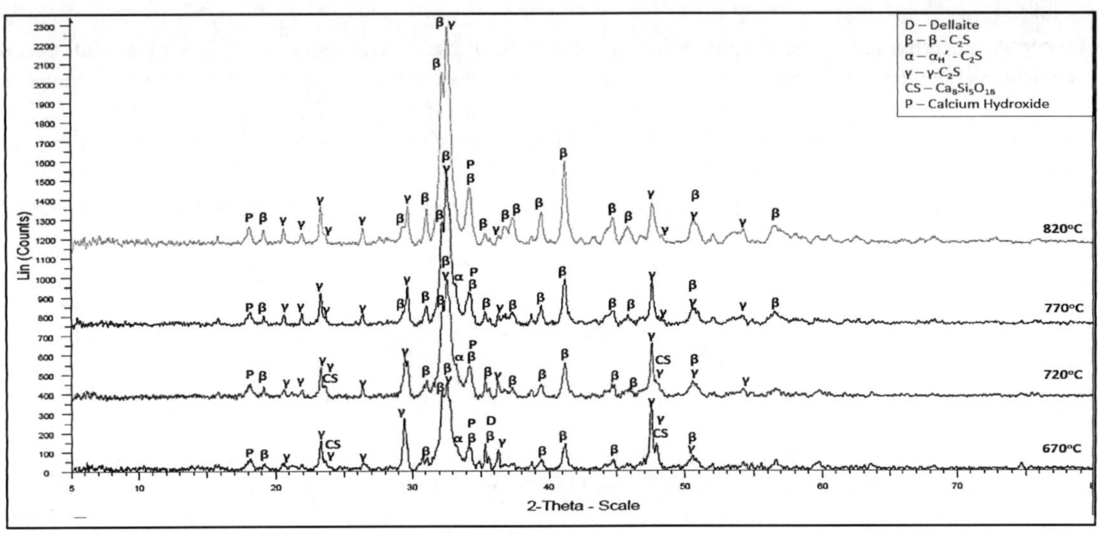

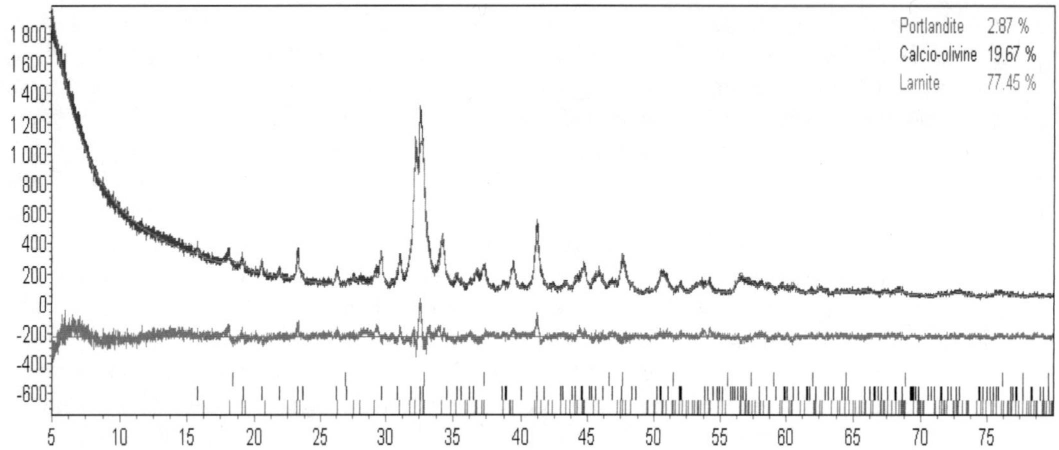

Figure 3. XRD patterns of products obtained after thermal treatment of hydrothermally synthesized sample –A) at 670°C, 720°C, 770°C and 820°C and B) at 820°C with Rietveld phase quantification.

4. Conclusions

In this work, we have utilizedthe hydrothermal process to synthesize calcium silicate hydrate phases with a Ca: Si ratio of 2:1. The synthesis reaction has been found to be favored at a water to solid ratioof 5/1 at a temperature of 250°C and a dwell time of 5 hours with a stirring rate of 400 RPM. At lower temperatures and dwell times, degree of reaction has been found to be low with unreacted intial components present in the product. At lower water to solid ratios, formation of calcium silicate hydrate phases with lower Ca: Si ratio is favored.

The desired β-C_2S phase is formed by heating the hydrothermally synthesized sample. At and above 820°C, the product comprised of77.45% ofβ -C_2S along with 19.67 % of γ-C_2S and 2.87 % of calcium hydroxide. Another known reactive phase, α_H' – C_2S was observed to be formed on heating the sample at 670°C, 720°C and 770°C.The BET surface area of the thermally treated sample at 820°C was found to be 8 m^2/g.

Hence,a hydrothermal assisted calcination route can be used to synthesize a β-C_2S phase with a high specific surface area. This phasecan be utilized by blending with ordinary Portland cement to render a higher early age strength contribution than what is contributed by the β-C_2S phase in ordinary Portland cement. This can be further leveraged by reducing the alite/belite ratio in cement and therefore reducing the overall energy consumption as well as the carbon dioxide emissions.Waste materials rich in lime and silica such as lime sludge, fly ash etc. can be explored further to synthesize this active belite phase.

References

[1] Cement technology roadmap 2009- carbon emissions reductions up to 2050, international energy agency (http://www.iea.org/publications/freepublications/publication/name,3861,en.html)

[2] Popescu CD, Muntean M and Sharp JH.Industrial trial production of low energy belite cement.2003 *Cement Concrete Comp.***25** [7], pp 689-93

[3] Ishida H, Mabuchi K, Sasaki K and Mitsuda T. Low temperature synthesis of β-Ca_2SiO_4 from Hillebrandite. 1992 *J. Am. Ceram. Soc.***75** [9], pp 2427 - 32

[4] Ishida H, Sasaki K and Mitsuda T. Highly reactive β-dicalcium silicate: I, Hydration behavior at room temperature. 1992 *J. Am. Ceram. Soc.***75** [2], pp 353-58

[5] Tantawy MA, Shatat MR, El-Roudi AM, Taher MA and Abd-El-Hamed M. Low temperature synthesis of belite cement based on silica fume and lime. 2014 *International scholarly research notices*Article ID 873215

[6] Shibata S, Ito S and Fujii N. Preparation and hydration of β-C_2S without stabilizer. 1984 *Cem.Concr. Res.***14**, pp 323–28

[7] Ishida H, Yamazaki S, Sasaki K, Okada Y and Mitsuda T. α-dicalcium silicate hydrate: preparation, decomposed phase and its hydration. 1993 *J. Am. Ceram. Soc.***76**, pp 1707-12

IOP Conf. Series: Materials Science and Engineering **216** (2017) 012028 doi:10.1088/1757-899X/216/1/012028

Effects of SBS Content on the Performance of Modified Asphalt

Chuanyi Zhuang[1], Ning Li[1], Wei Zhao[2] and Chuanfeng Cai[2]

1 School of Transportation and Civil Engineering, Shandong Jiaotong University. No.5001 Haitang Rd, University Science and Technology Park,China
2 Shandong Taihe Highway Engineering Company, No.36, Dongtai Road, Huantai New District, China

Email: 289580166@qq.com; 2512835041@qq.com; Skyangel001@126.com; 2895789724@qq.com;

Abstract. The effects of different dosage SBS（Styrene-butadiene-styrene）modifiers on the properties of asphalt were compared and analyzed from the temperature sensitivity, high temperature performance and low temperature performance of modified asphalt. The results show that with the increase of the content of SBS modifier, the softening point and kinematic viscosity of SBS modified asphalt are increased and the high temperature performance is improved. The low temperature ductility is improved; the penetration of modified asphalt is reduced, the temperature sensitivity of the asphalt is reduced. However, when the content of modifier is more than 4%, the penetration index of SBS modified asphalt decreases linearly with the increase of the content of modifier, and the temperature sensitivity increases. The engineering application should be determined according to the specific technical requirements.

1. Introduction

SBS modified asphalt is taken matrix asphalt as raw material, adding a certain proportion of SBS modifier, through shearing, stirring and other methods to make SBS evenly dispersed in the asphalt, at the same time, adding a certain proportion of the exclusive stabilizer to blend Materials, the use of SBS good physical properties of the modified asphalt treatment[1]. At present, SBS modified asphalt processing methods generally were used mechanical blending method, in the factory for mixing or site modification in China. The plant premix production through the swelling, shear, development of three stages, the SBS modifier by strong shear, dispersed into the matrix asphalt to achieve the modified effect, and then stored in a storage tank for using.

In this paper, by comparing the performance test and technical characteristics of modified asphalt with different dosage, the difference of modified asphalt and matrix asphalt was compared, and the change rule of SBS modified asphalt was explored, and reasonable SBS modifier Mixed dosage.

2. Raw materials and testing methods

2.1 Raw materials

90-A asphalt and linear SBS was used, the dosage of the agent was 4%, 4.5%, 5%, 5.5% and 6% respectively. FLUKO-FA25 high-speed shear dispersion emulsifier was used [2].

2.2 Sample preparation method

The process of SBS modified asphalt was shown in Figure 1, 90-A asphalt in the 135 °C oven was heated to a free flow state, adding SBS modifier and high-speed stirring, placed in the oven at 180 °C development expansion, then shearing machine under high-speed for 40 minutes, and then in the oven at 160 °C in the mixer for 2 hours of continuous development, and then remove the sample to test.

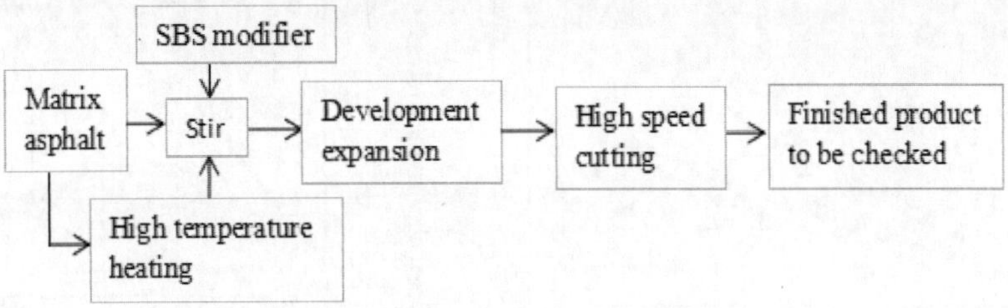

Figure 1. SBS modified asphalt process flow chart

3. Performance Index and Analysis of SBS Modified

Different proportion of SBS modifier to asphalt was produced in accordance with JTG E20-2011[3] for testing. The basic indicators of matrix asphalt are to meet the technical requirements of JTG F40-2004 [4] in the road asphalt 90, the matrix asphalt and SBS modified asphalt index as shown in Table 1.

Table 1. Performance index of SBS modified asphalt with different content.

Content	Penetration/0.1mm			PI	Softening Point /℃	Kinematic viscosity(135℃) /Pa·s	Ductility(5℃, 5cm/min) /cm	Ductility after Ageing (5℃, 5cm/min) /cm
	15℃	25℃	30℃					
0.0%	22.6	86.6	118.4	-1.36	45.1	0.32	0	0
4.0%	22.7	59.1	86.6	0.14	52.1	0.82	25	15
4.5%	21.7	58.3	85.6	-0.03	57.6	0.86	35	14
5.0%	20.2	57.0	82.2	-0.21	59.6	1.06	41	25
5.5%	19.5	52.7	80.2	-0.21	63.5	1.22	46	26
6.5%	18.4	50.9	79.7	-0.43	78.6	1.52	47	28

3.1 Temperature sensitivity

SBS as a rubber elastomer, itself and the asphalt material has different physical properties, which would cause significant changes in the nature of asphalt. The temperature sensitivity of asphalt is closely related to the paving and performance of asphalt pavements, which is the core indicator of the performance of asphalt, so it is very important to evaluate the temperature sensitivity [5].

It can be seen from Figure 2 that the penetration of SBS modified asphalt decreases with the increase of the modifier content at the same temperature, and the slope of the penetration-temperature relationship curve decreases, indicating that the asphalt temperature sensitivity reduce. It can be seen that the addition of modifier increases the penetration index of asphalt and reduces the temperature sensitivity of asphalt from Figure 3. When the amount of modifier is more than 4%, with the increase of the content of modifier, the penetration index of SBS modified asphalt decreases linearly and the temperature sensitivity increases, indicating that the amount of modifier has the best dosage .In order to meet the requirements of SBS modified asphalt in the technical specification for construction of highway asphalt pavements, lower modifier dosage was recommended so as to reduce costs.

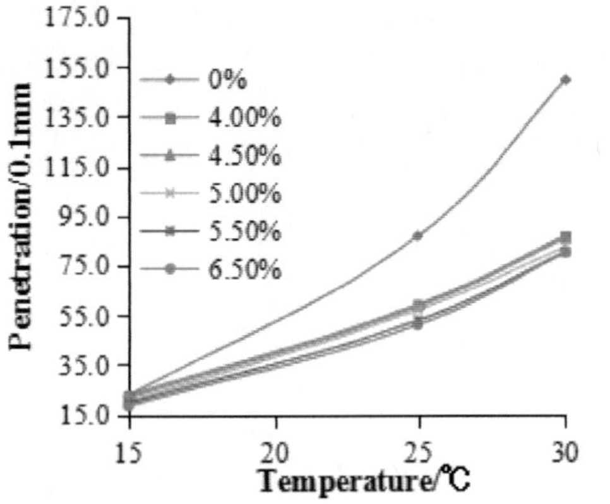

Figure 2. The Relationship between penetration degree and temperature of SBS modified asphalt

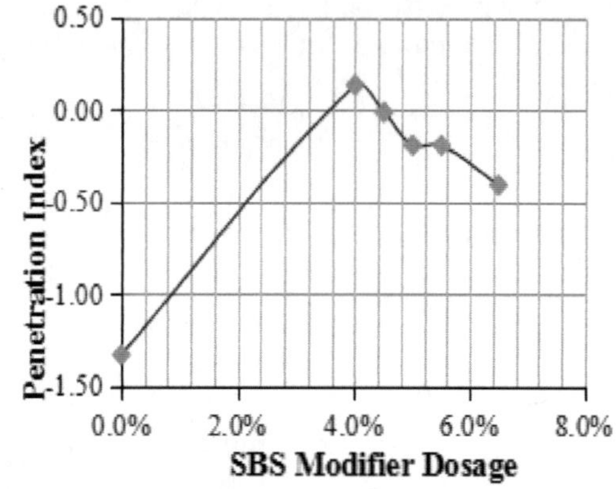

Figure 3. The Relationship between penetration index and modifier content of SBS modified asphalt

3.2 High temperature performance

The high temperature performance of asphalt pavement road performance evaluation is an important indicator [6].The Modified asphalt is a viscoelastic material, the deformation of the performance of elasticity, delay elasticity and viscous deformation, the former two in the external force will immediately after the recovery or gradually complete recovery, while the latter cannot be restored and become permanent deformation. Asphalt high temperature performance instability of the road, reflected in the summer hot season prone to rut, push and other permanent deformation, not only affect the driving comfort, but also a threat to traffic safety [7]. The addition of the modifier increases the elasticity and delayed elastic deformation of the asphalt, reducing the viscosity deformation. Modified asphalt standard is usually used to soften the point and viscosity of these two indicators to reflect the use of asphalt in high temperature conditions.

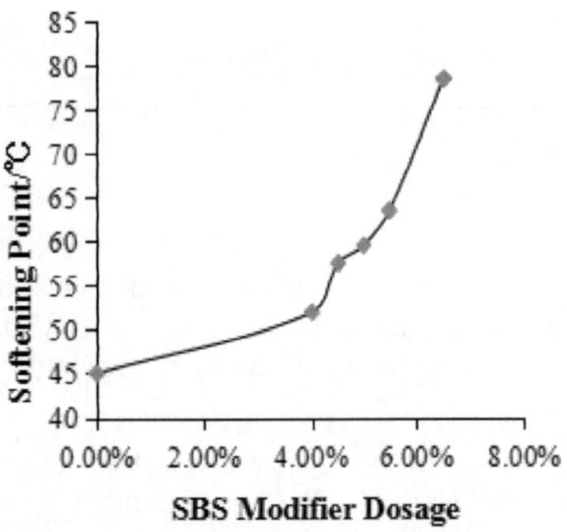

Figure 4. The relationship between the softening point of SBS modified asphalt and the content of modifier.

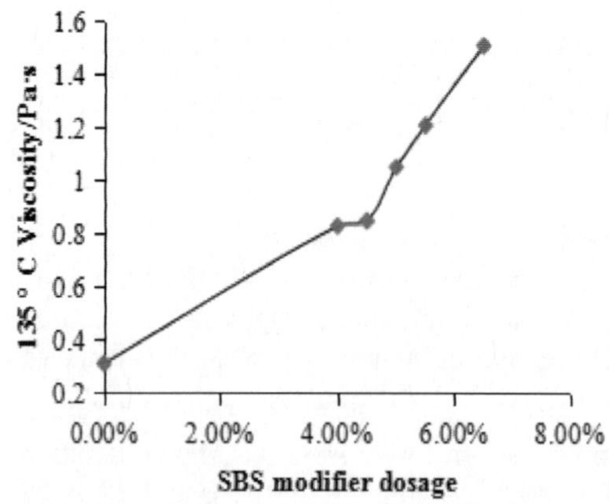

Figure 5. The Relationship between the Viscosity of SBS modified asphalt and the content of modifier.

From Figure 4 and 5 can be seen the viscosity increases gradually and does not exceed the viscosity of

SBS modified asphalt in the technical specification of asphalt pavement construction not more than 3.0 Pa·s limit. The addition of modifier can improve the high temperature performance of asphalt and meet the requirements of on-site pumping and mixing[8].The study shows that the incorporation of SBS modifier will greatly improve the high temperature performance of asphalt, thus improving the high temperature stability of asphalt mixture and the anti-rutting ability of asphalt pavement.

3.3 Low temperature performance and ductility

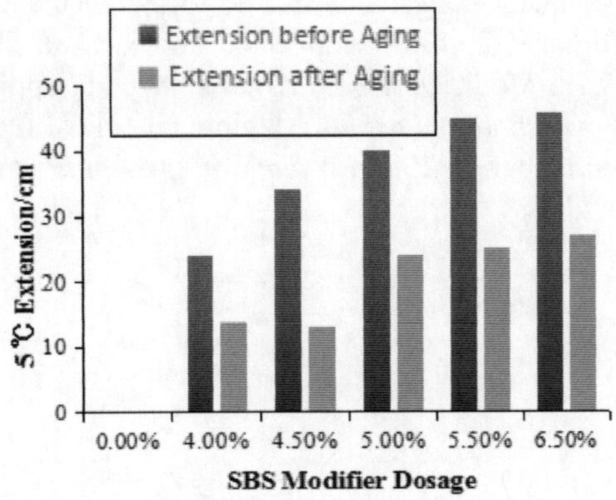

Figure 6. The Relationship between the ductility of 5 °C and the modifier content before and after aging of SBS modified asphalt

Asphalt pavement low temperature cracking resistance depends mainly on the low temperature tensile deformation of asphalt; China developed the polymer modified asphalt technical requirements [9]. The influence of the addition of modifier on the low temperature and anti-aging properties of asphalt was studied by measuring the ductility of SBS modified asphalt before and after aging [10].
With the increase of the content of the modifier, the value of 5 °C ductility of the asphalt before and after aging was gradually increased from Figure 6. The addition of the agent improves the low temperature ductility of the asphalt, the low temperature tensile properties of the asphalt was enhanced, improves the low temperature performance and anti aging function of the asphalt.The results show that the DSR test can accurately describe the effect of aging on the performance of modified asphalt.

4. Conclusions

(1)The addition of SBS modifier can improve the temperature sensitivity of asphalt, but the optimum dosage should be selected according to technical requirements and engineering conditions.
(2)SBS modifier can not only improve the high temperature performance of asphalt, but also meet the requirements of field pumping and mixing.
(3)The addition of SBS modifier has improved the low temperature ductility of the asphalt, enhanced the low temperature tensile properties of the asphalt, and improved the low temperature performance and anti-aging properties of the asphalt.

5. Acknowledgements

The paper was financial supported by Ministry of Transport Applied Basic Research Project (No. 2015 319 817 110) and Natural Science Foundation of Shandong (No. ZR2014EEQ035).

6. References

[1] Jie Lin 2011 *Study on Performance Evaluation and Quality Control of SBS Modified* (China: Changan University)
[2] Xiaoming Huang 2002 *Asphalt and asphalt mixture* (China: Southeast University Press)
[3] Highway Science Research Institute 2011 *Test Code for Asphalt and Asphalt Mixture for Highway Engineering* (China: Ministry of Transport of the People 's Republic of China)

[4] Ministry of Communications Highway Science Research Institute 2005 *Technical specification for highway asphalt pavement construction* (China: China Communications Press)
[5] Lizhi Wang, Jianming Wei and Yuzhen Zhang 2008 *Analysis and Discussion on Temperature Sensitivity Index of Road Asphalt* (China: East China University of Petroleum)
[6] Shuangyan Bao2007 *SBS Modified Asphalt Technical Performance and quality Evaluation Index* (China: Changsha University of Science and Technology)
[7] Liang Liu 2005 *Experimental Study on Evaluation Index of High Temperature Performance of Asphalt Binder* (China: Changsha University of Science and Technology)
[8] Dashang Peng, Weijian Guo, Zhongda Chen and Jie Lin 2013 Study on High Temperature Performance of SBS Modified *Shanxi Architecture* (12):pp 96-98
[9] Jun Yang 2007 *Polymer modified asphalt* (China: Chemical Industry Press)
[10] Linjiang Yang and Jingxuan Li 2001 *Production and Application of SBS Modified* (China: China Communications Press)

Performance of Granite Asphalt Mixture Modified by Silane Coupling Agent

Zhihang Liu, Xia Li, Li Wang and Rongling Kang

Weihai Highway Survey and Design Institute, No.20 Dongshan Rd, Huancui District of Weihai, China.

Email: 1158230266@qq.com; 1637955639@qq.com; designingwang@163.com; kangrongling@163.com;

Abstract. In order to improve pavement performance of granite asphalt mixture, the surface of granite mineral powder was organic modified by silane coupling agent. The water stability and high temperature stability of the asphalt mixture were analyzed by Marshall tests, immersion Marshall test, freeze-thaw splitting test and rutting test. The results show that the mixing amount of silane coupling agent in the range from 0.5% to 2.5% can significantly improve the high temperature stability and water stability of the asphalt mixture. Taking into account the performance and economic factors, 2.0% silane coupling agent on the surface of granite filler was recommended.

1. Introduction

Weihai is located in Shandong Province which is riched in granite and diorite. Due to the role of the ocean, the weather of rain or snow happened more than other areas in Shandong Province, the water stability requirements of the pavement are higher. Choose local granite mineral powder as filler of asphalt mixture, it is difficult to guarantee the water stability of the granite mineral powder as the filler of asphalt mixture, because of granite as a typical acidic material, and asphalt adhesion is poor. Experts and scholars at home and abroad in recent years, also the application of acidic stone in asphalt mixture was carried out by research, such as adding antistripping agent, the use of hydrated lime to replace some or all of the mineral powder, improved results have been achieved, but still happened a serious water damage of asphalt pavement [1, 2]. The improvement of the performance of the acid stone asphalt mixture is still an important issue. In this paper, the principle of modification of inorganic filler in the field of polymer materials. Using the modification mechanism of silane coupling agent [3], which make the inorganic surface organic matter of granite, thereby improving the compatibility with the asphalt organic surface with the asphalt adhesion [4].For granite stone road construction in the future of our country provides practical guidance to the use of the mass.

2. Composition of asphalt mixture

70-A road asphalt was selected, asphalt technology test results were shown in Table 1. Aggregate are made of granite gravel in Weihai, fine aggregate selection of natural sand and granite stone, filler selection of granite mineral powder. The apparent density of the filler was 2.73 $g \cdot cm^{-3}$, the water content was 0.1% and the hydrophilic coefficient was 0.7. According to the relevant test procedures to test the physical and mechanical indicators of aggregate in Table 2, its performance are to meet the technical requirements of the specification [5].Silane coupling agent KH-550 as a modifier, it is amino functional silane, was alkaline, the appearance of colorless or yellowish transparent liquid, versatile, soluble in organic solvents.

Table 1. Technical indexes of 70-A road asphalt.

Types of asphalt	Penetration of 25 degrees centigrade /(0.1mm)	Penetration index	Softening Point /^{0}C	Ductility of 5 degrees centigrade/cm	Density of 25 degrees centigrade ($g\cdot cm^{-3}$)
Matrix asphalt	71	-0.86	47.0	>100	1.038
Skills requirement	60～80	-1.50～+1.00	≥46	≥100	Recorded records

Table 2. Physical and mechanical indexes of coarse aggregate.

Crushed stone value /%	Los Angeles abrasion loss /%	Polished stone value /BPN	Sturdiness/ %	Apparent density ($g\cdot cm^{-3}$)
17.9	18.6	48	3.6	2.73

3. Molecular structure and mechanism of silane coupling agent

Silane coupling agent is a class of molecules containing both organic functional groups (Y group) and hydrolyzable inorganic functional groups (X group). The typical molecular structure is shown in Figure 1.

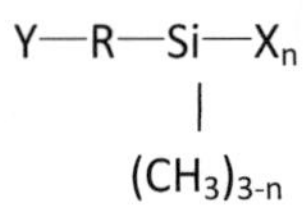

Figure 1. Schematic diagram of typical silence coupling agent.

Silane coupling agent modification of inorganic fillers is using the theory of chemical synthesis, the silane coupling agent forms a chemical bond with the active group on the surface of the material, the X group hydrolyzed into silanol alcohol with water layer attached to the surface of the granite mineral powder, Silanol alcohol forms hydrogen bonds with hydroxyl groups on the surface of granite mineral powder, but the silicon-hydroxy would condense to lose water and form covalent bond of dry curing, so that the surface of the granite mineral powder to form a coupling layer and its surface organic modification [6].Coupling layer block water immersion asphalt and stone interface, thus raising the granite water damage resistance of asphalt mixture.

4. Granite mineral powder surface modification method

Using the pretreatment of the modified method, not only the amount of coupling agent is small, and the silane coupling agent is distributed more uniformly on the surface of the material. In this paper, the quality of silane coupling agent is 0.5%, 1.0%, 1.5%, 2.0% and 2.5% of the mineral powder. Modified granite mineral powder by wet process，choose water and ethanol as solvent, the effect of the coupling agent was better when the ratio of silane coupling agent, ethanol and water was 5:45:50, through comparative tests. According to the corresponding proportion, adding the silane coupling agent to the solution of water and ethanol, stirred uniformly, the mixed solution is stirred at room temperature for 20min.Through comparing the test, taking into account the effect of coupling agent modification, economy, practical operation, the optimum operating conditions are that the rotational speed of cement glue sand mixer is 140r/min, total stirring time was 1.5h, and every 300 seconds pause 20 seconds, after mixing the mixture was dried in a 90 ^{0}C oven for 3 hours. To provide a reference for modification process of silane coupling agent modify inorganic fillers in the future. The granite mineral powder was introduced into the mixing pan, and the machine was started. After stirring for 5 seconds, the diluted silane coupling agent solution was added slowly while stirring, coupling agent in 20s ~ 30s plus finished, after completion of the stirring, the mixture was dried in a 90 ^{0}C for 3 hours [7].

5. Road performance test of modified granite mineral powder asphalt mixture

5.1 Mix proportion design of modified granite mneral powder asphalt concrete
AC-16 asphalt mix was used in the experiment [8]. Graphic design to obtain mineral aggregate with the ratio of 10～20mm: 5～10mm: natural sand: stone chips: mineral powder = 19: 44: 18: 13: 6. With the Marshall test, the optimum bitumen aggregate ratio is 4.6%.

5.2. Flooding Marshall test
According to the requirements of JTG E20-2011, the water stability of the mixture was evaluated by the residual stability of the Marshall test and the freeze-thaw splitting strength ratio of freeze-thaw splitting test. Immersion Marshall test using 6 groups of different silane coupling agent content of the comparative test (silane coupling agent content of 0% to 2.5% of the modified granite filler), the test results in Table 3.

Table 3. Results of Marshall test and immersion Marshall test.

The amount of silane coupling agent /%	Stability /KN		Residual stability /%
	0.5h	48h	
0.0	10.45	8.21	78.6
0.5	12.62	10.25	81.2
1.0	14.78	12.50	84.6
1.5	15.56	13.80	88.7
2.0	16.20	14.48	89.4
2.5	14.42	12.78	88.6

It can be seen from Table 3 that the silane coupling agent with 0.5% ~ 2.5% content can improve the surface properties of the granite mineral powder [9], in which the silane coupling agent content is 2.0 %, The modification effect of granite filler is the most prominent. Compared with the asphalt mixture with unmodified granite filler, the Marshall stability, immersion Marshall stability and residual stability are significantly improved.

5.3 .Freeze-thaw splitting test
The standard Marshall specimens were subjected to splitting tests under normal conditions and freeze-thaw cycles, respectively, and the residual strength ratios of the freeze-thaw splitting tests were calculated. The test results are shown in Table 4.

Table 4. Results of freeze-thaw splitting test.

The amount of silane coupling agent /%	Void ratio /%	Splitting strength before freezing and thawing /MPa	Splitting strength after freezing thawing /MPa	Freeze-thaw splitting strength ratio /%
0.0	6.1	0.86	0.52	60.5
0.5	6.3	1.04	0.81	77.9
2.0	6.5	1.07	0.92	86.0
Skills requirement	–	–	–	>75.0

It can be seen from Table 3 and Table 4 that the asphalt mixture without silane coupling agent does not meet the water stability requirements of asphalt mixture in wet areas such as Weihai [10]. When the dosage of silane coupling agent is 0.5% and 2.0%, the water stability is not only the requirement, but the TSR value is 17.4% and 25.5% higher than that of the unmodified limestone filler asphalt mixture after freeze-thaw cycle. Indicating that the use of silane coupling agent modified granite mineral powder can significantly improve the performance of asphalt mixture water damage.

5.4 .Rutting test

The rutting test results are shown in Table 5.

Table 5. Results of asphalt mixture rutting test.

The amount of silane coupling agent /%	Displacement of 45 minutes /mm	Displacement of 60 minutes /mm	Dynamic stability /(Times·mm^{-1})
0.0	5.200	5.834	994
0.5	5.022	5.553	1186
2.0	4.642	5.154	1230

From the Table 5 can be shown that compared with the unmodified granite mineral powder, the dynamic stability of the asphalt mixture with silane coupling agent modified granite mineral powder was increased by 19.3% and 23.7%. The high temperature stability of the asphalt mixture with modified granite mineral powder is significantly improved.

6. Conclusions

(1) To improve high temperature stability and water stability of the asphalt mixture synthesized, the amount of the coupling agent 2.0% was recommended by using the silane coupling agent.

(2) Compared with the unmodified granite mineral powder, the residual stability of the asphalt mixture increased by 10.8% and the freeze-thaw splitting strength ratio increased by 25.5%, the dynamic stability increased by 23.7%. Silane coupling agent modified granite mineral powder significantly improved the pavement resistance of asphalt mixture and other pavement performance.

(3) The modification of granite mineral powder by silane coupling agent, not only achieves the road performance requirement of the asphalt mixture with granite mineral powder as filler, but also significantly improved the water resistance of the asphalt mixture. So that rich granite stone Weihai area to make use of local materials, and granite mineral powder as filler of asphalt mixture has an extended significance and economic benefits.

7. Acknowledgements

The paper was financial supported by Ministry of Transport Applied Basic Research Project (No. 2015 319 817 110) and Natural Science Foundation of Shandong (No. ZR2014EEQ035).

8. References

[1] Changfa Ai, Tianbing Xu and Jong Cheng 2008 Experimental Study on Improving Performance of Granite Asphalt Mixture *Journal of Highway and Transportation Research and Development* (7) pp 96-98

[2] Shuyin Wang and Chunxiu Zhou 2003 Experimental Study on Improving Water Resistance of Asphalt - Acid Aggregate *Journal of China Highway* (1) pp 6-9

[3] Huixiang Du, Huoyang Huang and Wenpeng Wang 2013 Coupling Mechanism of Silane Coupling Agent and Its Application in Sealan *Chemistry and bonding* (02) pp 63-65

[4] Yiqiu Tan, Xiaolin li and Xingye Zhou 2009 Study on the Interaction Ability of Granite Mineral and Asphalt *Journal of Harbin Institute of Technology* (11) pp 95-98

[5] Ministry of Communications Highway Science Research Institute 2005 *Test Methods of Aggregate for Highωay Engineering* (China: China Communications Press)

[6] Benhui Gong, Feng Zhu, Aiqin Zhang 2012 Experimental Study on Surface Modification of

Limestone Filler *Chinese and foreign roads* (03) pp 1-3
[7] Hongchao Zhang and Lijun Sun 2002 Study on Evaluation Method of Water Stability Performance of Asphalt Mixture *Journal of Tongji University* (4) pp 2 -3
[8] Baosheng Yuan, Yuhua Peng and Linzhong Wang 2010 Application of Granite in Asphalt Overlay of Cement Concrete Pavement *Chinese and foreign roads* (5)
[9] Zhenzhen Kong, Manya Zhang and Aiqin Zhang 2014 Effect of granite slag packing on water stability of asphalt concrete, *Chinese and foreign roads* (05) pp 1-2
[10] Pan Lu, Jun Xie and Xiaojun Fan 2014 Application of Granite in Asphalt Pavement of Ma wu xpressway *Chinese and foreign roads* (01) pp 2-3

IOP Conf. Series: Materials Science and Engineering **216** (2017) 012030 doi:10.1088/1757-899X/216/1/012030

Effects of Aggregate Gradation on Drying Micro-Surfacing added by Waste Rubber Powders

Yali Ye, Jinke Guo, Fengjian Hou

School of Transportation and Civil Engineering, Shandong Jiaotong University. No.5001 Haitang Rd, University Science and Technology Park, China

Email: 75704581@qq.com;864603084@qq.com;1256048753@qq.com;

Abstract. In order to research the effects of aggregate gradation on micro-surfacing, the control variable method was used, the effects of different aggregate gradation on drying micro-surfacing added by waste rubber powders were studied, through a series of laboratory testing, indexes such as mixing time, cohesion torque, wet track abrasion value, load wheel value were studied. The results showed that intermediate gradation with 10% and 11% of the bitumen-aggregate ratio had the optimal performance. The mixing time and cohesion torque could all meet specification requirements; its anti-wear value and the adhering sand amount were smaller than conventional. So the micro-surfacing layer should form the internal compact, but also the formation of rough surface.

1. Introduction

The rubber powder dry micro-surfacing had significantly effect on anti-cracking, anti-wear, anti-fatigue, driving comfort and other performances. However, the quality of material was good, after paved still there were lots of problems in micro-surfacing. The practice showed that aggregate gradation had great effects on mix performance of micro-surfacing. Therefore, in order to ensure the quality of the material used in the micro-surfacing, it was very important to select a suitable gradation to match the quality of the micro-surfacing.

2. Experiment material

2.1. Modified emulsified asphalt

Test used SBR modified emulsified asphalt, its technical parameters were satisfied with the specification of micro-surfacing and slurry seal technology guide [1, 2], the test results were shown in Table 1.

Table1. Technical parameters of modified emulsified asphalt.

Pilot projects	Technical requirements	Test results	Experiment method
Demulsification rate	Slow breaking	Slow breaking	T0658
The remaining amount on the sieve (1.18mm)/%	≤0.1	0	T0652
Electric charge	Cation+	Cation+	T0653

Evaporation residue content/%		≥60	62.5	T0651
Evaporation residue properties	Penetration (25℃,100g,5s)/0.1mm	40~100	88	T0604
	Ductility（5°C)/cm	≥20	80	T0605
	Softening point/°C	≥53	56	T0606
	Solubility (trichlorethylene)/%	≥97.5	99	T0607
Storage stability	1d/%	≤1	0.6	T0655

2.2. Aggregate
The test aggregates were basalt, with a small particle size of more than 9.5 mm, which had been sieved before the test [3].The main technical indicators were shown in Table 2 [4].

Table 2. Aggregate technical properties and test results.

Pilot projects		Technical requirements	Test results	Experimental method
Coarse aggregate	Stone crushing value/%	≤26	12	T0316
	Los Angeles wear loss/%	≤28	13.6	T0317
	Stone polished value/BPN	≥42	46	T0321
	Ruggedness /%	≤12	0.9	T0314
	Stone needle-like content/%	≤15	0.5	T0312
Fine aggregate	Ruggedness /%	≤12	2	T0340
Synthetic mineral sand equivalent/%		≥65	69.5	T0334

2.3. Filling
Filling mainly cement and waste rubber powder. The cement used in this test was an ordinary portland cement without any additives, cement strength of 42.5, the amount of 1.5% (with the mass ratio of aggregate).The waste rubber used in this test was 40 mesh, the amount of 1% (with the mass ratio).

2.4. Water
The amount of water should be minimized when the requirements were met. The test used drinking water, and the content of water was 6% [1].

3. Performance of micro-surfacing

3.1. Mineral aggregate gradation
According to the International Slurry Seal Association's design method, the MS-3 gradation was required to mix. According to the sieving results of aggregates, coarse, intermediate and fine gradations were designed, as was shown in Table 3[3-5].

Table 3. Aggregate gradation table.

Gradation type	Percentage of mass by the following sieve (mm)%							
	9.5	4.75	2.36	1.18	0.6	0.3	0.15	0.075
5-10mm	100.0	9.7	2.0	1.9	1.9	1.8	1.8	1.7
3-5mm	100.0	93.8	3.4	1.5	0.9	0.8	0.7	0.7
0-3mm	100.0	100.0	77.8	57.9	42.6	29.0	20.9	15.1
Fine gradation	100.0	88.6	62.0	46.1	34.0	23.2	16.8	12.2
Intermediate gradation	100.0	80.4	53.7	40.0	29.5	20.2	14.7	10.7
Coarse gradation	100.0	73.8	45.4	33.8	25.0	17.1	12.5	9.2

3.2. Mixing test

Through a large number of tests, the best ratio of the three mixes was determined as the aggregate: cement: waste rubber powder: water: modified emulsified asphalt = 100: 1.5: 1: 6: 14.4.The test results were shown in Table 4.

3.3. Cohesion torque test

The cohesion test was carried out using the same mixing ratio as the mixing test,the test results were shown in Table 4.

Table 4. Results of mixing time and cohesion test.

Gradation type	Mixing time/s	Non-construction time/s	30min Cohesion torque/(N•m)	60min Cohesion torque/(N•m)
Coarse	>180	350	1.3	2.0
Intermediate	140	300	1.6	2.3
Fine	135	240	1.8	2.3

It could be seen from Table 4, the mix of coarse gradation mixing time was the longest. Intermediate gradation was 140s, fine gradation with the shortest mixing time. Not the construction time was the same law in descending order. Mainly because the intermediate and fine with the powder more, absorbed moisture and emulsified asphalt emulsion, at the same pitch ratio and the amount of water used, the mixture was thicker than the coarse gradate mix, mixing time and no construction time was not long. As could be seen from table 4, coarse grading of 30min and 60min cohesion value of the smallest, the cohesion of the intermediate with 60 min cohesive force had no aggregate dissolution and good effect. The main reason for the difference was that the coarse gradation of the fine aggregate was less; the indenter was slipped on the surface of the specimen so that the cohesive force was too small. And intermediate and fine with the mixture in the powder more, could filled the gap between the coarse aggregate, so the specimen was not damaged and cohesion value to meet the requirements [6-8].

3.4. Wet track abrasion test and load wheel test

Wet track abrasion test used the same mixing ratio as the cohesion test; the load wheel test was still the same mixing ratio as the cohesion test, the test results were shown in Table 5.

Table 5. Results of wet track abrasion test and load wheel test.

Gradation type	Asphalt-aggregate ratio/%	Wear value 1h(g/m^2)	Wear value 6d(g/m^2)	Adhesive sand content(g/m^2)
Coarse	9	428	545	310
	10	457	694	397
	11	643	806	680
Intermediate	9	1108	1257	208
	10	280	490	274
	11	291	301	390
Fine	9	1204	1637	179
	10	418	1269	266
	11	467	525	284

It could be seen from Table 5, in the wet track abrasion test, the intermediate aggregate with 10% and 11% asphalt-aggregate ratio when the wear value was small, asphalt and mineral filler and wear resistance of the best, this was because the number of medium-sized mineral material in the powder could just fill the gap between the coarse aggregate, and powder and rubber powder together could absorb the excess moisture and asphalt emulsion, moreover, made the inside of the specimen after the more compact, to form a whole; while the intermediate level with 9% of the asphalt-aggregate ratio, the measured wear value was large, beyond the provisions of the range, this was because the middle with more powder, made the material than the surface area, and the rubber powder adsorption emulsified asphalt was very strong, coupled with low than the asphalt-aggregate ratio, in the three under the combined effect of the formation of the specimen asphalt film thickness was small, so wear loss was great[6-10].

4. Conclusions

(1)The wear value of coarse gradation was more than specifications whether high asphalt-aggregate ratio or not, under the condition of certain asphalt-aggregate ratio, the values of cohesion of fine and intermediate gradation were larger than coarse gradation, it helped to be open to traffic quickly.
(2) The aggregate powder was filled in the gap of the coarse aggregate skeleton; excess water and asphalt emulsion were obsorbed by aggregate powder and rubber powder, micro-surfacing mixtures after paving had good wear resistance and water stability.
(3) Wet abrasion test and adhesion test results showed that when asphalt- aggregate ratio was 10% or 11%, the performance of intermediate gradation of MS-3 micro-surfacing was the best.

5. Acknowledgements

This paper was financial supported by Ministry of Transport Applied Basic Research Project (No. 2015 319 817 110) and Natural Science Foundation of Shandong (No. ZR2014EEQ035).

6. References

[1] Ministry of Communications Highway Science Research Institute S 2006 *Micro-Surfacing and Slurry Seal Technical Guide* (China: People's Communications Press)
[2] Songchang Huang, Jian Xu, Yongchun Qin M 2010 *Modified Emulsified Asphalt and Micro-Surfacing Technology* (Beijing: People's Communications Press)
[3] Yiqiu Tan M 2007 *Asphalt and asphalt mixture* (Harbin: Harbin Institute of Technology Press)
[4] Ministry of Communications Highway Science Research Institute S 2005*Test Methods of Aggregate for Highway Engineering* (China: People's-Communications Press)
[5] Aiqin Zhang, Xia Zhu M 2009 *architecture material* (Beijing: People's Communications Press)

[6] Hao Ju, Xiaoming Huang J 2006 *Effect of Different Gradation on the Performance of Mixture* asphalt20 (4):55-60
[7] Wei Chen, Liang Fan, Lin Wang J 2010 *Study on Optimization of Micro-surfacing of Technology of Rubber Powder* asphalt(06):24-25
[8] Liang Fan, Wei Chen, Fengren Cui more than three authors J 2011 *Experimental Study on Performance of Waste Rubber Powder Micro-Surfacing* Road Traffic Technology (04):50-51
[9] Jian Xu, Songchang Huang M 2011 *Ideas and Techniques of Preventive Maintenance of Asphalt Pavement* (Beijing: People's Communications Press)
[10] Shun Wang J 2012 *Performance Evaluation of Cement Emulsified Asphalt Cold Recycled Mixture* Road construction 37(1):171-172

IOP Conf. Series: Materials Science and Engineering **216** (2017) 012031 doi:10.1088/1757-899X/216/1/012031

Effects of Fineness of Rubber Powder on Micro-Surfacing Performance

Jinke Guo[1], Yali Ye[1], Chuanfeng Cai[2] and Wei Zhao[2]

[1]School of Transportation and Civil Engineering, Shandong Jiaotong University. No.5001 Haitang Rd, University Science and Technology Park, China
[2]Shandong Taihe Highway Engineering Company, No.36 Dongtai Rd, Huantai New District, China

Email: 864603084@qq.com; 75704581@qq.com; 2895789724@qq.com; Skyangel001@126.com;

Abstract. In order to research the effects of fineness of rubber powder on micro-surfacing, the control variable method was used. Through a series of basic performance tests, performance indexes such as mixing time, cohesion torque, wet track abrasion value and load wheel sand value were studied. The results showed that waste rubber powder with 40 mesh had good effects on non-construction time, anti-wear performance, the density and water sealed performance of micro-surfacing.

1. Introduction

Micro-surfacing as one of pavement preventive maintenance technologies, it had good effects on the repairments of pavement rutting and other diseases, in particular, the waste rubber powder dry micro-surfacing, it had great advantages in anti-cracking, anti-wear, anti-fatigue, riding comfort and other aspects. However, the quality of material was good, after paved still there were lots of problems in micro-surfacing. The practice showed that aggregate gradation had great effects on mix performance of micro-surfacing. The practice showed that the fineness of rubber powder in the mixture had a significant effect on the mix performance and micro-surfacing layer after paving. Therefore, in order to ensure the quality of the material used in the micro-surfacing, how to used a reasonable particle size of the rubber powder on the quality of micro-surfacing was essential. The effected of fineness of the waste rubber powder on the micro-surfacing performance was studied.

2. Properties of materials

2.1. Modified emulsified asphalt

SBR modified emulsified asphalt was used, the parameters met the requirements of micro-surfacing and slurry seal technology guide, the test results were shown in Table 1[1, 2].

Table 1. Technical parameters of modified emulsified asphalt.

Pilot projects	Technical requirements	Test results	Experiment method
Demulsification rate	Slow breaking	Slow breaking	T0658
The remaining amount on the sieve	≤0.1	0	T0652

(1.18mm)/%				
Electric charge		Cation+	Cation+	T0653
Evaporation residue content/%		≥60	62.5	T0651
Evaporation residue properties	Penetration (100g,25°C,5s)/0.1mm	40~100	88	T0604
	Ductility (5°C)/cm	≥20	80	T0605
	Softening Point/°C	≥53	56	T0606
	Solubility (trichlorethylene)/%	≥97.5	99	T0607
Storage stability	1d/%	≤1	0.6	T0655

2.2. Aggregate

The test aggregates were basalt, with a small particle size of more than 9.5 mm, which had been sieved before the test. The main technical indicators were shown in Table 2[3].

Table 2. Aggregate technical properties and test results.

Pilot projects		Technical requirements	Test results	Experimental method
Coarse aggregate	Stone crushing value/%	≤26	12	T0316
	Los Angeles wear loss/%	≤28	13.6	T0317
	Stone polished value/BPN	≥42	46	T0321
	Ruggedness /%	≤12	0.9	T0314
	Stone needle-like content/%	≤15	0.5	T0312
Fine aggregate	Ruggedness /%	≤12	2	T0340
Synthetic mineral sand equivalent/%		≥65	69.5	T0334

2.3. Filling

Filling mainly cement and waste rubber powder. The cement used in this test was an ordinary portland cement without any additives, cement strength of 42.5MPa, the amount of 1.5% (with the mass ratio of aggregate).The role of waste rubber powder was mainly to improve the road surface flexibility, improve the mixture of water damage and anti-wear performance. The waste rubber used in this test was 40 mesh, 60 mesh, 80 mesh, the amount of 2% (with the mass ratio).

2.4. Water

The amount of water should be minimized when the requirements were met. The test used drinking water, and the content of water was 6% [1].

3. Performance of micro-surfacing

3.1. Mineral aggregate gradation

According to the design method of the International Slurry Seal Association, the MS-3 micro-surfacing gradation was used to design the mix ratio. According to the sieving results of the aggregates, the gradation used was shown in Table 3[4].

Table 3. Aggregate gradation table.

Gradation type	Percentage of mass by the following sieve (mm)%							
	9.5	4.75	2.36	1.18	0.6	0.3	0.15	0.075
5-10mm	100.0	9.7	2.0	1.9	1.9	1.8	1.8	1.7
3-5mm	100.0	93.8	3.4	1.5	0.9	0.8	0.7	0.7
0-3mm	100.0	100.0	77.8	57.9	42.6	29.0	20.9	15.1
MS-3Upper limit	100	90	70	50	34	25	18	15
MS-3Lower limit	100	70	45	28	19	12	7	5
Median of gradation	100.0	80.0	57.5	39.0	26.5	18.5	12.5	10.0
Synthetic Gradation	100.0	80.4	53.7	40.0	29.5	20.2	14.7	10.7

Notes: 5-10mm: 3-5mm:0-3mm=21%:11%:68%

3.2. Mixing time test and cohesion torque test

Through number of tests, it was determined that the best ratio of three different mesh rubber powders were used as aggregate: cement: used rubber powder: water: modified emulsified asphalt = 100: 1.5: 2: 6: 16, the test results were shown in Table 4 [1], [5-9].

The cohesion test was carried out using the same mixing ratio as the mixing test, and the value of 300g aggregate was used. The test results were also shown in Table 4 [1], [5-9].

Table 4. Results of mixing time and cohesion test.

The fineness of rubber powder	Mixing time/s	Non-construction time/s	30min Cohesion torque/(N•m)	60min Cohesion torque/(N•m)
40	>180	350	1.8	2.3
60	140	300	1.6	2.3
80	120	220	1.15	1.8

It could be seen from table 4 that the mixing time of the mixture with 40 mesh powder was more than 180s, the mixing time of the mixture of 60 mesh powder was 140s, the mixing time of 80 mesh powder was 120s, and the non-construction time was the same law in descending order. Mainly becaused the rubber powder adsorption emulsified asphalt capacity than the aggregate was strong, 40 mesh powder particle size larger, so relative to 60 mesh and 80 mesh powder specific surface area were much smaller, adsorption of asphalt emulsion was also very less, so there were enough emulsified asphalt and intermediate with more powder combination, the mixture would not become sticky, so the mixing time was longer.The 80 mesh powder was opposite. Mixed with 40 mesh powder 30min and 60min cohesion torque were in line with the requirements, and mixed with 80 mesh powder 30min and 60min cohesion torque were not idealed. The reason was that 40 mesh powder surface area was small, the adsorption of emulsified asphalt less, thus ensured a sufficient emulsified asphalt and

IOP Conf. Series: Materials Science and Engineering **216** (2017) 012031 doi:10.1088/1757-899X/216/1/012031

powder combination, made the specimen after the internal compact,80 mesh powder was the opposite[5], [7-10].

3.3. Wet track abrasion test

Wet track abrasion test used the same mixing ratio as the cohesion torque test, and value of 800g of the aggregate was used. The test results were shown in Table 5.

Table 5. Results of wet track abrasion test and load wheel test.

The number of powder mesh	Asphalt-aggregate ratio /%	Abrasion value 1h(g/m^2)	Abrasion value 6d(g/m^2)	Adhesion sand amount(g/m^2)
40	9	451	735	269
	10	280	499	288
	11	274	437	350
60	9	937	1022	208
	10	481	562	260
	11	293	521	327
80	9	1204	1522	179
	10	548	984	246
	11	467	535	305

3.4. Load wheel test

The load wheel test was still the same mixing ratio as the cohesion torque test, and value of 500g of the aggregate was used. The test results were shown in Table 5.

The 40 mesh powder particle size was larger, the specific surface area was small, the adsorption of asphalt emulsion less, 10% and 11% of the asphalt stone ratio there was enough emulsified asphalt and powder combination, and the specimen dense. The 80 mesh powder was the opposite [5], [7-10].

4. Conclusions

(1) The use of 40 mesh of the rubber powder could properly absorb the mixture of excess emulsified asphalt, and asphalt emulsion was not reduced too much.

(2) The mixing time of waste rubber powder dry micro-surfacing was greatly affected by its fineness of rubber powder, the finer the powder, the shorter the mixing time.

(3) When the mesh of powder was greater than 60, the cohesion torque of micro-surfacing was lower than the specification requirements, rubber powder of 60 mesh or less was recommend. The powder mixture of 40 mesh powder had good abrasion resistance, and the asphalt emulsion by powder was less, and the specimen of 40 mesh rubber powder micro-surfacing was stabled and compacted.

5. Acknowledgements

This paper was financial supported by Natural Science Foundation of Shandong (No. ZR2014EEQ035) and Ministry of Transport Applied Basic Research Project (No. 2015319817110).

6. References:

[1] Ministry of Communications Highway Science Research Institute S 2006 *Micro-Surfacing and Slurry Seal Technical Guide* (China: People's Communications Press)

[2] People's Republic of China Industry Standard S 2011 *Test Code for Asphalt and Asphalt Mixture for Highway Engineering* (Beijing: People's Communications Press)

[3] Ministry of Communications Highway Science Research Institute S 2005 *Test Methods of Aggregate for Highway Engineering* (China: People's-Communications Press)
[4] Aiqin Zhang, Xia Zhu M 2009 *architecture material* (Beijing: People's Communications Press)
[5] Tianqing Ling, Yaonan Li, Wei Li J 2011 *Effect of Rubber Particles on Micro-Surfacing Performance* Road construction 36(2) pp 174-175
[6] Jianwei Fan, Xinzhan Liu J 2007 *Study on the Performance of Slurry Seal of Rubber Powder Modified Asphalt Mixture* Road construction machinery and mechanization of construction (1) pp 33-36.39
[7] Xiaochen Li,Lijuan Zhang J 2016 *Study on the Influence Mechanism of Rubber Powder Content and Head Number on Rubber Asphalt Performance* Guangdong building materials(2) pp 22-23
[8] Qiang Li, Wei Huang, Hang Ren J 2010 *Study on Modification of Waste Tire Rubber Powder* Journal of Highway and Transportation Science and technology (3) pp 67-68
[9] Jinfeng Li, Luoying Wang, Wenjie Li more than three authors J 2011 *Application of Waste Rubber Particles in Road Engineering* Chinese and foreign roads 31(4) pp 241-244
[10] Jian Xu, Songchang Huang M 2011 *Ideas and Techniques of Preventive Maintenance of Asphalt Pavement* (Beijing: People's Communications Press)

IOP Conf. Series: Materials Science and Engineering **216** (2017) 012032 doi:10.1088/1757-899X/216/1/012032

Full-scale Model Test for Anchorage Zone of Pylon: Error Analysis on Elongation of Annular Pre-Stressing Steel Strand

Xijun Ye[1], Zhuo Sun[1], and Weifeng Wang[2]
[1] School of civil engineering, Guangzhou University, Guangzhou, China, 510006
[2] School of Civil Engineering and Transportation, South China University of Technology, Guangzhou, China, 510507.

Email: xijun_ye@gzhu.edu.cn

Abstract. In order to study the mechanism of the annular Pre-stressing steel strands applied to cable-pylon anchorage zone, a full-scale model test and finite element analysis were conducted. Through the research on the full-scale model test of the U-shaped annular Pre-stressing steel strands for the Anchorage Zone of Guangzhou Bridge, the prestressing technique for the U-shaped annular Pre-stressing steel strands was investigated, and the stress distribution in the model were measured and analyzed. It can be found that the elongation of the U-shaped pre-stressing steel strands is greater than the actual theoretical limit, and the results of some project cases also showed that it is a common phenomenon. Finally, the reasons for thus problems and relative suggestions are proposed, they can be used as a reference to engineering practice.

1. Introduction

Guangzhou Bridge, with span length of (139m+106m), is an asymmetric concrete cable-stayed bridge with single pylon and single cable plane of 42 pair of stay-cables, whose pier, pylon and beam are consolidated together (as shown in Figure 1).

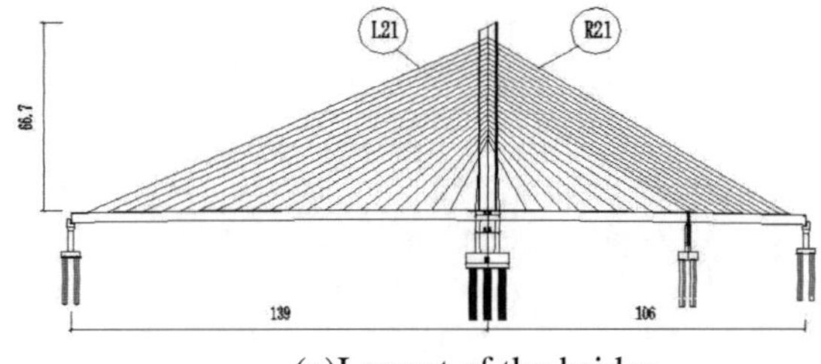

(a)Layout of the bridge

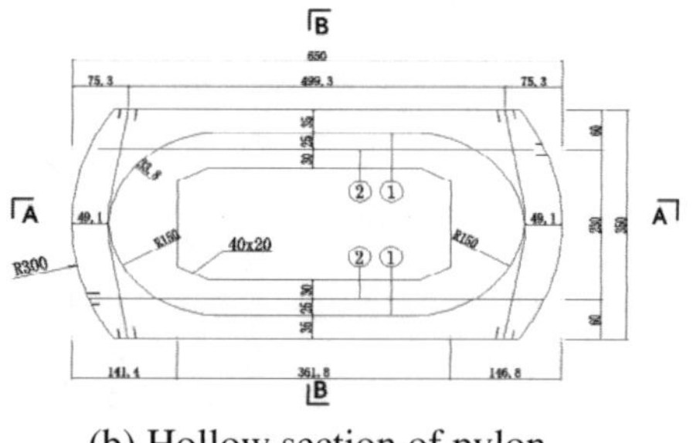

(b) Hollow section of pylon

Figure 1. Layout of Guangzhou Bridge(Unit: m)

Anchorage zone of the cable-stayed bridge uses approximate rectangular hollow section. Cable anchorage zone on the tower is stiffened by U-shaped annular pre-stressing steel strands（1#） and straight steel strands（2# (as shown in Figure 2). 1# steel cable includes 17 steel strands with diameter of 15.24mm; 2# steel cable embraces 9 steel strands with diameter of 15.24mm. Tension control stress is 1395MPa and standard value of tensile strength (f_{pk}) is 1860MPa and elasticity modulus (E_p) is 1.95×105MPa. Pre-stressed duct utilizes plastic corrugated pipes with inner diameter of 15×19mm.

2. Field test of the full-scale model of anchorage zone

Normal pitch of cable tower anchorage zone of Guangzhou Bridge is 1.8m. So the standard height of the full-scale model is 1.8m and the anchorage zone of L21 and R21 cables is selected, who has the maximum cable force. Each anchorage zone is equipped with 6 U-shaped annular pre-stressing steel strands and 6 straight steel strands. (As shown in Figure 2).

In order to mitigate the loss of pre-stressing, firstly, U-shaped steel strand is used in the form of two-end tension and then straight steel strand is used in the form of one-end tension. For purpose of synchronizing tension when both ends of a jack are tensioned, the bridge adopts grading tension, and loading rate is strictly controlled, which maintains 30s after loading. In strict accordance with tension procedures of post-tensioned pre-stressing regulated in 12.10.3 section of "Technical Specifications for Construction of Highway Bridges and Culverts(JTJ041—2000)"(hereafter referred to as JTJ041—2000)[1]: installation of steel strands, anchorage device and jacks, initial stress, double initial stress, triple initial stress, quadruple stress and σcon (sustaining load in 2 min and anchoring). σcon refers to tensile control stress, and the stress of this project is 1395MPa and the initial control stress is 15% σcon. The full-scale model can be seen in Figure 2.

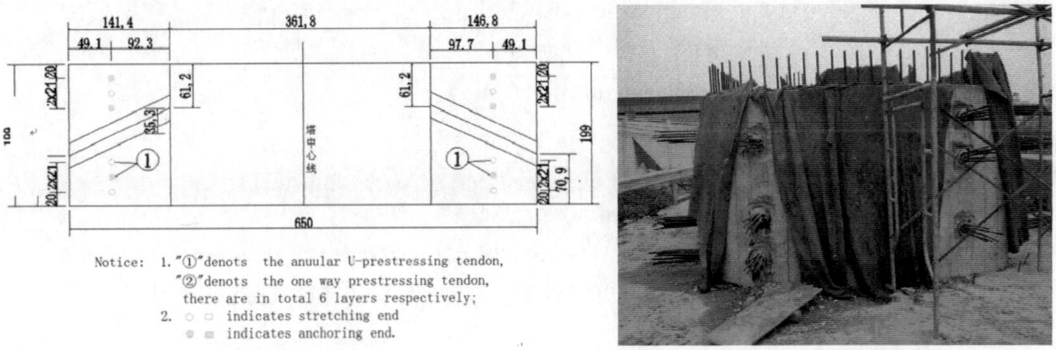

(a) Section A (b) Full-scale model

Figure 2. Field test (Unit: cm)

2.1. Calibration test.

Before the model test, a calibration test was conducted to verify the mechanical property of steel strands. A steel strand with diameter of 15.24mm was randomly selected, whose standard value of tensile strength (f_{pk})is 1860MPa and elasticity modulus (E_p) is 1.95×105MPa. The 1# steel cable includes 17 steel strands with diameter of 15.24mm, tension control stress is 3320MPa, the 2# steel cable embraces 9 steel strands with diameter of 15.24mm, tension control stress is 1750MPa. So the average tension control stress of each steel strand is 195kN.

Table 1. Test Result of calibraion test

Loaing grade	load (kN)	Theoretical strain (με)	Practical strain (με)	Relative Error (%)	Theoretical elongation (mm)	Practical elongation (mm)	Relative Error (%)
1	29.7	1087.9	1047.4	-3.7	4.188	4.236	1.3
2	58.7	2151.3	2060.1	-4.2	8.283	8.632	4.2
3	88.2	3231.5	3110.5	-3.7	12.441	12.786	2.8
4	117.3	4295.4	4150.8	-3.4	16.537	16.883	2.1
5	147.2	5392.5	5255.7	-2.5	20.761	21.113	1.7
6	178.3	6531.1	6368.6	-2.5	25.145	25.492	1.4

In the calibration test, 6 loading grades are applied: 1:30kN -260kN- 3:90kN- 4:120kN- 5:150kN- 6:180kN. The calibration results are shown in Table 1. It can be seen that the Theoretical values agree well with practical ones, the relative error between the two values is less than 5%. The steel strands selected meet the requirements of the full-scale model test.

2.2. Theory and Test Result.

According to 12.8.3 section of JTJ041—2000, as pre-stressed tendon is tensioned in the method of stress control method, proofreading shall be based on elongation. Meanwhile, the difference between actual elongation and theoretical elongation shall meet the requirements of design[1]. In the case of no design requirements, the difference shall be controlled within 6%, otherwise tension shall be suspended and can return to normal until problems have been found and solved. Meanwhile, corresponding design formulas of elongation (ΔL) of pre-stressing tension can be seen in JTJ041—2000:

$$\Delta L = {P_p \cdot L}/{A_p \cdot E_p} \quad .(1)$$

$$P_p = {P\left(1-e^{(kL+\mu\theta)}\right)}/{(kL+\mu\theta)} \quad (2)$$

Where P_p - average tension of pre-stressing tension (N)

L - the length of pre-stressing tension (mm)

A_p - sectional area of pre-stressing tension (mm^2)

E_p - elasticity modulus of pre-stressing (N/mm2)

P - stress under the anchorage of tensioned-end pre-stressing steel strands

μ - friction coefficient of the wall of pore passage

k - coefficient of influence of local deviation of pore passage in every meter on friction

θ - the sum of included angle of curves and tangent lines from tensioned-end anchor to calculation section.

Table 2. Test Result of elongation of U-shaped pre-stressing steel strands of Guangzhou Bridge

No. of steel strands	Theoretical elongation value (mm)	Elongation value (mm)	loss of due to draw in of clip (mm)	Practical elongation value (mm)	Relative Error (%)
C3-1	75	110	12	98	30.7
C3-2	75	106	12	94	25.3
C3-3	75	100	12	88	17.3
C3-4	75	102	12	90	20.0
C3-5	75	106	12	94	25.3
C3-6	75	112	12	100	33.3
C4-1	75	111	12	99	32.0
C4-2	75	105	12	93	24.0
C4-3	75	113	12	101	34.7
C4-4	75	109	12	97	29.3
C 4-5	75	106	12	94	25.3
C4-6	75	108	12	96	28.0

Actual elongation of steel strand begins at initial stress. Actual pre-stressing strand shall include estimated elongation based on initial value besides measured elongation. Actual elongation of pre-stressing tension (ΔL (mm)) can be calculated according to a formula:

$$\Delta L = \Delta L_l + \Delta L_2 \quad (3)$$

Where ΔL_l - actual elongation ranging from initial stress to maximized stress. (mm)

ΔL_2 - estimated elongation based on initial stress. ΔL_2 adopts grading tension from 15% σ_{con} to 60% σ_{con} and average elongation of steel strands while in grade 3 loading to this bridge. Formula (3) embraces withdrawal value of anchorage clip in the both ends of steel strands. Elongation value can be obtained from deducting withdrawal value. In accordance with given value in JTJ041—2000, deduction value of setback amount of each clip is 6mm. two-end tension method is applied to this bridge, deducting setback amount of 12mm from both ends of clips. Test results are shown in Table 2.

2.3. Error analysis.

It can be seen from Table 1 that actual elongation of U-shaped pre-stressing steel strands of Guangzhou Bridge are all greater than its theoretical values. And values of each steel strands exceed the standards regulated in JTJ041—2000, namely, deviation of actual elongation and theoretical elongation shall be controlled within 6%. The author has conducted an investigation in China of the comparison of actual elongation of U-shaped pre-stressing steel strands and its theoretical value (Table 3). And the results reveal that all the actual elongation test results are greater than its theoretical values which are within 6%. Deviation values with obvious discreteness range from 20% to 30%, sometimes reach to 50%.

Table 3. Error Comparison of Elongation of Steel Strands in China

Project cases	Average error	Maximum error
Guangzhou Bridge	27.0%	34.7%
Shenzhen bay bridge [2]	25.0%	35.9%
North channel of Hangzhou Bay sea crossing bridge [3]	16.2%	25.3%
A highway interchanges [4]	15.1%	25.9%
Bridge 4[5]	26.0%	37.5%
Tie Luo Ping Bridge 5[6]	25.7%	39.0%
Wuhan Junshan Yangtze River Highway Bridge [7]	26.0%	53.0%
Bridge 8[8]	33.3%	54.3%

By comparison and analysis on above data and combined with actual conditions, the author concludes following reasons that may lead to actual elongation of U-shaped pre-stressing steel strands are greater than its theoretical values:

(1) U-shaped pre-stressing steel strands with minor radius will generate huge radial pressure under the circumstance of huge tensile load so that steel strands will be stiffened, as well as, steel strands in corners sticks close to wall of plastic corrugated pipe. In this case, actual arc length is reduced and steel strands in pipes have extra length, which is transformed into piston stroke under the action of tension, thus making actual piston stroke greater than its theoretical elongation value.

(2) Considering that steel strands are loose before tension, according to calculation method of JTJ041—2000, generally, initial stress is 10-15% σ_{con}. Then elongation under initial stress can be back calculation through grading loading so that actual elongation can be calculated. However, when initial stress is 10-15% σ_{con}, U-shaped steel strands may be loose due to minor radius, twist of steel strands, friction loss, deviation of pipes and the like. Hence, partial unstressed extra elongation is calculated, which leads to a bigger actual value.

(3) Generally, sectional area (Ap) and elasticity modulus (Ep) of steel strands are given through design in engineering to calculate theoretical elongation of steel strands in the course of tension , rather than actual sectional area(*Ap*) and elasticity modulus (*Ep*) obtained from field measurement. In fact, during the process of tension, diameter of finished steel thread is not a constant value, but a variate that varies as tension proceeds. Meanwhile, correspondent *Ap* and *Ep* also change, thus having an impact on accurate value of theoretical elongation of steel strands.

3. Conclusions

(1) Comparatively accurate μ and k can be obtained from friction drag experiment which conducts through duct on site. In this paper, μ is 0.1929 and k is 0.0069. Precise Ap and Ep of steel strands can be obtained from calibration test so as to ensure the accuracy of theoretical elongation.
(2) Initial stress shall be increased so that the greater initial stress will slash extra elongation produced in the process of tension pre-stressed steel strands. The paper proposes initial stress shall range from 25% to 30% at best.
(3) The control of elongation plays an important role in describing overall tension state of pre-stressed steel bars and tension safety. It's essential to calibrate elongation, but it shall not be confined to ±6%. During construction, it would be better not to regard elongation as the main element of pre-stressed steel bar tension, otherwise pre-stressing may be inadequate.

Acknowledgment
This work was supported in part by the National Natural Science Foundation of China under Grant No.51608136 and the New Talent Project of Guangzhou University(No.2608). The authors confirm that this article content has no conflict of interest.

References
[1] Industry standard of the People's Republic of China 2000 *Technical Specifications for Construction of Highway Bridges and Culverts (JTJ041-2000).*
[2] Wuhan Bridge Science Research Institute of China Railway Bridge Bureau Group Co., Ltd., 2005 *Test report of Shenzhen bay bridge.*
[3] Wuhan Bridge Science Research Institute of China Railway Bridge Bureau Group Co., Ltd., 2004 *Model test report of the navigable span bridge about the north channel of Hangzhou Bay sea crossing bridge.*
[4] Xiaogang Chen, Wenqing Wu 2008 Experiment Research on the Friction Resistance of the Ring Prestressed Steel Bundle in the Anchorage Zone of Cable-stayed Bridge Tower *Modern Transportation Technology* vol.5 no.5 pp. 50-52,.
[5] Dongping Bu 2010 Experiment Research on Prestressed U-shaped Pipe Friction of 190 m Cable-stayed Bridge Main Tower Anchorage Zone *Modern Transportation Technology* vol.7 no.5 pp. 47-50.
[6] Huafu Ruan, Binxuan Wu 2009 Experiment Research on the tension of the Ring Prestressed Steel Bundle in the Anchorage Zone of Cable-stayed Bridge Tower *Urban Roads Bridges & Flood Control*, no.11, pp.105-107.
[7] Weijian Yi, Wangxi Zhang 2001 Experiment Research on the full-scale anchor block segment model of the anchorage zone of the Wuhan Junshan Yangtze River Highway Bridge *Central South Highway Engineering*, vol.26, no.4, pp. 33-35.
[8] Wei Shan 2005 Study on structural behavior for Anchorage Zone Segment of Cable-stayed Bridge Tower *Ph.D. dissertation*, Northeast Forestry Univ., Harbin, China.

IOP Conf. Series: Materials Science and Engineering **216** (2017) 012033 doi:10.1088/1757-899X/216/1/012033

Structural Performance's Optimally Analysing and Implementing Based on ANSYS Technology

Na Han[1,4], Xuquan Wang[2],Haifang Yue[3],Jiandong Sun[1] and Yongchun Wu[5]

[1]Structure Department, Shandong Academy of Building Research, Jinan, China
[2]Feicheng City Department for Quality Supervision of Building Engineering, Feicheng, China
[3]Guangrao County Brueau for Garden Greening, Guangrao, China
[4]Shandong Jianke Architectural Design Co.Ltd, Jinan, China
[5]Business School, Shandong Jianzhu University, Jinan, China

Email: hannah_hn@163.com;

Abstract. Computer-aided Engineering (CAE) is a hotspot both in academic field and in modern engineering practice. Analysis System(ANSYS) simulation software for its excellent performance become outstanding one in CAE family, it is committed to the innovation of engineering simulation to help users to shorten the design process, improve product innovation and performance. Aimed to explore a structural performance's optimally analyzing model for engineering enterprises, this paper introduced CAE and its development, analyzed the necessity for structural optimal analysis as well as the framework of structural optimal analysis on ANSYS Technology, used ANSYS to implement a reinforced concrete slab structural performance's optimal analysis, which was display the chart of displacement vector and the chart of stress intensity. Finally, this paper compared ANSYS software simulation results with the measured results, expounded that ANSYS is indispensable engineering calculation tools.

1. Introduction to CAE

Computer-aided Engineering (CAE) is a new discipline which began to be developed in the 1960s. This process is a product which is combined with a numerical analysis methods and modern computer technology. With numerical analysis theory methods maturing and rapid development and wide application of computer technology, there is a broad prospects which commonly used CAE for engineering analysis improving design quality and shortening design circle. At that time, there were two fact that brought a great power in development of CAE [1]. One is basic idea of the finite element method which solves the problem of complex structures. In 1943, Courant, on his first attempt, applied the definition in the triangle area of piecewise continuous function and the principle of minimum potential energy to be solved St.Venant torsion problem. The second is wide application and rapid development of electronic computers. In the 1950's, J.Turner for the Boeing company by the United States, and J.H.Argyris in the university of London , puts forward the structural matrix analysis method, applied electronic computers as a major means of operation. The finite element method was developed on basis of this new method.

A typical procedure to implement CAE generally consists of five steps, which is showed in Figure 1: establishing mathematical model, selecting optimal algorithm, designing program, developing objectives and requirements, and computer automatic screening optimal analysis scheme.

2. The Necessity for Structural Optimal Analysis

IOP Conf. Series: Materials Science and Engineering **216** (2017) 012033 doi:10.1088/1757-899X/216/1/012033

Traditional analysis process is generally completed by the structural engineer with continuously testing according to past experience. Although structure can be improved to a certain extent after modification, but this kind of analysis has certain blindness and uncertainty. The final result is different from person to person. Because of the complexity of the structure, how to adjust analysis and design in order to get better, which is often a difficult problem to solve. Structural optimal analysis is effective method which makes a computer to replace human laborious work. Its research content is the combination of numerical theory and mechanics analysis method. It automatically improves and optimizes the bearing structure analysis in various conditions by establishing a set of scientific, systematic, reliable and efficient methods and software [2].

2.1. Lack of dynamic response to a change of parameters relating to design in the process of construction

In the process of structural analysis, there is a change of parameters relating to design in each step of the site construction, such as excavation depth, soil conditions, surrounding environment, the additional load, bending moment and shear force acting on the structure, these factors in affecting pouring quality of concrete, and so on. Computer aided structural optimal analysis can provide internal force analysis under various operating conditions.

2.2. Lack of scientificity and systematicness in traditional analysis method

The traditional analysis was finished generally by the structural engineer according to past experience. The result is different from person to person. In structural optimal analysis, the amount participating in calculation, one is constant, and other is variable, that forms a whole structure analysis solution domain. The analysis organically unites mechanics concept and optimization technology. Practice proves that the structural optimal analysis can shorten design cycle, save manpower, improve the quality and level of structure design. Economic benefit and social benefit is completely remarkable.

2.3. Lack of coordination between the informationization construction and design

We try to establish the concept of work with a computer data process and site data monitor at the core in the process of informatization construction. Traditional analysis cannot provide the data of a whole site construction, unable to form the management value. This make the structural analysis can't realize informatization with the result that its security should be greatly reduced. Structural optimal analysis can provide management data to achieve the design and construction informatization.

3. The Framework of Structural Optimal Analysis Based on ANSYS Technology

There are three typical analysis process including Preprocessor, Solution Processor and Post Processor.

3.1. Preprocessor

The module has two main parts: the entity modelling and meshing. There are two kinds of entity modelling method: top-down and bottom-up. The users define a model primitives, thereby, to directly structure geometry model, such as circular , rectangular, 3D block, ball, taper and column. When Bottom-up modelling entity, users structure model upwards from the lowest figure yuan ,namely, users first define key point, and then in turn the line, surface, body related to it. No matter using which kinds of method, the user can use Boolean operation to combine data sets, and "sculpture" a solid model. ANSYS program provides convenient and high quality of grid partition function for CAD model. Four meshing are included: extended meshing, image meshing, free classified meshing and adaptive meshing.

3.2. Solution Processor

The user can define analysis type, analysis options, load data and load options, then began to finite element solution. Loading is that the boundary condition data descript the situation of actual structure , and analyze the interaction between structure and outside world. There is the meaning of load such as the constrained displacement of freedom degree, nodal force load (force, torque), surface pressure, and inertial load (gravity acceleration, angular acceleration).

3.3. Post Processor

A graphic form of the front analysis results can be displayed and output in post-processing. Different lines colors represent different values, different values, and clearly reflect the regional distribution. Different line colors represent different values, numerical areas, and clearly reflect resultant regional distribution. In addition, post-processing can also check results in a time period or a substep process .These results can be viewed by drawing curves or list, in order to help to visualize analysis results [3]. The data flow of analysis is shown in Figure 2.

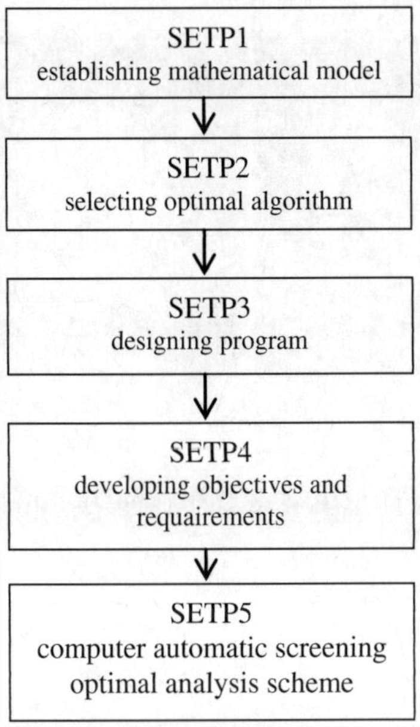

Figure 1. A typical procedure to implement CAE

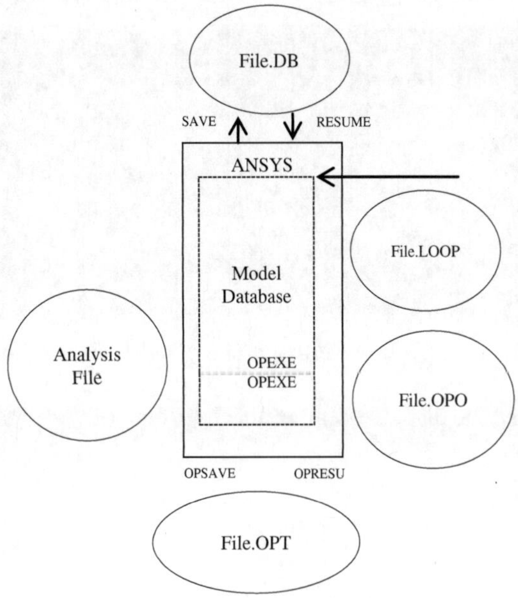

Figure 2. Data flow diagram in optimal analysis

4. The Implementing of Structural Optimal Analysis Based on ANSYS Technology

4.1. The experiment survey

There is a reinforced concrete slab of 1000*1000*180mm.Four levels exert concentrated load is applied respectively in the centre of the plate, which is 4kN, 6kN, 8kN, 10kN. Two long sides support slab, and two short sides node are constrained. The node doffs of UY constraints are applied to the slab bottom of Z = 1900 ~ 2000, at the same time, The node doffs of three directions constraints are applied to the slab bottom of Z = 0 ~ 100[4]. The change state of slab 's displacement and stress in the loading stage is observed and recorded. The accuracy of the numerical simulation analysis is verified by compared with the ANSYS software's calculation cases.

4.2. The simulation analysis of deflection and stress in the loading stage

4.2.1. The cloud charts of displacement vector sum.

A concentrated load is imposed on the center of slab. With concentrated load is gradually given an increase of 2kN, slab deflection value begins to expand gradually, however, the deflection value expands greatly slowly. The cloud charts of displacement vector sum are shown in Figure 3. From the figure, the deflection limit distributes the center on the bottom of the slab, whose value is the biggest, while the deflection value decreases in annular state trend on either side of slab [5]. There are four deflection cloud maximum values in the process of per level load, which are shown in Table 1.

4.2.2. The cloud charts of stress intensity.

A concentrated load is imposed on the center of slab. As the concentrated load is given an increase of 2 kN, stress intensity concentrates in the center of the slab bottom. The red part is dangerous stress area, which is also the largest stress intensity area of 1.09304 Pa. And then, you can draw a conclusion that stress intensity satisfies the requirement of the strength of material, if it is in the allowed range which is compared with the allowable stress of material. The minimum value is 0.051724 Pa, which is located on the support constraint of slab. There are shown in Figure 4. There are four stress intensity cloud maximum values in the process of per level load, which are shown in Table 2.

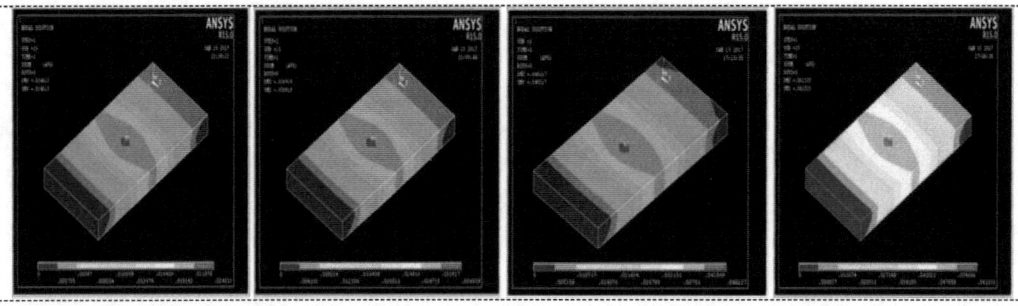

Figure 3. The cloud charts of displacement vector sum

Table 1. The displacement vector maximum of per level load.

Load(kN)	Displacement(mm)
4	24.613
6	36.919
8	48.227
10	61.533

Table 2. The stress intensity maximum of per level load.

Load(kN)	Stress(Pa)
4	0.437193
6	0.655802
8	0.875206
10	1.09304

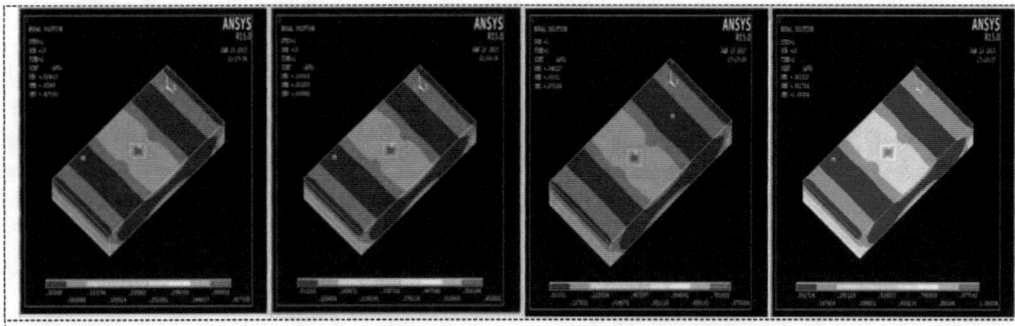

Figure 4. The cloud charts of stress intensity

4.2.3. Comparing ANSYS software simulation results with the measured results.

In the laboratory, the reinforced concrete slab is loaded with hydraulic jack. There are measured values of maximum deflection for leach level load which are shown in Table 3.

As using ANSYS software to simulate results, which comparing with the experimental measured results, we can see that the deflection values of computer simulation are consistent with the experimental result, which is shown in Figure 5. The chart is generally linear in the load - deformation curve chart [6].

The error in process is mainly manifested these great stress concentration in support of slab , which is with the corresponding concrete node displacement coordination; There are inaccurate dockings on working condition, which affect analysis model incorrect constraint and load.

Table 3. Comparison load with measured deflection

Jack readings (kN)	Measured deflection values(mm)
4	20.13
6	40.12
8	44.24
10	65.42

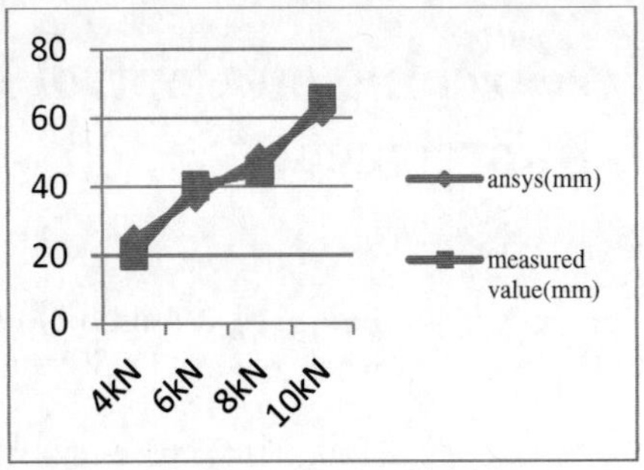

Figure 5. Comparison ANSYS simulation results with the measured results in deflection

5. Conclusions

The deflection and stress intensity of slab is simulated and analyzed by ANSYS soft. Although there is error in the calculating result of finite element model, it is basically in accordance with the measured results of experiment on the same change trend. The study shows that ANSYS is indispensable engineering calculation tools which simulate the stress process of slab better.

There are some problems in the finite element analysis of this study. (1) Some factors affecting convergence, such as mesh count, convergence criteria, bearing problem, appeared in the process of reinforced concrete slab's nonlinear calculation . (2) the reason of some error in process of finite element software calculation is that material parameters reflecting the calculation model and the working condition of docking is difficult to determine.

6. Acknowledgments

The authors would like to thank Shiqi Cui and Bo Cheng from Structure Department at Shandong Academy of Building Research, as well as the anonymous referees for valuable comments and critique during the review process of this paper. None besides the authors, however, can be held responsible for the result.

7. Reference

[1] Wei Li 2007 *ANSYS Civil Engineering Application Examples* Water Conservancy and Hydropower Press

[2] Mingwei Li 2012 *Based on ANSYS Finite Element Analysis of Deep Excavation Dynamic Optimization Design Method* Master Degree Dissertation of Xihua University

[3] Hao Liu, et al 2014 *ANSYS15.0 Finite Element Analysis from Entry to the Master* Mechanical Industry Press

[4] Cimian Zhu, et al 2016 *Strength of Structure* Higher Education Press

[5] James M.Gere and Barry J.Goodno 2011 *Strength of Materials* Mechanical Industry Press

[6] China Academy of Building Research 2014 *GB50204-2015 Code for Acceptance of Constructional Quality of Concrete Structures* the Ministry of Construction of the People's Republic of China

Researching on Control Device of Prestressing Wire Reinforcement

Jianhui Si, Yangbo Guo, and Maoshe Liu
[1]School of Civil Engineering and Architecture, Xi'an University of Technology, China

Email: sjhfr@163.com; 810979737@qq.com; liumaoshe@xaut.edu.cn

Abstract. This paper mainly introduces a device for controlling prestress and its related research methods, the advantage of this method is that the reinforcement process is easy to operate and control the prestress of wire rope accurately. The relationship between the stress and strain of the steel wire rope is monitored during the experiment, and the one - to - one relationship between the controllable position and the pretightening force of the steel wire rope is confirmed by the 5mm steel wire rope, and the results are analyzed theoretically by the measured elastic modulus. The results show that the method can effectively control the prestressing force, and the result provides a reference method for strengthening the concrete column with prestressed steel strand.

1. Introduction

According to the experience of earthquake damage in recent years at home and abroad, The design concept of "strong column and weak beam" is put forward. At present, the demand for improving the bearing capacity and ductility of the column has been gradually clear, combined with the actual situation of the existing building life in China, the reinforcement column technology has been widely used. At present, there are two kinds of reinforcement technology: Active and passive reinforcement; Passive reinforcement methods include: Enlarging section method, Concrete replacement method, Bonded steel structure strengthening method, Fiber reinforced plastic reinforcement method and so on. There is a common problem of stress hysteresis in these methods, which makes it necessary to have a certain degree of expansion cracks before the reinforcement materials play a role. The active reinforcement method, such as the external prestressing reinforcement method [1-3], to some extent, eliminates the problem of stress hysteresis, but there are still some problems. Guo Zixiong [4-5] et al carried out the experimental study on the axial compressive behavior of RC column strengthened with prestressed steel plate hoop, and the control of the prestress level is based on the average strain of 4 measuring points on the steel hoop. The method of reinforcing concrete column with wire winding is widely praised for its low cost, and it can improve the bearing capacity of the column, especially the ductility performance, especially in the case of only a small space. With the progress of the research on the reinforcement of reinforced concrete columns, the prestressed size monitoring [6-7] and the control have become a bottleneck in the further development of the wire.

2. Testing

The device is welded with the same size nut, and the nut sleeve can be matched with the bolt after welding. Bushing holes in the matching bolts to facilitate the rope through, The size of the apertures and bolts is determined by the actual use of wire ropes, The wire rope uses a 6 × 10 cord wire rope with a diameter of 5 mm, so the aperture is set to 12 mm.

IOP Conf. Series: Materials Science and Engineering **216** (2017) 012034 doi:10.1088/1757-899X/216/1/012034

The steel wire rope is wound to be reinforced, and both ends of the steel wire rope pass through the hole of the device and are reinforced by buckle, To ensure the reinforcement effect, the appropriate distance between buckle and bolt should be reserved, the reinforcement process continues to expand bolts, the reinforcement device is shown in Figure 1.

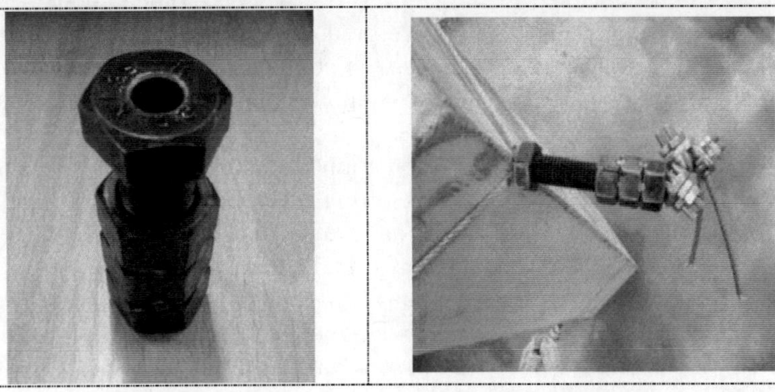

Figure 1. Reinforcement Device Diagram

3. Prestress Testing

The tensile test was carried out by the servo test machine of the material laboratory of Xi'an University of Technology， Before the start of the test, the rope fixed at both ends in the test machine and found the rope tension value equal to the actual value of prestressing. Mark the initial position of the bolt and set the initial pull value， After the test began, the bolts were rotated by 60 degrees each time， record the tensile force of the sensor and the wire rope deformation value. The effective length of the wire rope (equivalent to the distance between the two ends of the test machine) is 0.8m.The specific device shown in Figure 2。Easy to calculate and compare the data, adjust the test machine to the initial tension value of 300N.In order to study the instability of the stretching process, make a set of wire rope tensile test， Measured obtaining the stress - strain curve shown in Figure 3：

Figure 2. Tensile Testing

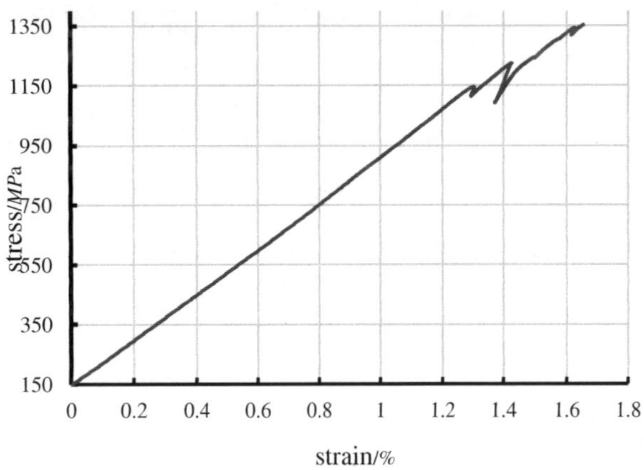

Figure 3. stress-strain curve

The following data can be obtained from the test：

①The actual measured elastic modulus of wire rope is 77.08Gpa，Strain is verified by resistance strain extensometer.
②The ultimate tensile force of wire rope is close to 9300N，And when the tension reaches 3000N, the stress-strain curve gradually becomes straight, so the data has high accuracy.
Based on the above data, the suitable scheme is used to solve the problem of stress relaxation. Wire rope was tightened to 3000N, until its tensile force gradually reduced to stable and then fine-tune to 300N, in this way, the error due to factors such as the combination of the device can be eliminated. The average of the set of data is shown in Table 1 .Combined with the Table 1, draw a curve as shown in figure 4.
As can be seen from the above figure.4, the deformation increment in the prestressing process is not uniform. The reason for this phenomenon is that during the tightening process, the nut sleeve ends have a friction effect on the wire rope, when the nut sleeve rotation, the friction is accumulated to a certain extent will be released, resulting in strain increment of uneven changes, And with the increasing stress of the wire rope, the wire rope on the nut pressure is also growing, resulting in increased static friction and the increased strain of the divergence situation. In spite of this instability, the change is relatively small relative to the tensile value, indicating that the prestressing control of the method is feasible. The corresponding relationship in the reinforcement process is not equal to the modulus of elasticity measured by the tensile test and is less than the value calculated by the stress-strain of the wire rope. The reason for the analysis is that the device makes the wire rope not as fully capable of material as the tensile test.

Table 1. Wire Rope Deformation - Rally Value Data Sheet

Wire rope deformation value(mm)	Tensile force(N)	Wire rope deformation value(mm)	Tensile force(N)
0	300	6.21	2025
0.51	480	6.73	2155
1.03	620	7.25	2240
1.55	795	7.77	2450
2.07	965	8.28	2655
2.59	1110	8.8	2730
3.1	1305	9.32	2920
3.62	1515	9.84	3015
4.14	1515	10.3	3220
4.66	1650	10.8	3240
5.18	1770	11.4	3525
5.69	1850	-	-

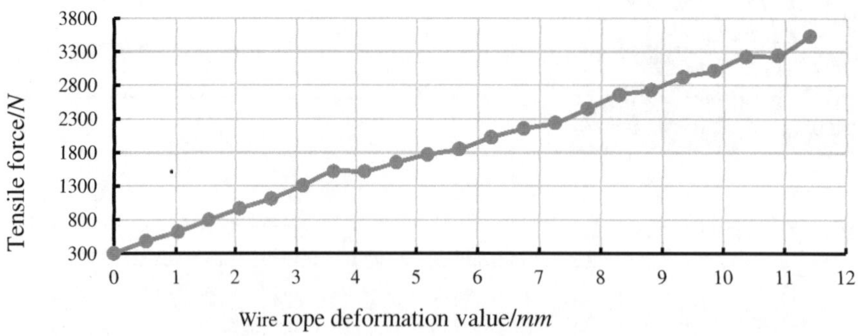

Figure 4. Wire rope deformation - tension curve

4. Theoretical Analysis and Error Analysis

The stress-strain relationship of the wire rope conforms to the elastic deformation. The section of steel strand: This kind of specifications every six strands wire rope, 10 per share, the specifications of wire rope cross-sectional area was 6.8mm2.Check from the specification table, the bolt thread spacing is a fixed value, so each rotation of a circle, the device expansion distance is fixed. However, according to the test results, the deformation of the wire rope is not equal to the bolt spacing, the actual result is that each rotation of the screw 60 °, the deformation of the wire rope is 0.13mm. And rotate once, both ends at the same time shrink, so the deformation should be multiplied by 2.

In accordance with Chinese Code for Design of Concrete Structures (GB 50010-2010),The theoretical calculations are as follows[8-10]:

$$\sigma = E \times \varepsilon \quad 1)$$

$$\frac{F}{S} = E \times \frac{\Delta L}{L}$$

$$\frac{\Delta F}{6.8mm^2} = 77.08Gpa \times \frac{0.13mm \times 2}{800mm}$$

$$result\text{:}\ \Delta F = 170N$$

Because of the interference by the friction, we have made improvements to the actual device and polished a small angle in the steel wire rope face contact with the nut sleeve, also smear a small amount of lubricant in the course of the experiment, and the coefficient of the theoretical value of the amendment. Three steel wires with the same specifications and the same batch were selected as the experimental results, and compared with the theoretical calculation values. tension - wire rope deformation curve shown in Figure 5.

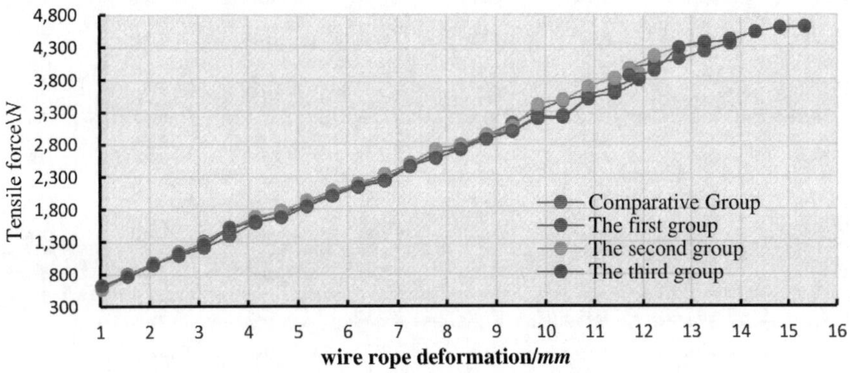

Figure 5. Total curve of deformation and tensile force of wire rope

As is clearly shown in the Figure 5, the tension value can be kept uniform, and the phenomenon that the prestress peak becomes larger, the improvement of the device can reduce the influence of friction on the experiment to a certain extent, but with the increase of the friction force, the friction between the corner and the device will inevitably increase, so we modify the theoretical value of friction coefficient γ =0.93.From the 30% of the ultimate tensile stress, the results are in good agreement with the actual values. The test data and error analysis are shown in the following table 2.

After correction by coefficient, the error between theoretical value and actual value can be kept within ±8%.It is proved that this method can not only ensure the accurate value of prestress, but also can better correspond to the relationship between the rotation angle of bolt and the prestress value of steel wire rope.

Table 2. The data of deformation and tension of wire rope

Deformation (mm)	*Comparative Group(N)*	*The first group(N)*	*The second group(N)*	*The third group(N)*	*Average value(N)*	*Theoretical value(N)*	*Modification value(N)*	*Error(N)*	*Rate of Deviation*
0	300	300	300	300	300	300	300	0	0
0.51	480	455	453	470	459	470	470	10.6	0.02
1.03	620	615	560	617	597	641	641	43.6	0.07
1.55	795	785	780	760	775	811	811	36	0.04
2.07	965	956	955	940	950	981	981	30.6	0.03
2.59	1110	1087	1140	1110	1112	1152	1152	39.6	0.03
3.11	1305	1210	1287	1255	1251	1322	1322	71.3	0.05
3.63	1515	1390	1487	1490	1456	1492	1492	36.3	0.02
4.14	1650	1590	1680	1610	1627	1663	1546	-80	-0.05
4.66	1770	1705	1775	1675	1718	1833	1704	-13	-0.01
5.18	1850	1890	1930	1845	1888	2003	1862	-25	-0.01
5. 698	2025	2040	2080	2005	2042	2174	2021	−19	−0. 01
6. 212	2155	2175	2195	2140	2170	2344	2179	9. 9	0
6. 736	2240	2280	2335	2235	2283	2515	2338	55. 6	0. 02
7. 251	2450	2470	2505	2455	2477	2685	2497	20. 3	0. 01
7. 777	2655	2635	2715	2580	2643	2855	2655	11. 8	0
8. 288	2730	2765	2790	2715	2757	3026	2814	57. 5	0. 02
8. 806	2920	2935	2945	2875	2918	3196	2972	53. 9	0. 02
9. 328	3015	3120	3070	3000	3063	3366	3130	67. 1	0. 02
9. 842	3220	3310	3405	3194	3303	3537	3289	−13	0
10. 36	3240	3465	3485	3215	3388	3707	3447	59. 2	0. 02
10. 88	3525	3559	3680	3495	3578	3877	3605	27. 6	0. 01
11.41	-	3680	3815	3585	3693	4048	3764	71.3	0.02
11.91	-	3850	3885	3790	3842	4218	3922	81.1	0.02
11.71	-	3961	3960	3855	3925	4388	4080	155	0.04
12.23	-	4035	4155	3935	4042	4559	4239	198	0.04
12.75	-	4120	4285	4280	4228	4729	4397	169	0.04
13.27	-	4230	4370	4355	4318	4899	4556	237	0.05
13.78	-	4355	-	4410	4383	5070	4715.1	332	0.07
14.3	-	-	-	4525	4525	5240	4873.2	348	0.07
14.82	-	-	-	4595	4595	5410	5031.3	436	0.08
15.34	-	-	-	4610	4610	-	-	-	-

5. Conclusion

This paper puts forward a convenient and feasible device used for wire rope reinforced concrete member pre-tightening force control. The feasibility of the device was verified by field test and theoretical calculation, and correction coefficient of friction was introduced to the theoretical calculation result has carried on the correction, through the revised coefficient, the error of the theoretical value and practical value to keep within the plus or minus 8%, Which provides the basis for the application of controllable preloading force for the reinforcement of concrete members. There is still a lot of space to explore the steel wire rope reinforcement technology. In the future, with the development of the construction technology, the research of wire rope reinforcement technology will be of great value.

6. References

[1] Wu Gang, Jiang Jianbiao, Wu Zhishen, et al. Experimental study on flexural strengthening of RC beams with prestressed high strength steel wire ropes[J].China Civil Engineering Journal,2007,40(12):17-27.
[2] Liao Xiaochun, He Tinghui, Tang Liqun, et al. Prestress Lose Calculation and Analysis of Long Span Bridge Base on Monitoring of Strain[J].JOURNAL OF EXPERIMENTAL MECHANICS,2006,21(6):742-746.
[3] Lan Chunguang, Liu Huang, Zhou Zhi, et al. Experimental investigation of prestress lose in PC beams based on BOTDA/FBG smart steel strands[J].China Civil Engineering Journal,2013,46(9):55-61.
[4] GUO Zixiong, Zhang Jie, Li Chuanlin. Seismic strengthening of rectangular RC columns using prestressing steel jackets[J].China Civil Engineering Journal,2009,42(12):112-117.
[5] GUO Zixiong, ZENG Jianyu, HUANG Qunxian, et al. Experimental study on axial compression behavior of RC columns retrofitted by prestressed steel jackets [J]. Journal of building Structures, 2012, 33(11):124-131.
[6] DENG Zongcai, Li Hui.Seismic Behavior of RC Columns Strengthened with Prestressed Steel Wires [J].JOURNAL OF BASIC SCIENCE AND ENGINEERING,2014,22(5):941-951.
[7] Guo Junping, Deng Zongcai, Lin Jinsong, et al. Experimental study of reinforced concrete slabs strengthened with prestressed high strength steel wire mesh[J].China Civil Engineering Journal,2012,45(5):84-92.
[8] GB 50010-2010, Code foe design of concrete structures[S].2010.
[9] Liu Zhaoqing, Qian Jiaru, Xu Yan, et al. Investigation on stress increments of unbounded pre-stressed strands of cast-in-situ concrete hollow floor slabs[J].Building Structure,2013,43(6):82-85.
[10] LUO Xiaoyong, Li Zheng. Mechanical Performance of Corroded Unbonded Prestressed Steel Stranded Wires [J]. JOURNAL OF THE CHINA RAILWAY SOCIETY, 2008, 30(2):108-112.

Experimental Study for Structural Behaviour of Precast Lightweight Panel (PLP) Under Flexural Load

W I Goh[1,*], N Mohamad[1], Y L Tay[1], N H A Rahim[1], A A Jhatial[1], A A A Samad[1], R Abdullah[2]

[1] Jamilus Research Center, Faculty of Civil and Environmental Engineering, Universiti Tun Hussein Onn Malaysia, Batu Pahat, Johor, Malaysia
[2] Universiti Teknologi Malaysia, Skudai, Johor Malaysia.

Email: wigoh@uthm.edu.my

Abstract. Precast lightweight concrete slab is first fabricated in workshop or industrial before construction and then transported to site and installed by skilled labour. It can reduce construction time by minimizing user delay and time for cast-in-situ to increase workability and efficiency. is environmental friendly and helps in resource reduction. Although the foamed concrete has low compressive strength compared to normal weight concrete but it has excellent thermal insulation and sound absorption. It is environmental friendly and helps in resource reduction. To determine the material properties of foamed concrete, nine cubes and six cylindrical specimens were fabricated and the results were recorded. In this study, structural behaviour of precast lightweight panel (PLP) with dry density of 1800 kg/m^3 was tested under flexural load. The results were recorded and analysed in terms of ultimate load, crack pattern, load-deflection profiles and strain distribution. Linear Voltage Displacement Transducers (LVDT) and strain gauges were used to determine the deflection and strain distribution of PLP. The theoretical and experimental ultimate load of PLP was analysed and recorded to be 70 and 62 kN respectively, having a difference of 12.9%. Based on the results, it can be observed that PLP can resist the adequate loading. Thus, it can be used in precast industry for construction purposes.

1. Introduction

Lightweight foamed concrete (LFC) is comprised of Ordinary Portland cement (OPC), fine aggregates and foamed agent without using any coarse aggregate. This helps in reduction in density. The density of lightweight concrete ranges from 1440 to 1840 kg/m^3 which is approximately 80% of normal weight concrete. It is lightweight, free-flowing, easy to pump and does not required compaction [1]. It also has good thermal insulation and sound absorption as compared to normal weight concrete [2]. Although the Industrialised Building System (IBS) was introduced in Malaysia in early 1960s but the usage level of IBS in the local construction industry stands at only 15% [3] due to lack of knowledge about properties of lightweight concrete. Thus, this study was carried out to achieve better understanding of the structural behaviour of PLP subjected to flexural load.

2. Literature Review

2.1 Structural Behaviour of Panel.

Each construction has its own structure part which is responsible to sustain the loading and prevent it from failure. Structures are designed to sustain loads and types of building to be constructed. Thus, ultimate load, cracking pattern and deflection of structures have a relationship to the failure of structure. The ultimate flexural load of panel will be influenced by compressive strength, thickness [4] and density [5]. According to Mustaffa, higher density of foamed concrete also influences the attained ultimate load of slab under test. According to Mohamad et al., two types of crack patterns were discovered on a panel which are flexural crack and shear crack. Flexural cracks started to occur between the point of loading (mid-span) where the concrete experienced high local tensile stresses as loading applied to the slab followed by a shear crack propagated along the slabs. Further increasing the load, the flexural crack propagated upwards reaching top of the slab with increased deflection. Finally, the panels failed by crushing of concrete at the spot nearest to the point of loading [5].
According to the study on the use of additives to enhance properties of preformed foamed concrete [6], it was found that the compressive strength of foam concrete decreased with a reduction of density but the compressive strength increased with the use of additives. This is due to the reduction of water content with the formation of less porous interfacial zone and provided a better interlocking between the paste and the aggregates. The results showed that the mixes with density 1300 and 1600 kg/m^3 are not suitable for structural purposes due to the low compressive strength obtained. The tensile strength and splitting of conventional mixes is higher than mixes with additives. For a given density, the sand content is lower in mixes with additives and degraded the shear capacity.
A research was conducted to study the flexural behaviour of structural fibre composite sandwich beams in flatwise and edgewise positions [7] in which composite sandwiches with nominal thicknesses of 18 mm and 20 mm were tested under flexural load. In the flatwise position failed with sudden brittle type failure and the presence of fibre composite skins increased the ultimate strength prevented the sudden failure of the composite sandwich in edgewise. Strength is significantly improved by the introduction of the non-horizontal fibre composite skins, thus could be used in the inner portion of the laminated beams to carry shear.

3. Materials and Methods

The materials used in foamed concrete are Ordinary Portland cement, fine aggregate, water and foam. 2:1 sand-cement ratio was used while the foam-water was 1:20 and 0.55 water-cement ratio was taken. In order to test the mechanical properties of foamed concrete and flexural behaviour of panel, cube specimens of 100 x 100 x 100 mm, cylinder specimens of 300 x 150 mm and PLP specimens of 1400 x 500 x 200 mm dimensions were cast. The detailing of the panel is showed in figure 1.

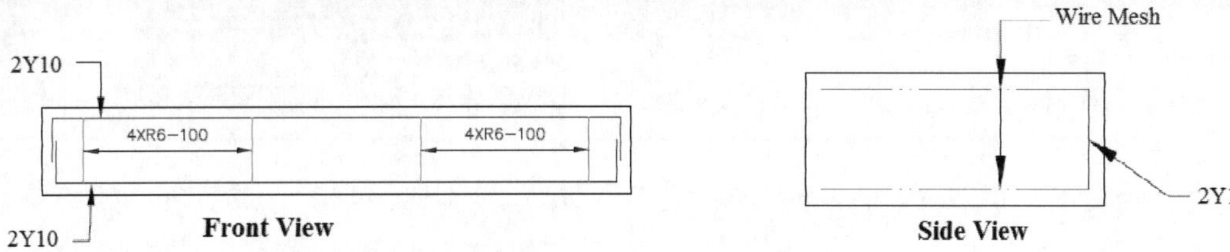

Figure 1. Detailing of PLP

Test Set-up and Procedure.

The flexural strength was performed on specimens of 1000 × 500 × 150 mm dimensions using the four-point loading procedure. PLPs were cast and were tested after 28 days curing. The test specimen was placed in the Magnus Frame, hydraulic jack of machine was loaded to the mid-span of the panel and the distance between two point loads was 400 mm as shown in figure 2.
A total of 24 electrical strain gauges of 30 mm length were placed on top, bottom, front and rear surface. Each surface consisted of 6 strain gauges. To measure the deflection of the panel three Linear Voltage Displacement Transducers (LVDT) were used. LVDT 1 was placed at the center while LVDT 2 and LVDT 3 were placed at the bottom of panel where the load was applied. All the strain gauges and

LVDTs were connected to data logger and computer for data recording purposes. The crack pattern was marked on the surface and noted at each load stage.

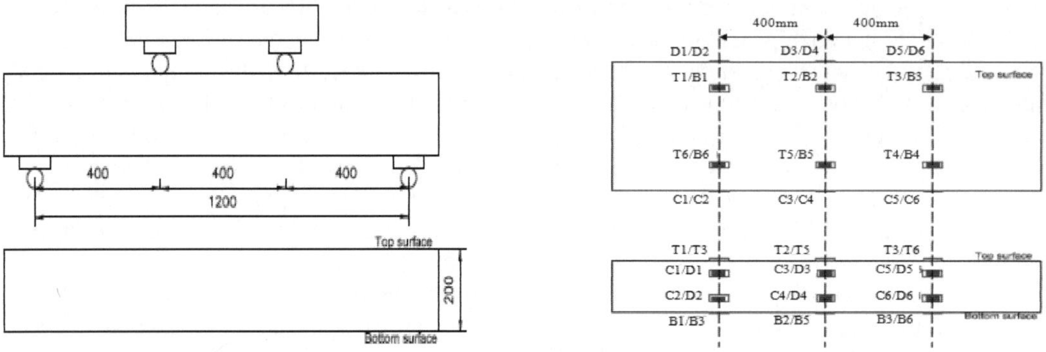

Figure 2. Test setup specimen and location of strain gauge and LVDT

4. Results and discussion

The material properties of foamed concrete and ultimate load of PLP are shown in table 1.

Table 1. Properties of Foamed Concrete of PLP

Compressive strength	Tensile strength	Elastic modulus	Ultimate load	
			Experimental	Theoretical
13.93 kPa	1.34 kPa	8.509 kN/mm^2	62 kN	70 kN

4.1 Crack Pattern.

Figure 3 shows the cracking pattern of PLP at different location. The load was applied on panel incrementally but it was stopped before failure but after the ultimate load occurred. The first crack occurred when 30 kN load was applied. The flexural crack started at the bottom and propagated to top of panel. All the cracks developed at the middle part of the panel where the concrete experienced highest tensile stresses. However, no shear crack occurred, this is due to the reinforcement and stirrups provided in the panel. The reinforcement provided had gripped the concrete together and reduced the cracking. The major crack happened at mid-span of the panel.

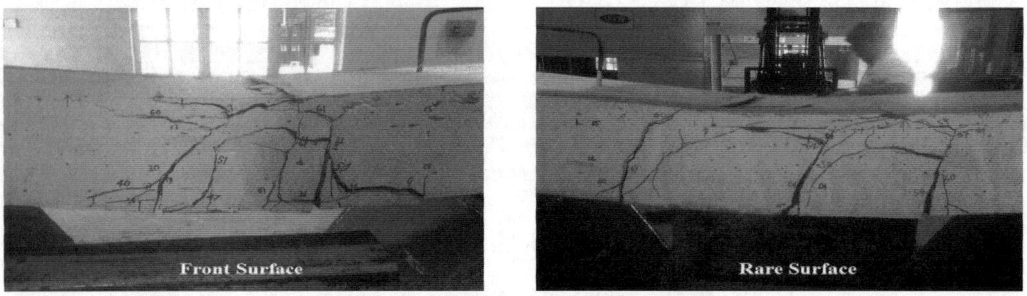

Figure 3. Cracking pattern of PLP

4.2 Load Deflection Profile

The deflection happened before first crack meaning that the panel had deformed as the loads were applied. Load-deflection curves were approximately linear which can be assumed as elastically deformation but the curve become non-linear and deflection increased significantly until reaching ultimate load after first crack as shown in figure 4. The deflection for LVDT2 is higher compared to LVDT1 and LVDT3 at ultimate load. It can be observed that as the loads increase, the deflection of panel also increases as shown in figure 5. At initial stage, the deflection of right part of panel deformed more as compared to other parts but at the ultimate load, mid-span deflected most indicating critical area of panel.

IOP Conf. Series: Materials Science and Engineering **216** (2017) 012035 doi:10.1088/1757-899X/216/1/012035

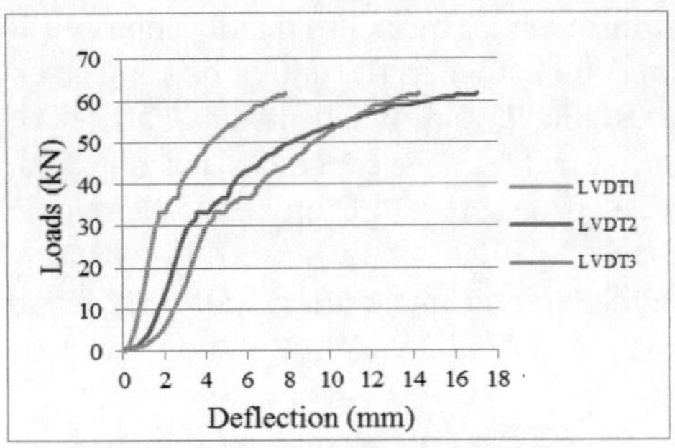

Figure 4. Loads versus deflection

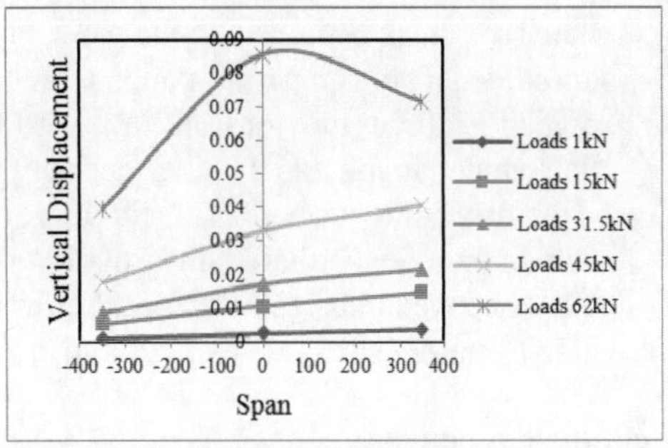

Figure 5. Vertical displacement along span

4.3 Strain distribution

There are total of 12 strain gauges in front and rear surface. Each surface consists of two rows of strain gauges and each row has 3 strain gauges. It is distributed into three parts (left, right and middle) and labelled as top (left or right) and bottom (left or right) as shown in figure 6, while front (up or down) and rear (up or down) as shown in figure 7. load-strain curve at upper part tends to move to the negative along the applied load. Therefore, all the strain gauges at right part experienced compressive strain at ultimate load and this indicated that the right part of the panel undergoes compressive strain. In addition, some of the curve is not completed compared with the others. This is because the data had been removed since this inconsistence results is affected by the cracks which had passing through the location of strain gauge during testing.

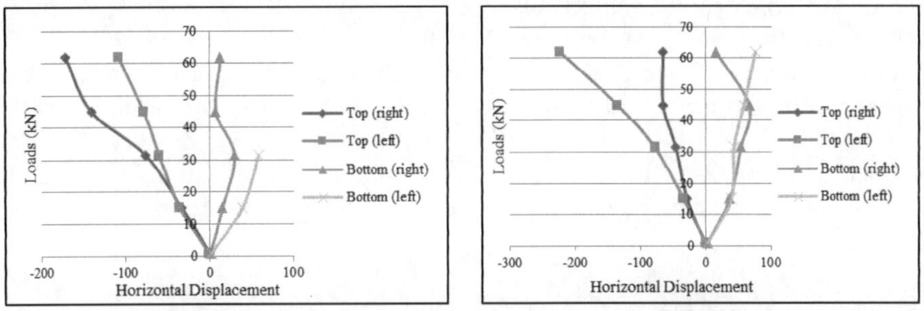

Left Side **Right Side**

Figure 6. Comparison of loads versus strain graph between top and bottom surface of panel

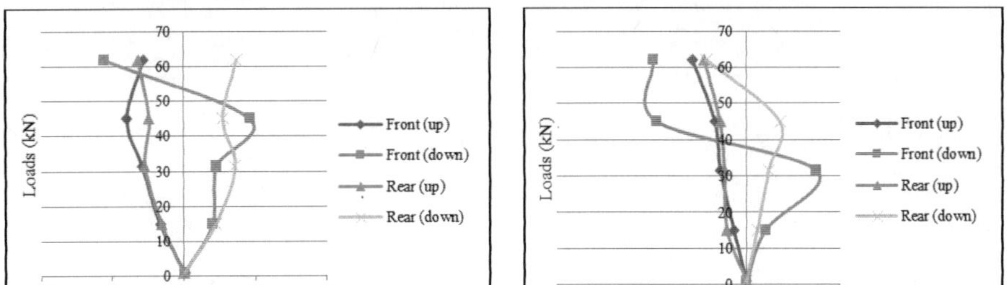

Left Side **Right Side**

Figure 7. Comparison of loads versus strain graph between front and rear surface of panel

5. Conclusion

This study was carried out to investigate the flexural behaviour of Precast Lightweight Panel (PLP) under flexural load. Based upon the results, the following were concluded:

i. The ultimate load tabulated according to BS 8110 is 70 kN which has a difference of 12.9% from experimental data, 62 kN. Although the loading sustained by PLP is not as much as normal weight concrete but it is able to be applied in low rise building which carry lesser loading.

ii. The first crack occurred on the panel at 50% of the ultimate load. The cracks mainly appeared at middle and bottom part of panel. Cracks which appeared are known as flexural or bending crack which start at the tension face and extend upwards. Steel reinforcements that had been introduced in the panel had effectively reduced the shear crack as no shear cracks had appeared.
iii. The larger the applied load, the larger the deflection of panel. The PLP proved to be ductile which experienced large deformation before failure.
iv. PLP behaved elastically before the first crack but it turned to non-linear afterwards. PLP failed due to tensile stress with excessive cracks appearing in the tension zone of the bottom surface.

6. Acknowledgements
The authors would like to thank Office for Research, Innovation, Commercialization, Consultancy Management, Universiti Tun Hussein Onn Malaysia for the financial support for this project.

References
[1] Mohamad N, Goh W I, Abdullah R, Samad A A A, Mendis P and Sofi M 2017 Structural performance of FCS wall subjected to axial load *Construction and Building Materials* **134** p 185 - 198.
[2] Goh W I, Mohamad N, Abdullah R and Samad A A A 2014 Compression test and finite element analysis of foamed concrete cube. *Journal of Engineering and Technology* **5 (1)** p 1-9.
[3] IBS survey 2003 *Lembaga pembangunan industri pembinaan Malaysia.* CIDB Malaysia.
[4] Mohamad N, Khalil A I, Samad A A A and Goh W I 2014 Structural behaviour of precast lightweight foam concrete sandwich panel with double shear truss connectors under flexural load *ISRN Civil Engineering.*
[5] Mustaffa N K 2008 Structural behaviour of reinforced foamed concrete slab.
[6] Hilal A A, Nicholas H T and Andrew R D 2015 The use of additives to enhance properties of preformed foamed concrete *International Journal of Engineering and Technology* **7 (4)** p 286–293.
[7] Manalo A C, Aravinthan T, Karunasena W and Islam M M 2010 Flexural behaviour of structural fibre composite sandwich beams in flatwise and edgewise positions *Composite Structures* **92 (4)** p 984 – 995.

IOP Conf. Series: Materials Science and Engineering **216** (2017) 012036 doi:10.1088/1757-899X/216/1/012036

Passive Vibration Control of Existing Structures by Gravity-Loaded Cables

E Alvis, H H Tsang, and M J Hashemi

Department of Civil and Construction Engineering, Swinburne University of Technology, Melbourne, Australia, 3122

E-mail: 100033624@student.swin.edu.au, htsang@swin.edu.au, jhashemi@swin.edu.au

Abstract. Structures with high concentration of mass at or close to the top such as highway bridge piers are vulnerable in earthquakes or accidents. In this paper, a simple and convenient retrofit strategy is proposed for minimizing vibrations and damages, extending service life and preventing collapse of existing structures. The proposed system comprises of tension-only cables secured to the sides of the structure through gravity anchor blocks that are free to move in vertical shafts. The system is installed in such a way that the cables do not induce unnecessary stress on the main structure when there is no lateral motion or vibration. The effectiveness of controlling global structural responses is investigated for tension-only bilinear-elastic behaviour of cables. Results of a realistic case study for a reinforced concrete bridge pier show that response reduction is remarkably well under seismic excitation.

1. Introduction

One of the major challenges facing structural engineers today is to develop creative ways to reduce the risk of catastrophic damage due to extreme events, and to enhance the resiliency of urban infrastructure. Particularly, earthquake is a hazard with severe consequences on civil infrastructures that ranges from direct and indirect economic losses due to structural and non-structural damages as well as business interruption; to social effects such as injuries and fatalities; and environmental impacts such as energy consumption and carbon dioxide emissions due to subsequent repair/replacement, among others [1] .

The need for retrofitting existing infrastructures that are prone to earthquakes is common all around the world. Possible reasons include aging and/or recognition of their unsatisfactory performance as a result of updates in seismic code requirements. There are also changes in the design performance objectives under the emerging performance-based seismic design framework.

Apart of seismic motion, non-redundant structures are liable to suffer imminent collapse due to unexpected loads. Figure 1 shows the same pattern of failure for single rowed pier bridges under earthquake motion or truck collision. In both scenarios the strength and stiffness of the reinforced concrete columns were not enough to prevent excessive lateral displacement and P-Delta effects disregarding the high and low level of seismic hazard in where they are located.

Figure 1. Left: Collapse of bridge in the Kobe earthquake 1995 [2], Right: Pedestrian bridge collapse after being hit by a truck [3].

Traditionally, approaches taken by engineers to mitigate the earthquake hazard have involved methods with the assumption of seemingly limitless natural resources, and with little regard to the impact of their rehabilitation techniques on the environment, and local disruption. Davidson et al. [4] suggest that the next generation of engineers must be able to design with a narrowing and diminishing set of natural resources for a wider variety and greater number of end users. The development of rapid rehabilitation strategies has sustainability benefits in terms of providing more resilient infrastructure for the community as well as minimizing environmental and economic impacts and social consequences during the rehabilitation project and after occurrence of earthquakes.
The primary objective of this study is to design and evaluate an innovative, rapid retrofitting strategy for structures with high concentration of mass at or close to the top such as highway bridge piers that are susceptible to P-Delta effects and therefore vulnerable in earthquakes or collisions. Also, the proposed supplemental system is to be consistent with the following principles of sustainable design. The systems must: 1) result in robust and resilient infrastructure; 2) be efficient in the use of materials, with minimal energy requirements and emissions; 3) be able to achieve its goals with only minor on-site construction and disruption; 4) require minimal modification to existing structural elements; and 5) require little maintenance. To achieve these goals, a novel approach for a supplemental seismic retrofitting system is proposed, which includes the use of gravity-anchored cable-bracing system consisting of tension-only cables that are attached to the top of the structure from one-side and secured to gravity anchor blocks from the other side. Utilizing this method, cables begin to carry load immediately after a deformation reversal through a stable tension-only bilinear-elastic behaviour. The level of accepted damage in the cable (e.g. no-damage, yielding, rupture) can be controlled by assigning appropriate gravitational force to the anchorage system. In this paper, the concept and installation of this retrofitting system are discussed and the effectiveness of implementing this system to control the global response of a bridge pier subjected to sequential ground motions with increasing intensities is investigated. Relative movements between the bridge deck and the pier are not considered in this study. The simulation results show promise for effective vibration control and collapse prevention of existing structures.

2. Gravity-anchored cable-bracing system

2.1. Concept of the proposed retrofitting system

Cable-bracing systems are commonly used in conjunction with linear/rotational friction-based dampers [5] or hysteretic dampers [6] and [7] while the cable-bracing system proposed here can also be used in combination with energy-dissipating devices, this study is focused only on the gravity-anchored cable-bracing system and its effectiveness for vibration control of structures.
The concept of the proposed gravity-anchored cable-bracing system is depicted in figure 2 through a single-degree-of-freedom (SDOF) structure. The installation of the system requires drilling vertical shafts on opposite sides of a structure to accommodate the gravity anchor blocks. In the first stage of the installation (Figure 2(A)), the tension-only cables are placed as braces for the structure with one side attached to the top of the structure, and the other side secured with the gravity anchor blocks.

Initially, light weights are used to develop low-level tensions in the cable. In the second stage (Figure 2(B)), clamp elements are installed at the top of the shaft (i.e. intersection of Lo1 and Lo2) to restrain the downward motion of cable (Figure 2(C)). After installation of the clamps, gravity blocks are added to produce pre-tension forces only in the underground part of the cable. This allows the cables to be effective only with lateral movements of the main structure without applying unnecessary forces during normal service. This system can be comparable to seatbelts in vehicles where levels of tension can be adjusted, allowing passengers to move freely, while if the acceleration exceeds certain pre-defined level, system raises its tension resistance, proportionally.

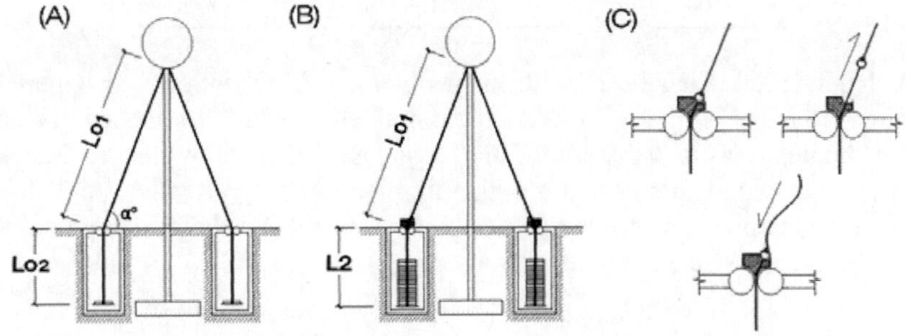

Figure 2. Concept and installation stages of the proposed gravity-anchored cable-bracing system.

The behaviour of the cables in the proposed system follows stable tension-only bilinear-elastic behaviour. As shown in figure 3(A), while the tension forces developed in the cable are less than the gravitational force of the blocks, they behave in linear-elastic range. The cables are placed on opposite side of the structures and are activated as opposed to direction of the external forces for controlling the excessive deformations. As this stage, the cables behave as if they are fixed to the ground using a normal pin connection.

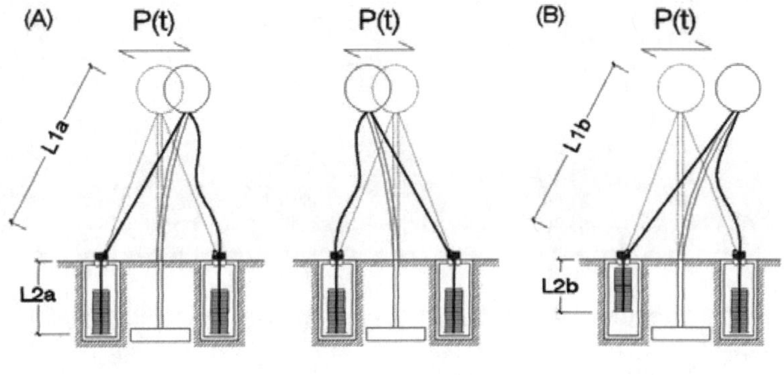

Figure 3. Seismic behaviour of a SDOF retrofitted structure.

Once the cable tension force reaches the gravitational force of the blocks, they will lift the blocks and displace them freely in vertical shafts (Figure 3(B)). This allows to calibrate the weight of gravity blocks based on preventing different levels of damage in the cables (e.g. no damage, yielding, rupture); making this retrofitting system a sustainable solution, without the need for repair/replacement of the cables.

3. Sesimic retrofit of A reinforced concrete bridge pier

3.1. Description of the model

In order to assess the effectiveness of the gravity-anchored cable-bracing system in a realistic scenario, a reinforced concrete bridge pier was considered as a case study. The bridge pier is 15m high and has a circular cross-section with 1m diameter. The longitudinal column reinforcement consists of 22 bars of N24. The pier contains transverse reinforcement throughout the entire length with 20mm cover

IOP Conf. Series: Materials Science and Engineering **216** (2017) 012036 doi:10.1088/1757-899X/216/1/012036

thickness that adds 5% to maximum compressive stress of the concrete in the confined region. The material properties of the bridge pier are presented in table 1.

Table 1. Material properties of the bridge pier.

	Concrete (cover)	Concrete (core)		Steel
ε_c	0.002	0.002	ε_y	0.00275
f'_c	25 MPa	26.25 MPa	f_y	550 MPa
ε_{cu}	0.006	0.006	ε_u	0.05
f'_{cu}	0.0 MPa	5.0 MPa	f_u	594 MPa

The bridge pier is modelled using beam-with-hinges element in OpenSees [8], where the nonlinear behavior is assumed to occur within a finite-length at both ends using fiber-based plasticity approach. The tension-only bilinear-elastic behavior of the cable is modelled by taking 2500kN/m for the horizontal component of the axial stiffness and the ultimate force is equivalent to 30% of the gravity load applied on the bridge pier. The combined dead and live loads on the bridge pier is assumed to be 10% of the ultimate compressive axial load capacity.

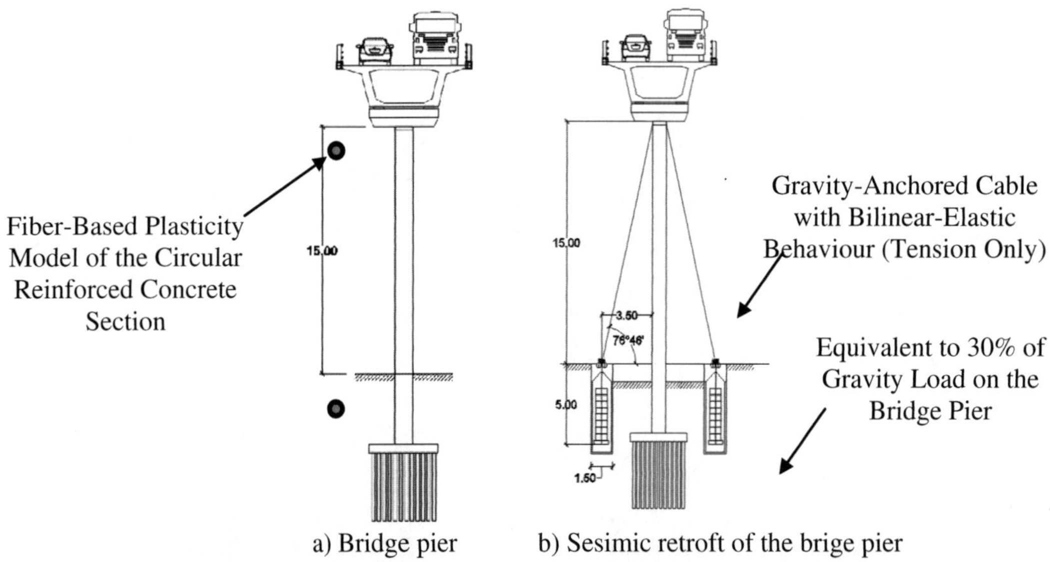

Figure 4. Seismic retrofit of the bridge pier using gravity-anchored cable-bracing system.

Figure 4 shows installation of the retrofitting system on opposite sides of the bridge pier. Note that, no friction-based or energy-dissipating devices are used and it is assumed that the cables cannot be installed with the optimized angle of 35° due to the limited available space. The slope ratio of the cable is set to 3.5/15, resulting in an installation angle of approximately 75°. The bridge pier for the cases with and without seismic retrofitting system was then subjected to four sequential ground motion increasing intensities. The El Centro earthquake is the nominated response spectra input force (Figure 5) and the selected scale factors are 0.5, 1.0, 2.0 and 4.0 to capture the full range of response from linear-elastic to the state of near-collapse.

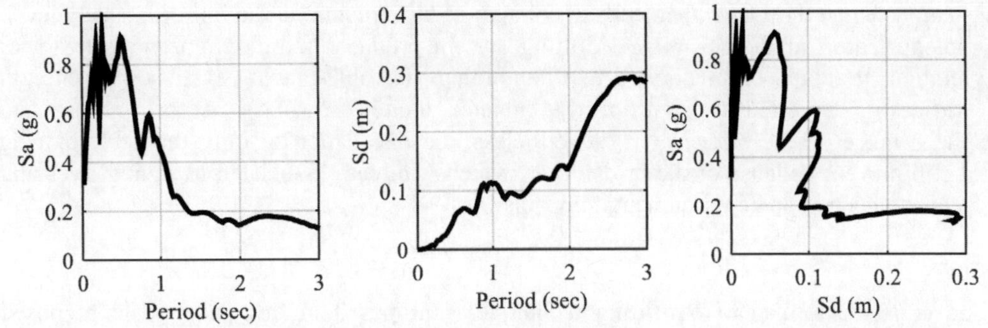

a) Acc. Response Spec. b) Disp. Response Spec. c) Acc.-Disp. Response Spec.

Figure 5. Response spectra for the El Centro ground motion.

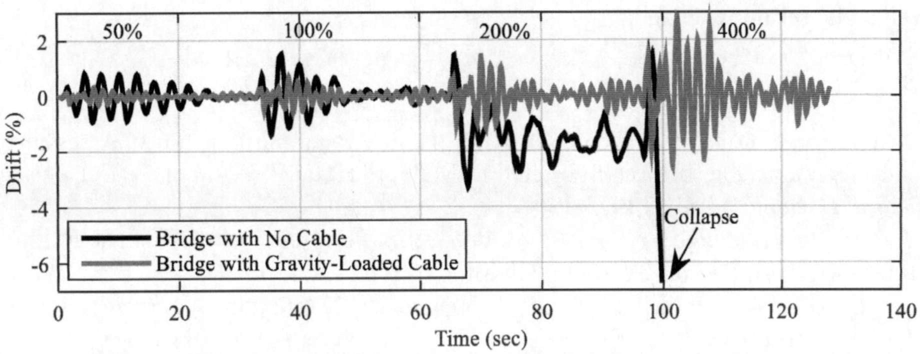

a) Drift time-history for four sequential ground motions with increasing intensities

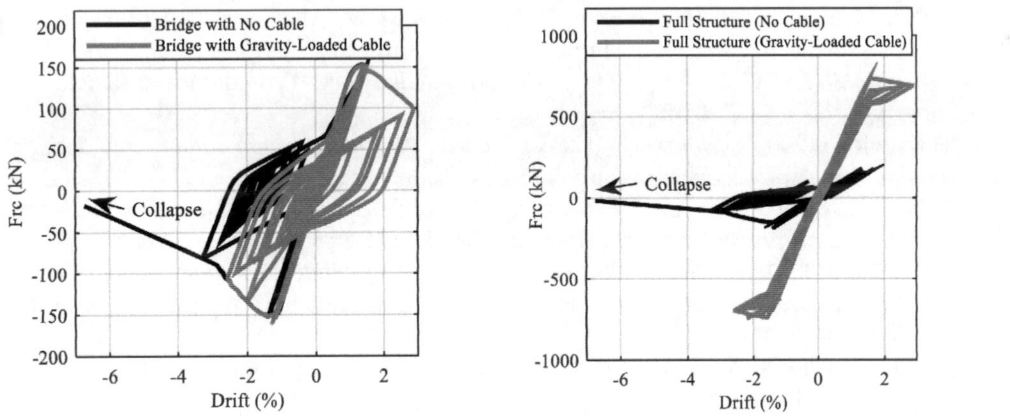

b) Comparison of hysteretic response of the bridge pier c) Comparison of hysteretic response of the full structure including the bridge pier and cables

Figure 6. Comparison of structures response for cases with and without seismic retrofitting system.

3.2. Results and discussion

Figure 6 shows the comparison of the structural response for both cases, including the lateral-drift response time history, the hysteretic response of the bridge pier element and the hysteretic response of the full structure including the cable-bracing system. As observed from the figures, the bridge pier without the use of cable-bracing system shows larger drifts in the first two intensity levels and large plastic deformations in the third level that leads to collapse of the structure in the fourth intensity level. However, implementation of the seismic retrofitting system produces more symmetric with lower amplitude vibration. It should be mentioned that combining the cable-bracing system with energy-dissipating devices is expected to significantly enhance the effectiveness of this system and minimizing the damage on the bridge pier. Nevertheless, the installation of only the cable-bracing system could still prevent collapse of the bridge pier, which is directly translated to saving lives and reducing the economic costs and environmental impacts.

4. Conclusion

A simple and convenient seismic retrofitting technique is proposed in this paper. The proposed gravity-anchored cable-bracing system comprises of tension-only cables attached to opposite sides of the structure through gravity anchor blocks that are free to move in vertical shafts. The effectiveness of controlling global structural responses has been demonstrated based on tension-only bilinear-elastic behaviour of cables, and the optimisation of cable angle has also been discussed. Results of a realistic case study for a reinforced concrete bridge pier show that reduction of displacement response is remarkably well under seismic excitation.

5. REFERENCES

[1] Dong Y, Frangopol D M & Saydam D 2013 Time-Variant Sustainability Assessment of Seismically Vulnerable Bridges Subjected to Multiple Hazards, *Earthquake Engineering & Structural Dynamics*, 42(10), 1451-1467.

[2] Scheer J 2010 *Failed bridges, case studies, causes and consequences*. (Ernest and Sohn Wiley Company). Berlin, Germany Chapter 10 p p20.8

[3] Lawrence E, Allen R and Stafford K (2014, September, 27). Southfield Freeway Pedestrian Bridge Collapse. Detroit Free Press. Retrieved from: http://www.freep.com/story/news/local/michigan/detroit/2014/09/26/southfield-freeway-pedestrian-bridge-collapse/16252769/

[4] Davidson C I, Matthews H S, Hendrickson C T, Bridges M W, Allenby B R, Crittenden J C, Chen Y S, Williams E, Allen D T, Murphy C F & Austin S 2007 *Adding sustainability to the engineer's toolbox: A challenge for engineering educators*, Environmental Science & Technology, 41(14), 4847-4850.

[5] Mualla I H & Belev B 2002 *Performance of steel frames with a new friction damper device under earthquake excitation*, Engineering Structures, 24(3), 365-371.

[6] Phocas M C & Pocanschi A 2003 *Steel frames with bracing mechanism and hysteretic dampers*, Earthquake Engineering & Structural Dynamics, 32(5), 811-825.

[7] Kurata M, Leon R T & DesRoches R 2012 *Rapid Seismic Rehabilitation Strategy: Concept and Testing of Cable Bracing with Couples Resisting Damper*, Journal of Structural Engineering-Asce, 138(3), 354-362.

[8] McKenna F 2011 Opensees: A Framework for Earthquake Engineering Simulation, *Computing in Science & Engineering*, 13(4), 58-66.

IOP Conf. Series: Materials Science and Engineering **216** (2017) 012037 doi:10.1088/1757-899X/216/1/012037

Analysis of Lateral Buckling of Bar with Axial Force Accumulation in Truss

Nuttapon Wattanamankong, Arthit Petchsasithon and Suwat Dhirasedh

Faculty of Engineering, King Mongkut's Institute of Technology Ladkrabang, Thailand

E-mail: civil_ct25@hotmail.com; art_arthit@hotmail.com and kdsuwat@kmitl.ac.th;

Abstract. This research studies the lateral buckling behavior in truss and lateral buckling coefficient of truss. Lateral buckling analysis of truss is performed by simulating the structural model with both end supports being pinned and roller-supports. The analysis is indirectly conducted using Elastic Theory to evaluate the length of lateral buckling by calculating the determinant of the Matrix [K]. Results from the analysis are marginally different from those obtained from finite element program and are considerably less than those obtained from Eurocode standard. This can be concluded that using elastic theory to evaluate lateral buckling coefficient of truss member will result in more economical section.

1. Introduction

For decades, Truss has been one of the most popular structures used for long span roof structure. Since the internal forces of truss structure are purely axial force (both tension and compression), buckling of compressive members is the most important aspect in the structural design. Nevertheless, structural engineers generally concern only in-plane buckling of truss member and ignore the lateral buckling effect. In most case, the failure mode of truss is out-of-plane or lateral buckling. Lateral buckling shape in truss without and with lateral bracing at the center of truss is illustrated in fig. 1(a) and 1(b), respectively

In the Eurocode3 [1], lateral buckling coefficient, *k*, is suggested to be 0.9 for truss structures using hollow section members no matter what the magnitude of the applied forces and the number of elements in truss are. Nevertheless some researchers have studied the effect of the number of elements in truss and magnitude of the applied force on lateral buckling coefficient such as Joanna [2] and Nattapon *et al.* [3]. Joanna [2] studies the effect of position of applied load, unbraced length and type of restraint on lateral buckling coefficient. Tada *et al.* [4] behavior post-buckling by dynamic loads and Chan *et al.* [5] the methodology has been experimentally verified with full scale test of one-bay truss. It has been shown that taking into consideration torsional stiffness of the both chords can considerably decrease the out-of-plane chord's buckling length.

From both Biegus [6] and Joanna [2] researches, it was found that the coefficient of lateral buckling suggested in the Eurocode3 [1] is too conservative.

Thus, this research aims to directly study the behavior of lateral buckling of truss structures in order to efficiently evaluate the lateral buckling coefficient. The analysis is performed using Buckling of a Bar with Intermediate Compressive Forces [7]. The above theory can be applied to truss structure by applying axial forces along the length of the member.

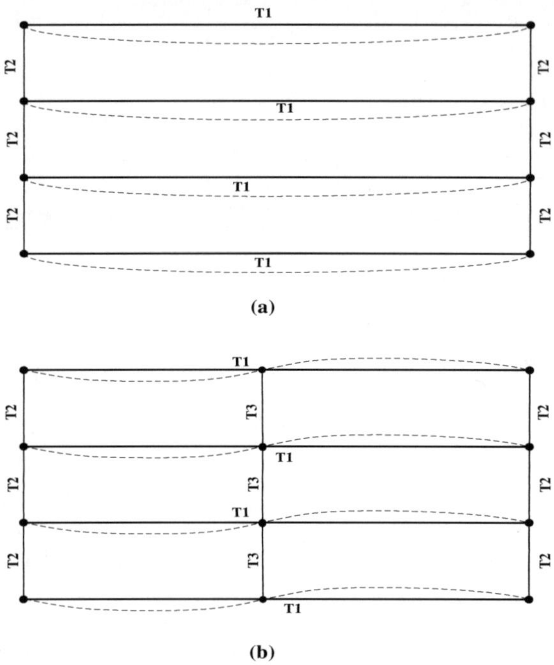

Figure 1. The plan lateral buckling of truss.

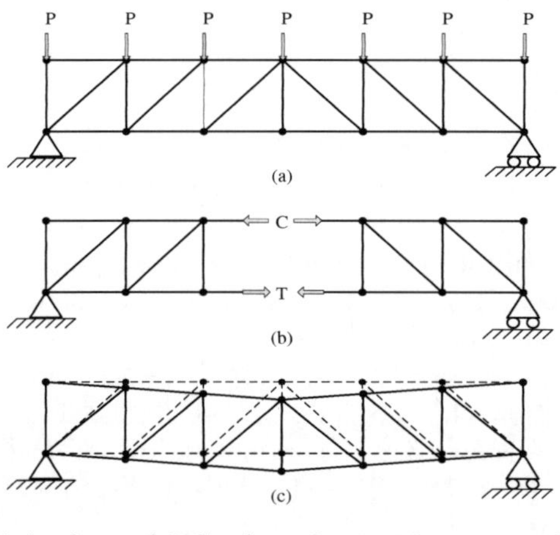

Figure 2. The behavior and deflection of truss when external forces are applied.

2. Modeling of Member in truss

To simplify the analysis of truss structure, 2-D model with external loads acting at the joints is simulated as illustrated in fig. 2(a). It can be seen from the structural analysis that top chord members resist compressive force and transfer loads to bottom chord members through diagonal members as shown in fig.2(b). This causes deflection in truss and therefore the behavior of truss is like non-homogeneous beam structure.

Analysis of truss structure in fig. 2(c) illustrated that the internal force in top chord members are compression. Therefore they behave like structural columns. The analysis is performed by simulating stability of bar member in truss model using Elastic Theory shown in fig. 2(a). The column like member is divided into n elements with equal length and constant stiffness. The external load is

IOP Conf. Series: Materials Science and Engineering **216** (2017) 012037 doi:10.1088/1757-899X/216/1/012037

applied at each element. This would increase the accumulated force shown in fig. 3 and consequently, increase stress in each element shown in fig. 3(c).

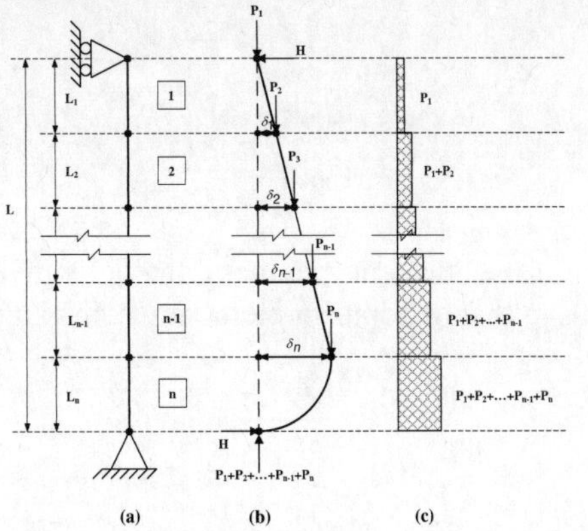

Figure 3. Accumulation of compressive force in bar member model with single curvature buckling shape

3. The Analysis of Bar Member in Truss

Analysis using Elastic Theory is performed under the assumptions that all axial forces are in compression, member stiffness is constant throughout the whole member and lastly buckling shape of the member is single curvature shown in fig. 4.

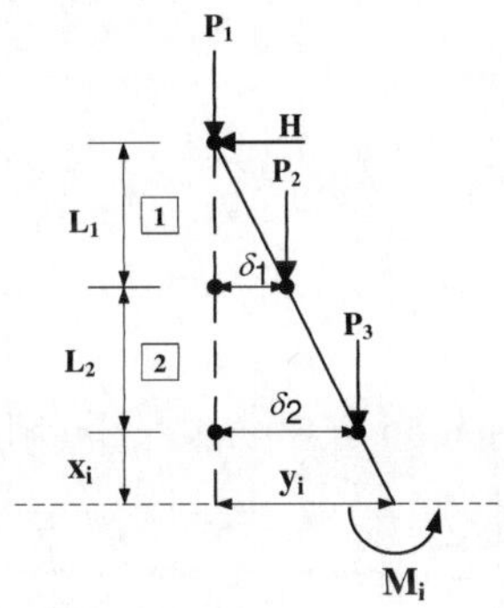

Figure 4. Analysis of the accumulating axial force in member

The second-order differential equation for beam-like element can be written as

$$EI\frac{d^2y}{dx^2} = -M \tag{1}$$

Using the following notations

$$k^2 = \frac{P}{EI} \tag{2}$$

The general solution for the above differential equation is

$$y = A\sin(kx) + B\cos(kx) + Cx + D \tag{3}$$

The critical force (P_{cr}) in curved member can be obtained from the analysis of bar member model. The critical force is carried out by setting the determinant of the matrix $[K]$ to be zero in (4) and solving equation by numerical analysis [8,9]. Consequently, coefficient of lateral buckling (k) is obtained and therefore critical force can be evaluated as illustrated in (5).

$$[K]\{\delta\}=0 \tag{4}$$

$$P_{cr}=\frac{\pi^2 EI}{(kL)^2} \tag{5}$$

With the continuity condition of the adjacent members, the derivative of lateral displacement *(y)* with respect to x of element i equals to that of adjacent element which can be written as

$$\left(\frac{dy_i}{dx_i}\right)_{x_i}=\left(\frac{dy_{i+1}}{dx_{i+1}}\right)_{x_i} \tag{6}$$

In which $x_i = L-\sum_{i=1}^{n} L_i$

Applying the lateral reaction at support

$$H=\frac{1}{L}\sum_{i=2}^{n} P_i\delta_{i-1} \tag{7}$$

Using the second order differential equation

$$EI\frac{d^2 y_1}{dx_1^2}=P_1 y_1 - H(L-x_1) \tag{8}$$

$$EI\frac{d^2 y_n}{dx_n^2}=\sum_{i=2}^{n} -P_1 y_i - P_i(y_i-\delta_{i-1})-H(L-x_i) \tag{9}$$

Under the condition $L-\sum_{i=1}^{m} L_{i-1}\le x_m \le L-\sum_{i=1}^{m} L_i$ By $m=1\ 2\ 3\ ...n$

The general solution for differential equation at any lateral buckling position can be demonstrated as,

$$y_1=-\left[\frac{\sin(k_1\cdot x_1)\cdot\cos(k_1\cdot L)}{Lk_1^2\cdot\sin(k_1\cdot L_1)}\left[\left(k_1^2\cdot\delta_1\right)\cdot L\right]+\left[\left(\sum_{i=2}^{n} k_{2i-2}^2\cdot\delta_{i-1}\right)\cdot L_1\right]\right]$$
$$+\left[\frac{\sin(k_1\cdot L)\cdot\cos(k_1\cdot x_1)}{Lk_1^2\cdot\sin(k_1\cdot L_1)}\left[\left(k_1^2\cdot\delta_1\right)\cdot L\right]+\left[\left(\sum_{i=2}^{n} k_{2i-2}^2\cdot\delta_{i-1}\right)\cdot L_1\right]\right]+\left[\frac{x_1-L}{Lk_1^2}\sum_{i=2}^{n} k_{2i-2}^2\cdot\delta_{i-1}\right]$$

(10)

$$
\begin{aligned}
y_t = {} & \left[\frac{\sin(k_{2t-1} \cdot x_{t-1})}{L k_{2t-1}^2 \sin(k_{2t-1} \cdot L_{t-1})} \left[\begin{aligned} & \left(\begin{aligned} & \left(k_{2t-1}^2 \cdot \delta_{t-1} - \sum_{i=2}^{t} k_{2i-2}^2 \cdot \delta_{i-1} \right) \cdot L \\ & + \left(\sum_{i=2}^{n} k_{2i-2}^2 \cdot \delta_{i-1} \cdot \sum_{i=1}^{t} L_{i-1} \right) \end{aligned} \right) \cdot \cos\left(k_{2t-1} \cdot L - \sum_{i=1}^{t} L_i \right) \\ & + \left(\begin{aligned} & - \left(k_{2t-1}^2 \cdot \delta_{t} - \sum_{i=2}^{t} k_{2i-2}^2 \cdot \delta_{i-1} \right) \cdot L \\ & + \left(\sum_{i=2}^{n} k_{2i-2}^2 \cdot \delta_{i-1} \cdot \sum_{i=1}^{t} L_{i} \right) \end{aligned} \right) \cdot \cos\left(k_{2t-1} \cdot L - \sum_{i=1}^{t} L_{i-1} \right) \end{aligned} \right] \right] \\
& - \left[\frac{\cos(k_{2t-1} \cdot x_{t-1})}{L k_{2t-1}^2 \cdot \sin(k_{2t-1} \cdot L_{t-1})} \left[\begin{aligned} & \left(\begin{aligned} & \left(k_{2t-1}^2 \cdot \delta_{t-1} - \sum_{i=2}^{t} k_{2i-2}^2 \cdot \delta_{i-1} \right) \cdot L \\ & + \left(\sum_{i=2}^{n} k_{2i-2}^2 \cdot \delta_{i-1} \cdot \sum_{i=1}^{t} L_{i-1} \right) \end{aligned} \right) \cdot \sin\left(k_{2t-1} \cdot L - \sum_{i=1}^{t} L_i \right) \\ & + \left(\begin{aligned} & - \left(k_{2t-1}^2 \cdot \delta_{t} - \sum_{i=2}^{t} k_{2i-2}^2 \cdot \delta_{i-1} \right) \cdot L \\ & - \left(\sum_{i=2}^{n} k_{2i-2}^2 \cdot \delta_{i-1} \cdot \sum_{i=1}^{t} L_{i} \right) \end{aligned} \right) \cdot \sin\left(k_{2t-1} \cdot L - \sum_{i=1}^{t} L_{i-1} \right) \end{aligned} \right] \right] \\
& + \left[\frac{x_t}{L k_{2t-1}^2} \sum_{i=2}^{n} k_{2i-2}^2 \cdot \delta_{i-1} \right] - \left[\frac{1}{k_{2t-1}^2} \left[\left(\sum_{i=2}^{n} k_{2i-2}^2 \cdot \delta_{i-1} \right) - \left(\sum_{i=2}^{t} k_i^2 \cdot \delta_{i-1} \right) \right] \right]
\end{aligned}
$$

(11)

$$
\begin{aligned}
y_n = {} & \left[\frac{\sin(k_{2n-1} \cdot x_n)}{L k_{2n-1}^2 \sin(k_{2n-1} \cdot L_n)} \left[\left(k_{2n-1}^2 \cdot \delta_{n-1} - \sum_{i=2}^{n} k_{2i-2}^2 \cdot \delta_{i-1} \right) \cdot L + \left(\sum_{i=2}^{n} k_{2i-2}^2 \cdot \delta_{i-1} \cdot \sum_{i=1}^{n} L_{i-1} \right) \right] \right] \\
& + \left[\frac{x_n}{L k_{2n-1}^2} \sum_{i=2}^{n} k_{2i-2}^2 \cdot \delta_{i-1} \right]
\end{aligned} \tag{12}
$$

Which $t = 2\ 3\ 4\ \ldots\ n-1$

4. Result and Conclusion

Analysis of the lateral buckling coefficient of bar members with 2 to 5 elements have been carried out using Elastic Theory and compared with finite element program[10]. Lateral buckling coefficients, k, of bar member with 2, 3, 4 and 5 elements obtained from Elastic theory are 0.90, 0.83, 0.81, 0.79 respectively. Results obtained from Elastic theory are slightly higher than those obtained from finite element program [10] (less than 3% higher). It is found that bar members with more elements tend to have smaller value of lateral buckling coefficient as depicted show in fig. 5.

According to the results from this study, lateral buckling coefficient of compressive members in truss structure suggested from Eurocode 3 [1] is 0.9 which is considerably higher than that obtained from Elastic Theory. This means that using lateral buckling coefficient from Eurocode 3 [1] in designing truss structures may result in larger truss members compared with using Elastic Theory or Finite Element program. Although lateral buckling coefficient using Elastic Theory in this study may not be applied for every truss structure since there are various factors which affect lateral buckling coefficients, authors hope that this study could be useful for structural designer to use appropriate value of lateral buckling coefficient.

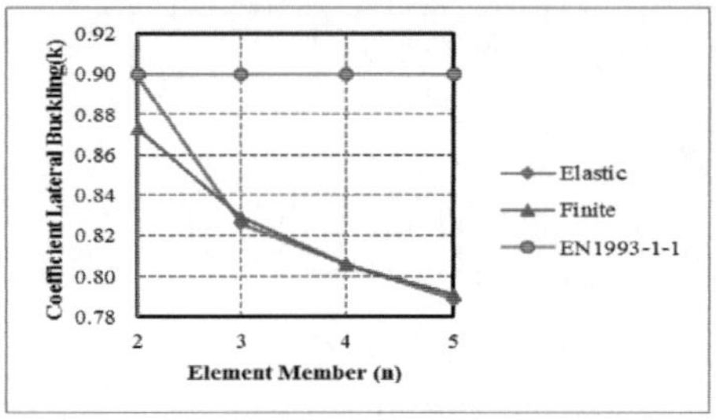

Figure 5. The relation between number of member and lateral buckling coefficient

5. Appendix A Notations

E = Elastic Modulus of member

I = Moment of inertia of member

k = Coefficient of Lateral Buckling

y = Displacement of Bar at i-segment

δ = Displacement of Bar at Apply force (node)

$\{\delta\} = \{\delta_1, ..., \delta_{n-1}\}^T$ = Displacement in Coordinate System

$[K] = [kL]$ = Stiffness Matrix of Effective Lengths

6. References

[1] European Standard (Eurocode 3), *Design of steel structure-Part 1-1: General rules and rules for building*. EN1993-1-1:2005, CEM, Brussels: CEM, 2005, pp. 82.

[2] J.S. Joanna and K. Jaroslaw, "Experimental Study of steel truss lateral-torsional buckling," *Engineering Structure*., vol.46, pp.165-172, 2011.

[3] L. Nattapol., S. Tanawat., S. Sakchai. and Suwat T., The Stability of Axial Force Accumulation. *Proceeding Nation Convention on Civil Engineering*, Thailand, 2005, pp. 203-209.

[4] J. Tada and A. Suito, "Experimental Study of steel truss lateral-torsional buckling," *Engineering Structure*., vol.20, pp.384-389, 1998.

[5] SL. Chan, CM. Koon and FG. Albermani, "Theoretical and experimental studies of unbraced tubular trusses allowing for torsional stiffness," *Steel and Composite Structures*., vol.2, pp.209-222, 2002.

[6] A. Biegus and D. Wojczyszyn, "Studies on buckling lengths of chords for out-of-plane instability," *Archives of civil and Mechanical Engineering*., vol.11, pp.507-517, 2011.

[7] S.P. Timoshenko and J.M. Gere., *Theory of Elastic Stability*, 2nd ed. New York: McGraw-Hill, 1961, pp. 98-100.

[8] G.Z. Dennis, *A First Course in Differential Equations with Modeling Applications,* California: Cengage Learning, 2009.

[9] H.V. Laksminarayama, *Finite Element Analysis (Procedures in engineering),* India: Vistas, 2004.

[10] MIDAS Information Technology Co.,Ltd., *Midas Engineering Software: Integrated Solution system for building and General structures*., Korea: MIDAS GEN, 2015.

IOP Conf. Series: Materials Science and Engineering **216** (2017) 012038 doi:10.1088/1757-899X/216/1/012038

Anchorage Behaviors of Frictional Tieback Anchors in Silty Sand

Shih-Tsung Hsu[1], Wen-Ta Hsiao[1], Ke-Ting Chen[1], Wen-Chi Hu[2] and Ssu-Yi Wu[3]

[1] Department of Construction Engineering, Chaoyang University of Technology, 168, Jifeng E. Rd., Wufeng Dist., Taichung, Taiwan.
[2]Department of English, Taichung University of Education, No. 140, Minsheng Rd., West Dist., Taichung, Taiwan.
[3]Department of Civil and Construction Engineering, National Taiwan University of Science and Technology, No.43, Keelung Rd., Sec.4, Da'an Dist., Taipei, Taiwan.

Email: sthsu@cyut.edu.tw;winniehu@gmail.com;

Abstract. Soil anchors are extensively used in geotechnical applications, most commonly serve as tieback walls in deep excavations. To investigate the anchorage mechanisms of this tieback anchor, a constitutive model that considers both strain hardening and softening and volume dilatancy entitled SHASOVOD model, and $FLAC^{3D}$ software are used to perform 3-D numerical analyses. The results from field anchor tests are compared with those calculated by numerical analyses to enhance the applicability of the numerical method. After the calibration, this research carried out the parameter studies by numerical analyses. The numerical results reveal that whether the yield of soil around an anchor develops to ground surface and/or touches the diaphragm wall depending on the overburden depth H and the embedded depth Z of an anchor, this study suggests the minimum overburden and embedded depths to avoid the yield of soils develop to ground surface and/or touch the diaphragm wall. When the embedded depth, overburden depth or fixed length of an anchor increases, the anchorage capacity also increases. Increasing fixed length should be the optimum method to increase the anchorage capacity for fixed length less than 20m. However, when the fixed length of an anchor exceeds 30 m, the increasing rate of anchorage capacity per fixed length decreases, and progressive yield occurs obviously between the fixed length and surrounding soil.

1. Background

Currently, deep excavations are supported by diaphragm walls with tieback systems or internal braces. A complicated internal brace system in an excavation with an irregular area or an asymmetrical cross-section usually causes difficulty in operating within the space and increases construction expenditures. By contrast, a tieback system is more cost-effective and provides a larger excavation working space.

Virtually all of the fixed anchors constructed around the Taipei Basin are typically installed in silty sand layers. Several in-situ tests have revealed that the increment of ultimate load per fixed length decreases as the fixed length increases over a threshold value [1], which shown that soils around a fixed anchor exhibit progressive yielding.

Figure 1(a) illustrates the regulation of installing tieback systems by BSI [2]. With regard to design of a tieback anchor, the fixed anchor must be installed deep enough to avoid the localized passive failure of the soil associated with the failure condition for shallow anchors, and placed far enough away from the wall to ensure against a slip failure beneath the toe of the wall and beyond the fixed anchor zone. As illustrated in Figure 2(a), an inaccurate tieback system in the deep excavation in Zhubei City,

Taiwan which constructed by Feng-Yi Construction Company had induced site disaster. Figure 2(b) exhibits that because all fixed anchors almost placed in the hatch area displayed in Figure 1(a), the tieback systems fail. Therefore, it is important to install tieback anchors in appropriate location, and the behavior of tieback anchors demand to investigate.

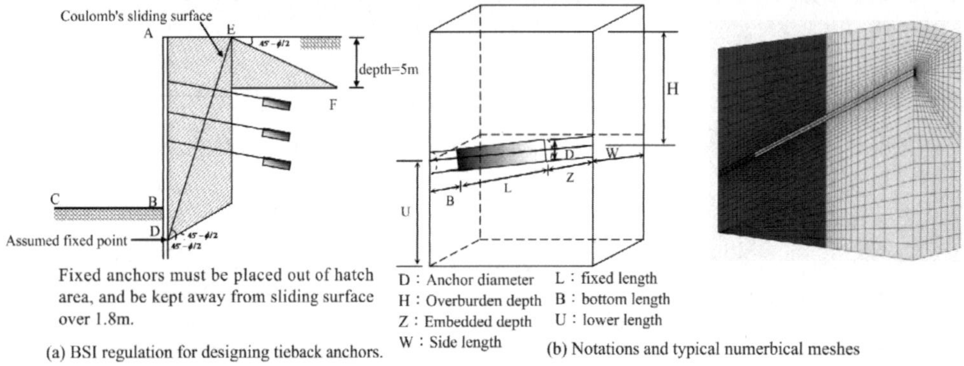

(a) BSI regulation for designing tieback anchors. (b) Notations and typical numerbical meshes

Figure 1. BSI regulation and numerical study for a tieback anchor

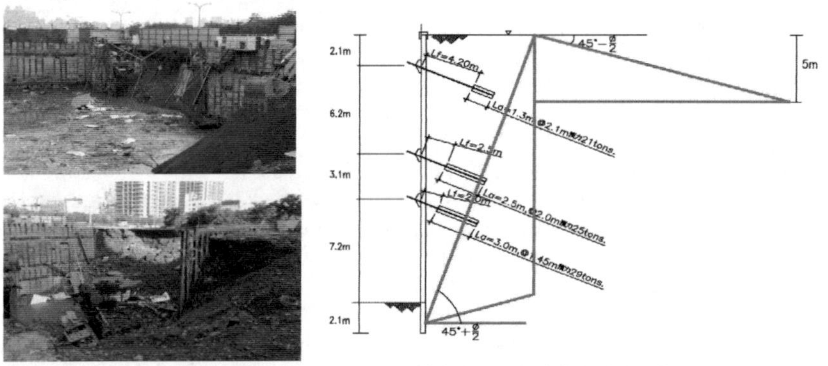

(a) Disaster of the tieback system (b) An inaccurate designs of tieback anchors

Figure 2. An inaccurate design of tieback system in a deep excavation in the Zhubei City, Taiwan

For numerical analyses, most studies have adopted techniques of 2-dimensional analyses to study the anchor behavior [3, 4]; however, the tieback anchor is an inclined anchor, the 3-D model must be established to obtain an elaborated result.

To manage these issues, this study has applied a constitutive model of sandy soils entitled SHASOVOD model and the FLAC^{3D} software to examine the anchorage behavior of tieback anchors in silty sand. To demonstrate the suitability of the numerical model in evaluating the load-displacement behavior of anchors in silty sand, the calculated results are compared with the results of in-situ anchor tests. Based on both the numerical and test results, a recommendation was presented for the design of tieback anchors in silty sand. Parametric studies on factors that affect the anchorage behavior of tieback anchors were conducted to address the anchorage capacity of the anchors.

2. Constitutive Model and Numerical Analysis

Hsu [5] suggested a constitutive model with non-associated flow rules that considers the strain hardening and softening and volume dilatancy (SHASOVOD model). Previous results demonstrate that the numerical results are in agreement with the field test results. This study used the SHASOVOD model and the FLAC^{3D} software to study the anchorage behavior of tieback anchors in silty sand.

The soil specimen used in the triaxial CD (Consolidated Drained) test was sampled from Taipei Basin. The soil is classified as silty sand, SM. The specific gravity of the sandy soil is 2.68; its maximum dry density is 16.1 kN/m^3, and its minimum dry density is 12.0 kN/m^3.

As mentioned above, this study adopted the constitutive model proposed by Hsu [5] to represent the stress-strain behavior of silty sand. This model is based on plasticity theory and the non-associated flow rule. The yield function *f* in this study is similar to Mohr-Coulomb criterion, and can be written as followings

$$f = \sigma_1 - \sigma_3 \frac{1+\sin\phi^*}{1-\sin\phi^*} \tag{1}$$

where σ_1 represents the major principal stress; σ_3 is the minor principal stress, and ϕ^* is the mobilized friction angle, which is a function of the accumulative plastic strain ε^p.

The plastic potential function g was required to describe the relationship between the plastic strain increment $d\varepsilon_{ij}^p$ and the stress tensor σ_{ij}. The plastic potential function can be expressed as

$$g = \sigma_1 - \sigma_3 \frac{1+\sin\psi^*}{1-\sin\psi^*} \tag{2}$$

where ψ^* is the mobilized dilatancy angle, which is also a function of accumulative plastic strain ε^p.

To exhibit the reliability of this constitutive model, the stress-strain behaviors of sand with a relative density of 30%, determined from the triaxial test, were compared with those calculated using the proposed model. As illustrated in Figure 3, the calculated results were perfectly consistent with the triaxial test results not only for the strain hardening stage but also for strain softening stage. Hsu [5] provided a detailed elucidation of this model.

Figure 1(b) plots the typical numerical meshes and notations used to define the tieback anchor. To reduce calculation time, one-half symmetric meshes were used for the tieback anchor.

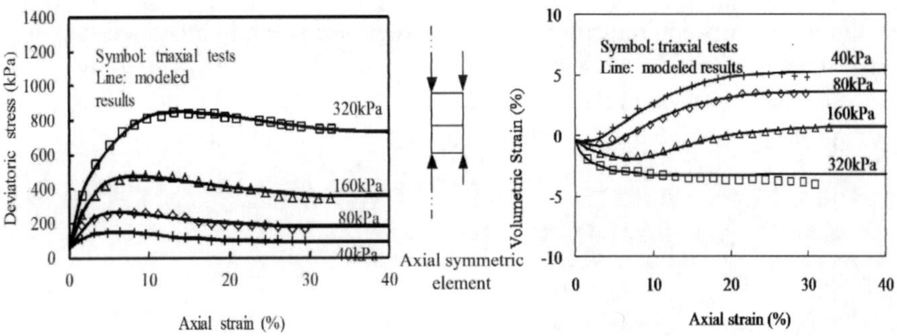

Figure 3. Comparison of measured and modeled stress-strain behavior of silty sand with a D_r of 30%

3. Verification of Numerical Results

As illustrated in Figure 4(a), two frictional anchors were placed and tested at same site to verify the suitability of the proposed numerical method. Figure 4(a) also presents information on the tested anchors and subsoil conditions.

The test anchors were pulled out following the proving test procedure, as recommended by Deutsche Industrie Norm, DIN [6]. First, an initial load was applied to the two anchors, which were subsequently stressed in five loading-unloading cycles. The test anchors were pulled out to failure after completing the five loading-unloading cycles.

The test anchors were almost anchored in silty soils, the relative density of in-situ sandy soil can be determined from the relationship between the relative density D_r and *SPT-N* value of sandy soil of the Taipei Basin. The *SPT-N* of 10 approximates to D_r of 30%.

As displayed in Figure 4(b), the calculated load-displacement curves of the test anchors were nearly consistent with those measured in the field tests. After the proposed model was calibrated using the results of the pullout test on the field anchor, the numerical method was applied to elucidate the anchorage behavior of tieback anchors in silty sand.

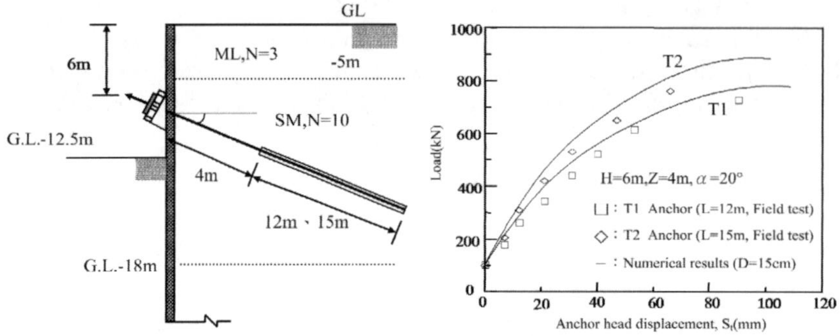

(a) Dimensions and location of test anchors

(b) Comparison of experimental and numerical load-displacement relationships of T_1 and T_2 anchors

Figure 4. Field anchor information and experimental/numerical results

4. Analysis and Discussion of results

4.1. The minimum overburden and embedded depths.

As mentioned before, for designing tieback anchors, the fixed anchor must be placed a considerable distance away from ground surface, Coulomb's sliding surface [2] and diaphragm wall to avoid the yield zones develop to these regions. As illustrated in Figure 5(a), when the overburden depth H=4m and embedded depth Z=2m, the yield zones develop to both the ground surface and diaphragm wall. Figure 5(b) displays that when the H is 6m and Z is 2m of an anchor, the yield zones merely develop to the diaphragm wall. This study uses same procedures for various overburdens and embedded depths to figure out the minimum requirements of both the overburden and embedded depths for a tieback anchor design. As demonstrated in Figure 6, the fixed anchor must be installed away from the hatch area. Figure 6 illustrates that, when the overburden H ≧ 3m, embedded depth Z ≧ 3m and H+Z ≧ 8m, then the yielding soils cannot develop to both the ground surface and diaphragm wall.

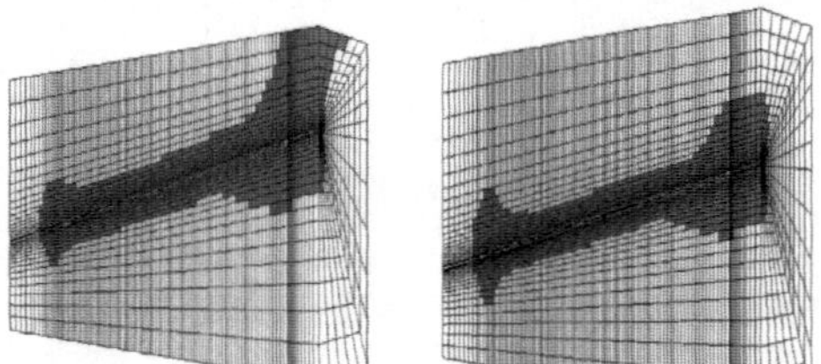

(a) H=4m, Z=2m (b) H=6m, Z=2m

Figure 5. Yielding soil around an tieback anchor

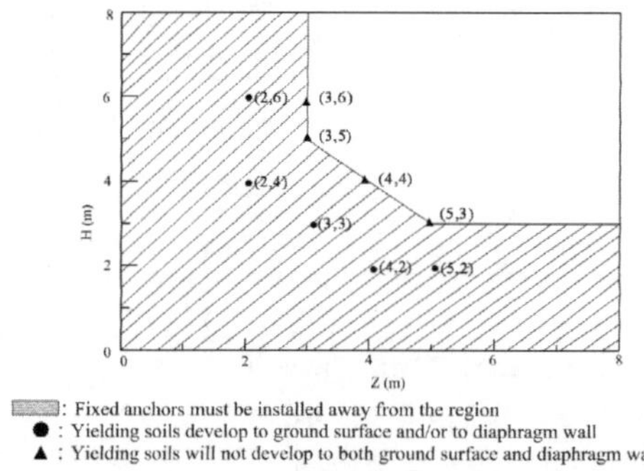

Figure 6. The minimum requirements of overburden and embedded depths of a tieback anchor

4.2. Parametric effects on the ultimate load.

Theoretically, when the embedded depth Z increases, then the shear strength of silty sand increases, the ultimate load of an anchor also increases. Figure 7(a) depicts the increment of ultimate load per embedded depth is approximately 15kN/m.

Because increasing the overburden H could raise the shear strength of sandy soil, increasing the overburden could raise the ultimate load of an anchor. Figure 7(b) indicates the increasing rate of ultimate load per overburden is around 45kN/m.

Figure 7(c) plots the relationship of ultimate load and fixed lengths. The ultimate load increases in an increasing rate with the fixed length when L ≦ 20m, the increasing rate of ultimate load per fixed length exceeds 60 kN/m, because no progressive yield occurs between the fixed length and sandy soil; when 20m≦L≦30m, the ultimate load increased in a decreasing rate because progressive yield occurs. However, when L exceeds 30m, the progressive yield occurs obviously, thus the increasing rate of ultimate load per fixed length decreases clearly.

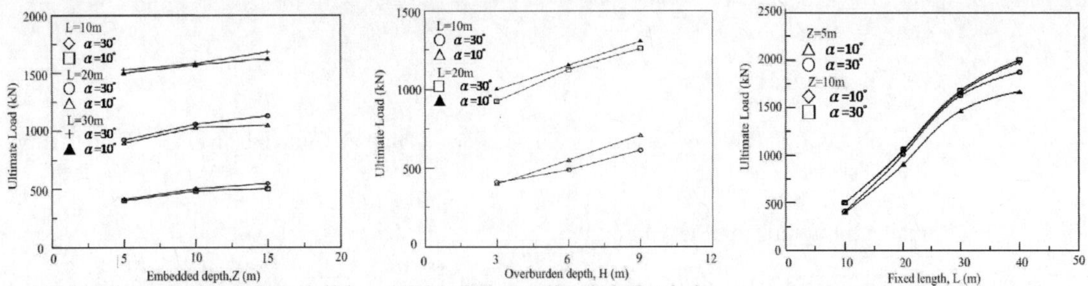

(a) Relationship between ultimate load and embedded depth (H=3m)

(b) Relationship between ultimate load and overburden depth(Z=5m)

(c) Relationship between ultimate load and fixed length (H=3m)

Figure 7. Parametric effects on the ultimate load for tieback anchors

5. Conclusions

The results of the numerical analyses and in-situ tests lead to the following conclusions.

(1) Numerical results shown that the load-displacement behaviors calculated by the numerical model were consistent with those measured by experimental anchor tests.

(2) As the overburden H≧3m, embedded depth Z≧3m, and H+Z≧8m, then the yielding soils could not develop to both the ground surface and diaphragm wall.

(3) When the embedded depth, overburden depth or fixed length of an anchor increased, the anchorage capacity also increased. Increasing fixed length should be the optimum method to increase the anchorage capacity for fixed length less than 20m.

(4) For an anchor with long fixed length (says, L=40m), a progressive yield occurred obviously between the interface of the fixed length and sandy soil, the increasing rate of the ultimate load per fixed length decreased clearly.

6. Acknowledgments

The authors would like to thank the Ministry of Science and Technology of the Republic of China, Taiwan, for financially supporting this study under Contract No. MOST-105-2221-E-324 -004.

7. References

[1] Hsu S T and Hu W C 2014 Uplift behavior of shaft anchors in silty sand in Taipei Basin. *Journal of the Chinese Institute of Engineers* 37(2) pp 175-188.

[2] British Standards Institution-BSI (8081) 1989 *British Standard Code of Practice for Ground Anchorages* London.

[3] Hsu S T and Tang Y G 2013 Anchorage behavior of mechanically under-reamed anchors in silty sand. *Disaster Advances* 6(10) pp 31-44.

[4] Hsu S T and Liao H J 1998 Uplift behavior of cylindrical anchors in sand. *Canadian Geotechnical Journal* 35(1) pp 70-80.

[5] Hsu S T 2005 A constitutive model for uplift behavior of anchors in cohesionless soils. *Journal of Chinese Institute of Engineers* 28(2) pp 305-317.

[6] Deutsche Industrie Norm (DIN4125) 1988 *Verpressanker fur vorubergehende zwecke im lockergestein* Bemessung, Germany.

IOP Conf. Series: Materials Science and Engineering **216** (2017) 012039 doi:10.1088/1757-899X/216/1/012039

Modeling of Batter Pile Behavior under Lateral Soil Movement

C Y Chen and H Q Hsu

Dept. of Civil and Water Resources Engineering, National Chiayi University, Chiayi City 60004, Taiwan

Email: chienyuc@mail.ncyu.edu.tw

Abstract. Pile foundation is frequently used when structures are located on weak sublayers or are at risk from lateral loadings such as earthquakes. The design of pile foundations has recently become crucial to stop slope movement. To understand the behavior of pile foundations subjected to lateral soil movement, the three-dimensional Fast Lagrangian Analysis of Continua (FLAC3D) program was used to perform numerical simulations, which can reduce the cost of field testing. Vertical piles and batter piles were combined into 3 × 3 pile groups, and the response of batter piles to soil movement was analyzed. The outer batter piles led to an increased bending moment in the middle, vertical pile row. Increasing the pile spacing and the presence of battered piles reduced the pile group's displacement. The batter pile group's maximum bending moment was smaller than the vertical pile group's in sand soil, but 5–8 times higher in clay soil.

1. Introduction

Pile foundations are common used to reduce settlement and displacement. The main function of pile foundations is to transfer the load of a superstructure onto a hard layer through a soft upper layer. Batter piles are used to support horizontal loads, for example bridge piers, wharf piles, and retaining walls beneath the sloping areas, and they reduce displacement of piles and surrounding soil settlements, thus enhancing the stability of a superstructure. Numerous documented studies and models for the use of batter piles have been published [1-6].

Pile response is affected by soil–pile interaction, the space and arrangement of piles, and interaction with the pile cap. It is also influenced by liquefied surrounding soil, pile deformation, and kinematic response. Batter piles subject to lateral or seismic loadings have decreased pile cap displacement and are subject to axial forces up to eight times higher than vertical piles [1-4]. The presence of batter piles in a foundation can reduce the bending moment of a vertical pile and horizontal displacement under seismic loading. However, a batter pile was shown to cause non-negligible residual moments and higher rotation at the pile cap after seismic loadings [7]. Inclined batter piles were reported to have increased seismic resistance during numerical modeling using three-dimensional Fast Lagrangian Analysis of Continua (FLAC3D) [5]. Experimental and practical studies of batter pile foundations are still being conducted, but their interactions with superstructures and soft soil require further study to optimize their safe and economic design.

2. Material and method

The response of batter piles within a group of piles subjected to soil movement was studied. Specifically, finite difference analysis using FLAC3D was validated [6] and then employed to model the pile–soil interaction.

The numerical analysis assumed that soil was deposited under gravity (9.8 m/s^2) in an initial state with a high water table at ground level, to model a riverside or wharf scenario. The concrete pile was modeled as an elastic material following the Mohr–Coulomb failure criteria, and soil movement induced by liquefaction was replaced at a constant horizontal rate (-1×10^{-6} m/s) at the boundary. Pile responses in sand soil were considered and discussed.

Grid size and the boundary conditions used in the mesh are crucial control factors that determine the efficiency of the numerical calculations. For example, a smaller grid size and larger study area increases the precision of the modeling compared with a coarse mesh used with boundary conditions. To avoid boundary effects, a minimum horizontal distance to the boundary of 15–25 times the pile space is required [8,9]. A distance to the boundary 25 times the pile space was used in the present study (20 m vs 0.75 m). The boundary conditions used were roller for the surrounding soil, hinged for the base, and free at the soil surface with high water table (Figure 1). Single batter pile response under soil movement was modeled first and compared with two pile groups in a 3 × 3 scheme: one with nine vertical piles, and one with three vertical and six 20°-inclined batter piles (Figure 2).

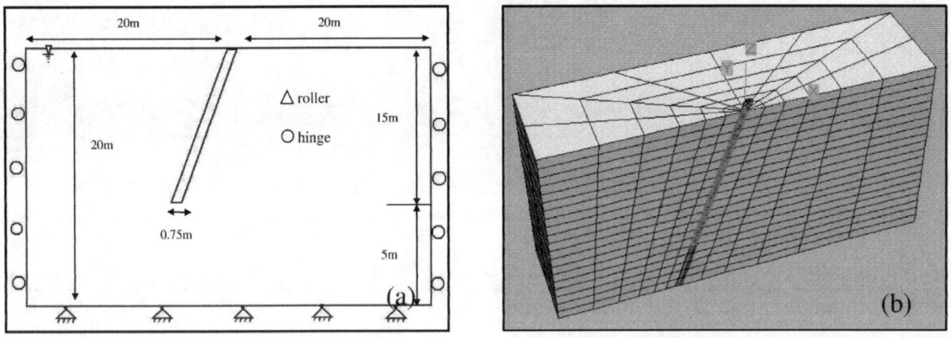

Figure 1. Numerical scheme for (a) boundary conditions and size used, and (b) the design mesh for a single batter pile.

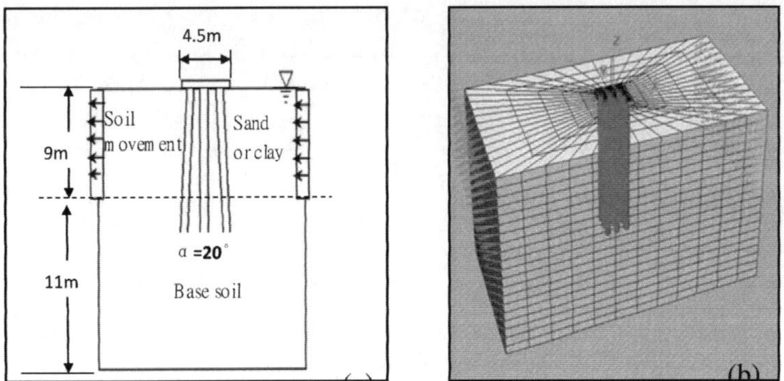

Figure 2. Modeling of the batter pile group subjected to soil movement (a) soil profile, and (b) designed 3D mesh.

Several elasticity and strength parameters were required for the numerical analysis, namely bulk modulus, shear modulus, internal friction angle, dilation angle, and cohesion strength (Table 1).

An interface element was used to model slippage at the soil–pile interface. This element was single sided with zero thickness, and had normal stiffness K_n and shear stiffness K_s to transfer loadings from soil to the pile shaft. The parameters are suggested be ten times larger than the stiffness of the surrounding soil [10], as described by

$$K_n = K_s = 10 \times \max\left[\frac{(K+\frac{4}{3}G)}{\Delta Z_{min}}\right] \tag{1}$$

where K is the bulk modulus, G is the elastic modulus, and ΔZ_{min} is the minimum mesh size in the interface element. Parameters for friction angle, cohesion strength, and tension strength are also required to model slip movement and tension failure in the plastic state. The friction angle between the pile shaft and soil has a large effect on the bearing capacity of the frictional pile and has been previously estimated as 0.6–0.7 times the soil internal friction angle [11,12]. Table 2 presents a list of the parameters used for the interface element.

Table 1. Material parameters used during numerical analysis.

Material/ parameter	Bulk modulus (MPa)	Shear modulus (MPa)	Cohesion (kPa)	Friction angle (°)	Dilation angle (°)	Density (kg/m^3)
Sand	15	10	0	30	0	2000
Base soil	50	30	0	40	0	2000
Concrete pile	27800	20800	--	--	--	2500

Table 2. Parameters for the interface element used during numerical analysis.

Parameters/ Soil types	ΔZ_{min} (m)	K_n (MPa/m)	K_s (MPa/m)	Cohesion (kPa)	Friction angle (°)
Sand	0.25	1100	1100	0	21
Base soil	0.25	1100	1100	0	28

The pile shaft moment was calculated using the shear forces along the pile section. These shear forces were measured at the mesh nodes in a horizontal direction using the FISH language (the language of FLAC3D) and modified to the normal direction of the pile shaft. The top of the pile, which was connected to the pile cap, was free to rotate but could have been fixed to model a real field situation. Pile space is the distance between the centers of two adjacent piles; thus, piles in a given group were coded in a numerical sequence, as portrayed in Figure 3.

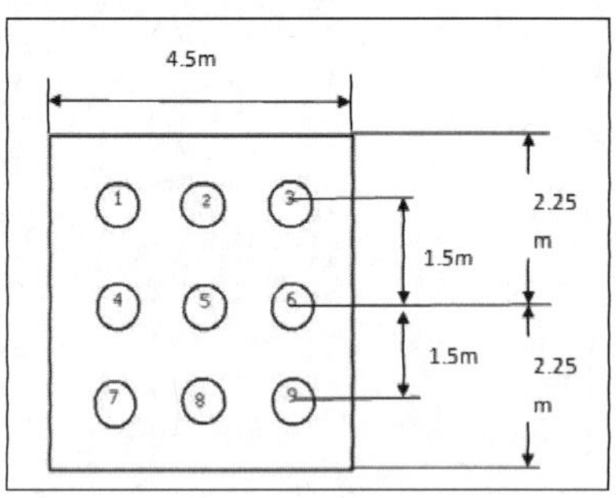

Figure 3. Code numbers of 3 × 3 pile group (top view).

3. Results and analysis

A 3 × 3 pile group arranged with two rows of three inclined batter piles and one middle row of three vertical piles was modeled (Figure 3), and the results were compared with a 3 × 3 pile group containing nine vertical piles. The moment distribution along the pile shaft subjected to soil movement in two times (2d) and four times (4d) the pile diameter space is illustrated in Figs 4 and 5, respectively. The maximum moment of the pile shaft occurred at a depth of 6 m for the vertical piles and at 12 m for the 20° inclined piles. The 2d space of the vertical piles' maximum moment due to sand soil movement was approximately two times larger than that for the inclined batter piles (Figure 4); in this scenario, pile #2 in the batter pile group was subjected to the largest moment. When the pile space increased to 4d, the middle row piles #2 and #5 had larger moments in the batter pile group. This demonstrates that the presence of surrounding batter piles did not protect the vertical middle pile during sand soil movement. The pile displacement curve illustrates that the affected area was less covered, and that the beneficial grouping effect tends to diminish as pile space is increased (Figure 5). Moreover, the batter piles were less displaced by the sand soil movement than were the vertical piles when the pile space was increased (Figure 6). Pile displacement decreased from 33% for the 2d batter pile space to 80% for the 4d batter pile space. Finally, the presence of the batter piles reduced the displacement of pile #5 in the pile group, whereas the pile moment caused by sand soil movement was not reduced.

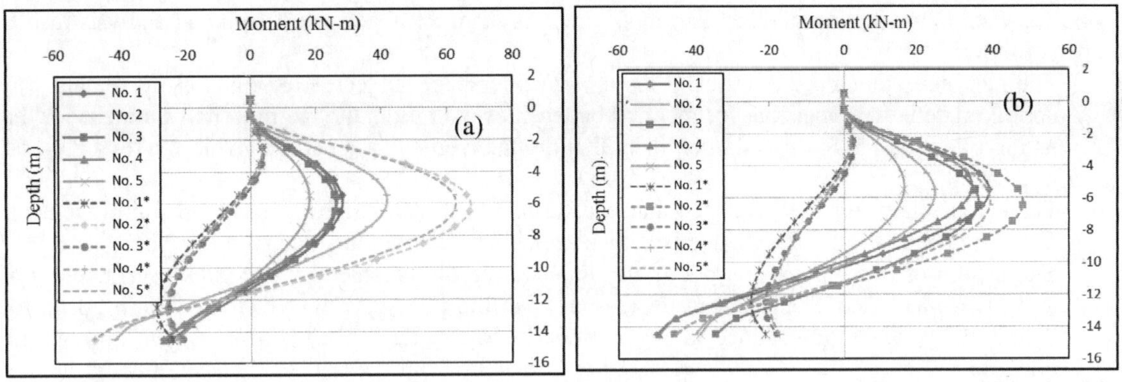

Figure 4. Pile shaft moment distribution in two 3 x 3 pile groups (a) in the 2d pile space (b) in the 4d pile space that were subjected to sand soil movement (*batter pile).

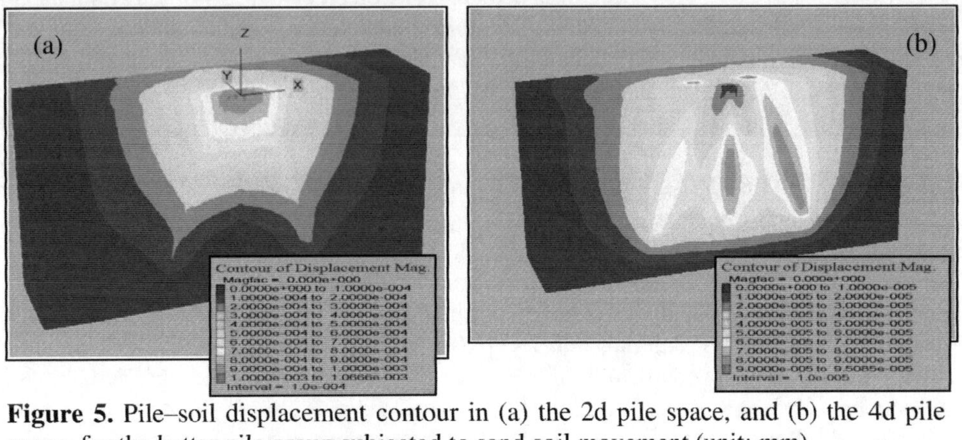

Figure 5. Pile–soil displacement contour in (a) the 2d pile space, and (b) the 4d pile space, for the batter pile group subjected to sand soil movement (unit: mm).

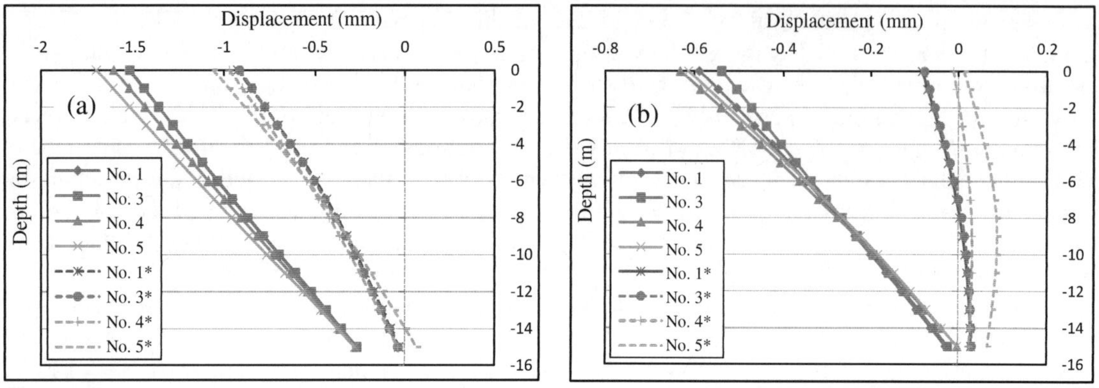

Figure 6. Lateral pile displacement in (a) the 2d pile space, and (b) the 4d pile space of the pile group subjected to sand soil movement (*batter pile).

4. Conclusions

The response of batter pile groups under lateral soil movement was modeled using finite difference analysis. Pile shaft moment, lateral displacement, and group effects were evaluated by adjusting the pile space and introduction pile inclination. The comprehensive batter pile design and analysis results are as follows:

1. The maximum moment of vertical piles in a 2d pile space subjected to sand soil movement was approximately two times that for inclined batter piles. Additionally, the pile shaft moment for the batter piles was 5–8 times larger than that in the vertical pile group subjected to clay soil movement.
2. The batter piles were less displaced than the vertical piles when subjected to sand soil movement. The presence of batter piles reduced the displacement of the middle pile in a pile group.
3. The design of batter piles in a pile group reduced the displacement of the pile group under soil movement, and their displacement decreased when the pile space was increased from 2d to 4d. Additionally, a soil arch formed in the first row of batter piles to halt soil movement through the pile group.

5. References

[1] Sadek M and Shahrour I 2004 Three-dimensional finite element analysis of the seismic behavior of inclined micropiles *Soil Dynamics and Earthquake Engineering* **24** 473–485

[2] Sadek M and Shahrour I 2006 Influence of the head and tip connection on the seismic performance of micropiles *Soil Dynamics and Earthquake Engineering* **26** 461–468

[3] Deng N, Kulesza R and Ostadan F 2007 Seismic soil-pile group interaction analysis of a battered pile group *Proc. 4th Int. Conf. on Earthquake Geotechnical Engineering* CD-ROM Greece.

[4] Poulos H G 2006 Raked piles-virtues and drawbacks *Journal of Geotechnical and Geoenvironmental Engineering* **132** 795–803

[5] I Shahrour, A Hassan and S Mhamed 2012 3D elastoplastic analysis of the seismic performance of inclined micropiles *Soil Dynamics and Earthquake Engineering* **42** 275–291

[6] Chen C Y and Tsai C X 2014 Batter pile behavior modeling using finite difference analysis, Applied Mechanics and Materials **566** 199–204

[7] Escoffier S 2012 Experimental study of the effect of inclined pile on the seismic behavior of pile group *Computers and Geotechnics* **39** 1–7

[8] Trochanis A M, Bielak J and Christiano P 1991 Three-dimensional nonlinear study of pile *Journal of Geotechnical Engineering* **117** 429–447

[9] Lim C H, Chow Y K and Karunaratne G P 1993 Negative skin friction on single piles in a layered half-space *International Journal for Numerical and Analytical Methods in Geomechanics* **17** 625–645

[10] Itasca Consulting Group Inc. 2005 FLAC3D Fast Lagrangian Analysis of Continua in 3 Dimensions, user manuals

[11] Potyondy J G 1961 Skin friction between various soils and construction materials *Geotechnique* **11** 339–353
[12] Acer Y B, Durgunoglu H T and Tumay M T Interface properties of sands *J. Geotech. Engrg. Div.* **108** 648–654

IOP Conf. Series: Materials Science and Engineering **216** (2017) 012040 doi:10.1088/1757-899X/216/1/012040

Applying the Artificial Neural Network to Predict the Soil Responses in the DEM Simulation

Z LI, J K Chow and Y H Wang

The Hong Kong University of Science and Technology, Clear Water Bay, Hong Kong

Email: zlibi@connect.ust.hk; junkangchow@gmail.com; ceyhwang@ust.hk;

Abstract. This paper aims to bridge the soil properties and the soil response in the discrete element method (DEM) simulation using the artificial neural network (ANN). The network was designed to output the stress-strain-volumetric response from inputting the soil properties. 31 biaxial shearing tests with varying soil parameters were generated using the DEM simulations. Based on these 31 training samples, a three-layer neural network was established. 2 extra samples were generated to examine the validity of the network, and the predicted curves using the ANN were well matched with those from the DEM simulations. Overall, the ANN was found promising in effectively modelling the soil behaviour.

1. Introduction

The soil behaviour of is significantly regulated by the soil properties and testing condition, such as the initial void ratio [1] and particle shape [2]. These factors interact during the testing process and collectively determine the soil response. However, the associated mechanism is not complicated and an effective way to quantify the effects of soil properties is needed. One of the feasible approaches is to implement the artificial neural network (ANN). In Ellis et al. [3], the ANN was first applied in modelling the stress-strain response of sands with varying grain size distribution and stress history. Penumadu and Zhao [4] considered more parameters in their ANN model, e.g., the grain shape and size distribution. Besides, this technique has also been adopted to evaluate the soil liquefaction resistance [5] and slope stability [6].

Inspired by these pioneer works, in this study, an attempt was made to effectively bridge the soil properties and the soil behavior in the discrete element method (DEM) simulation; that is establish an artificial neural network, which can output the mechanical response from the input soil parameters. This paper will first introduce the details of preparing training samples in the DEM simulation. Then, the network will be established using the training samples. The validity and efficiency of this ANN-based model will be demonstrated by its performance in predicting the response of the untrained samples.

2. Details of the DEM simualtion

The discrete element method (DEM), can numerically generate soil samples and obtain the relevant mechanical responses. An ANN-based model was designed to predict the stress-strain-volumetric responses for the soil samples from the DEM simulation. A total number of 31 biaxial shearing tests have been conducted in the DEM simulation. In the simulations herein, the particles interact via the preset contact model, which includes the Hertz-Mindlin model in the normal and tangential direction and the rolling resistance model in the rotational direction. The normal and tangential contact stiffnesses, k_n and k_s, are determined as in equation (1),

$$\begin{cases} k_n = \dfrac{G\sqrt{2RU_n}}{(1-\upsilon)} \\ k_s = \dfrac{2(1-\upsilon)}{2-\upsilon}k_n \end{cases} \tag{1}$$

where G and υ are the shear modulus and Poisson's ratio of the particles, respectively; R is the equivalent particle radius; and U_n is the contact overlapping. Then, contact force/torque is aroused due to the contact movement, as expressed in equation (2),

$$\begin{cases} P = 2k_n U_n/3 \\ \Delta Q = k_s \Delta U_s \text{ and } |Q| \le \mu P \\ \Delta M = 2J_n P \Delta U_r \text{ and } |M| \le \eta RP \end{cases} \tag{2}$$

where P, ΔQ and ΔM are the contact normal force, incremental tangential force and incremental torque, respectively; μ is the friction coefficient; J_n is the rolling coefficient; and η is the rolling sliding coefficient; ΔU_s and ΔU_r are the incremental tangential and rolling displacement.

Table 1. Parameters used in the DEM simulation for ANN training.

No.		G (GPa)	p_c (kPa)	υ	J_n	η	μ_c †	μ
Ref	1	10	50	0.25	0.0	0.0	0.0	0.5
G	2-6	0.5, 1, 5, 50 and 100	50	0.25	0.0	0.0	0.0	0.5
p_c	7-11	10	10, 20, 100, 200 and 500	0.25	0.0	0	0.0	0.5
J_n	12-16	10	50	0.25	0.05, 0.1, 0.5, 1 and 2	0.25	0.0	0.5
η	17-21	10	50	0.25	0.5	0.05, 0.1, 0.5, 1 and 2	0.0	0.5
μ_c	22-26	10	50	0.25	0.0	0	0.1, 0.2, 0.3, 0.4 and 0.5	0.5
μ	27-31	10	50	0.25	0.0	0	0.0	0.1, 0.2, 0.8, 1 and 2

†μ_c is the friction coefficient in the consolidation stage. A lower value makes the initial packing denser and it is equivalent to effect of initial void ratio.

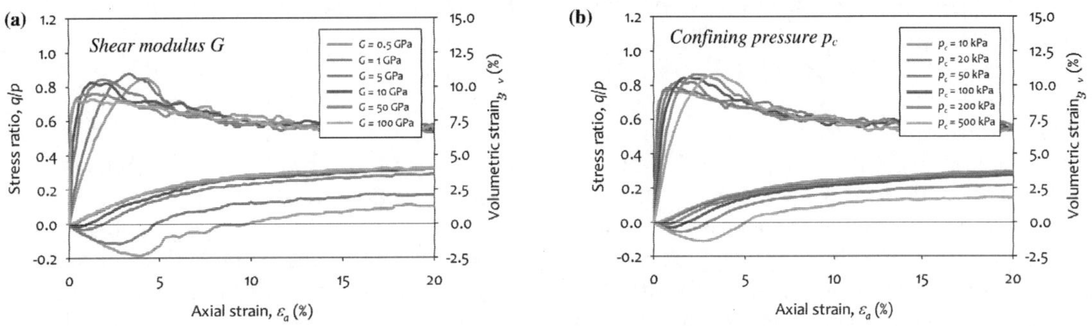

IOP Conf. Series: Materials Science and Engineering **216** (2017) 012040 doi:10.1088/1757-899X/216/1/012040

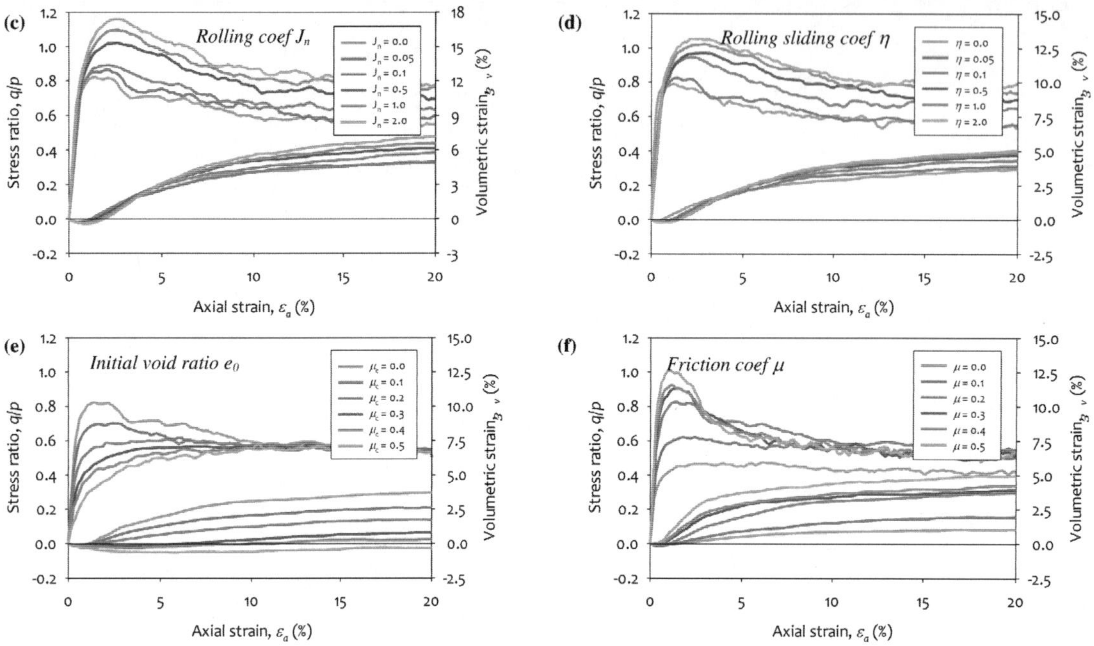

Figure 1. Stress-strain-volumetric responses with varying parameters: (a) shear modulus; (b) confining pressure; (c) rolling coefficient; (d) rotational sliding coefficient; (e) initial void ratio; and (f) friction coefficient.

In these 31 tests, the influences of particle shear modulus, confining pressure, rolling coefficient, rolling sliding coefficient, initial void ratio and friction coefficient on the soil response were examined. Parameters used in the simulations are summarized in Table 1. A reference test was set and six sections were classified, in which only one parameter was changed compared to the reference one. Figures 1(a) to 1(f) illustrate the stress-strain-volumetric responses for each section. Each parameter significantly contributes to the response, but their influences are different. In figure 1(a), with the shear modulus increasing, the peak stress came at a smaller strain with a lower value, and the initial contraction was reduced whereas the dilation enhanced. On the contrary, when the confining pressure was raised, as in figure 1(b), the peak stress then came larger and later and the dilation was inhibited. Obviously, the soil parameters intricately interact during the shearing stage. Such influences from the soil properties may not be quantifiable properties in the conventional way, i.e., the constitutive modeling. Thus, a better modeling method, i.e., the artificial numerical network was adopted.

3. Artificial Neural Network Modeling

Soil parameters of these 31 samples and the associated responses were used to train the artificial neural network (ANN), which was designed to output the stress-strain-volumetric curve by inputting the soil parameters. The ANN is a 3-layer model, including 6 input parameters, 1 hidden layer with 41 neurons and 2 output curves. The associated toolbox in Matlab was utilized and the data are randomly divided for training (87%, 67 samples), validation (6%, 2 samples) and testing (6%, 2 samples) uses. When training the ANN, the "Scaled Conjugate Gradient" algorithm was used since it requires less memory and training stops when generalization stops improving. 112 iterations were carried out and parameters for connecting the neurons in the network finally converged. The ANN was well trained in this process since the fitting quality, i.e. mean square error, has reached to 0.0009. In order to verify the validity and efficiency of this trained ANN, two extra DEM simulations were carried out. The parameters in these simulations were different from those of the training samples, but within the range scope of them. Parameters for ANN verification are summarized in table 2. These sets of parameters were then input into the ANN model and the predicted stress-strain-volumetric curves were obtained. Meanwhile, the responses were also obtained from the DEM simulation using these parameters.

Table 2. Parameters used in the DEM simulation for ANN verification.

No.	G (GPa)	p_c (kPa)	υ	J_n	η	$\mu_c(e_0)$	μ
32	7.5	0.25	0.45	0.3	25	0.15	0.55
33	25	0.25	0.6	0.4	50	0.5	0.5

Comparison is plotted in figure 2(a) and figure 2(b) and good agreements between the actual responses from the DEM simulations and the predicted responses from the ANN can be observed. Despite the minor discrepancies, the general trends between the curves from DEM and ANN are quite similar. Therefore, by using this ANN model, the repetitive DEM simulation work can be avoided since the results are predictable now. This ANN model again demonstrates that soil parameters collaboratively contribute to the soil response during the test process and it provides a promising tool for effectively and efficiently modeling the soil behavior. Also, it may be an effective way for the model calibration in the simulation since the ultimate parameters can be obtained by try and error.

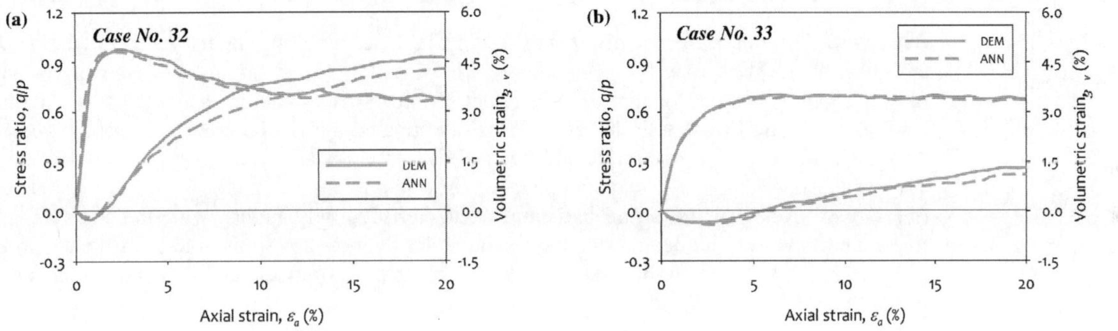

Figure 2. ANN verification using two extra DEM simulations: (a) Case No. 32 and (b) Case No. 33.

4. Conclusions

In this paper, the artificial neural network (ANN) was applied to predicting the soil mechanical response in the discrete element method (DEM) simulation. 31 DEM samples were prepared for the ANN model training. The soil parameters were the inputs and the stress-strain-volumetric response were the output in this network. This ANN model was 3-layer model with 6 input, 41hidden and 2 output neurons. It was well trained and the predicted results using this network matched well with the actual ones from the extra two DEM simulations. Overall, the ANN is a promising tool for effectively and efficiently modeling the soil behavior and thus repetitive simulation work can be reduced.

Acknowledgments
This research was supported by the Hong Kong Research Grants Council (project no. T22-603/15N) and Hong Kong PhD Fellowship Scheme (HKPFS).

References

[1] Salot C, Philippe G, Pascal V, 2009 Granul. Matt. **11**(4) 221-236.
[2] Wood M, Maeda K, 2008 Acta Geotech. **3**(1) 3-14.
[3] Ellis GW, Yao C, Zhao R, Penumadu D, 1995 J. Geotech. Engng **121**(5) 429-435.
[4] Penumadu D, Zhao R, 1999 Comput. Geotech. **24**(3) 207-230.
[5] Goh ATC, 1994 J. Geotech. Engng **120**(9) 1467-1480.
[6] Ni S.H, Lu PC, Juang C.H, Comput. Aided Civ. Infrastruct. Engng **11**(1) 59-66.

IOP Conf. Series: Materials Science and Engineering **216** (2017) 012041 doi:10.1088/1757-899X/216/1/012041

Experimental Study on the Effect of freeze-thaw cycle on Silty Clay Stress-Strain Behaviour

Rong Chen, Nan Xue, Dong-xue Hao and Yu-cong Gao
School of Civil Engineering and Architecture, Northeast Electric Power University; Jilin, 132012, China

Email: xuenan9073@163.com;

Abstract. The area of seasonally frozen soil in China accounts for about 54% of the national territorial area. The property of seasonally frozen soil has a great influence on highway, railway and architectural foundation. The effect of freeze-thaw cycle for saturated silty clay in Northeast Area on stress-strain behaviour was investigated by triaxial test in this paper, where three different freezing temperature and five different freeze-thaw cycle times were considered. The results showed that silty clay exhibited strain softening characteristics for small number of freeze-thaw cycles, while strain hardening characteristics will appear with the increase in freeze-thaw cycles. The deviatoric stress-axial strain relationship shows the transitional form from strain softening to strain hardening. It will be helpful to understand the strain behaviour of soil under freeze-thaw cycle.

1. Introduction

The area of seasonally frozen soil in our country is about 5,137,000m^2, accounting for about 54% of the national territorial area [1]. After the seasonally frozen soil experience several times of frost heaving in winter and thawing in spring, the thermodynamic transformation of temperature may cause migration of water content [2]. The physical and mechanical property may also change, the corresponding parameters will change obviously [3, 4] along with the cycle number of freezing and thawing so as to affect the engineering characteristic of seasonally frozen area [5]. With the rapid development of China's economy and the implementation of the old industrial base in Northeast China, the development of highway, railway, water conservancy and hydropower in large scale determines the physical and mechanical property of soil in seasonally frozen area under various environmental and human factors, which is extremely necessary and valuable to guarantee the stability and safety of the project [6].

Xiaojie Ma [7] studied the elastic behaviour of sandy soil and silty soil under the triaxial strain. The result indicated that the maximum compactness corresponded to maximum resilient modulus, while it was false for silty clay. Xuesong Mao [8] studied the resilient modulus under different water contents and cycle number of freezing and thawing. The results indicated that for the sample under optimum water content, the change in modulus of resilience was greatest when the water content changed for the first time, but he didn't consider saturated sample.

Stress-strain curve of soil changes after freeze-thaw cycle and the change mode is greatly related with the soil quality and test type; The shearing strength of soil also appears different change trends to the freeze-thaw cycle. In the freeze-thaw cycle test of soil, the previous research only concerns about the influence of freeze-thaw cycle on the shear performance under optimum water content or nearby the optimum water content, while the mechanical property of saturated soil was not to paid enough attention; another research only focuses on the number of freeze-thaw cycle, but ignores the influence of freezing temperature on the mechanical property of soil. Therefore, the influence of saturated soil on the soil stress-strain behaviour under different freezing temperatures and freezing times is still an outstanding problem.

2. Sample preparation and test method

2.1 Basic nature of soil sample

The soil sample in the test was taken from the foundation soil (the burial depth is 20m and below the underground water level) of library under construction in new campus of Northeast Electric Power University，Jilin. Due to the fact that the original soil contains non uniform sand particles, which easily causes the difference in the sample. Therefore, the test adopted remoulded sample for research. Drying of undisturbed soil and crushes the original soil and then sieving it with 2mm screen. Then a series of physical tests and compaction test were carried out. The related parameters are shown in Table 1.

Table 1. Physical and mechanical parameters of test soil

Parameter of soil	Value	Parameter of soil	Value
Specific gravity of soil, *Gs*	2.72	Plastic limit, w_p	25.27%
Uniformity coefficient, C_u	58.75	Plasticity index	11.31
Bending coefficient, *Cc*	0.18	Optimum moisture content, w_{op}	21%
Liquid limit, w_l	36.58%	Maximum dry density, $\rho_{d,,max}$	1.74 kg/m^3

2.2 Specimen preparation

Vacuum suction method [9, 10] was used to prepare saturated samples. The specific procedures are as below: (1) Lay two layers' plastic film in the trough and then lay 3~5mm round gravel cushion and put 10mm thick geotechnical cloth on it; (2) After drying the excavated original silty clay, crush it and screen it so as to get rid of big gravel; Add distilled water and stir into mud and then pour it onto the geotechnical cloth; (3) Sealing the plastic film totally, turn on the vacuum pump, the water and gas in the soil samples were pumped into the pressure storage tank and make the soil consolidation for saturated soil sample.

Figure 1. Triaxial test soil sample

2.3 Test method

The prepared remoulded saturated soil sample was cut off in cylinder with diameter of 39.1 mm and height of 80 mm for triaxial test. Then these cylindrical soil samples were put in the constant-temperature and humidity freeze-thaw test tank for 0-9 times freeze-thaw cycle. In order to avoid moisture loss of the soil samples or deformation during the process of freezing and thawing, the metal plates with 39.1mm diameter were added on the top and at the bottom of the samples and then the rubber membrane were wrapped on the sample, as shown in Figure 1.

Based on the annual temperature curve data in Jilin, the average lowest temperature is found in January, and the average temperature is -17.8°C. Therefore, we control the freezing temperature of sample within such range and select the freezing temperature of -3°C, -7°C and -12°C. The thawing temperature of the sample is set as 20°C, and the freezing and thawing process experiences 12h. Meanwhile, in order to consider the influence of freeze-thaw cycle number on the mechanical property of soil, the samples were subjected to 0, 1, 3, 5, 7, and 9 freeze-thaw cycles respectively before the conventional triaxial test.

To consider the interaction of confining pressure, cycle number of freezing and thawing and freezing temperature, 61 samples were prepared and tested, shown in Table 2.

Table 2. Experimental program of triaxial test

Freezing temperature, t/°C	Group No.	Cycle number, n	Confining pressure, σ_3/kPa
20	A1	0	100, 200, 300, 400
-3	B1-B4	1	
	C1-C4	3	
	D1-D4	5	
	E1-E4	7	
	F1-F4	9	
-7	G1-G4	1	
	H1-H4	3	
	I1-I4	5	
	J1-J4	7	
	K1-K4	9	
-12	L1-L4	1	
	M1-M4	3	
	N1-N4	5	
	O1-O4	7	
	P1-P4	9	

3. Test result

Non consolidation and non-drainage (UU test) triaxial tests were carried out by TSZ-2 triaxial apparatus. Deviatoric stress-axial strain curves for different temperatures (-3, -7 and -12℃) under confining pressure of 100kPa are shown in Figure 2.

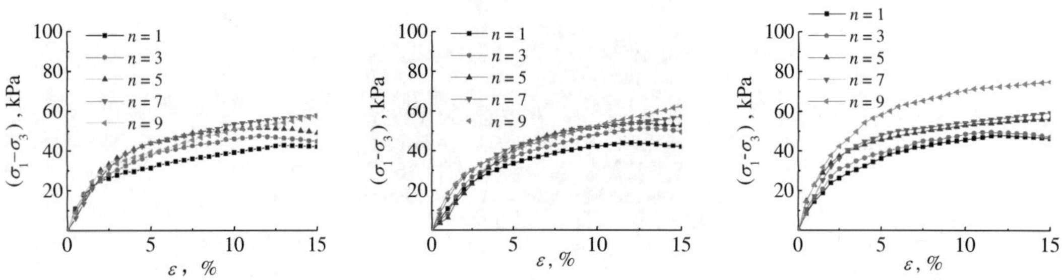

(a) t=-3°C (b) t=-7°C (c) t=-12°C

Figure 2. Deviatoric stress-axial strain curves under different freeze-thaw cycles at σ_3=100kPa

It can be seen that the silty clay has a certain softening characteristic after freeze-thaw cycles 1 ~ 5 times. However, the homogeneous soil presents a hardening curve form after 7 to 9 freeze-thaw cycles as Figure 3. It shows soils with more number of freeze-thaw cycles are more likely to behave strain hardening phenomenon.

Figure 3 shows the effects of different confining pressures and temperatures on the stress-strain curves at 9 times freeze-thaw cycles. The stress strain curves behave strain hardening when the confining pressures are 100, 200, 300 and 400kPa, respectively. But obviously when the confining pressure 300kPa and 400kPa, the $(\sigma_1 - \sigma_3) \sim \varepsilon$ curve is steeper and larger than the confining pressure at

100kPa and 200kPa. It can be found that the $(\sigma_1-\sigma_3)\sim\varepsilon$ curve hardening degree of high confining pressure is more obvious than the low confining pressure, the strain hardening phenomenon is more prominent.

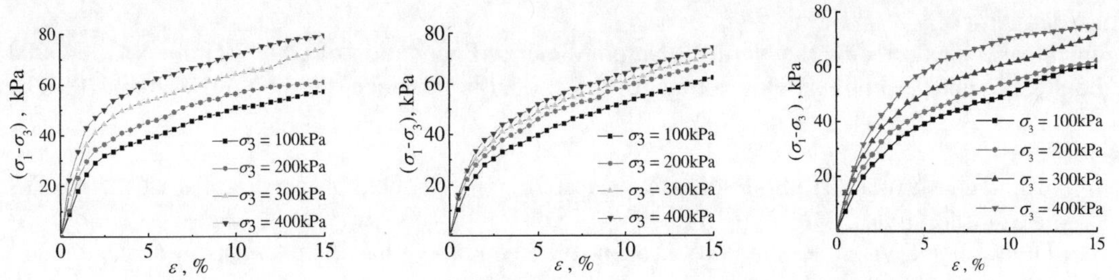

(a) t=-3°C (b) t=-7°C (c) t=-12°C

Figure 3. Deviatoric stress-axial strain curves under different confining pressure at *n*= 9

Figure 4 shows the deviatoric stress and strain curves for various freezing temperatures under confining pressure of 400kPa and different cycle times. Combine all curves to know that when the freeze-thaw cycle reaches 5 times, it appears strain hardening characteristic and the freezing temperature is -12°C. The process of soil freezing is from external to internal. There is a certain time gradient, which doesn't conduct at the same time. The temperature is low when the initial freezing temperature is −12°C and the time for soil precooling and congealing to ice is less than it at the initial temperature of −3 and −7°C. Moreover, after freeze-thaw cycle for many times, the strain hardening process is accelerated.

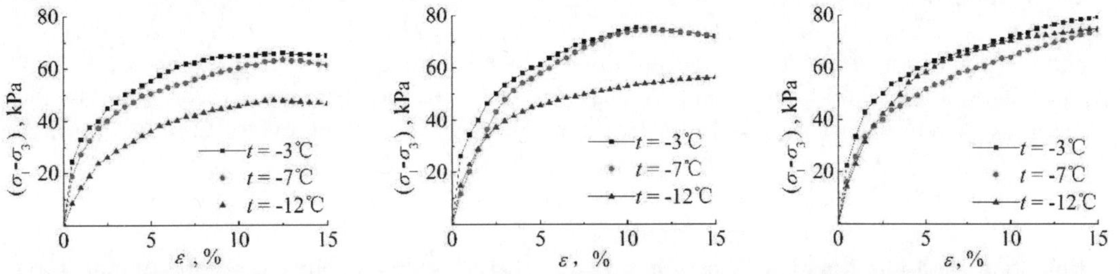

(a) n = 1 (b) n = 5 (c) n = 9

Figure 4. Deviatoric stress-axial strain curves under different freezing temperature at σ_3=400kPa

According to the $(\sigma_1-\sigma_3)\sim\varepsilon$ curve of the soil after freeze-thaw cycle, it can be known that under different confining pressures and at different freezing temperatures, when the number of freeze-thaw cycle is 1-3 times, 23 of 24 $(\sigma_1-\sigma_3)\sim\varepsilon$ curves appear strain softening and the occurrence rate of strain softening is 96%. When the number of freeze-thaw cycle is 5 times, stain softening happens 8 times among 12 $(\sigma_1-\sigma_3)\sim\varepsilon$ curves and the strain hardening occurs 4 times; and all of them happen at the freezing temperature -12°C. When the number of freeze-thaw cycle is 7~9 times, all the 24 curves appear strain hardening and the occurrence rate of strain hardening is 100%.

4. Conclusion

Through triaxial compression test of saturated clay and analysing and processing the test data, the paper summarizes the mechanical property of silty clay on different initial conditions as below:

(1) When the number of freeze-thaw cycle is low, the silty clay exhibits strain softening property. After experiencing freeze-thaw cycle for 5 times, the curve turns into strain hardening gradually. When the number of freeze-thaw cycles increases, all of them show strain hardening characteristics.

(2) The $(\sigma_1-\sigma_3)\sim\varepsilon$ curve hardening degree of high confining pressure is more obvious than the low confining pressure, the strain hardening phenomenon is more prominent

IOP Conf. Series: Materials Science and Engineering **216** (2017) 012041 doi:10.1088/1757-899X/216/1/012041

(3) The deviatoric stress axial strain relationship is characterized by the transition from strain softening to strain hardening at three different freezing temperatures. Moreover, after 7 cycles of freeze-thaw strain hardening is most obvious.

Acknowledgement
This financial is supported by the National Natural Science Foundation of China (Grant No. 51409045) and the Science and Technology Development Project of Jilin Province (Grant No. 20170520105JH).

5. References

[1] Qi Jilin and MaWei 2010 State-of-art of research on mechanical properties of frozen soils. *Rock and Soil Mechanics*. **31**(1), pp 133-143

[2] Li Dongwei and Wang Renhe 2008 Frozen soil creep model based on statistical damage theory. *Chinese Journal of Applied Mechanics*. **25**(1), pp 133-136

[3] Qi Jilin, Ma Wei and Song C X 2008 Influence of freeze–thaw on engineering properties of a silty soil. *Cold Regions Science and Technology*. **53**(3), pp 397-404.

[4] Zhang Xiangdong, Zhang Shuguang, Li Yongjing, et al. Testing study on triaxial creep properties of frozen soil to determine thickness of frozen wall. *Chinese Journal of Rock Mechanics and Engineering*. **23**(3), pp 395-400

[5] Chang Dan, Liu Jiankun, Li Xu, et al. 2014 Experiment study of effects of freeze-thawing cycles on mechanical properties of Qinghai-Tibet silty sand. *Chinese Journal of Rock Mechanics and Engineering*. **33**(7), pp 1496-1502

[6] Sun Xingliang, Wang Ren and Hu Mingjian 2005 Meso-scopic mechanism analysis of deformation properties of frozen silty clay under triaxial shear. *Acta Mechanica Solida Sinica,* **26**(1), pp 62-66

[7] Ma Xiaojie, Zhang Jianming, Chang Xiaoxiao, et al. 2007 Experimental study on creep warm and ice-rich frozen soil. *Chinese Journal of Geotechnical Engineering,* **29**(6), pp 848-852

[8] Mao Xuesong, Hou Zhongjie and Wang Weina 2010 Experimental research on resilient modulus of remolded soil based on water content and freeze-thaw cycle. *Chinese Journal of Rock Mechanics and Engineering,* **28** (S2), pp 3585-3590

[9] Qi Jianfeng 2007 A study on cycle shear behavior and softening deformation of saturated clay. *Dalian University of Techno*logy, Dalian

[10] Liu Gongxun, Luan Maotian, Tang Xiaowei, et al 2010 Critical cyclic stress ratio of undisturbed saturated soft clay in the Yangtze Estuary under complex stress conditions. *Transactions of Tianjin University,* **16**(4), pp 295-303

IOP Conf. Series: Materials Science and Engineering **216** (2017) 012042 doi:10.1088/1757-899X/216/1/012042

Influence of Water Content on Pullout Behaviour of Geogrid

Rong Chen, Yang-yang Song, Dong-xue Hao and Yu-cong Gao

School of Civil Engineering and Architecture, Northeast Electric Power University, Jilin, 132012, China

Email: 15144216791@163.com;

Abstract. The interaction between geogrid and soil is fundamental and crucial factor on safety and stability of geogrid-reinforced earth structure. Therefore, the interface index between geogrid and soil is of vital importance in the design of reinforced earth structures. The pullout behaviour of geogrid in soil is studied, an experimental investigation is conducted using geogrid in four groups of soil with 20%, 24%, 28%, 32% water contents, which correspond to normal stresses of 50, 100, 200 and 300 kPa respectively. The results indicate that the geogrid embedded in soil mainly represents pullout failure, and the ultimate pullout force is sensitive to water content. It decreases with the increase of the water content firstly. Besides, the water content influences the process of the pullout behaviour. The increase of water content leads to the ultimate pullout force soon.

1. Introduction

Geogrid is commonly used in many works such as soil reinforced walls, soft ground improvement, roads and railway embankments, slope stabilizations and bridge abutments constructed also in seismic areas. In all previous applications the main function of geogrid is reported to be the soil reinforcement and the high tensile resistance, and the long-term efficient tensile behaviour of these materials makes possible this function. Wilson [1-2] studied of the short-term and long-term pullout characteristics in detail, a set of methods for predicting the pullout resistance of geogrid in sand soil was obtained. Sugimoto M [3-4] used the stress and strain equation of the pull-out test to simulate the pullout characteristics of the whole geogrid and achieved good results fruit. Farsakh [5] studied the pullout behaviour of geogrid in clay and obtained the formula for calculating the pullout resistance. Yang Guang-qing [6] studied the interface characteristics of geogrid in clay and sand soil, the results show that the shear strength is high in sand soil. Shi You-zhi [7] studied the pullout process of geogrid in the soil and divided the test into 4 stages. Ma Cun-ming [8]and Zhang Bo[9]studied on interface friction of plastic geogrid reinforce earth At present, the studies about water content influences the pullout behaviour were relatively less, so this study was particularly important. This paper carried out a series of pullout tests for geogrid reinforcement soil, testing the relationship between the ultimate pullout capacity and normal pressure under the different water content.

2. Pullout test

2.1. Fill material

Soil used in pullout test is taken from the library construction site of Northeast Electric Power University in the northeast province of Jilin, China. The buried depth of undisturbed soil sample is about 1.5 m. It has well-graded, non-uniform grain size and less impurities. Soil samples after drying, crushing, grinding treatment is shown in figure 1. The conventional tests were conducted including the grain size distribution analysis, specific gravity test, liquid plastic limit test, compaction test. The soil properties are shown in table 1. Based on the grain size distribution analysis and plastic index, the soil is named as silty clay.

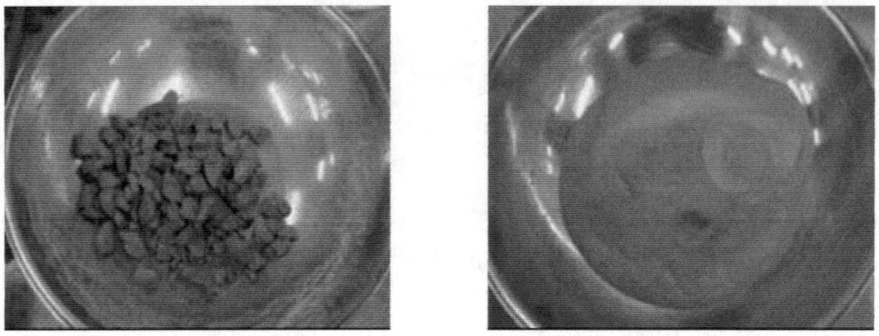

Figure 1. The soil sample in test

Table 1. Properties of test soil

Property	Value
Specific gravity, Gs	2.72
Plastic limit, w_p	25.27%
Liquid limit, w_l	36.58%
The maximum dry unit weight, $\rho_{d,\max}$	1.74g/cm 3
Optimum moisture content, w_{op}	16.5%

2.2. Geogrid

One geogrid, identified as TGSG3030, was used in this investigation. Specifically, the geogrid products were biaxial geogrid that was made of PP. It is manufactured by BOSTD Geosynthetics Qingdao Ltd. in China. The geometric features and main indexes of the geogrid were shown in figure 2 and table 2.

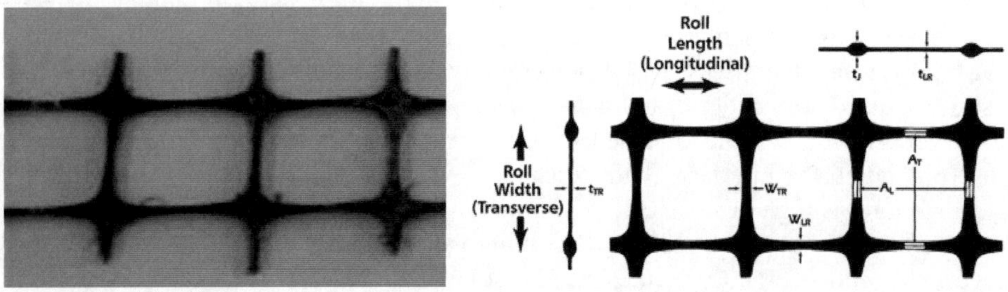

Figure 2. Biaxial geogrid

Table 2. The main indexes of the geogrid in test

Property	Value
Transverse value of aperture, A_T (mm)	32.08
Longitudinal value of aperture, A_L (mm)	36.12
Width of longitudinal ribs, F_{WL} (mm)	2.38
Width of transverse ribs, B_{WT} (mm)	2.14
Thickness of transverse ribs, t_B (mm)	2.25
Thickness of longitudinal ribs, t_F (mm)	1.23
Tensile strength (at 2% strain) (kN/m)	13.6
Tensile strength (at 5% strain) (kN/m)	24.9

2.3. Pullout test device

The integrated test system about geosynthetics was used in this investigation, which could do most conventional test about geogrid. The picture about device was shown in figure 3. The length, width and height of soil-bin are 210 mm, 200 mm and 115 mm. The normal pressure was accurately applied by airbag, and the pullout load was applied using a universal load frame, capable of recording the pullout load and the front displacement.

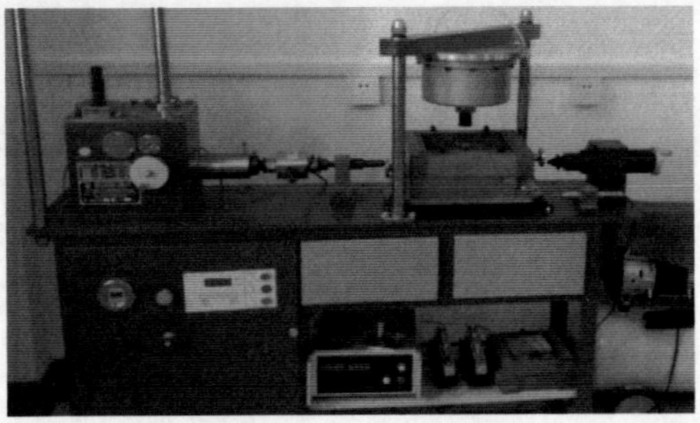

Figure 3. Integrated test system about geosynthetics

2.4. Experimental Method

The soil used in the pullout test program was placed at the lower shear box. The soil was compacted in three layers. The geogrid specimen was then positioned on top of the lower shear box and subsequently clamped on the front edge of the lower shear box. The pullout tests were conducted using normal stresses of 50, 100, 200 and 300 kPa respectively. Consistent with the test device, a pullout rate of 2.2 mm/min was used during testing. When the peak of the pullout forces appeared, continued the pullout test until pullout forces became stable, then stopped the experiment, Besides, The tests were terminated when the displacement reached approximately 30 mm, which corresponds to the maximum travel allowed in the equipment. The test program was shown in table 3.

Table 3. Pullout test program

Influence factor	Test No	Water content（%）	Normal pressure（kPa）
Water content	A1	20	50、100、200、300
	A2	24	
	A3	28	
	A4	32	

3. Experimental results and analysis

The pullout test results are analyzed in order to evaluate the influence on the peak pullout resistance and on the interface pullout friction coefficient of different water content investigated in the research. According to the formula (1) and (2), the pullout friction strength between the geogrid and soil as well as the pullout friction coefficient of the geogrid could be calculated. That is

$$\tau_p = \frac{T_p}{2LB} \quad (1)$$

$$f = \frac{\tau_p}{\sigma} \quad (2)$$

Where τ_p = pullout friction strength; T_p = ultimate pullout force; L = the length of geogrid buried in the soil; B = the width of geogrid buried in the soil; f = the pullout friction coefficient; σ = the normal pressure.

The load- displacement curves obtained from the pullout tests conducted on silty clay, the curves were shown in figure 4. The curves indicated that the pullout friction strength increased with the increase of displacement when the normal stress was 50kPa and 100kPa, the pullout friction strength achieved the limit value and remains stable when the pullout displacement continued to increase. The tests were terminated when the displacement reached approximately 30 mm. The geogrid broken in the process of pullout test when the pullout friction strength was greater than the tensile strength of geogrid, the $2\tau p$ didn't reach the limit value under this condition. From (a) to (d) of figure 4, when the normal stress was 300kPa, the $2\tau p$ was nearly 200 kPa at 20% water content, the $2\tau p$ was nearly 130 kPa at 32% water content, so the $2\tau p$ decreased with the increase of water content under different normal stress.

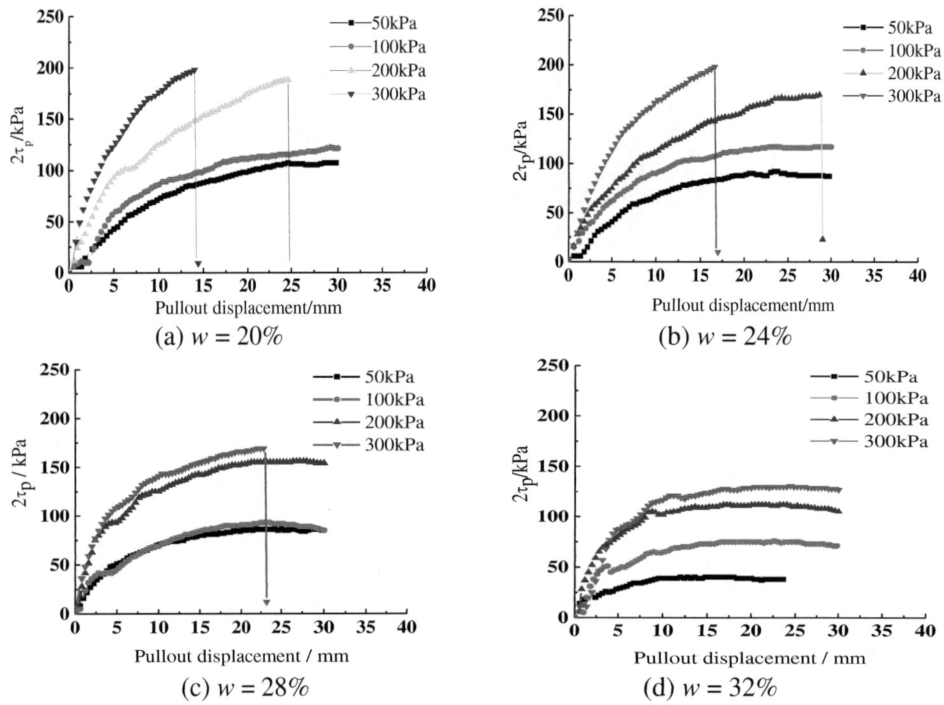

Figure 4. Relation load-displacement curves of pullout test under different water content

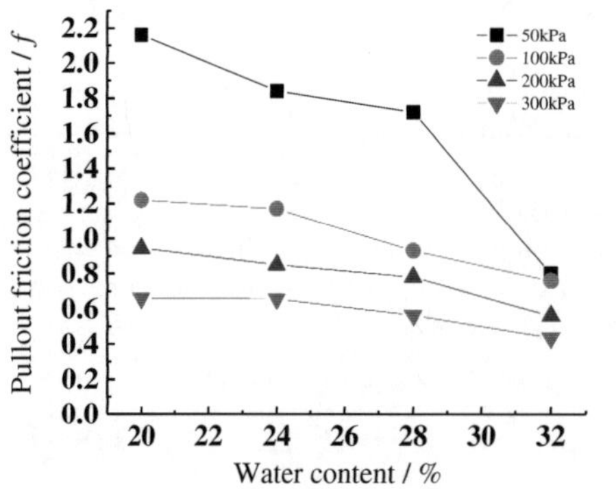

Figure 5. Relation between pullout friction coefficient and water content

On the basis of these formulas the friction coefficient under different water content and normal stresses can be achieved, the curve was shown in figure 5.

When the normal stress was 50kPa, the f was nearly 2 at 20% water content, the f was nearly 0.8 at 32% water content, the f had a great decrease. It indicated that the pullout friction coefficient between the geogrid and soil would decrease with the increase of water content.

It had charged effect on the surface of the clay particles, which adsorbed water molecules and formed water membranes. With the increase of the water content, the water membranes became thin gradually.

Therefore, the friction force between the soil and geogrid decreased gradually, in this case, the pullout friction coefficient would decrease with the increase of soil moisture content.

4. Conclusion

A series of pullout tests were conducted to evaluate the influence of water content on pullout behaviour of geogrid. The main conclusions that can be drawn from this investigation are as follows:

- The pullout friction strength of geogrid decreased with the increase of water content under different normal stress. The geogrid broken in the process of pullout test when the pullout friction strength was greater than the tensile strength of geogrid.
- The test results show that the influence of water content within the range of this experiment (20%~32%) seems to be significant, the friction coefficient between the geogrid and soil would greatly reduce with the increase of moisture content in soil.

5. Acknowledgement

This financial is supported by the National Natural Science Foundation of China (Grant No. 51409045) and the Science and Technology Development Project of Jilin Province (Grant No. 20170520105JH).

6. Reference

[1] Wilson Fahmyr F, Koerner R M and Sansone L J 1994 Experimental behaviour of polymeric geogrids in pullout *Journal of Geotechnical Engineering* 120(4) pp 661–677
[2] Wilson Fahmyr F, Koerner R M and Harpur W A 1995 Long-term pullout behavior of polymeric geogrids *Journal of Geotechnical Engineering* 121(10) pp 723–728
[3] Sugimoto M and Alagiyawanna A M N 2003 Pullout behavior of geogrid by test and numerical analysis *Journal of Geotechnical and Geoenvironmental Engineering* 129(4) pp 361–371
[4] Sugimoto M, Bueno B S and Zornberg J G 2007 Pullout resistance of individual longitudinal and transverse geogrid ribs *Journal of Geotechnical and Geoenvironmental Engineering* 133(1) pp 37–50
[5] Farsakh M, Farrag K and Almohd I 2005 Bearing and frictional contributions to the pullout capacity of geogrid reinfor-cements in cohesive backfill *Proceedings of the Geo-frontiers Congress* pp 1-11
[6] Yang Guang-qing, Li Guang-xin, Zhang Bao-xian 2006 Experiment studies on interface friction characteristics of geogrids *Chinese Journal of Geotechnical Engineering* 28(8) pp 948-952
[7] Shi You-zhi, Ma Shi-dong 2003 Test for interface characteristics of geogrids *Rock and Soil Mechanics* 24(2) pp 296-299
[8] Ma Cun-ming, Zhou Yi-tang, Liao Hai-ni, et al 2004 Experimental study on interface friction of plastic geogrid reinforce earth *Chinese Railway Science* 25(3) pp 36-39
[9] Zhang Bo, Shi Ming-lei 2005 Research on direct shear test and pullout test between clay and geotextile *Rock and Soil Mechanics* 26(5) pp 61-64

IOP Conf. Series: Materials Science and Engineering **216** (2017) 012043 doi:10.1088/1757-899X/216/1/012043

Numerical Simulation of Creep Characteristic for Composite Rock Mass with Weak Interlayer

Jian-guang Li, Zuo-liang Zhang, Yu-biao Zhang, Xiu-wen Shi, Jian Wei
College of Electromechanical Engineering, Qingdao University of Science and Technology, Qingdao 266061, China

Email: qingdaokeda@126.com; starabc123@163.com; 791840317@qq.com; 2859484611@qq.com; 1824611071@qq.com

Abstract. The composite rock mass with weak interlayer is widely exist in engineering, and it's essential to research the creep behavior which could cause stability problems of rock engineering and production accidents. However, due to it is difficult to take samples, the losses and damages in delivery and machining process, we always cannot get enough natural layered composite rock mass samples, so the indirect test method has been widely used. In this paper, we used ANSYS software (a General Finite Element software produced by American ANSYS, Inc) to carry out the numerical simulation based on the uniaxial compression creep experiments of artificial composite rock mass with weak interlayer, after experimental data fitted. The results show that the laws obtained by numerical simulations and experiments are consistent. Thus confirmed that carry out numerical simulation for the creep characteristics of rock mass with ANSYS software is feasible, and this method can also be extended to other underground engineering of simulate the weak intercalations.

1. Introduction

Layered structure is the most common type of structure in rock engineering. According to the survey, the sedimentary rocks with layered structure accounts for two-thirds of land area in nature, even many metamorphic rocks also have the characteristics of layered structure. Therefore, it was inevitable to encounter a great deal of layered rock mass in underground engineering. In various of rock engineering, the weak interlayer frequently appeared as a special form of layered rock mass. According to many research and practices show that the weak interlayer makes significant impact on the overall stability of rock engineering. And even control it under certain conditions. On the other hand, there are problems that effectiveness of mechanical characteristic, also namely the rheological characteristic of rock mass in longer service life of rock engineering should not be neglected, especially it has brought a series of special problems to design and construction in underground engineering with weak interlayer, due to the significant rheological characteristic of interlayer, which even would cause safety accidents. Therefore, it is necessary to study the creep characteristic of rock mass with weak interlayer. All kinds of rheological experiments within the scope of layered rock mass were carried out by ji Chen [1], bin Hu etc. [2], Ming-rong Shen etc. [3], Pinglao Zhao etc. [4], Zeqi Zhu etc. [5], Baoping Xi etc. [6], Deyi Jiang etc. [7], and the creep characteristic of layered rock mass were simulated by Liang-xiao Xiong, Lin-de Yang [8], An-ming Wang etc.[9] [10], Bing Han etc. [11], Huoming Zhou etc. [12], and so on, thus obtained some valuable conclusions.
In this paper, based on creep experiments for composite rock mass with weak interlayer, numerical simulation was carried out, to compare the simulation results with the experimental phenomena and

data, to explore the feasibility of using ANSYS software simulate the creep characteristics of composite rock mass.

2. The creep experiments for composite rock mass with weak interlayer

In this paper, according to the similarity principle, the soft interlayer and the surrounding rock material are made of sand and cement, and the composite cylindrical rock specimen with 30 degree of inclination is prepared, as shown in figure 1(a).The uniaxial compressive creep experiments was carried out after maintenance period of time. Mechanics experiment that adopts hierarchical load method was carried out by the TAW- 200 microcomputer control rock triaxial high temperature triaxial test machine of mechanical research center in Qingdao university of science and technology, the duration of each load was determined according to the variation of the strain rate of the specimen, which was finally loaded into the specimen, as shown in figure 1 (b). Compared with the conventional uniaxial compression experiment, damage form was nearly same interlayer still peeling destruction, but the main rock mass was not involved.

(a)pre-experiment (b) pro-experiment

Figure 1. The weak interlayer of rock specimen.

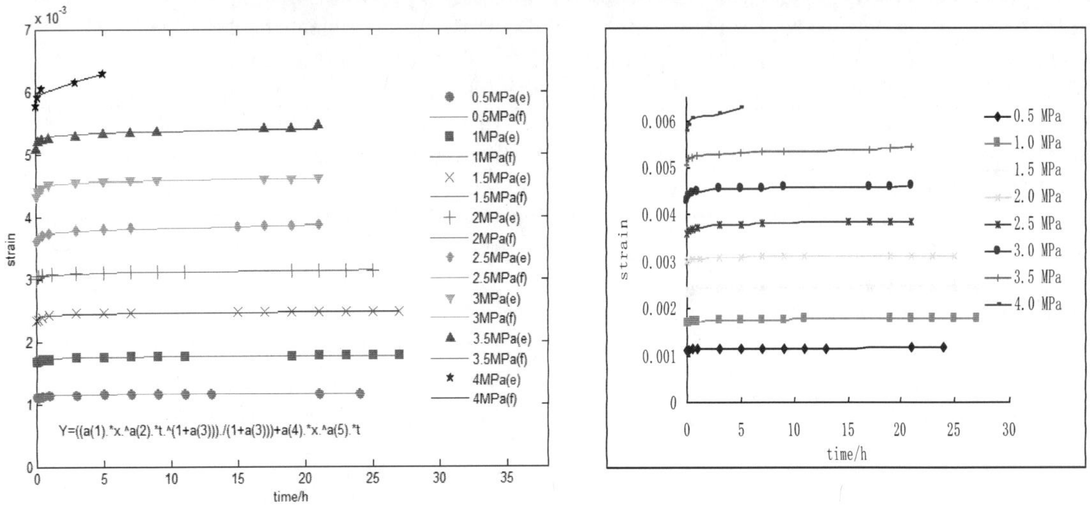

(a) The experimental curve (b) The fitted curve

Figure 2. The uniaxial compression creep curves of class rock mass.

IOP Conf. Series: Materials Science and Engineering **216** (2017) 012043 doi:10.1088/1757-899X/216/1/012043

Figure 2 was the creep axial curves obtained from experiments. In curves, the measured displacement was transformed into strain. The characteristics of loading process from composite rock mass with weak interlayer are shown as follows:

(1)It can produce stain form each level loading, the higher the stress level is, the larger the instantaneous strain absolute value is.

(2) Under the low stress level, the creep curves was straight line, it not occur creep or attenuation creep; However, under the high stress level, the creep curves showed a state of rising, reached steady-state creep, even accelerated creep stage.

(3) The interval of the adjacent two levels of stress curves was different, there are no laws that it can increase or decrease of consistence.

Compared the above characteristics with the predecessors' study for rock mass without interlayer the laws are consistence, which means the existing creep models can be used directly when it was built and used for rock mass with interlayer.

3. Based on ANSYS software, the creep characteristic was simulated for composite rock mass with weak interlayer.

3.1. The choice of creep model

There are 13 kinds of creep models which provided by ANSYS software, some of them are suitable for the creep stage of simulating attenuation, such as, TBOPT = 1, 2, 3, 4, 5, 6, 7, 13; Some of them are suitable for the steady creep stage of simulation , such as TBOPT = 8, 9,10; At the same time, there only two models are suitable to simulate for attenuation stage and steady stage, one is the time hardening model of TBOPT = 11, another is rational polynomial model of TBOPT = 12, the former contains seven parameters and the latter contains 12 parameters. The time hardening model is simple and easy to use relative to rational polynomial model, and if it can accord with the experimental data well, it is the ideal creep model. The form as following:

$$\varepsilon_{cr} = \frac{C_1\sigma^{C_2}t^{C_3+1}e^{-C_4/T}}{(C_3+1)} + C_5\sigma^{C_6}te^{-C_7/T} \tag{1-1}$$

When the temperature impact was not considered, the seven parameters can be degraded to five.

Exporting the creep experimental data, with the aid of matlab software, making the time hardening model be fitted through the least squares, at least the correlated coefficient reached 0.98, and impact was good, therefore, it was shown that experimental data follow the time hardening model.

As figure 2 (b), the fitting comparison illustration between the time hardening model and the measured results of the specimen is given. In illustrations, there are experimental data points with the suffix of letter "e", the fitting curves with the suffix of letter "f". Obviously, for the same series of experimental data, the time hardening model can be fitted perfectly, therefore, it illustrates that the feasibility and reliability of simulate calculation carried out for creep in ANSYS software directly, and it maybe provide more accurate and quantitative reference for the stability of engineering.

3.2. The numerical simulation based on ANSYS software

Geometric model that it was cylinder of Φ 50 mm x 100 mm and interlayer with 30 °, and the bottom of all constraints, the surface load of two Mpa was borne on top. The models were built respectively for interlayer and the main rocks, and input two kinds of material parameters.

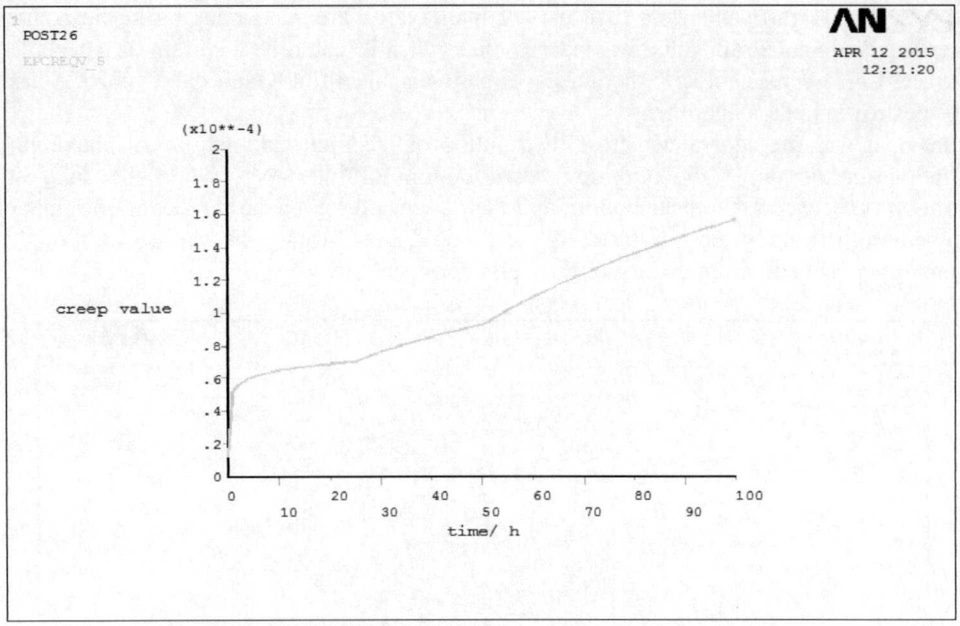

Figure 3. The creep experiment curves under step loading

Figure 3 is the curves for loading experiment courses. It can be seen that there is a corresponding instantaneous strain produced at all levels of load loading moment, and deformation would appear step change. In the process of constant load, the instantaneous strain is no longer change, but the creep variables would growth with time, and the steady creep rate would greater with the stress level higher, the creep variable also greater, finally, it also would cause the macroscopic deformation increasing with time. It was consistent with the process of real experiments.

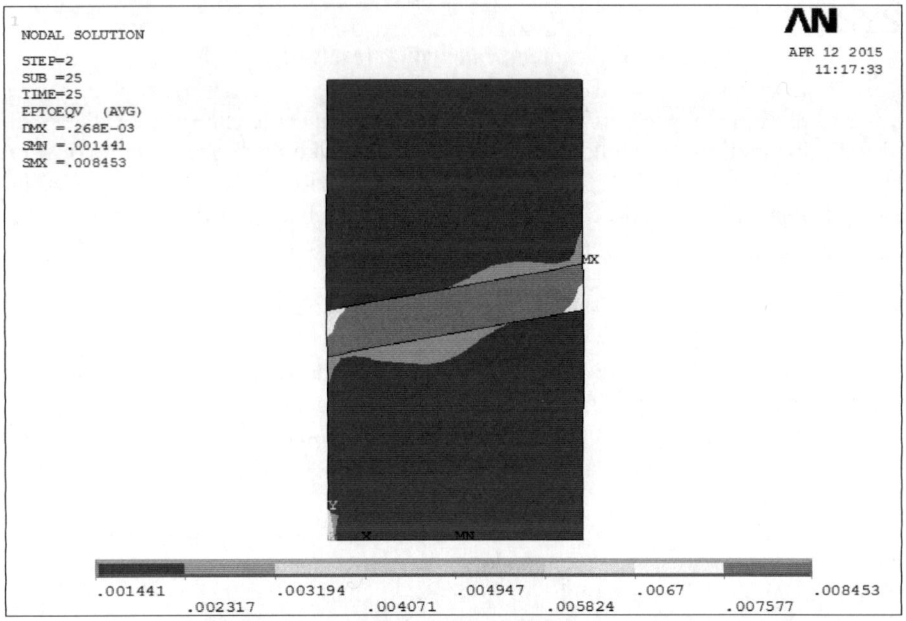

Figure 4. Displacement distribution of sample

Figure 4 provided that the whole strain nephogram of specimens under the pressure 2 Mpa after 25 hours. Total strain was made up by two parts of creep and the elastic strain of instant. The numerical of main rock mass significantly less than part of interlayer, there was changing in the both sides of rock interface, and total strain value was nearly equal when it would be a certain distance away from the interface. It was consistent with the interlayer parts damaged firstly and the main rock mass could nearly be destroyed in the experiment.

As Figure 5, it was the equivalent stress distribution of the final calculation, the maximum stress concentration on the upper top and downward bottom of interlayer, while the smallest stress concentration on between the upper bottom and the downward top of the interlayer, the final damage of specimens maybe along the interface slip, or it could also through the two position of the stress largest and produce the fracture surface of through interlayer.

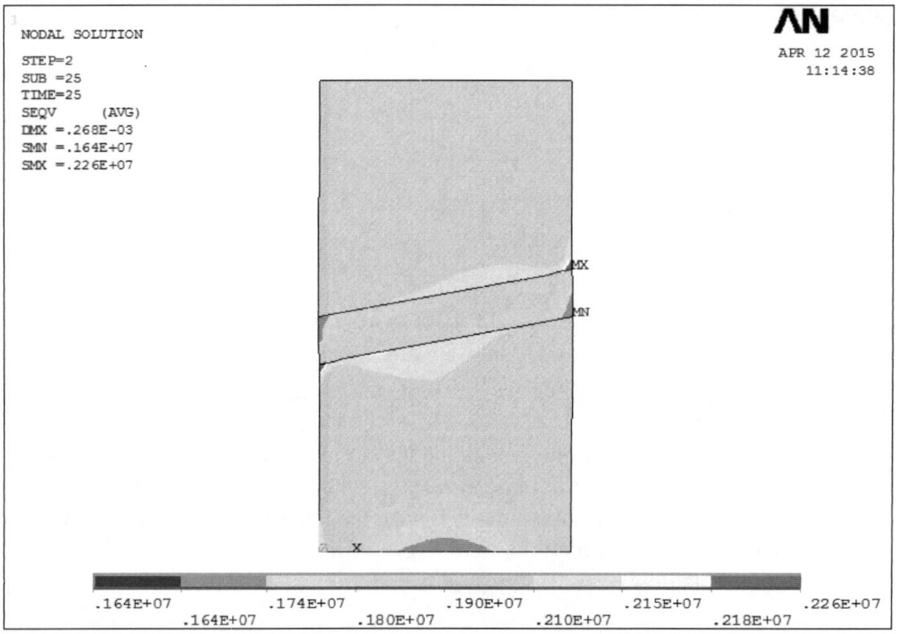

Figure 5. The creep calculation results of sample

However, due to the limit of the software function, in this paper, it can only on the basis of stress analyzing, strain distribution and value within the object to make damage and numerical predictions, and it can't observe the process of the material burst and gradually drop. In view of the diversity and discreteness of materials, we should combine the mechanical characteristics of material to make the actual judgment more proper after obtaining the simulation results, and thus extending the areas of application.

4. Conclusion

In this paper, a typical specimen with inclined interlayer as the representative, through experiments and numerical simulation, mutual verification, the following conclusions:

(1) It is feasible that using the software of ANSYS to simulate the creep characteristic for the composite rock mass with weak interlayer. The experimental phenomena and data were fitted well, so it is a good study method.

(2) This simulated method can be applied to the engineering research of long-term stability, such as simulation of cavity with weak interlayer and slope etc.

In this paper, although only lists a sample of the relevant chart, but it does not mean that only doing one experiment, in fact, it is based on a large number of experimental conclusions and also applicable to other test pieces. For specific engineering structures and projects, the details and the state of the combination of the structural surface need for further study.

5. Acknowledgement

I gratefully acknowledge the financial support from the National Natural Science Fund project (51374134) and (51674149), and the Natural Science Foundation of Shandong Province (ZR2014EL20).

6. References

[1] Ji Chen. Experimental Study of Shear Rheological for Rock Joint Surface [J]. Journal of Huaihai Institute of Technolog (Jiangsu). 2004, 13(4) pp 74-77.

[2] Bin hu, Haifei Jiang, Xinli Hu,etc. Mechanical Characteristics Analysis of Shear Rheological for Purple Mudstone [J]. Journal of Rock Mechanics and Engineering (Wuhan). 2012, 31 (supp.1) pp 2796-2802.

[3] Mingrong Shen, Qingzhao Zhang. Study of Shear Creep Characteristics for Green Schist with Weak Structure Plane [J]. Journal of Rock Mechanics and Engineering (Wuhan). 2010, 29(6) pp 1149-1155.

[4]Pinglao Zhao. Experimental Study on Anisotropy of Strength of Layered Rock [J]. Journal of Lanzhou University (Lanzhou). 1990, 26(4) pp 135-139.

[5] Zeqi Zhu, Yawu Zeng. Study on uniaxial compressive strength and shear failure surface of layered rock mass[J]. Geotechnical Engineering World (Beijing). 2005, 8 (4) pp 27-29.

[6]Baoping Xi,Suguo Xu, Yanlin Zhao,etc.Experimental Research and Theoretical Analysis for Gypsum Layered Creep characteristics[J]. Journal of Taiyuan University of Technology(Taiyuan). 2006,37(2) pp 123-126.

[7]Deyi Jiang, Tao Ren, Jie Chen, etc. Experimental Study for Mechanical Characteristics for Gypsum Rock Mass with Weak Interlayer [J]. Journal of Rock Mechanics and Engineering(Wuhan). 2012, 31(9) pp 1797-1803.

[8] Liangxiao Xiong, Linde Yang. Numerical Analysis of Adhesion Elastic-plastic Rheological Characteristics for Alternative Stratified Rocks [J]. Journal of Rock Mechanics and Engineering (Wuhan). 2011, 30 (supp.1) pp 2803-2809.

[9] Anming Wang, Chunhe Wang, Jianwen Chen, etc. Creep Constitutive Model of Bedded Salt Rock with Nonlinearity [J]. Journal of Rock Mechanics and Engineering (Wuhan). 2009, 28 (supp.1) pp 2708-2714.

[10] Anming Wang, Xiaogen Li,Chunhe Yang, etc. Study of Creep Deformation Interaction for Layered Salt Rocks [J]. Rock and Soil Mechanics (Wuhan). 2010, 31(12) pp 3964-3970.

[11] Bing Han, Zhiyin Wang, Xiuli Ding, etc. Numerical Simulation of Rheological Characteristics for Rock Mass with Hard and Soft Layer [J]. Journal of Yangtze River Scientific Research Institute (Wuhan). 2007, 24(2) pp 25-29.

[12] Huoming Zhou, Qian Sheng, Suwei Chen, etc. Numerical Simulation of Size Effect for Layered Composite Rock Mass Deformation Test [J]. Journal of Rock Mechanics and Engineering (Wuhan). 2004, 23(2) pp 289-292.

IOP Conf. Series: Materials Science and Engineering **216** (2017) 012044 doi:10.1088/1757-899X/216/1/012044

Floodplain Modelling of Malaking-Ilog River in Southern Luzon, Philippines Using LiDAR Digital Elevation Model for the Design of Water-Related Structures

J. R. Ternate, M. I. Celeste, E. F. Pineda, F. J. Tan, and F. A. A. Uy
School of Civil, Environmental, and Geological Engineering, Mapúa Institute Technology, Manila, Philippines

{jasperternate, maigceleste, efpineda2000, fibortan}@gmail.com, {fjtan, faauy}@mapua.edu.ph

Abstract. Hydrologic and hydraulic model applications for floodplain analysis are very limited in the Philippines due to the insufficiency of topographic, river geometric and hydrologic data. However, with Light Detection and Ranging (LiDAR) technology, the country is able to overcome this problem resulting to much more accurate mapping and modeling possibilities. With the use of a modern hydrologic modelling software, rainfall runoff model based on LiDAR data was created for Malaking-Ilog River in Batangas-Quezon with precipitation input of five, ten, twenty-five, fifty and a hundred-year return period. In this study, the significance of the hydrologic model and the selection of return period in the design of various water related structures were discussed. Moreover, river analysis software was utilized in a case study for design of a dike. The communities which are politically grouped as "barangays" surrounding the river that are directly affected when the river overflows were identified. With the help of the hydrographs generated from the rainfall runoff model in this study, the design parameters of various water related structures are easily determined leading to a more efficient design process.

1. Introduction

On average, there are about twenty (20) typhoons that hit the Philippines annually. These have consistently resulted in widespread flooding, casualties, and destruction of structures and properties on the typhoon's path as it traverses the archipelago. In line with this, the Disaster Risk and Exposure Assessment for Mitigation (DREAM) Program was formed to help the country and its people prepare for such natural disaster. DREAM is part of the Nationwide Operational Assessment Hazards (NOAH) project of the Department of Science and Technology (DOST). The said program involves different universities in the Philippines one of which is the Mapúa Institute of Technology (MIT) with its Mapua -Phil LiDAR 1 Project.

LiDAR uses light detection and ranging equipment to gather elevation data of an area. This remote sensing technology is done through an airborne mechanism which has the ability to gather geographic information of a terrestrial structure through recording of laser pulses which hit the land and reflected back to the sensors. Compared to the conventional photogrammetry and field surveying, gathering topographic information through LiDAR technology is not time consuming and it is accurate enough to provide the required elevation data [1]. The use of the LiDAR technology is now acknowledged by the government sectors in the Philippines. The project aims to produce an up-to-date 3-D flood and hazard maps as well as predict, in a more accurate manner, the extent of flooding in Southern part of Luzon using LiDAR DEM.

Malaking-Ilog River is located in the southern part of Luzon Island particularly in the boundaries of Batangas and Quezon provinces. The river is surrounded by developing municipalities with potentials for both social and economic growth. This development may include construction of water-related structures that can change the course of the river, which can affect an entire area or community. With this in mind, accurate urban planning for the future of the city is critical. Being one of the largest watersheds in southern Luzon, Malaking-Ilog drains an area of about seven hundred and thirty eight (738) square kilometers that includes most of Malepunyo Volcano, two-thirds of San Cristobal Volcano and the southwest quadrant of Banahaw Volcano. This extensive watershed causes flooding and flash floods in the lower reaches of the river. On October 12, 2012 the municipality was put under state of calamity due to flooding caused by overflowing of this river when typhoon Ofel (International name: Son-Tinh) hit the city. According to the mayor, 4,100 families or 20,000 individuals were affected while five died due to landslide and drowning. An estimated 125 million pesos worth of structures were damaged and 140 million pesos worth in agriculture. This problem, could have been prevented or at least controlled if only there were more accurate information regarding the geographical features of the river.

2. Floodplain Modelling

LiDAR-derived digital elevation models (DEMs) are important part of the floodplain modelling procedure. LiDAR data can provide a more accurate topographical information of rivers such as slopes, elevations, and cross-sections which are useful in the design aspect of water resources engineering. The process starts with the data acquisition through air-borne light detection and ranging. The gathered LiDAR data are pre-processed in a special computer software to generate the DEM. The data collected is then validated through conducting actual field surveys. Once validated, the data are processed using hydrologic and hydraulic modeling software of the namely: the Hydrologic Engineering Center–Hydrologic Modeling System (HEC-HMS) and the Hydrologic Engineering Center - River Analysis System (HEC-RAS). These are computer generated program developed by the US Army Corps of Engineers designed for runoff simulation and river analysis [2], [3]. The production of flood models and maps is done through incorporating generated digital elevation models with rainfall data. With the aid of these models, water resources engineers can provide information for activities such as the development of new hydraulic structures and flood control measures, evaluation of the performance of existing water facilities, and preparation for disasters that may arise due to floods.

3. Methodology

The researchers gathered DEM files of Malaking-Ilog as shown in Fig. 1 which is processed through the aid of Geographical Information System (GIS) and the existing HEC-HMS parameters. The precipitation data for different return periods (5, 10, 25, 50, and 100-Year Return Period) were gathered from the Philippine Atmospheric, Geophysical and Astronomical Services Administration (PAGASA). The researchers created a rainfall-runoff model based on the collected data and this were simulated using the HEC-HMS software. The outputs are graphs and time series which were analyzed to produce applications for water structures.

Figure 1. Digital Elevation Model of Malaking-Ilog.

3.1. Rainfall Data Acquisition

Precipitation data were gathered from PAGASA in the form of Rainfall Intensity Duration Frequency (RIDF) curve from Tayabas Synoptic rainfall station which is the closest station to Malaking-Ilog watershed.

In order to get the amount of rainfall from the given RIDF curve, interpolation and re-arrangement of values were done to produce a synthetic storm as seen in Fig. 2 where peak values are achieved at a particular time. The extreme values shown in Table I were computed based on a 41-year record.

TABLE I. RIDF VALUES FOR TAYABAS STATION GENERATED BY PAGASA

COMPUTED EXTREME VALUES (in mm) OF PRECIPITATION						
T (yrs)	**1 hr**	**2 hrs**	**3 hrs**	**6 hrs**	**12 hrs**	**24 hrs**
2	59.3	83	99.9	128.2	161.5	195.9
5	77.3	116.1	143	192.6	232.3	279.5
10	89.2	138	171.5	235.2	279.3	334.9
15	96	150.3	187.6	259.3	305.7	366.1
20	100.7	159	198.9	276.1	324.3	388
25	104.3	165.7	207.5	289.1	338.5	404.8
50	115.5	186.2	234.3	329.1	382.5	456.7
100	126.6	206.6	260.8	368.8	426.2	508.3

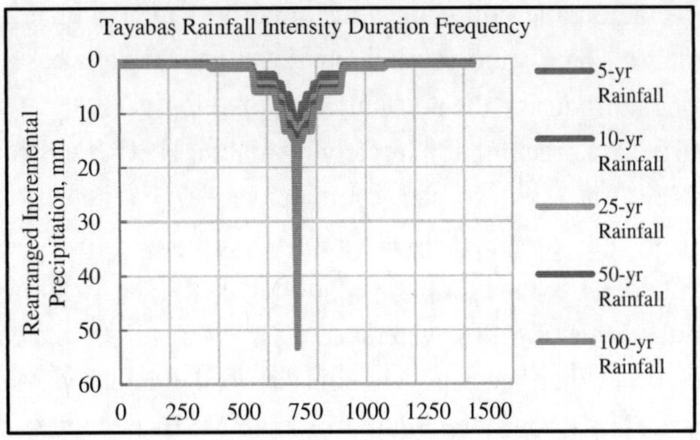

Figure 2. Synthetic storm for a 24-hr period generated at various return periods.

3.2. Data Processing

The researchers utilized HEC-HMS to compute for the discharge hydrograph of each return period. The incremental precipitation is processed in such way that the researchers would be able to identify the discharge hydrograph and the peak discharge on every return period [4].

The basin model shown in Fig. 3 for the Malaking-Ilog includes 68 subbasins, 34 River reach, and 34 junctions. Four important parameters were considered namely, SCS Loss method; Transform - Clark Unit Hydrograph; Baseflow – Recession; and Routing – Muskingum-Cunge. For each return period, there is a corresponding frequency storm. These are stored in the Meteorologic component of the HEC-HMS model with the computed extreme values shown in Table I as inputs After all the inputs were done, the model was simulated to obtain the discharge hydrograph and generate the rainfall-runoff model.

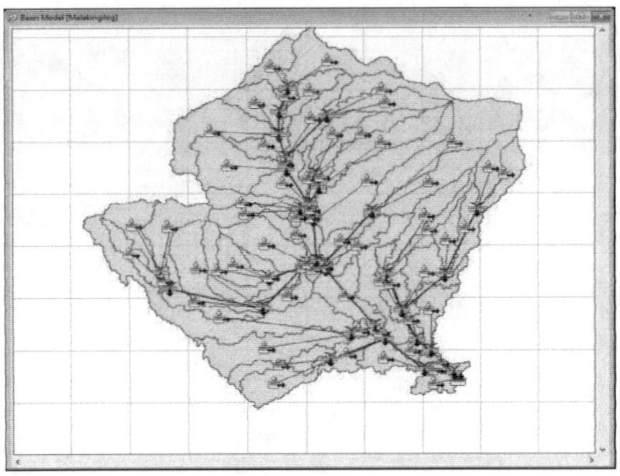

Figure 3. Malaking-Ilog Basin Model in HEC-HMS.

3.3. Model Calibration

The four parameters used in the HEC-HMS model were calibrated by comparing actual discharge to simulated discharge produced by the model. In order to collect actual discharge, flow velocity from a rainfall event is needed. The discharge can be computed by multiplying the river cross sectional area to the accumulated velocity. Furthermore, precipitation data is also needed in the model calibration.

These were taken from the automatic rain gauges installed by DOST which is located on or near the watershed. These precipitation data were rainfall observations that are based on the actual rainfall event or typhoon which occurred during the accumulation of velocity.

With the observed discharge and precipitation already available, HEC-HMS can perform an automatic calibration by optimization trials which can estimate model parameters and other preliminary conditions. After every trial, the computed hydrograph is compared to the observed hydrograph to determine how well the model fits to the observed flow. Fig. 4 shows the observed and simulated discharge hydrographs in the Malaking-Ilog watershed.

The accuracy of the model was then measured in contrast to the actual values. Different methods of comparison were employed. The Root Mean Square Error (RMSE) method sums up the differences of the two values. It was identified at 4.4 m^3/s. The Pearson correlation coefficient (r^2) measures the linear relationship between the actual and computed value. A value close to 1 indicates that the two values have the best fit. It is identified to be 0.8708. The Nash-Sutcliffe (E) method was used to measure the certainty of the model with also an optimal value of 1. Here it is computed as 0.85. The Observation Standard Deviation Ratio, RSR, is an error index. A value of zero means the two values have a perfect match. The model has an RSR value of 0.39 [5].

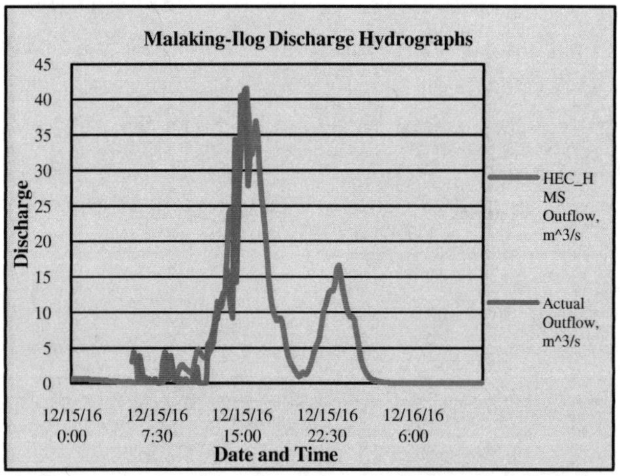

Figure 4. Malaking-Ilog HMS model outflow compared to actual outflow.

4. Results and Discussion

4.1. HEC-HMS Discharge Hydrograph

The HEC-HMS simulation for 5, 10, 25, 50, and 100-year return period reveals the time and value of peak discharge for each hydrologic element in a precipitation event. From this graph, it can be observed how discharge varies over time such that as the rain continues to fall the runoff discharge also continues to rise until it reaches its peak discharge in that case it gradually falls into recession.

During an intense rainfall event, the discharge of the river is high since there is an additional amount of water to the stream flow of the river. Precipitation has a great effect on the discharge of the river. More precipitation means more water will be flowing in the river. Precipitation highly depends on the recurrence interval (return period) of a particular rainfall event [6]. Observing the design hyetograph in Fig. 5, precipitation increases depending on the return period being considered. The highest rainfall is 53.4 mm which occurred at 100-year return period. Based on the hydrograph, the highest discharge occurred also at 100-year return period with a flow of 11905.3 m³/s implying that return period, precipitation, and discharge has a direct relationship with each other.

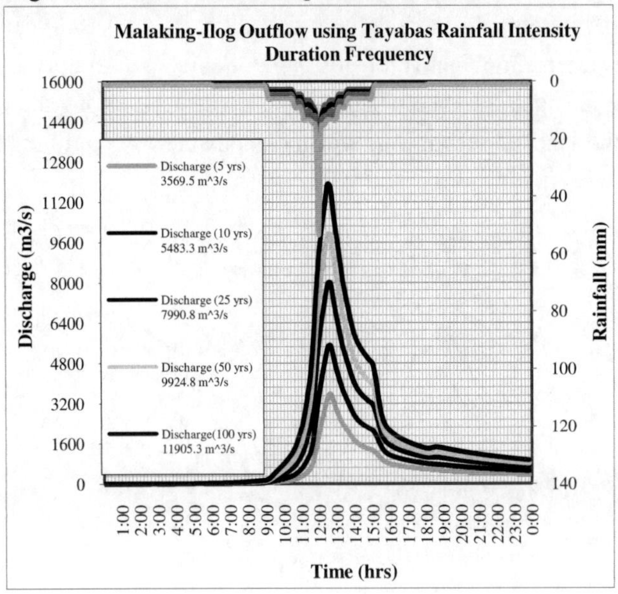

Figure 5. Simulated Discharge Hydrograph with Corresponding Rainfall

4.2. Application of Design Discharge

Considering the following structures, the role of a rainfall runoff model, return period, and accurate LiDAR DEM in the design of water related structures are discussed. Ponce [7] provided some guidelines for the selection of return period. For drainage systems, 25 to 50-year return period is needed wherein peak discharge from hydrograph determines the design capacity and the cross-section of the pipe needed. For dams, 25 to 100-year return period is required. Considering its complex routing process, frequency analysis of variables such as flood peak and volume (discharge) are necessary. As for bridges, necessary data is based on the maximum design discharge corresponding to 100 to 500-year return period. With these data, the substructures of the bridge such as abutment and pier could be designed. Fishways are hydraulic structures usually built on disrupted path of flow of fishes caused by construction other hydraulic structures such as dam or dikes.

Other water related structures, like revetment and bulkhead, also requires the discharge value in order to be designed. Revetments and bulkheads are structures capable of protecting riverbanks from eroding and sliding. It can be constructed in areas where some parts of the riverbank can be drag away due to high discharge rate. Although, revetment and bulkheads are not mainly designed to prevent flooding but rather control the discharge and prevent it from breaking the banks. As for design, hydraulic parameters like local-depth average velocity, velocity near the bed, waves, level of turbulence in the flow, and various water levels, are necessary to determine the stable sizes of both structures. The values that will be adopted for each of the hydraulic parameters mentioned above will depend on the return period for which the structures will be designed [8].

In the design of a dike, one has to consider the maximum water surface level in critical cross-sections of the river. If the maximum water surface level (design flood level) surpasses the elevation of the riverbanks, the river overflows and construction of dike must be initiated. The height of the dike is based on the design flood level which can be calculated using available flow hydrographs. In determining the design hydrographs on medium risk dikes, 50 to 100-year return period can be considered. In line with the main objective of this paper, the researchers present an application of the rainfall runoff model by designing a dike which can be constructed along the banks of a river for the purpose of protecting the landslide from overflowing floodwater by confining the stream flow in the regular channel.

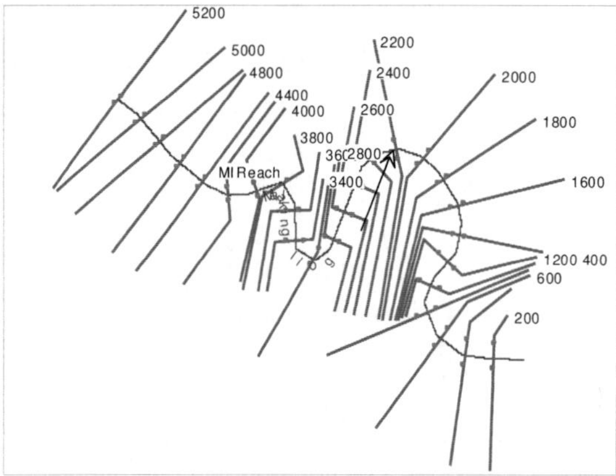

Figure 6. Stationing of Malaking Ilog

In HEC-RAS, cross-sections for 26 river stations were created as shown in Fig. 6. Unsteady flow simulation was performed. Boundary conditions for the upstream were selected as the flow hydrograph of Station 5+200. For the downstream, normal depth as boundary condition was chosen which was measured to be 0.0003. The flow hydrograph considered for a medium risk dike is of 100-year return period. Taking that as an input to HEC-RAS, the design discharge for every station and the elevation of the maximum water surface that the river can reach are determined. Whereas, the design flood level is calculated by subtracting the elevation of the edge of the river from the water surface elevation. Consequently, it is identified that 25 out of 26 stations are susceptible to overflow.

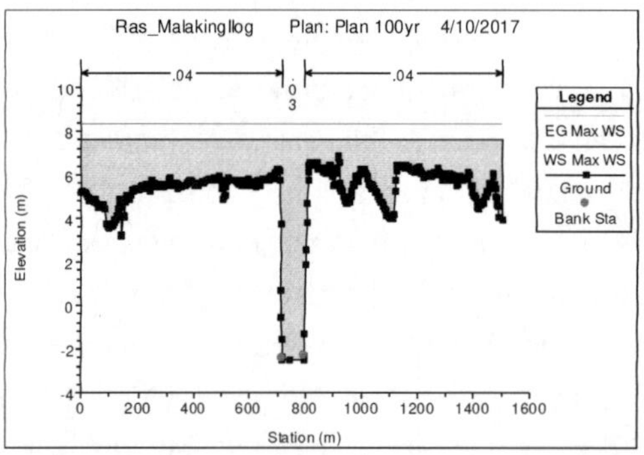

Figure 7. Cross-sectional area of Station 5+000 based on 100-Year Return Period.

The cross-section of each station within the river can also be viewed upon simulation of HEC-RAS. The stations where water overflows start from Station 5+000 to the downstream which is at Station 0+200. Fig. 7 shows the water surface level at Station 5+000 based on 100-year return period.

The general dike design considerations are specified in Table II. The freeboard and crest width are computed based on the acquired value of discharge, Q. For discharge 2,000 m^3/s up to 5,000 m^3/s, the freeboard height is 1.2 meters while crest width is 5 meters. For 5,000 m^3/s up to 10,000 m^3/s, the freeboard height is 1.5 meters while crest width is 6 meters. For 10,000 m^3/s and over the freeboard height is 2 meters while crest width is 7 meters. As for the slope, since the distance from the crest to the river bed is more than 6 meters, the slope is 3:1 [9].

TABLE II. DIKE DESIGN CONSIDERATIONS OF CRITICAL AREAS

River Sta	Discharge Total (m^3/s)	Design Flood Level (m)	Free-board (m)	Dike Design Height (m)	Crest Width (m)	Slope (H: V)
5+000	8066.85	1.41	1.5	2.91	6	3:1
4+800	11746.97	2.02	2	4.02	7	3:1
4+600	11692.55	2.47	2	4.47	7	3:1
4+400	11630.32	2.29	2	4.29	7	3:1
4+200	11563.59	2.41	2	4.41	7	3:1
4+000	11499.72	2.91	2	4.91	7	3:1
3+800	11447.66	3.04	2	5.04	7	3:1
3+600	10778.64	2.62	2	4.62	7	3:1

5. Conclusion

The latest technology from Mapúa-DOST-Phil LiDAR1 allowed the researchers to define highly accurate topographical parameters of Malaking-Ilog for a rainfall runoff model considering 5, 10, 25, 50, and 100 year return period. This model easily computes discharge for various return periods and may be utilized to develop a floodplain analysis to design flood control structures such as dikes, embankments, dikes, drainage systems etc. The results can also provide information for the design of other hydraulic structures such as weirs and dams.

Based on the results, the researchers came up with the following findings: Malaking Ilog experiences an overflow when strong typhoons hit the province of Batangas. In an ocular visit to the study area, the researchers confirmed these theoretical outputs by conducting an interview with the residents living beside the river. According to them, due to strong typhoons in the past the river tends to overflow resulting to floods in the area.

So far, with the coupled simulation of the hydrologic and hydraulic modeling programs, the water level in Malaking-Ilog overflows especially to the areas where people reside. Residences were observed just beside the river and when the river overflows, the lives of the residents are in danger. From this the researchers conclude that flood control facilities/measures must be instigated in order to reduce the risks. With the help of the hydrographs from the rainfall runoff model in this study, the design parameters of various water related structures could easily be determined.

This study can serve as a reference for water resources engineers and designers who decide to pursue construction of flood control facilities. With the knowledge on return periods considered for individual water structures, engineers could utilize the rainfall-runoff model developed by the researchers to determine the design discharge. Furthermore, the use of river analysis software, HEC-RAS, is recommended in this study for identifying the areas that needs flood control measures and facilities.

Acknowledgments
The authors extend their sincerest gratitude to the Mapua-DOST Phil-LiDAR 1 Project funded by Mapua Institute of Technology and the Department of Science and Technology (DOST) thru its Philippine Council for Industry, Energy, and Emerging Technology Research and Development (PCIEERD). This project is jointly done with the Training Center for Applied Geodesy and Photogrammetry (TCAGP) of the University of the Philippines – Diliman thru its Phil LiDAR 1 Program.

References
[1] A. Fernando, and J. Mills, "Accuracy Assessment of LIDAR-Derived Digital Elevation Models," The Remote Sensing and Photogammetry Society and Blackwell Publishing Ltd, 2008.
[2] US Army Corps of Engineers Hydrologic Engineering Center, "HEC-HMS – Hydrologic Modeling System: Technical Reference Manual," Davis, CA., 2000.
[3] US Army Corps of Engineers Hydrologic Engineering Center, "HEC-RAS – River Analysis System: User's Manual," Davis, CA., 2010.
[4] Department of Public Works and Highways Japan International Cooperation Agency, "Manual on Runoff Computation with HEC-HMS. Project for the Enhancement Capabilities in Flood Control and Sabo Engineering of the DPWH," 2003.
[5] MIT Phil-LiDAR 1, "Malaking-Ilog HEC-HMS and HEC-RAS Model Report, Hazard Mapping of the Philippines using LiDAR (Phil-LiDAR 1)," DOST-Grant-In-Aid Program, 2017.
[6] H. Q. Urias, H. Garcia, J. S. P. Mendoza, "Determination of Relationship Between Precipitation and Return Periods to Assess Flood Risks in the City of Juarez, Mexico," 2007.
[7] V. M. Ponce, "Q&A on the return period to be used for design," 2008.
[8] M. Escarameia, "River and Channel Revetments: A Design Manual," Thomas Telford Publications.K. Elissa, 1998.
[9] Department of Public Works and Highways Japan International Cooperation Agency, "Technical Standards and Guidelines for Design Flood Control Structures. Project for the Strengthening of Flood Management Function of the DPWH.J," 2010.

IOP Conf. Series: Materials Science and Engineering **216** (2017) 012045 doi:10.1088/1757-899X/216/1/012045

Estimation of River Discharge at Ungauged Catchment using GIS Map Correlation Method as Applied in Sta. Lucia River in Mauban, Quezon, Philippines

Monjardin, Cris Edward F., Uy, Francis Aldrine A., Tan, Fibor J.
School of Civil, Environmental and Geological Engineering, Mapua Institute of Technology, Manila, Philippines

cefmonjardin@mapua.edu.ph, faauy@mapua.edu.ph, fjt@mapua.edu.ph

Abstract. This paper presents use of GIS Map Correlation Method, a novel method of Prediction of Ungauged Basin, which is used to estimate the river flow at an ungauged catchment. The PUB Method used here intends to reduce the time and costs of data gathering procedure since it will just rely on a reference calibrated watershed that has almost the same characteristics in terms of slope, curve number, land cover, climatic condition, and average basin elevation. Furthermore, this utilized a set of modelling software which used digital elevation models (DEM), rainfall and discharge data. The researchers estimated the river flow of Sta. Lucia River in Quezon province, which is the ungauged catchment. The researchers assessed 11 gauged catchments and determined which basin could be correlated to Sta. Lucia. After finding the most correlated basin, the researchers used the data considering adjusted parameters of the gauged catchment. In evaluating the accuracy of the method, the researchers simulated a rainfall event in the said catchment and compared the actual discharge and the generated discharge from HEC-HMS. The researchers found out that method showed a good fit in the compared results, proving GMC Method is effective for use in the calibration of ungauged catchments.

1. Introduction

Daily stream flow is essential before designing such structures because it is the one that can determine if a certain area or watershed is feasible or not for the project. Daily stream flow can be obtained easily from gaged catchments, however to those ungauged catchments, gathering stream flow information is still a challenge in hydrology [9]. This study will basically focus on the determination of stream flow from ungaged basins using PUB (Prediction at Ungage Basin) Method which is said to be more accurate than The basin of Sta. Lucia which covers Mauban has a total land area of 10,059 ha. Sta. Lucia River covers Mauban Municipality in the Province of Quezon, the area is an ideal watershed where it is not yet overpopulated by people and many parts of it are still full of vegetation. And in the vicinity, it can be seen that large area is available for use if ever we want to put up a hydropower plant or reservoir.

Phil-LiDAR 1 is the project of DOST given to the SUC's and HEI's in the Philippines, and Mapua Institute of Technology is one of the selected schools and actually the only private HEI. This project gave opportunity to the researchers to acquire lots of data.

The study will predict the water availability of a certain ungage catchment using PUB Methods which were widely used in other countries but still not here in the Philippines. PUB Methods don't require expensive equipment and data, most of the data that will be using here were all free such as Google Earth, which is a free satellite image by Google so it will really be much easier to conduct the study since all the data need are available. Determination of water availability is very important since not all the watersheds in the country are accessible; some are located in remote areas which are difficult to reach. It will be much better if there are certain studies that can predict the hydrograph without even going on the place itself because it can save time and money, since people don't need to travel just to study the watershed and money since people don't need to buy expensive equipment to validate the model as all that is needed is to rely on the previous available data [10]. This study can also help water industries that wants to put up a hydropower plant or a reservoir on a certain watershed.

Water availability is one of the major concerns of the people related to water engineering, and these industries have many projects in mind but they can't pursue it since they don't have the data about the watershed.[6] Nowadays, the government is getting very strict when it comes to the environment, industries can't easily do a project without consent or permit. Many of the approved projects by the government were in the remote areas since it can't affect many people in that location but the problem is they don't have any data about that place. Water availability is one of the major things they need to know before the industries can conduct a certain project, and through PUB methods, without even going on the site and those expensive equipment the researcher can easily determine the hydrograph of a certain watershed.

Equipment and accessibility are one of the major things people need to consider when studying a certain watershed [8]. But since there is no such budget and time some become less interested in planning projects to the watersheds. With the use of PUB Method to determine water availability, many industries and even the Phil-Lidar 1 of DOST will benefit from it, since it won't be needing any other expensive equipment just to calibrate a certain watershed model.

Determination of water availability has seen a great impact in many studies [3]. It is very convenient if this can be utilized using PUB methods in determining such parameters. The industries and government can save time and money since they won't be needing expensive equipment; all they need to do is to apply and understand the proposed method. PUB Method was already widely used in other countries, it's a big step for the Philippines if it also start using this because this is really a good thing to many future environmental projects.

1.1. Objectives

The main objective of this study is to use PUB Method in determining the water availability of ungaged catchments such as the Sta. Lucia Watershed. Specifically, the project seeks to:

a. Determine the water availability of ungaged catchments.

b. Come up with own PUB Method (GMC Method)

c. Suggest a method to Mapua Phil-Lidar 1 to calibrate their model at ungage catchment

It is expected that this study would help many industries and government agencies in solving environmental issues. Water availability is one of the main concerns of everyone who are into water industry since it is the main requirement before a project can be established or not, and for the government agencies it is very helpful in terms of disaster preparedness.
Also, with the said methodology not only the cost will be lessened but also the safety of the people who will be conducting a study or project won't be a problem because they don't need to visit the site every time they can even rely on the image provided by Google Earth or if they want a clearer image they can ask an orthophoto in NAMRIA (National Mapping and Resource Information Authority). Since most of the projects must be done on the upstream and as everyone already knows what the situations are there, many of the said areas are being occupied by New People's Army and other Rebel groups.
Moreover, with the goal of determining water availability using PUB (Prediction at Ungaged Basin) Method it would be a good solution to the problems. The Philippines is encountering nowadays in the environment, having a model of a watershed is very important to predict the ever changing behavior of the environment.[11] Government agencies and water industries will be the ones who would be benefiting a lot from this study.

2. Methodology

The objective of this study is to estimate the water availability of a certain watershed without the use of rain gage and stream gages. Arc GIS and HEC HMS are the primary programs to use in the estimation since it will involve the study of maps, specifically land cover maps and the hydrologic cycle in the area. PUB Methods will be used to determine the flow hydrograph of Sta. Lucia, Quezon. Most of the data to be used here came from Phil-Lidar 1 and from open sources, then these data will be processed. Next thing is study the characteristics of the watershed and compare it to others then decide what PUB Method will be used. With the use of PUB Method the researcher can predict the hydrograph of an ungaged basin that can be used to calibrate the said river.[1] Calibration is done using HEC-HMS' optimization function. After calibration, next thing to do is to validate the accuracy of the model, this will determine the effectiveness of PUB Method used. Once the model is validated, it will now be the output of the study. Finally, an ungage basin calibrated using PUB Method without the use of modern technologies is produced.
Actual flow data will be gathered on site to test if the calibrated model using PUB Method is effective. If the actual data is almost similar to the simulated what the researcher can say that this PUB Method is efficient. The calibration and validation will be done with the help of HEC-HMS. To predict the flow at ungage basins is a challenge to many hydrologists and water resources engineers, so coming up with a certain method and proving its effectiveness is a milestone for them.[9] There were already

many studies from different countries about this PUB Methods but still no one conducted it here in the Philippines; it will be a good thing that with this paper the Philippines won't be left behind in terms of this kind research study.

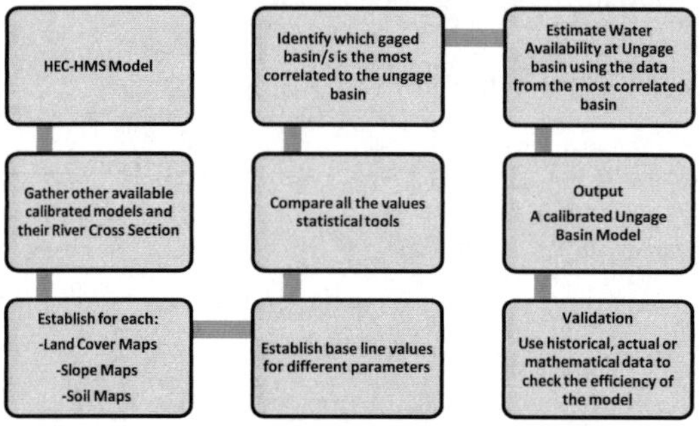

Figure 1. Flow chart for GMC Method.

In this PUB Method as illustrated in Figure 1, the researcher will make use of the other calibrated gaged basin model. The calibrated basin models' properties will be compared to the property of the ungage catchment using land cover maps for land use, slope map for contours and soil maps for soil types. Using Pearson's r correlation, the researcher will compare all the properties of the basins then identify which is/are the most correlated catchment that can be used to calibrate the ungage basin. Using these criteria, it will be the basis that can tell if a certain gage basin/s is correlated to an ungage basin. After identifying which is/are the most correlated basin, make use of that catchment as reference to calibrate the model. And, when the calibration is done, test the accuracy of the model using Nash-Sutcliffe test for goodness of fit.

3. Results and Discussion

In the study, the researcher considered 11 watersheds to know which among those basins is the most correlated one to Sta. Lucia Basin. These 11 watersheds were Lian, Maragondon, Canas, Ilang-Ilang, Binabang Molino, Bolbok, Rosario-Lobo, Catanauan, Pandanan, Vinas and Kilbay Catabangan. These watersheds were all located in Region IV-A "CALABARZON" where the Mapua Phil-Lidar 1 conducting their study. On the other hand, parameters considered in the correlation were the following: land cover, slope, curve number and average basin. These parameters are the ones that might affect the river flow in a certain basin, therefore if the researcher can identify all these parameters and compare to each river basin they can easily identify which river basin is most correlated to Sta. Lucia River.

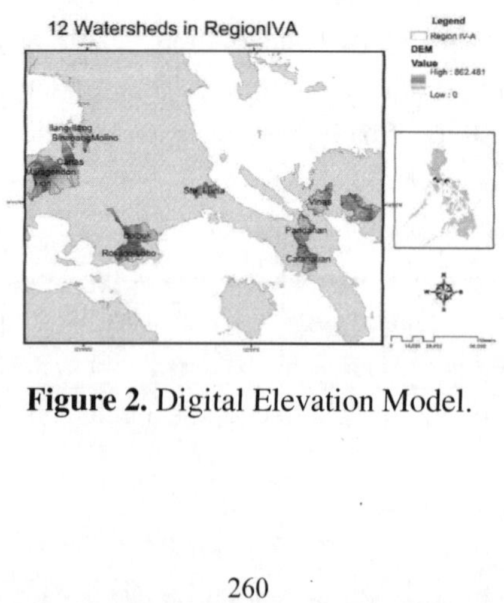

Figure 2. Digital Elevation Model.

3.1. Slope

First parameter to be considered is slope since this determines how fast the water inside the watershed will flow, the steeper the slope the faster the travel will be. The researcher utilize the data they have specifically their SAR-DEM that has a cell size of 10m by 10m which is very efficient for watershed modelling. Using this DEM, the researcher created slope maps that helps to identify which river basin is the most correlated to Sta. Lucia. Region IV-A is mountainous that is why slope inside each watershed is very significant for the watershed's behaviour and must be considered.

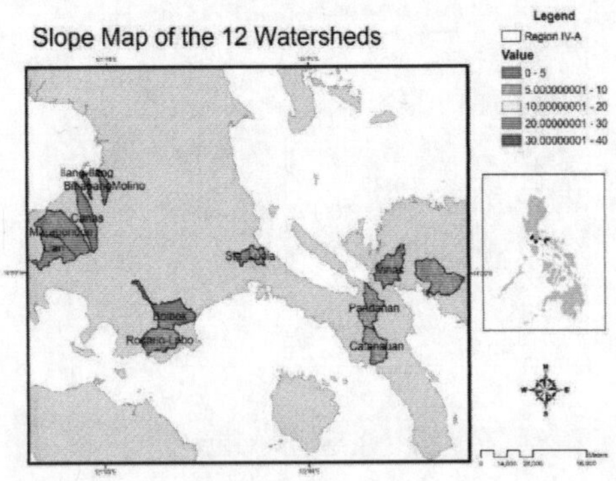

Figure 3. Slope Map.

All percentages were given above and compared with respect to Sta. Lucia. Out of the 12 watersheds, two showed significant relationship with Sta. Lucia that was Catanauan which gave an average percent error of 14.13%, and the other was Rosario-Lobo which showed 8.72% percent error. Since the two said watersheds is the most correlated with respect to Sta. Lucia; they will be prioritized for comparison to the next parameters that will be considered.

3.2. Curve Number

Curve number is a parameter that mainly determines the infiltration rate of a certain watershed. Curve numbers are dependent to what type of soil is present in the watershed. Determining the curve number of the watersheds is not that difficult because such data were already gathered by other researchers and shared online at PhilGIS.org there is already an available shape file for the entire Philippines indicating different types of soil a certain area has. There is also another data made by other researchers as well that gives the curve number for every type of soil. This data were very helpful in different kinds of study like this one. With the help of Arc GIS, the researcher easily created a Soil Map to determine the correlation of the watersheds to Rosario-Lobo.

As reflected in the calculation above, other watersheds show a significant difference with Sta. Lucia that it goes beyond 100%. Only Rosario-Lobo gives a percent difference below 100%. That's why, one can easily say that in terms of curve numbers Roasrio-Lobo is the most correlated basin to Sta. Lucia.

3.3. Land Cover

IOP Conf. Series: Materials Science and Engineering **216** (2017) 012045 doi:10.1088/1757-899X/216/1/012045

Land cover is one of the important parameters that must be considered when comparing watersheds. Land cover determines the land use of different areas inside the watershed. This must be considered because land use plays an important role to what will be the response of a watershed when rainfall occurs. Many studies show that as land cover of a certain area changes, it shows significant effect to its river flow. The data used in here came from Philgis.org a reliable source resources which is very helpful to different studies. The researcher downloaded the shapefile from Philgis. This shapefile contains digitized boundary for different types of land use of the whole Philippines.

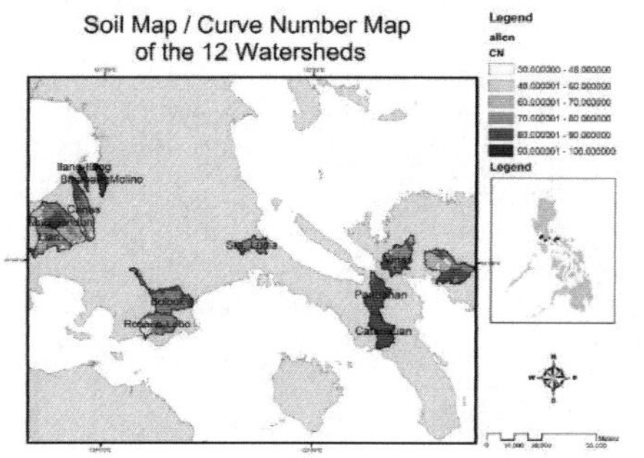

Figure 4. Soil Map.

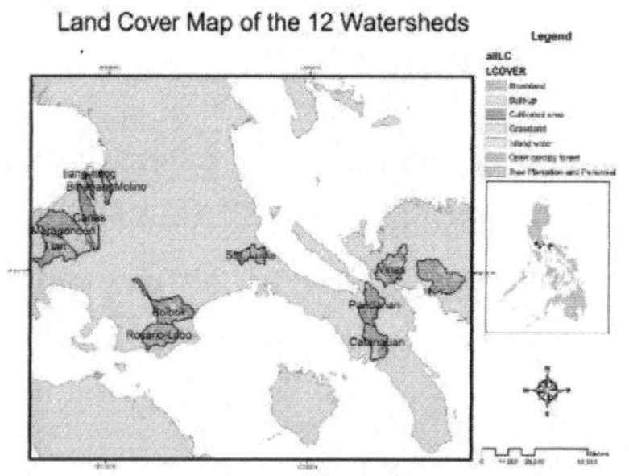

Figure 5. Land Cover Map of the 12 Watersheds.

By figures, Sta. Lucia is mostly brushland which is 89.61% of the watershed area, and 9.93% of it is cultivated area. Comparing Sta. Lucia to other watershed, Rosario-Lobo shows a significant similarity with Sta. Lucia having 81.35% of the total watershed area as brushland and 10.04% of it is cultivated area. Therefore, the researcher found out that among the 11 watershed Rosario-Lobo shows significant similarity in terms of its land cover to Sta. Lucia.

3.4. Average Basin elevation

Table 1. Average Basin Elevation % Difference with Sta. Lucia

River Basin	Average Basin Elevation	% difference with Sta. Lucia as Reference Gage
Sta Lucia	128.8861601	0
Maragondon	248.476991	92.78795396

Kilbay	103.6618015	19.5710374
Bolbok	100.9212203	21.69739543
Babang Molino	52.21346079	59.4886986
Lian	183.9925036	42.75582689
Canas	233.2969405	81.0100792
Rosario-Lobo	180.7300661	40.22457176
Vinas	39.394054	69.43500065
Pandanan	44.52822871	65.4515049
Ilang Ilang	130.7022922	1.409097854

The average basin elevation of the 11 watersheds were also computed and compared with Sta. Lucia. Considering its % difference, there were four basins that can be compared to Sta. Lucia which were Ilang-Ilang with 1.41% difference, Kilbay-Catabangan with 19.57% difference, Bolbok with 21.7% difference and Rosario-Lobo with 40.22% difference. With the figures shown those four were the ones close to Sta. Lucia in terms of average Mean Sea Level.

Considering the four most important parameters of a watershed the researcher found out that among the 11 watersheds the most correlated one to Sta. Lucia is Rosario-Lobo. Rosario-Lobo showed a significant homogeneity with Sta. Lucia in terms of its slope profile, curve number, land cover and average basin elevation. With this, the researcher can use the data gathered in Rosario-Lobo, which is a gaged basin to calibrate the watershed model at ungaged site which is Sta. Lucia.

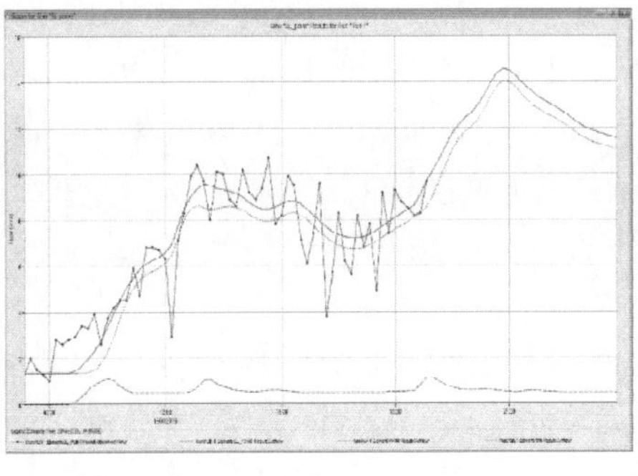

Figure 6. Calibration of Sta. Lucia using data from Rosario-Lobo.

To validate the accuracy of the data, the researcher gathered actual flow rate in the river during an event (Rainfall). The simulated hydrograph by HMS must coincide with the actual hydrograph gathered from the river.

The efficiency of the model with respect to the actual flow were tested using the three formulas used for determination of goodness of fit (Nash Sutcliffe, Percent Bias and RMSE-observation Standard Deviation Ratio). The results were significant, in the calculation above Nash-Sutcliffe had 0.5 which is satisfactory, Percent Bias was 2.16% which is Very good and RSR was 0.71 which is satisfactory. All values were above the minimum requirements therefore the researcher can conclude that the model simulation has a good fit with respect to the actual data.

Test for goodness of fit was made using the following equation: Nash-Sutcliffe, Pbias and Root Mean Square Error.
The efficiency of the model with respect to the actual flow were tested using the three formulas used for determination of goodness of fit (Nash Sutcliffe, Percent Bias and RMSE-observation Standard Deviation Ratio). The results were significant, in the calculation above Nash-Sutcliffe had 0.5 which is satisfactory, Percent Bias was 2.16% which is Very good and RSR was 0.71 which is satisfactory. All values were above the minimum requirements therefore the researcher can conclude that the model simulation has a good fit with respect to the actual data.

4. Conclusion

In this study the researcher came up with a method that will help in the calibration of the watersheds that don't have its own rain gages and automatic water level sensors, it is called GIS Map Correlation (GMC) Method that uses base maps and digital elevation models such as SAR and LiDAR data to compare the physical/geographical characteristics of the watersheds and determine which were correlated to each other.
Sta. Lucia Watershed Model was calibrated using the data from Roasrio-Lobo basin, this method can be used as long as there is an available data, specifically the DEMs which is one of the important parameters that needs to be considered. This method can help Mapua Phil-LiDar 1 to accomplish their project Flood Hazard Modeling of Region IVA.
To check the efficiency of the model calibrated with the use of GMC method, the researcher gathered actual discharge data. This data was compared with the simulated discharge of the HMS model. For the actual discharge, the peak flow was 0.3 m^3/s while the simulated peak discharge was 0.3 m^3/s. The researcher test the goodness of fit of the model using Nash-Sutcliffe, PBIAS and RSR. The results were the following Nash-Sutcliffe had 0.5 which is satisfactory, Percent Bias was 2.16% which is Very good and RSR was 0.71 which is satisfactory. Having those values indicate that the simulated data had a good fit with respect to the actual data.

5. Recommendation

The research was all about watershed models, in the study the researcher consider five parameters to determine which watershed is most correlated to Sta. Lucia and the result is Rosario-Lobo. Sta. Lucia watershed model were already established using the data gathered from a gaged basin which is the Rosario-Lobo since the two watershed were correlated to each other. For the duration of the study, the Philippines is experiencing El niño which became a challenged for the researcher in gathering data. The researcher is recommending that the model's accuracy in forecasting discharge flow must be checked further, checking the reliability of the model will increase the proof that it can be used for Disaster Management programs. Checking the accuracy of the model will just give an idea how effective it is. Since all the data used in this study specifically the parameters used, came from other study done in the past therefore it may not be up to date considering the recent condition therefore the researcher is recommending that data must be updated so accurate correlation can be done.

6. Acknowledgement

This material was inspired by the work done by University of the Philippines – Diliman Phil-LiDAR 1 Program. The authors are thankful to the people behind it, to the Department of Science and Technology - Philippine Council for Industry, Energy and Emerging Technology Research and Development (DOST-PCIEERD) and to Mapúa Institute of Technology Phil-LiDAR 1 Project for all the support to this research.

References

[1] Kentel, E., (2009). Estimation of River Flow by Artificial Neural Networks and Identification of input vectors susceptible to producing unreliable flow estimates. Journal of Hydrology, 375 (2009) 481–488.

[2] Kentel, E. and Ergen, K. (2015). An Integrated Map Correlation Method and Multiple-Source sites Drainage-Area Ratio Method for Estimating Streamflows at Ungauged Catchments: A Case Study of the Western Black Sea Region, Turkey. Journal of Environmental Management, 166 (2016) 309e320.

[3] Yoshida, K. and Ishikawa, T. (2014). Flood Hydrograph Estimation Using An Adjoint Shallow-Water Model. Journal of Hydro-environment Research 9 (2015) 429e440..

[4] Mutie S.M., Mati B, Home P., Gadain H. and Gathenya J. (2006). Evaluating Land Use Change Effects On River Flow Using Usgs Geospatial Stream Flow Model In Mara River Basin, Kenya. Center for Remote Sensing of Land Surfaces, Bonn, 28-30 September 2006.

[5] Getachew, H. and Melesse, A. (2012). The Impact of Land Use Change on the Hydrology of the Angereb Watershed, Ethiopia. International Journal of Water Sciences.

[6] Mazvimavi, D. (2003). Estimation of Flow Characteristics of Ungauged Catchments Case Study in Zimbabwe. Ph.D. Thesis, Wageningen University.

[7] Granato, G. (2012). Estimating Basin Lagtime and Hydrograph-Timing Indexes Used to Characterize Stormflows for Runoff-Quality Analysis. U.S. Geological Survey, Reston, Virginia.

[8] Cooper, R. (2002). Determining Surface Water Availability in Oregon. State of Oregon, Water Resources Department.

[9] Archfield, S. A. and Vogel, R. M. (2010). Map Correlation Method: Selection of a Reference Streamgage to Estimate Daily Streamflow at Ungaged Catchments. Water resources research, vol. 46, w10513.

[10] Gamble, C. (1989) Techniques For Simulating Flood Hydrographs And Estimating Flood Volumes For Ungaged Basins In East And West Tennessee. U.S. Geological Survey Water-Resources Investigations Report 89-4076.

[11] Gamble, C. (1989) Techniques For Simulating Flood Hydrographs And Estimating Flood Volumes For Ungaged Basins In East And West Tennessee. U.S. Geological Survey Water-Resources Investigations Report 89-4076.

IOP Conf. Series: Materials Science and Engineering **216** (2017) 012046 doi:10.1088/1757-899X/216/1/012046

Analysis of Manning's and Drag Coefficients for Flexible Submerged Vegetation

Khamaruzaman Wan Yusof[1], Muhammad Mujahid Muhammad[1,2], Muhammad Raza Ul Mustafa[1], Nor Azazi Zakaria[3] and Aminuddin Ab. Gahani[3]

[1]Department of Civil & Environmental Engineering, Universiti Teknologi Petronas (UTP), Seri Iskandar, Malaysia.

[2]Department of Water Resources & Environmental Engineering, Ahmadu Bello University, Zaraia – Nigeria.

[3]River Engineering and Urban Drainage Research Centre (REDAC), Universiti Sains Malaysia (USM) Pinang, Malaysia.

khamaruzaman.yusof@utp.edu.my

Abstract. Accurate determination of flow resistance is of great significance in modelling of open channels that will convey water efficiently. Although, resistance or drag induced by vegetation have been systematically studied for several decades, estimating of the resistance remain as a challenge. This is because most of previous studies use artificial vegetation to investigate flow – vegetation interactions. To overcome this, the present study evaluates the vegetation resistance in terms of Manning's roughness coefficient and drag coefficient using a natural flexible vegetation (cow grass) under submerged condition. From the experimental result obtained, it was observed that the Manning's and drag coefficients decreased with the increasing in average velocity. Also, graphical relationship between Manning's coefficient, n and drag coefficient, C_D has been developed with $R^2 = 0.9465$, which indicate that there exist a strong correlation between n and C_D, and one can use the proposed graphical model to predict the n - values corresponding to the C_D – values.

1. Introduction

Flow resistance induced by vegetation have been systematically studied for several decades, estimating of the resistance remain as a challenge. This is because most of previous studies use synthetic vegetation to investigate flow – vegetation interactions.

Thus, for accurate determination of flow resistance, the present study employed the use of natural flexible vegetation, as it is of great significance in modelling of open channels that will convey water efficiently. Also, if the estimated roughness is too low, this can result in over – estimating the discharge, under – calculating the flood levels, and consequently, over - design of drainage facilities with excessive expenditure in erosion control works and vice versa [1]. However, the estimation of this roughness, n, depends on several factors as illustrated by Chow [2], that the value of n is highly variable which can be affected by the following: surface roughness, vegetation, channel irregularity, channel alignment, silting and scouring, obstruction, and stage and discharge. From these factors, vegetation resistance is of very significant in predicting n, as it has large Manning's coefficient, n, compared to other factors, depending on the height, density, distribution and type of vegetation.
Furthermore, another important parameter that can be related with flow resistance is the drag on vegetation, this is because the vegetal drag increases the overall flow resistance and decreases the bed shear stress of the channel, causing in reduction for bed-load transport capacity. Hence, the drag improved tendency for trapping, deposition, and stabilization of sediments. Numerous methodologies have been proposed to model the effects of vegetation on open-channel flow and sediment transport [2-7]. Also, vegetation-induced drag force has been comprehensively investigated in recently by Kothyari, Tang, Wu and others [8-10].
The evaluation of the vegetation resistance in terms of Manning's roughness coefficient and drag coefficient is therefore the main objective of this study. A laboratory experiments was conducted to assess the flow resistance as well as the drag coefficient using a natural flexible vegetation (cow grass) under submerged condition.

2. Flow resistance due to flexible vegetation

The flow resistance in channels with flexible vegetation can be determined using relative roughness method similar to the commonly used resistance relationships derived for rigid roughness in pipes and channels [11, 12]. In this regard, Kouwen and Li [13] suggested that the Darcy-Weisbach friction factor, λ, can be found using a semi-logarithmic resistance equation also known as the Colebrook-White equation:

$$\frac{1}{\sqrt{\lambda}} = a + b \log(R/k) \tag{1}$$

Where k is the roughness height of the vegetation, R = hydraulic radius of the channel, a and b two fitted parameters that are found to be dependent on the relative magnitude of the shear velocity *U and a critical value U*crit. Also, based on numerical model test, Darby [5] used a general approach to evaluate the roughness of vegetation and movable bed:

$$\frac{1}{\sqrt{\lambda}} = 2.03 \log\left(\frac{a_s R}{k}\right) \tag{2}$$

$$a_s = 11.0(R/h_{max})^{-0.314} \tag{3}$$

Where, as is a non-dimensional shape correction factor and hmax being the maximum flow depth in the cross section. For flexible submerged vegetation, the roughness height, k, is a function of the amount of drag exerted by the flow and the flexural rigidity MEI [13]:

$$k = 0.14h_v \left[\frac{(MEI/\tau)^{0.25}}{h_v}\right]^{1.59} \tag{4}$$

Where, τ = bed shear stress, MEI is the flexural rigidity and hv is vegetation height. Using laboratory experiments, Kouwen [14] and Temple [15] developed empirical equations to compute the MEI with vegetation height, hv, for a variety of growing and dormant grass species as follows:

$$MEI = \begin{cases} 319h_v^{2.3} & \text{for green, growing grass} \\ 25.4h_v^{2.26} & \text{for dormant or dead grass} \end{cases} \tag{5}$$

Also, the vegetation roughness can be determined in terms of the popularly known Manning's roughness, n, using Manning's equation. Thus, in this work the Manning equation would be adopted for being the usual method for estimating energy losses in hydraulic modelling:

$$Q = \frac{1}{n} AR^{2/3} S_0^{1/2} \tag{6}$$

Where, Q = discharge (m^3/s), A = cross-sectional area (m2), R = hydraulic radius (m) = A/P, P = wetted perimeter (m), and S = slope of gradient of the bed.Please follow these instructions as carefully as possible so all articles within a conference have the same style to the title page. This paragraph follows a section title so it should not be indented.

3. Drag coefficient for submerged vegetation

The drag in vegetated channel is determined based on the concepts suggested by Wu et al. [10]. Noting that, it is challenging to represent a vegetation element with simple geometry because of its irregularity in shape. Thus, as an approximation, a vegetation stem is conceptualized as a cylinder with height, h, and a representative diameter, D. The procedure to calculate the vegetative drag coefficient due to submerged vegetation is represented as follows:

$$F_G = F_D + F_S \tag{7}$$

IOP Conf. Series: Materials Science and Engineering **216** (2017) 012046 doi:10.1088/1757-899X/216/1/012046

Where, FG is the gravitational force, FD is the drag force exerted on the vegetation and FS is the surface friction of the sidewalls and bottom.
In which the gravitational force FG is expressed as:

$$F_G = \rho . g(B.H.L).S \tag{8}$$

Where, ρ is the mass density of water (Kg/m^3), g is the gravity constant (m2/s), H is depth of water above the vegetation, B is channel width (m), S is the bed slope (m/m) and L is length of vegetated channel reach. For uniform flow in vegetated channel, FS is negligible compared to FD. The drag force FD is given by:

$$F_D = C_D(\lambda . h_v . B.L)\frac{\rho . U^2}{2} \tag{9}$$

Where, hv is height of vegetation, and C_D is the drag coefficient, λ is the vegetal area coefficient representing the vegetation density per unit channel length (m-1). Equating FG and FD gives:

$$C_D = \left(\frac{H}{h_v}\right)\frac{2.g.S}{U^2\lambda} \tag{10}$$

Equation (10) can be used to evaluate the vegetal drag coefficient C_D, which accounts for the features of vegetation, as indicated above. Also, it should be noted that vegetation density, λ, was determined based on concept of Jihong and Launia [16], that vegetation density can be generally considered as the momentum absorbing area per unit water volume taking into account the whole part of the plant, which is more elaborate than defining it as the projected area of rods per unit volume.

4. Experimental work

The laboratory experiments were conducted in a man - made concrete rectangular channel of dimension 16 m overall length, 1.5 m width and 1 m depth respectively, located at the physical modeling laboratory of River Engineering and Urban Drainage Research Center (REDAC), Universiti Sains Malaysia (USM). It consists of three sections which were the inlet sump, the test flume (12 m length) and the outlet sump. Fig. 1 shows the plan view of the test channel with the Cow grass planted on a bed slope of 1 in 1000. Also, the figure shows the various measurement locations.

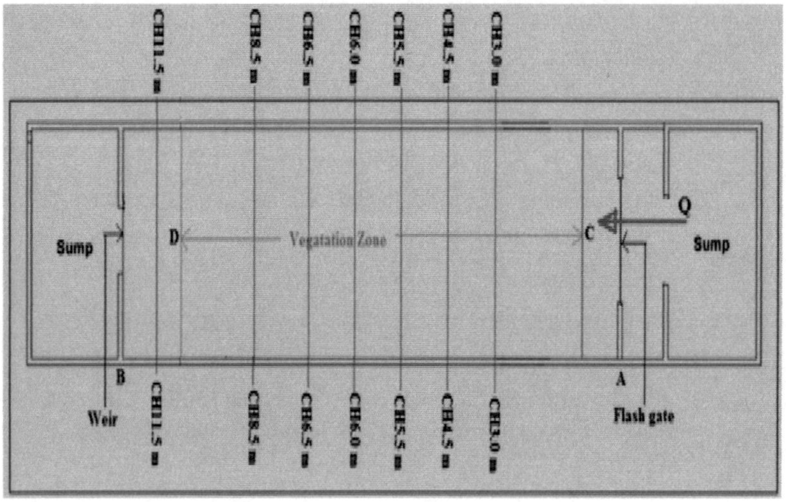

Figure 1. Plan view of test channel

From Figure 1, the Cow grass planted over a length of 10 m in the vegetated zone labeled CD. The average height of the grass was maintained at 50 mm. Afterwards the experimental testing started using three different flow depths with the grass under submerged conditions. The varied flow depths were y = 0.15 m, y = 0.20 m and y = 0.40 m, respectively. Velocity Distributions were determined at five different vertical points (that is width of channel 1.5 m divide by 5 equals 0.3 m) measured along 7 – cross sections labelled CH3.0 m as the first cross section, meaning 3.0 m away from the inlet, similarly cross section CH4.5 m, means 4.5 m away from inlet and so forth, up to CH11.5 m as shown in Fig. 2. Also, at each point, 8 different depths were measured in fractions of the flow depths, that is, 0.2y, 0.25y, 0.3y, 0.4y, 0.5y, 0.6y, 0.7y and 0.8y, respectively. Acoustic Doppler Velocimeter (ADV) was used for the velocity measurement which can determine the velocity in x, y and z – directions as stream-wise, lateral and vertical, respectively. However, only the average vertical velocities were used for the analysis of the flow resistance. Also, the measured velocities are influenced by the particles moving along with the water which help in obtaining a better result. As such turbid water was used in this experiment.

5. Results and discussion

Tables 1 - 3 show the average vertical velocity profiles of the various flow depths. From these tables, variations of Manning's Roughness, n and drag coefficient, C_D, with Velocity were shown, where both n and C_D values decrease with increase in velocity and vice versa.

It should be noted that vegetation density, λ, in Tables 1 -3 was obtained based on the following criteria: Length of vegetation reach L_v = 10 m; Area of vegetation reach, Av = B*Lv = 15 m2, where B is the width of the channel = 1.5 m; Volume of water, $V_w = A_v*h$; and vegetation density, $\lambda = A_v/V_w$.

Table 1. Variation of manning's and drag coefficients for the flow depth, y = 0.15 m

Total Flow depth (m)	Fraction water depth, h (m)	Average Velocity, V (m/s)	Flow Area A (m^2)	Vegetation density λ (m^{-1})	Drag Coefficient C_D	Manning n
	0.03	1.42	0.05	33.33	0.000	0.001
	0.04	1.11	0.06	26.67	0.001	0.001
	0.05	0.57	0.07	22.22	0.005	0.002
	0.06	0.50	0.09	16.67	0.011	0.004
y = 0.15	0.08	1.15	0.11	13.33	0.003	0.002
	0.09	0.99	0.14	11.11	0.007	0.003
	0.11	1.07	0.16	9.52	0.008	0.003
	0.12	0.67	0.18	8.33	0.026	0.005

Table 2. Variation of manning's and drag coefficients for the flow depth, y = 0.20 m

Total Flow depth (m)	Fraction water depth, h (m)	Average Velocity, V (m/s)	Flow Area A (m^2)	Vegetation density λ (m^{-1})	Drag Coefficient C_D	Manning n
	0.04	0.37	0.06	25.00	0.009	0.003
	0.05	0.48	0.08	20.00	0.009	0.003
	0.06	0.32	0.09	16.67	0.029	0.006
	0.08	0.63	0.12	12.50	0.013	0.004
y = 0.20	0.10	0.26	0.15	10.00	0.118	0.011
	0.12	0.17	0.18	8.33	0.402	0.019
	0.14	0.16	0.21	7.14	0.590	0.023
	0.16	0.31	0.24	6.25	0.210	0.013

Table 3. Variation of manning's and drag coefficients for the flow depth, y = 0.40 m

Total Flow depth (m)	Fraction water depth,	Average Velocity,	Flow Area	Vegetation density λ (m^{-1})	Drag Coefficient	Manning n

	h (m)	V (m/s)	A (m^2)		C_D	
	0.08	0.45	0.12	12.50	0.025	0.005
	0.10	0.44	0.15	10.00	0.042	0.006
	0.12	0.32	0.18	8.33	0.112	0.010
	0.16	0.34	0.24	6.25	0.182	0.012
y = 0.20	0.20	0.09	0.30	5.00	3.955	0.056
	0.24	0.10	0.36	4.17	4.299	0.056
	0.28	0.14	0.42	3.57	3.300	0.047
	0.32	0.12	0.48	3.13	5.787	0.060

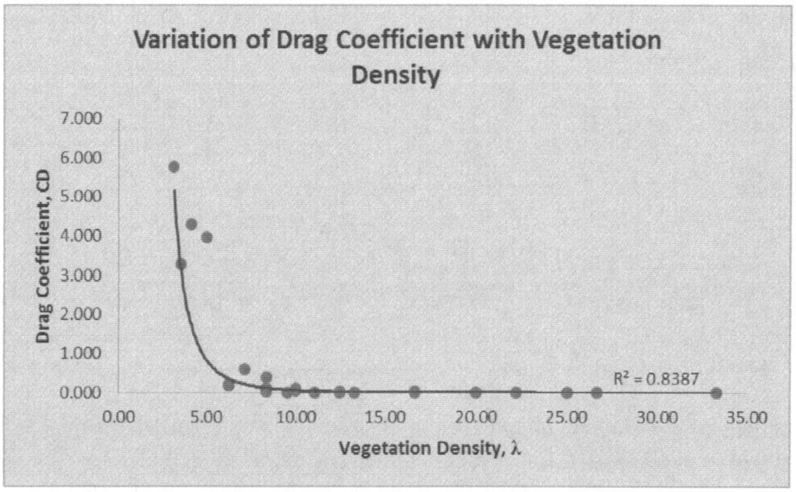

Figure 2. Variation of drag coefficient with vegetation density

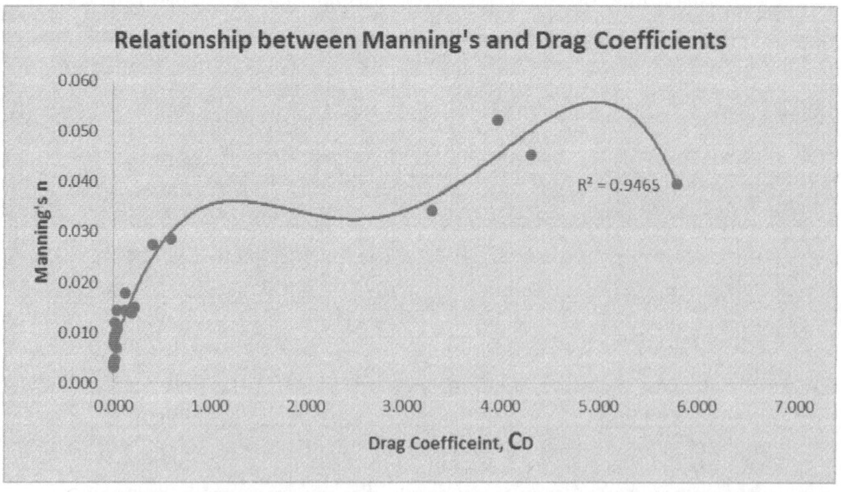

Figure 3. Relationship between manning's and drag coefficient

Also, using the above tables, Figures 2, 3 and 4 were plotted to illustrate the overall relationship between C_D and vegetation density λ, Manning's n with C_D and n-V relationships respectively, for the combined depths. From these Figures, it shows that the drag C_D, has a very good relation with the Manning's n, and λ, as the correlation coefficient R^2 is generally more than 80%. Similarly, from Figure 4 there exist a strong relations as R^2 is more than 90%. And Figure 5 shows the overall velocity distributions with good correlation of R^2 = 70%. Hence, these curves obtained can be used for the design of channel stability or its capacity in vegetated waterways [17].

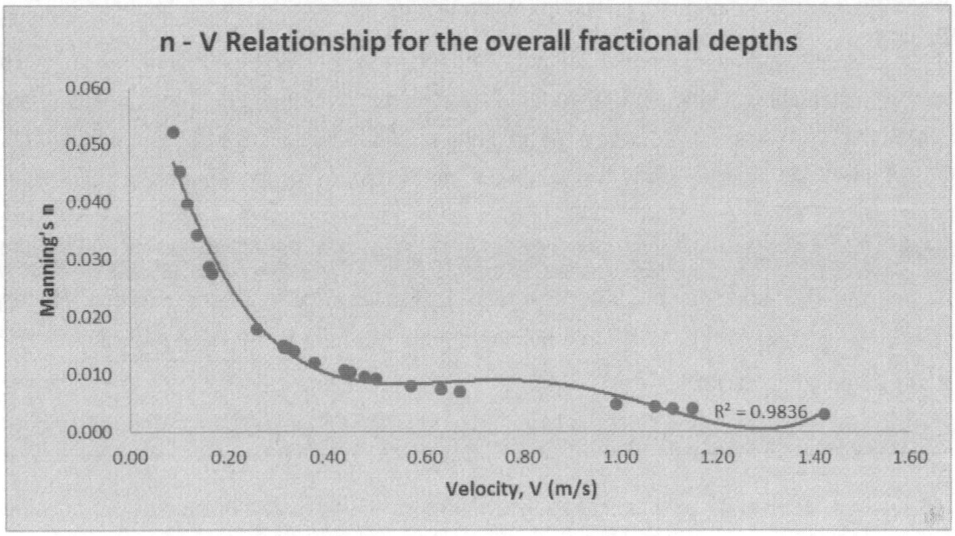

Figure 4. Relationship between manning's and velocity

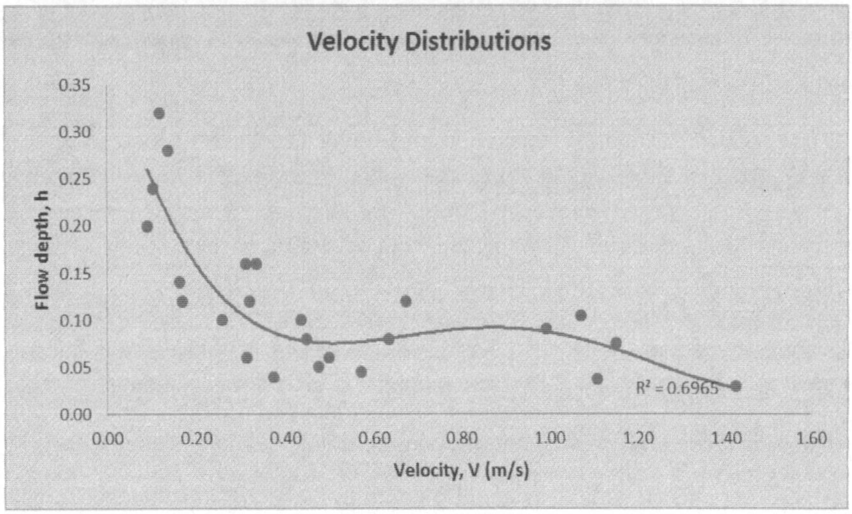

Figure 5. Overall velocity distributions

6. Conclusion

An investigation on flow resistance and drag coefficient has been carried experimentally using natural vegetation to study the variation of vegetation resistance based on Manning's coefficient with flow velocity, depth and drag coefficient, for submerged condition only. It was deduced that the Manning's and drag coefficients decreased with the increasing in average velocity. Also, graphical relationship between Manning's, n and drag, C_D has been developed with $R^2 = 0.9465$, which indicate that there exist a strong correlation between n and C_D. Thus, the curves can be applied in construction of vegetated channels.

Acknowledgments

The authors gratefully acknowledge the financial support from Ministry of Education under HiCOE's niche area Sustainable Urban Stormwater Management, Universiti Sains Malaysia (Grant No.311.PREDAC.4403901). And, the second author also acknowledge the support of Universiti Teknologi PETRONAS for the fellowship scheme to undertake the postgraduate study.

References

[1] A. Ladson, B. Anderson, I. Rutherfurd, and S. van de Meene, "An Australian Handbook of Stream Roughness Coefficients.How hydrographers can help," in Proceeding of 11th Australian Hydrographic conference, Sydney, 2002.

[2] V. Te Chow, "Open channel hydraulics," 1959.

[3] H. Gotoh, T. Tsujimoto, and H. Nakagawa, "Numerical simulation of transition from saltation to sheet flow in sediment transport," Proc. of Coastal Engrg, vol. 40, pp. 326-330, 1993.

[4] Y. Shimizu and T. Tsujimoto, "Numerical analysis of turbulent open-channel flow over a vegetation layer using a k-ε turbulence model," Journal of hydroscience and hydraulic engineering, vol. 11, pp. 57-67, 1994.

[5] S. E. Darby, "Effect of riparian vegetation on flow resistance and flood potential," Journal of Hydraulic Engineering, vol. 125, pp. 443-454, 1999.

[6] F. Lopez and M. H. Garcia, "Risk of sediment erosion and suspension in turbulent flows," Journal of Hydraulic Engineering, vol. 127, pp. 231-235, 2001.

[7] W. Wu and S. S. Wang, "A Depth - Averaged Two - Dimensional Numerical Model of Flow and Sediment Transport in Open Channels with Vegetation," Riparian vegetation and fluvial geomorphology, pp. 253-265, 2004.

[8] H. Tang, Z. Tian, J. Yan, and S. Yuan, "Determining drag coefficients and their application in modelling of turbulent flow with submerged vegetation," Advances in Water Resources, vol. 69, pp. 134-145, 2014.

[9] U. C. Kothyari, K. Hayashi, and H. Hashimoto, "Drag coefficient of unsubmerged rigid vegetation stems in open channel flows," Journal of Hydraulic Research, vol. 47, pp. 691-699, 2009.

[10] F.-C. Wu, H. W. Shen, and Y.-J. Chou, "Variation of roughness coefficients for unsubmerged and submerged vegetation," Journal of Hydraulic Engineering, 1999.

[11] N. Kouwen, T. Unny, and H. M. Hill, "Flow retardance in vegetated channels," Journal of the Irrigation and Drainage Division, vol. 95, pp. 329-344, 1969.

[12] J. Pethick, D. Leggett, and L. Husain, "Boundary layers under salt marsh vegetation developed in tidal currents," Vegetation and erosion, pp. 113-124, 1990.

[13] N. Kouwen and R.-M. Li, "Biomechanics of vegetative channel linings," Journal of the Hydraulics Division, vol. 106, pp. 1085-1103, 1980.

[14] N. Kouwen, "Field estimation of the biomechanical properties of grass," Journal of Hydraulic Research, vol. 26, pp. 559-568, 1988.

[15] D. M. Temple, "Velocity distribution coefficients for grass-lined channels," Journal of Hydraulic Engineering, vol. 112, pp. 193-205, 1986.

[16] J. Xia and L. Nehal, "Hydraulic Features of Flow through Emergent Bending Aquatic Vegetation in the Riparian Zone," Water, vol. 5, pp. 2080-2093, 2013.

[17] Chen, Y.-C., Kao, S.-P., Lin, J.-Y., and Yang, H.-C. Retardance coefficient of vegetated channels estimated by the Froude number. *Ecological Engineering, 35* (7), pp. 1027-1035. 2009.

IOP Conf. Series: Materials Science and Engineering **216** (2017) 012047 doi:10.1088/1757-899X/216/1/012047

Measuring Tyre Rolling Noise at the Contact Patch

P Kozak[1], R Matuszkova[1], M Radimsky[1] and J Kudrna[1]

[1]Faculty of Civil Engineering, Brno University of Technology, Veveri 331/95, Brno 60200, Czech Republic

E-mail: matuszkova.r@fce.vutbr.cz

Abstract. This paper deals with noise generated by road traffic. A focus is concentrated solely on one of its sources related to tyre/road interaction referred as rolling noise. The paper states brief overview of various approaches and methods used to measure this particular source of road traffic noise. On the basis of literature reviews, a unique device has been designed. Development of the measuring device and possibilities of its usage are described in detail in this paper. Obtained results of noise measurements can then be used to design measures that increase safety and a lead to better comfort on the road.

1. Introduction

Road traffic brings not only benefits for society but also some negatives. Roadway noise pollution and air pollution are definitely some of these negatives. The way how can we mitigate the negatives is, due to rising vehicles per capita, an important topic and also an actual topic for several last years.

When we focus just to roadway noise pollution, we still have quite complex problem to solve. The source of roadway noise is not one specific phenomenon, but several phenomena occurring together. We can identify many sources of noise pollution on each vehicle, like an engine, a gearbox, breaks, an exhaust system, a vehicle body or vehicle tyres. This paper aims to describe the phenomenon of tyre/road contact, where rolling noise is generated [1].

2. Rolling noise - background

Principles of tyre/road noise generating are fairly complex. A contact patch is an area where tyre treads and a road interact. While tyre tread blocks travel around the tyre, they hit the road on the contact patch. Energy created by this impact together with other noise source generating mechanisms is radiated as sound. Other several aspects further enhance the radiated sound. Rolling noise is dominant part of roadway noise, especially when vehicles are moving medium to high speeds. For passenger cars it is above 40 kmh^{-1}, for trucks above 65 kmh^{-1}. In case of a vehicle moving faster than 200 kmh^{-1} aerodynamic noise of a vehicle body starts to be dominant [2, 3].

3. Noise measurement approaches

Rolling noise generated at the tyre/road interface is studied since the 1970s. Results from numerous studies done since then were used as the basis to identify and understand the mechanisms of generating and radiating rolling noise. These results were then used for the development of materials and technologies which affect rolling noise on both sides of the contact patch [1, 4].

Due to that it was necessary to develop exact methods for noise measuring of tyre/road interaction. Generally, there are 3 known approaches [5].

3.1 Wayside noise measurements

The static approach by the wayside limits measuring to one point, one cross section, of a road. In addition to that measuring methods define further requirements for the surrounding of measured cross section or for the traffic flow. Measuring higher number of cross sections, or to obtaining high number of measurements, with this approach is time consuming.

3.2 On-board noise measurements

On the other side the on-board approach (measuring the contact patch of a tyre of a moving vehicle/trailer on a road) allows examination of any part of a road network, which is limited just by capacity of a measuring device or range of a measuring vehicle (trailer). This approach can be easily and repeatedly used to specific sections of a road exposed to surrounding conditions. In case of occurrence of deviations or long-term trend of measured values, it is possible to localize the problem and realize follow-up examination, which will help to reveal causes.

3.3 Laboratory Drum noise measurements

Last described approach uses laboratory drum methods. The significant drawback of this approach is that a real wearing course is not used, testing is done on the recreation of a wearing course with similar texture. That is also the reason why the approach cannot be applied to the real road structures. This approach is more suitable for researching causes and mechanisms of rolling noise on the tyre side. By using laboratory drum it is possible to follow exact same track, which is unreal in actual traffic.

Based on the description of approaches and due to the research planning, the on-board noise measurement approach was selected.

4. Onboard measuring methods

For qualification and evaluation of rolling noise at a tyre/road interface are mostly used two variables. Sound pressure level (SPL) and sound intensity level (SIL). In Europe is for measuring noise of tyre/road interface used Close-ProXimity method (CPX), where sound pressure level is measured. However, in the United States is used On-Board Sound Intensity method (OBSI), which is based on measuring sound intensity level. Both methods are described in certified standards [6, 7]. While both methods use same units (decibels – dB), they are not directly comparable. Both methods use different acoustic variables and also the sensors on measuring devices are placed in the different distance from a noise source.

4.1 On-board Sound Intensity (OBSI) method

Some text. The method is described in standard AASHTO TP 76-13 [7]. It is based on the method developed to assessing vehicle noise by General Motors Company in the 1980s in USA. Measuring is done by 2 acoustic probes placed in required distances in relation to the front and rear edge of the tyre/road contact patch (Figure 1). Acoustic probes contain two microphones placed in described distances, which measure sound intensity level (SIL). Sound intensity is a vector dimension, with value and direction. Both probes are oriented towards the contact patch.

4.2 Close-ProXimity (CPX) method

This method (CPX), defined by standard ISO 11819-2 [6], was designed as a complement to the Statistical Pass-By method (SPB), to compare different properties of different wearing courses. CPX method is based on measuring with at least 2 (mostly 3 or 5) microphones placed in close proximity to the tyre/road contact patch (Figure 2). During measurement Sound Pressure Level (SPL) is measured. Sound pressure level describes different conditions at different locations. Sound pressure doesn't have any direction; it is a scalar. Due to that measuring method CPX is more sensitive to interfering sound sources.

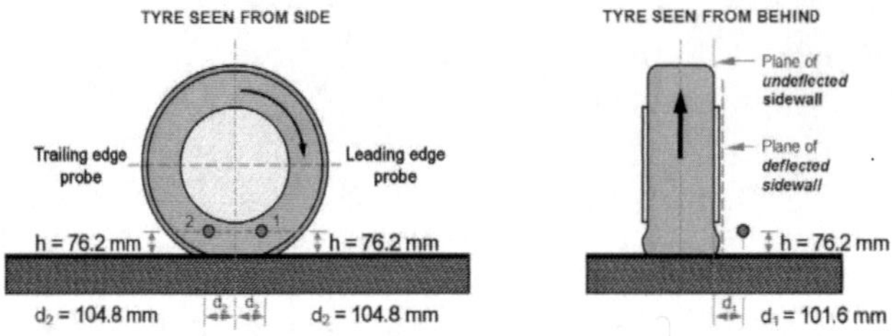

Figure 1. Sound intensity probe positions showing leading and trailing edge measurements according to AASHTO TP 76-13.

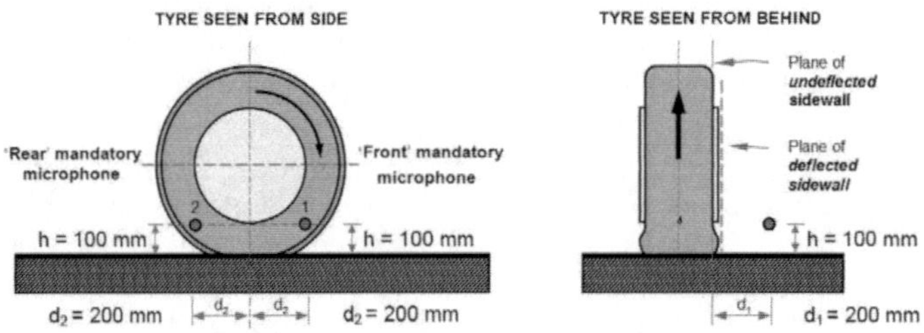

Figure 2. Mandatory microphone positions for the measurements according to ISO-11819-2.2.

5. Design and development of the measuring device

The first roadway noise measuring attempts at our institute were done in the 1990s. Since the very beginning the development was aimed to on-board measuring. The first developed measuring device can be described as "behind the tyre" (BTT). An acoustic sensor was attached to the measuring vehicle, nearby a rear mudflap, and in that position it measured sound pressure (SP).

5.1. Suction cups device

The first version of a device measuring on both ends (front and rear) of the contact patch was designed in 2012. The approach to measuring the sound pressure level was preserved. For measuring sound pressure were chosen couple of microphones placed according to a draft used in the standard for Close Proximity method (CPX) issued in 2000 [8]. For a construction of the measuring device was used a mounting system RK Rose+Krieger. Aluminium beams and joints were used to create the structure which holds measuring microphones in the prescribed positions (Figure 2). Whole structure is attached to a vehicle body by two vacuum suction cups (Figure 3a).

The measuring structure was equipped with the couple of microphones iSEMcon type EMM-13D082/H-P48/RM connected to an external 2-channels audio card M-AUDIO Fast Track Pro. The card was connected via USB to a laptop with SpectraPLUS software. The software saved audio recording and also was used for the evaluation. The structure was also equipped with an acoustic transmitter using 2500 Hz frequency. The transmitter was used for marking important points (start, end, change, etc.). An independent GPS device was used to record a position and a speed of the measuring vehicle.

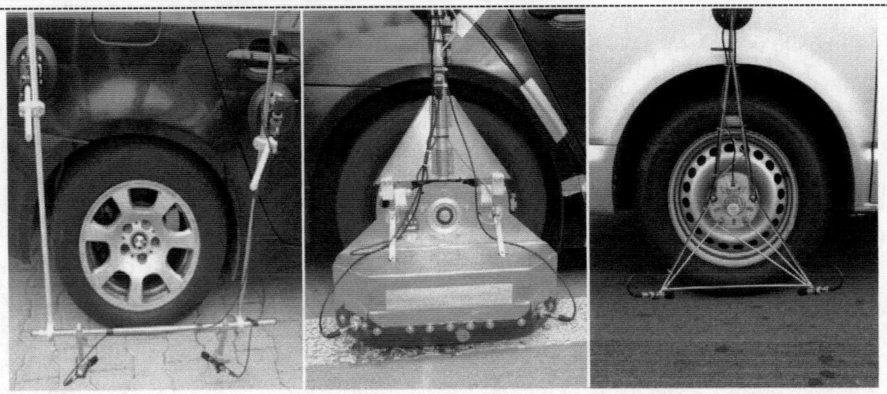

Figure 3. Measuring devices: a) suction cups device; b) custom-made wheel rim device; c) collet-sleeves & truss device.

5.2. Custom-made wheel rim device

In 2014 upgrade of measuring device was done, its target was to eliminate some of the shortcomings. Construction of the measuring device was changed, adjusted was also the device mounting to a vehicle and the process of evaluation. The measuring device was attached directly to one of the spinning wheels and not to a vehicle body. Into the custom made rim an aluminium disc was inserted, later the disc was changed to a performed aluminium hoop. Before measurement a bearing was installed into the hoop, on the other side of the bearing were attached microphones. These microphones were placed under cover, which hold them in position and which also should help with an elimination of background noise from unwanted sources (Figure 3b). Due to a custom made software, evaluation was done immediately after measuring. The software was based on National Instruments – Labview. Every noise level entry was registered together with an actual time and a distance. Both values were provided by Hall probe which was also attached to the bearing. The software allowed inserting sings during measuring ("Z" – start, "K" – finish, "Z" – change, etc.). Data stored in this specific format saved us time later with post-processing.

5.3. Collet-sleeves & truss device

Series of measurements demonstrated that the cover doesn't have positive impact to shielding interfering noise sources. Due to that another adjustment of the measuring device was realize in 2016. Concept of mounting the measuring device to the active wheel was preserved, but realization was redesigned. The bearing is now mounted by collect-sleeves to wheel nuts' heads. The cover with microphones was replaced with the subtle truss structure (Figure 3c). Hall probe was replaced by GPS/GNSS unit. That enables evaluating software assign precise location to measured noise levels, in addition to time and chainage. Measuring device was complemented by two temperature sensors. The infrared sensor Optris CS LT used to measure a temperature of a road in track of the measuring wheel and the temperature sensor Papouch TMU-USB to measure a temperature of surrounding air. Measured temperature values are used for correction of measured noise values to the reference temperature 20°C. Mentioned parameters can be observed in real time, including an actual sound pressure level measured by each microphone. Due to linking GPS location and signs registered by the device operator, it is possible to verify sing's location and importance on a map underlay.

6. Conclusions

The paper describes the development of the measuring device used to measure rolling noise at the tyre/road interface at the Institute of Road Structures, Faculty of Civil Engineering, Brno University of Technology. The measuring device combines a lightweight and a simple construction of American OBSI devices together with parameters of the CPX method used solely in Europe. This solution allows

easy and fast mounting and demounting onto the measuring vehicle. The measuring device is used to the noise comparison of different wearing courses and different road equipment, like road marking.
The paper describes the development of the measuring device used to measure rolling noise at the tyre/road interface at the Institute of Road Structures, Faculty of Civil Engineering, Brno University of Technology (VUT). The measuring device combines a lightweight and a simple construction of American OBSI devices together with parameters of the CPX method used solely in Europe. This solution allows easy and fast mounting and demounting onto the measuring vehicle. Results are verified before and after a measurement by a single-frequency sound calibrator ISEMcon SC-1 Microphone Calibrator. The unit generates a 1 kHz reference tone at 94 and 110 dB SPL. Using calibration results, calibration constants for each of microphones are changed in a user interface of the measuring device. Numerous measurements were performed during the development of the measuring device, using both CPX method and the measuring device. The CPX-lorry without anechoic chamber accredited for the rolling noise measurements according to ISO-11819-2 [8] was used. Comparing CPX results to the results of the on-board measuring device developed by VUT, differences on homogenous surfaces were constant. Absolute sound pressure level difference was between tenths and ones of decibels, depending on the generation of VUT's measuring device.

7. Acknowledgments

This paper has been worked out under the project No. FAST-J-16-3685 Acoustic characteristics of various types of road marking and project No. LO1408 AdMaS UP - Advanced Materials, Structures and Technologies, supported by Ministry of Education, Youth and Sports under the National Sustainability Programme I.

8. References

[1] Sandberg, U. 2001. Tyre/road noise – Myths and realities. The 2001 International Congress and Exhibition on Noise Control Engineering; Proc. intern. symp., The Hague, The Netherlands, 27-30 August 2001.
[2] Bernhard, R. J. & Watson, R. L. 2005. An Introduction to Tire/Pavement Noise, Final report SQDH 2005-1 HL 2005-2, Purdue University, West Lafayette, IN.
[3] Gail, A. & Wolfram B. 2014. Influence of surface textures of road markings on tyre/road marking noise. In: INTERNOISE 2014 - 43rd International Congress on Noise Control Engineering: Improving the World Through Noise Control. Melbourne: Australian Acoustical Society.
[4] Goubert, L. & Debroux P. 2014. Assessing the acoustic properties of audio-tactile road markings. In: INTERNOISE 2014 - 43rd International Congress on Noise Control Engineering: Improving the World Through Noise Control. Melbourne: Australian Acoustical Society.
[5] Świeczko-Żurek, B. & Ejsmont, J. & Ronowski, G. & Taryma, S. 2014. Comparison of road and laboratory measurements of tyre/road noise, Inter Noise 2014; Proc. intern. symp., Melbourne, Australia, 16- 19 November 2014.
[6] International Standard ISO-11819-2.2. 2015. Acoustics-Measurement of the Influence of Road Surfaces on Traffic Noise. Part 2: Close-Proximity Method; International Organization for Standardization: Geneva, Switzerland, 2015.
[7] AASHTO TP 76-13. 2013. Standard Method of Test for Measurement of Tire-Pavement Noise Using the on-Board Sound Intensity (OBSI) Method; American Association of State and Highway Transportation Officials: Washington, DC, USA, 2013.
[8] International Standard ISO-11819-2. 2000. Acoustics-Measurement of the Influence of Road Surfaces on Traffic Noise. Part 2: Close-Proximity Method; International Organization for Standardization: Geneva, Switzerland, 2000.

IOP Conf. Series: Materials Science and Engineering **216** (2017) 012048 doi:10.1088/1757-899X/216/1/012048

Moisture Content Numerical Simulation on Structural Damage of Hot Mix Asphaltic Pavement

O S Abejide and M M H Mostafa
Sustainable Urban Roads and Transportation (SURT) Research Group,
Department of Civil Engineering, Central University of Technology, Private Bag X20539, Bloemfontein, 9300, Free State, South Africa.

E-mail: bskrtell@gmail.com; mmostafa@cut.ac.za;

Abstract. Considering the merits of road transportation in the economy and communication activities of the modern societies, it is imperative to design a safe, stable, efficient and cost effective road that will lead to increased economic development and growth of the South African nation. Although, the overarching effect of failed roads has in many ways led to increased travel time, loss of life and property; leading to reduced driver control on failed road sections (riding quality). Thus, time rate delamination of flexible pavement is a major focus of this study. Since structural collapse in a flexible pavement structure is caused by the evolution of different types of damage mechanisms; fatigue cracking, advanced crushing, temperature variation, and delamination. The effect of moisture content on HMA was analysed. The analysis from the multi-layered elastic model indicates that increase in moisture content in the underlying layer of HMA pavement results to increase in the strain of the individual layers and culminates to a decrease in the structural carrying capacity of the pavement with respect to number of load cycles that can be carried on the HMA pavement. This study shows a clear relationship between the moisture/saturation coefficient and the Elastic Modulus of the underlying geometric material layer properties of the pavement during the service life of the pavement.

1. Introduction

Excessive moisture in the underlying granular base layer can result to lose of its structural capacity by reducing the area over which the traffic load will be distributed. The increased moisture level has a significant effect on pavement structure which decreases resilient modulus and increases permanent deformation [1]. Excessive moisture and fine particles can be transported by hydrostatic pressure to the surface layers, thus reducing the strength of the overlying HMA by increasing strain under service conditions. Understanding the environmental effects on HMA pavements allows better prediction of pavement performance and behavior under different environmental conditions [2]. Regardless of the strength of the HMA surface course, the stability and durability of a pavement depends on the strength of its subbase and subgrade [3]. A strong surface layer will undoubtedly fail prematurely if constructed over a weak foundation as the strength of a pavement may fluctuate with changes in moisture content of its sub layers [4]. It is well known that the performance of a pavement construction is adversely affected by the presence of moisture within underlying layers [1]. Design and construction of the pavement is generally carried out with the intention of keeping the layers unsaturated.

2. Stress and strain models

The response of a pavement structure to traffic loading is mechanistically modeled by computing stresses and strains within its layers. However, excessive stresses may cause pavement fatigue cracking and/or surface rutting. This will further result in both structural and functional failure, thus causing a safety hazard to motorists. These failure distresses are minimised among others by use of stochastic pavement designs. Pavement stress-strain analysis is an ideal tool for analytical modeling of pavement behavior and thus, constitutes an integral part of pavement design and performance evaluation.

2.1 Two-dimensional stress state

The analysis is based on a mechanistic design approach [5], [6], and a linear elastic two-layer pavement system [7]. The common analytical method to model pavement-traffic load response is stress-strain analysis.

2.2 Three-dimensional stress state

The development of FEM and its analysis presents description of three FE models in line with the set objectives with regards to moisture ingress. This model was used to validate the efficiency of using 3D FEM over a multilayered elastic design model in the design of pavement structure and examine the structural response of a stabilized base layer in terms of the stresses and strains at the surface of the subgrade and the tensile strain at the bottom of the Asphalt surface layer. Overall, all the models were developed using Abaqus® [8].

2.3 Traffic loading

In view of the current traffic regime, this study's emphasis was focused on high traffic loading which is considered as a major factor responsible for most pavement damage world-wide [9]. However, a design input value of 80,000N and a pressure distribution of 650Mpa is adopted in the design. The tire track is assumed to act along a radius of 195mm2; assuming a category "A" road with maximum traffic count at peak periods over a wet condition environment.

3. Methodology

3.1 Fatigue cracks and rutting

Fatigue cracking in asphalt layers is considered a major structural distress and is predominantly caused by traffic loading. In addition, ingress of rainwater through the cracks can lead to serious structural failure of the underlying layers particularly the unbound granular materials, subgrade and base course. [10] there exist two primary models used and would be considered in this study[11]. The strain values gotten from the FEM will be inserted into the equation for fatigue cracking and rutting [11] and the regression graph of safety will be plotted based on a statistical approach.

3.2. The NCHRP model

[11] introduced in the Mechanistic-Empirical Pavement Design Guide [12] estimates the resilient modulus using a generalized constitutive model for Level 1 analysis for the nonlinear stress-dependent modeling of both the unbound aggregates and fine-grained soils. Tests done by [13] also showed that the deviator stress is between 14 and 20 kPa (2 and 3 psi) and developed an expression for variation of resilient modulus obtained from laboratory tests on soil from the Ohio SHRP test section which will be adopted in this study to determine the resilient modulus of the UGB material when saturated.

3.3. Multi layered elastic theorem

The methodology adopted for this study makes use of a Multi-Layered Elastic Theory of pavement design and Analysis. The material properties and parameters of the pavement are as shown in Table 1. Consequently, the analysis was followed by a Finite Element analysis mesh optimization in Abaqus to determine the horizontal tensile strain at the bottom of the Asphalt pavement and also the vertical compressive strain at the top of the subgrade considering an Axis-symmetric modeling.

IOP Conf. Series: Materials Science and Engineering **216** (2017) 012048 doi:10.1088/1757-899X/216/1/012048

4. Results and discussion

Radial strain at the bottom of Asphalt layer and vertical strain at the top of Subgrade have been used to control fatigue and rutting of flexible pavements respectively using a multi-layered elastic design principle by mePADS Several fatigue and rutting models were developed to relate the asphalt modulus and the measured strains to the number of load repetitions to pavement failure. The result of the analysis as generated from the input values in (table 1) is shown in (table 2).

Table 1. Material properties of pavement layers

Layer	Material Code (Colto 2008)	Thickness (mm)	Elastic Modulus (Mpa)	Poisson's Ratio
Asphalt	AC	30	4000	0.40
Base C4	C2	200	2560	0.35
Unbound Granular Base	G4	350	Varied	0.35
Subgrade	G5	5000	200	0.35

From [11], the data in (table 2) was further analysed into to the Number of fatigue cracks to failure as well as the Number of Rutting to failure. A clear difference can be seen between the Normal strain at 20% Saturation Coefficient and 70% Saturation Coefficient; at this point, it is expected that the pavement has exceeded its yield point where it fails and deforms completely. The Figures 1-A and 1-B clearly reveal the behavior response of the soil with increasing saturation coefficient.

Table 2. Strain and resilient modulus for different percentage saturation coefficient levels

Saturation Coefficient (S %)	Elasticity Modulus (Mpa)	Horizontal Tensile Strain Et(10^{-6}) Bottom AC *MLED*	Vertical Compressive Strain *Top Subgrade* *MLED*	Horizontal Tensile Strain Et(10^{-6}) Bottom AC (Abaqus)	Vertical Compressive Strain Top Subgrade (Abaqus)
20	441	67.8	253	93.2	238.8
30	369.9	72.2	257	99.2	244
40	325.5	75.2	260	103.5	246.2
50	281.4	79.3	261	108.4	247.9
60	237.4	83.7	260	114.2	248.2
70	187	89.8	257	122.3	245.6
80	130	99.2	245	134.6	235.5

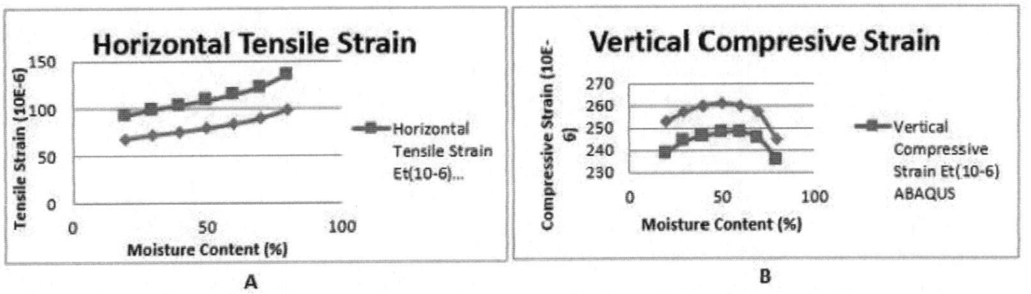

Figure 1. A - Horizontal Tensile Strain at Bottom of Asphalt, **B -** Vertical Compressive Strain at Top of Subgrade

IOP Conf. Series: Materials Science and Engineering **216** (2017) 012048 doi:10.1088/1757-899X/216/1/012048

5. Conclusion

A Multi-Layered Elastic design was used to determine the horizontal tensile strain at the bottom of the Asphalt and the vertical compressive strain at the top of the subgrade. The analysis from the multi-layered elastic model indicates that increase in moisture content in the underlying layer of HMA pavement results to increase in the strain of the individual layers and culminates to a decrease in the structural carrying capacity of the pavement with respect to number of load cycles that can be carried on the HMA pavement. The results obtained from the Finite Element Method indicates that; considering an Axis symmetric model, the percentage difference of the deformation response with an optimized meshing for Vertical Compressive Strain has an average percentage difference of 5.4% with the strain values obtained from the Multi-Layered Elastic Method, and the Horizontal Strain has an average percent difference of 31% compared with the results generated from the Multi-Layered Elastic Method. This study shows a clear relationship between the moisture/saturation coefficient and the Elastic Modulus of the underlying geometric material layer properties of the pavement during the service life of the pavement.

Furthermore, this study serves as a preface to an ideal and highly sophisticated field testing and monitoring model to be analysed with moisture instrumentation devices for an optimized and reliable analysis of moisture content effect in the flexible pavement which will be carried out on both old and newly constructed flexible pavement in South Africa.

6. References

[1] Saevarsdottir T and Elingsson S 2013 Water impact on the behavior of flexible pavement structure in an accelerated test. *Roads Materials and Pavement Design* 2013, Vol.14, No. 2,225-227. Taylor and Francis.

[2] Davies B O A 2004 A model for the prediction of subgrade soil resilient modulus for flexible pavement design: Influence of moisture content and climate change. *Master of Science Thesis in Civil Engineering*. University of Toledo December 2004.

[3] Kiehl J T and Briegleb B P 2011 Pavement Surface Condition Field Rating Manual for Asphalt Pavement, ARPN. *J. Sci. Technology*, 2(1):17-25, 2011.

[4] Ahmed M Y, Nury A H, Islam F and Alam M J B 2012 Evaluation of geotechnical properties and structural strength enhancing road pavement failure along Sylhet-Sunamganj Highway, Bangladesh. *Journal of Soil Science and Environmental Management*. Vol.3 (5), pp: 110-117, May 2012.

[5] Croney D and Croney P 1998 *The design and performance of road pavements*: 3rd edition, New York, NY 10011, 1998.

[6] Huang Y. H. 1993. Pavement analysis and design, Englewood Cliffs, NJ 07632, 1993.

[7] Abaqus 6.13 CAE 2013 Finite Element Design Drawing Environment for deformation of structures. *Simulia*. 2013.

[8] Walubita L F and Van de Ven M F C 2000 Stresses and strains in asphalt-surfacing pavements, *South African Transport Conference. Action in Transport for the New Millenium. SATC July 2000.*

[9] De beer M, Sadzik E M, Fisher C and Coetzee C H 2005 Tyre-Pavement Contact Stress Patterns from the test tyres of the Gautrans Heavy Vehicle Simulator (HVS) MK IV+, *Proceedings of the 24th Southern African Conference (SATC 2005).* ISBN: 1-920-01712-7, Pretoria, South Africa, pp 413-430, 2005.

[10] Huang Y. H. 2004. Pavement Analysis and Design. *2nd ed. (NJ: Pearson Education Inc)*, 2004.

[11] Guide for Mechanistic-Empirical Design of New and Rehabilitated Pavement Structures, *Part2. 2004 Design Inputs*, National Cooperative Highway Research Program, Washington, DC, USA, 2004.

[12] Mechanistic–Empirical Pavement Design Guide: *A Manual of Practice.* (2008). American Association of State Highway and Transportation Officials; Washington, DC, USA.

[13] Alvarez C A 2000 Study of Environmental Factors and Load Response in Flexible and Rigid Pavements at the Ohio Test Road. *Master's Thesis*, CWRU, OH, 2000.

IOP Conf. Series: Materials Science and Engineering **216** (2017) 012049 doi:10.1088/1757-899X/216/1/012049

The Status and Prospect of Research into Protective Structures of Bridge Piers against Rockfall Impact

***GAO Liang[1], ZHANG Shan[1], ZHANG Junfa[1] and WU Xiangnan[1]**
1. School of Civil Engineering and Architecture, Xi'an University of Technology, Xi'an 710048, China

Email: 156820773@qq.com;

Abstract: Rockfall impact on bridge piers threats severely the mountain bridge structures of lifeline engineering. Intended for mountain bridge pier protection against rockfall impact, the paper conducted comprehensive reviews on the research status of impact effects, anti-collision structure, impact response to rockfall, and protective design at home and abroad, and proposed a new-type protective structure against rockfall impact. In addition, the paper carried out deep studies on such key scientific issues as impact effect calculation, protective materials against rockfall impact, damage mechanism of protective units, and parameter optimization on the system of protective structures against rockfall impact as well, aiming to strength disaster prevention of mountain bridge structures.

1. Introduction

As a common geological disaster in mountainous areas, rockfall is characterized by high unexpectedness, severe destruction, and great harm to lifeline engineering. In recent years, the frequency and scales of rockfall have mounted up rapidly in China. With the implementation of the national strategy of the Silk Road Economic Belt and the 21st-Century Maritime Silk Road, domestic transportation infrastructure has been expanded enormously, and the ratio of bridges to tunnels has been increased as well. In this connection, it has become urgent to protect mountain bridges from rockfall impact. Considering that bridge piers are an important part of bridge engineering, bridge will malfunction once the bridge piers are damaged by rockfall.

To relieve rockfall impact on bridge piers, cushion materials or anti-collision systems are used on bridge piers during engineering design so that the prolonged period of rockfall impact is able to offset the impact effect partly by dissipating impact energy. However, given high randomness and complicated mechanical process of rockfall impact, the existing design of anti-collision system is far from effective. Thus, it has become the basic difficulty to determine rockfall impact effect and to optimize protective performance during the design of bridge pier protection.

2. Analysis of corresponding research status at home and abroad

Here are reviews of relevant research on rockfall impact force, protection of bridge structures against rockfall impact, and protection of bridge piers against rockfall impact as well.

2.1 Rockfall impact effect

As rockfall impact is a kind of pulse [1], the rockfall impact force can hardly be calculated with either Hertz elapse collision theory or Thornton elastic-plastic collision theory. Domestic references are limited to *the Code for Design of Highway Subgrades* (JTJ13-95) and *the Handbook for Design of Railroad Engineering (revised version),* in which it is only suggested to use empirical algorithms to

IOP Conf. Series: Materials Science and Engineering **216** (2017) 012049 doi:10.1088/1757-899X/216/1/012049

calculate rockfall impact force. Semi-empirical and semi-theoretical algorithms, which are based on in-situ tests of rockfall impact force, are commonly-used in foreign countries, the Japan Highway Public Corporation algorithm [2], for example. In addition, from the perspective of engineering practice, Yang Qixin et al. [3] simulated the impact load for shed-type open cut tunnels with cushion layers of different thicknesses, and also provided the calculation formulas of rockfall impact according to the law of rockfall acceleration change for the duration of impact. Ye Siqiao et al. [4] compared five representative algorithms of rockfall impact force at home and abroad, and concluded that the Chinese algorithm results were much lower than real values while the Japan Highway Public Corporation algorithm that was based on field measurement on rockfall impact conformed to reality better.

Yuan Jinke et al. [5] conducted tests on rockfall impact force, and established the calculation method of maximum impact force which covered various influential factors, such as rock weight, impact velocity, incident angles, and the property and thickness of cushion materials. With the help of impact tests of cone-type rockfall on gravel soil layer, Pichler B et al. [6] proved that the rockfall impact time increased with the increase of rock quality but decreased with the increase of impact velocity. He also considered rockfall impact as an act of collision in a moment, and held it that the calculation of rockfall impact time would affect the calculation results of impact force directly.

To summarize, basic theories of rockfall impact were introduced to China through the above research. By means of theoretical deduction, rockfall impact tests, and numerical simulation, and in view of both theoretical calculation of rockfall impact force and parameter analysis of influential factors on impact force, the aforementioned research expounded calculation formulas of rockfall impact force in aspects of rockfall and impacting objects.

2.2 The protective structure against rockfall impact

Passive protections are mostly used in engineering practice, including rock shed, open cut tunnel, safe net (or wall), protective dyke, and rockfall crater. The following is analysis of research status of existing passive protective structures against rockfall impact.

Wang Min, Shi Shaoqing, Yang Youkui et al. [7] conducted research on energy consumption of the rockfall SNS flexible safety net system with steel rope net as the main component, and established the calculation model of ring network with two forms of composition. What's more, by applying the above research results to rock sheds, they [8] proposed the flexible structure for rock sheds, and also conducted corresponding computed analysis on full-scale experiments, dynamic response, and energy absorption characteristics. Shi Shaoqing et al. [9] applied waste tires to new-type flexible safety wall structures. By researching on the energy consumption characteristics of materials for each part of waste tires, he pointed out that the use of waste tire could improve significantly the protective capacity of new-type flexible safety walls due to its obvious ability of energy consumption and buffer function.

He Siming et al. [10] proposed a new energy dissipation-seismic reduction shed. By installing energy dissipation-seismic reduction apparatus on the shed base, he built up a nonlinear mass-spring system model to simulate dynamic response of shed structures under the rockfall impact load. He also used the energy method to analyze the protective mechanism against rockfall impact for the new energy dissipation-seismic reduction shed. Moreover, He Siming et al. introduced the energy dissipation-seismic reduction materials of EPS slates [11], foam aluminum [12], and modified concrete to the protective structure against rockfall impact for the new shed. With static loading tests and dynamic finite elements, they analyzed dynamic response of the protective structure against rockfall impact, and further revealed the energy dissipation-seismic reduction mechanism for the new shed.

All in all, with the emergence of new structural materials and the flexible application of various protective apparatus in engineering practice, engineering technicians have developed multiple new protective systems against rockfall impact, and discuss relative technical problems concerning material selection for protection use and improvements on protective apparatus.

2.3 Impact response and protective design for bridge piers

Researchers have done a great deal of research work on rockfall impact on bridge piers from different perspectives. Luo Songnan et al. [13] simplified the constraints of upper parts and lower parts of high bridge piers, respectively, and established the basic nonlinear dynamic equation for high bridge piers under the action of impact load. Through numerical simulation, they also compared the displacement response under the action of triangular impact load and that under the action of rectangular impact load. Pei Xiangjun et al. [14] used the non-continuum theory to analyze the impact of peripheral unstable hill stones on bridge piers in meizoseismal areas, and found that under the impact and extrusion of rockfall, the maximum permanent displacement occurred on collar beams and bridge tops. With the help of HJC damage mechanism, Yuan Jinke, Pei Xiangjun [15] established a numerical model to analyze kinetically the rockfall impact on bridge piers, obtaining the bridge piers' change of stress fields and displacement fields. They denoted that rockfall impact force presented an obviously nonlinear trend, and that for passive anti-collision structures, the maximum impact force should be referred to when determining corresponding indices. Based on LS-DYNA software and explicit nonlinear analysis, Yu Zhixiang et al. [16] used HJC constitutive relations to simulate reinforced concrete. From the perspectives of volume damage ratio and pier stud displacement, they researched on how different impacting factors helped damage pier studs, and concluded that rock diameters and impact velocity contributed more to pier stud damage. The above studies on dynamic response, deformation laws, and volume damage ratios under impact load are of great significance to anti-collision of bridge piers.

Research on rockfall impact prevention has been conducted successively since Chediguan bridge was destroyed by rockfall during the 5·12 Wenchuan Earthquake. He Siming et al. proposed a protective structure for bridge piers against rockfall impact, and used the dynamic finite elements to compare the dynamic response before and after the bridge pier was intensified. In this regard, they revealed the energy dissipation-seismic reduction mechanism, based on which they designed a protective structure for Chediguan bridge pier against rockfall impact. Moreover, He Zilu and He Siming invented a tiled compound pier protective apparatus, with protective units that were fixed firmly onto the external pier wall. The units contained several protective components of the same structure, each of which was composed of internal fillings, external reinforcing ribs, and filled airtight cavities that were separated by the reinforcing ribs. With explicit nonlinear dynamic analysis, and in combination with model test results, Yu Zhixiang et al. [17] conducted numerical simulation on two protective materials of steel sleeves and thickened plain concrete, and proposed an improved type of separable steel sleeves to cover the shortage of bonded steel sleeves as well. The above computed research on energy consumption buffer laws, crashworthiness, and design parameters with respect to anti-collision bridge piers against rockfall impact is of great significance to the establishment of anti-collision systems for bridge piers.

All in all, theoretic analysis and the finite element method have been used for exploratory study on rockfall impact prevention in terms of rockfall impact, protective structures against rockfall impact, and structural response of bridge piers to rockfall impact together with corresponding protective design.

2.4 Scientific issues to be further studied

Certain research achievements have been attained on rockfall impact, and design and improvements of the protective system. However, with respect to the fundamental scientific issue of rockfall impact prevention of bridges, the following aspects still await studies:

2.4.1 Calculation methods of rockfall impact effect. Existing calculation formulas of rockfall impact force are mainly based on structural response backstepping and static equivalence. The process of rockfall impact is regarded as a typical inelastic collision, because it involves elastic-plastic deformation, impact energy conversion, and determination of impact force.

2.4.2 Damage mechanism of protective materials and protective units under rockfall impact. The dynamic analysis of protective structures and their materials under rockfall impact contains material deformation, structural deformation, structural dynamic characteristics, and structural dynamic response. The relation remains unclear between the damage law of protective systems for bridge piers and rockfall impact force.

2.4.3 Parameter optimization of the rockfall impact prevention system for bridge piers. The ability of protective system mainly depends on the cushion materials and protective performance of the structure, and the protective units of the protective structure is the key of the design, and therefore it's must be based on dynamic analysis of the structure under rockfall impact, and based on the multi-level protective goal to study the cushioning property, and then carry on its parameter optimization analysis.

3. The rockfall impact prevention system for bridge piers

According to engineering practice, the tiled compound pier protective structure helps relieve the impact force on bridge piers to a certain degree. Specifically, this structure was applied to Chediguan bridge that was destroyed by rockfall during the 5·12 Wenchuan Earthquake and proved to be effectively protective. This structure gives prior consideration to material buffer and multi-level protection. However, this paper proposed a new-type rockfall impact prevention system for bridge piers (Figure 1) as an improvement to it in lights of rockfall impact dynamic effect, damage modes of rockfall prevention material and component, protective structure, and connecting structure of bridge piers. For this new system, rockfall impact tests were mainly carried out as a way to improve the protective performance and to unveil the operating mechanism as well. This system has realized design optimization for both protective parameters and protective units, thus further upgrading performance of the protective system for bridge piers.

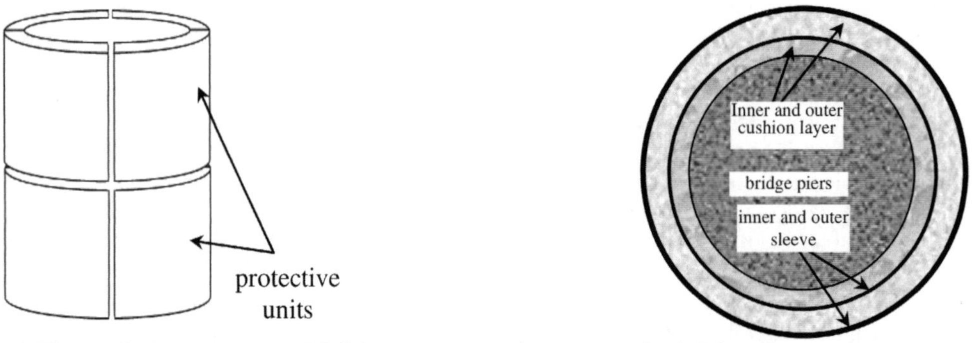

Figure 1. A new-type rockfall impact prevention system for bridge piers

Specific significances are summarized as follows:

3.1 Lay theoretical foundation for rockfall impact dynamic computation

The foundation of rockfall impact research includes rockfall impact force computation, motion curves analysis, and impact effect forecast. Relevant design theories are mostly experience-based or formula-based, whose formulas target mainly at soil cushion or sand cushion. Thus, the paper conducted dynamic analysis on rockfall impact from the aspects of rockfall itself and the stuff being impacted.

3.2 Provide means for parameter optimization of rockfall impact prevention structures for bridge piers

Existing protective structures for bridge piers are mainly against vessel collision and vehicle collision, whose methods and effects of collision are greatly different from those by rockfall impact. Therefore, the paper carried out dynamic buffer performance tests to research into the deformation and damage laws of the protective structures of bridge piers under rockfall impact. By analyzing changes of impact force, the paper analyzed the protective mechanism of both cushion materials and the protective components.

3.3 Provide reference for design of rockfall impact prevention systems for bridge piers

Based on cushion performance, the paper dynamically analyzed anti-collision performance of the protective structure against rockfall impact, and also introduced the multi-level prevention idea. The dynamic model for bridge piers under rockfall impact was established to obtain the operating principle of the protective structure against rockfall impact, which helped improve the protective performance for bridge piers.

4. Conclusion

In terms of protective structures for mountain bridge piers against rockfall impact, the paper reviews comprehensively relevant research status at home and abroad, and proposes a new-type rockfall

impact prevention system for bridge piers. Through impact test and finite element numerical, this paper shows the conclusions as follows:
1. Due to the flexible cushion layer under the rolling impact, the impact contact belongs to the point contact, limited by the thickness of cushion layer, the cushioning property of protective cushion structure is difficult to give full play. So, in order to improve the performance of buffer on the protective cushion structure, we can increase stiffness protective layer on the outside of flexible cushion layer, so as to make the point contact into surface contact, effectively improve the utilization rate of buffer material.
2. Project team in the study of impact protective measures for roof boarding in early, and found that the foam sandwich reinforced concrete plate has good shock resistance, applied to pier impact protective structure, from the change or block the shock wave propagation path and deformation energy dissipation, namely using the flexible buffering inner core and rigid surface layer protective measures, and draw lessons from the construction method of shield tunneling segment, the tube-plate type hard-soft laminated protection system for bridge piers against rockfall impact is put forward.

5. Acknowledgments
The research is supported by the National Natural Science Foundation of China (51279162).

6. References
[1] Zhang Guang-cheng, Xiang Xin and Tang Ming-hui. 2011 Field test and numerical calculation of restitution coefficient of rockfall collision. J. *Rock Mechanics and Engineering*, **30**(6): 1266-1273.
[2] Kawahara S and Muro T. 2006 Effects of dry density and thickness of sandy soil on impact response due to rockfall. J. *Terramechanics*, **43**(3): 329-340.
[3] Yang Qi-xin and Guan Bao-shu. 1996 Test and research on calculating method of falling stone impulsive forc. J. *China Railway Society*, **18**(1): 101-106.
[4] Ye Si-qiao, Chen Hong-kai and Tang Hong-mei. 2010 Comparative research on impact force calculation method for rack falls. J. *Hydrogeology & Engineering Geology*, **37**(2): 59-64.
[5] Yuan Jin-ke, Huang Run-qiu and Pei Xiang-jun. 2014 Test research on rockfall impact force. J. *Rock and Soil Mechanics*, **35**(1):48-54.
[6] Pichler B, Hellmich Ch and Mang H A. 2005 Impact of rocks onto gravel Design and evaluation of experiments. J. *Impact Engineering*, **31**(5):559- 578.
[7] Yang You-kui. 1998 The SNS Flexible Systems for Holding Rockfall. J. *The Chinese Journal of Geological Hazard and Control*, **S1**:321-329.
[8] Wang Min, Shi Shao-qing and Yang You-kui. 2014 Numerical simulation of a flexible rock-shed under the Impact of a rockfall. J. *Engineering Mechanics*, **31**(5): 151-157.
[9] Deng Li-yuan, Shi Shao-qing, Wang Min and Sun Jian-hu. 2015 Application of Waste Tire to New Type Flex Block Wall and Its Numerical Analysis. J. *Logistical Engineering University*, **31**(1):1-6.
[10] He Si-ming, Shen Jun and Luo Yu. 2011 Study on the characteristics of normal impact of post-earthquake rock-fall on slope. J. *Rock and Soil Mechanics*, **32**(3): 781-788.
[11] Wang Dong-po, He Si-ming, Wu Yong, et al. 2014 Cushioning effect of rock sheds with EPS cushion on rock-falls action. J. *Vibration and Shock*, **33**(4): 199-203+214.
[12] Wang Dong-po, He Si-ming and li Xin-po. 2016 Study on the Dissipating Effects of Aluminum Foam Sandwich Panel Under Rockfall Impact. J. *Sichuan University (Engineering Science Edition)*, **48**(1): 43-49.
[13] Luo Song-nan and Li Li. 2013 Elastic-plastic dynamic buckling of a high pier under impulse loads. J. *Vibration and Shock*, **32**(23):196-200.
[14] PEI Xiangjun, HUANG Runqiu and LI Shigui. 2011 Study of dynamic response of bridge pier shocked by falling rock induced by intensive earthquake. J. *Rock Mechanics and Engineering*, **30** (S2): 3995-4001.
[15] Yuan Jin-ke and Pei Xiang-jun. 2014 Modeling and Study of Impact Damage to Piers of Chediguan Bridge. J. *Highway*, **59**(2):22-27.
[16] Yu Zhi-xiang, Xu Hu. Influence on Damage of Bridge Pier in Mountain Area Under Impact Load of Rockfall [J]. Journal of Sichuan University(Engineering Science Edition), 2012, 44(6):86-91.
[17] Yu Zhi-xiang, Yang Chang and Luo Yan. 2013 Crashworthiness of the protective measures for mountain bridge piers under impact load of rockfall. J. *Sichuan Building Science*, **39**(5):11-15.

IOP Conf. Series: Materials Science and Engineering **216** (2017) 012050 doi:10.1088/1757-899X/216/1/012050

Wireless Smart Sensor Network System Using SmartBridge Sensor Nodes for Structural Health Monitoring of Existing Concrete Bridges

J R Gaviña[1,2] , **F A Uy**[1,2], **and J D Carreon**[1,2]

[1] School of Civil, Environmental, and Geological Engineering Department, Mapua Institute of Technology, Intramuros, Manila, Philippines
[2] SmartBridge Project, Mapua Institute of Technology, Intramuros, Manila, Philippines

Abstract. There are over 8000 bridges in the Philippines today according to the Department of Public Works and Highways (DPWH). Currently, visual inspection is the most common practice in monitoring the structural integrity of bridges. However, visual inspections have proven to be insufficient in determining the actual health or condition of a bridge. Structural Health Monitoring (SHM) aims to give, in real-time, a diagnosis of the actual condition of the bridge. In this study, SmartBridge Sensor Nodes were installed on an existing concrete bridge with American Association of State Highway and Transportation Officials (AASHTO) Type IV Girders to gather vibration of the elements of the bridge. Also, standards on the effective installation of SmartBridge Sensor Nodes, such as location and orientation was determined. Acceleration readings from the sensor were then uploaded to a server, wherein they are monitored against certain thresholds, from which, the health of the bridge will be derived. Final output will be a portal or webpage wherein the information, health, and acceleration readings of the bridge will be available for viewing. With levels of access set for different types of users, the main users will have access to download data and reports. Data transmission and webpage access are available online, making the SHM system wireless.

1. Introduction

Deterioration of civil infrastructure systems, such as bridges, buildings, and pipelines, is a complex, yet common problem. Currently, visual inspection is the most common practice used in monitoring the structural integrity of bridges. Visual inspections are highly variable, lack resolution, and fail to detect damage unless it is visible. Thus, structurally substandard, and even dangerous bridges may be left undiscovered, potentially placing the public at risk. Structural Health Monitoring (SHM), is used to combine all the inspection methods performed on a bridge. Alongside visual inspections, some systems include a sensor monitoring system to measure the response of a bridge (e.g. acceleration, displacement). Other SHM systems are using different techniques for structural health monitoring. However, in the Philippines, due to high-cost of instrumentation devices, large installation costs (e.g. wiring), or high maintenance costs, engineers and maintenance managers are limited to visual inspection. Introduced in the SHM system in this study is a sensor of considerably low price and efficiently programmed and built to read acceleration in real-time. The SmartBridge Sensor Nodes will be used in the detection of vibration of a concrete bridge. The changes to structural properties of a bridge caused by damage will change the way the bridge responds to ambient motions. For example, corrosion of steel members of a steel bridge results in a reduction in the cross-sectional area of those steel members, which then would lead to a reduction in the stiffness or other structural properties.It is crucial to inspect the bridge periodically and to assess its condition and evaluate its integrity.

1.1. Objectives

This study addresses the following: a) To develop a Wireless Smart Sensor Network (WSSN) System for SHM of an existing concrete bridge; and b) To develop a portal or webpage, wherein authorized users may input visual inspection data, download sensor data, and check the health of the bridge.

1.2. Significance

In compliment with visual inspection, SHM would be the best way to monitor the health of a bridge. SHM provides a way to continuously capture structural response and assess structural condition in real-time. SHM can also be utilized to characterize actual loading, which enables engineers to predict the response of a structure to certain extreme loading conditions. SHM can also be used in forensic engineering, wherein data collected and stored by the sensors during natural hazards or man-made disasters, can be analysed and used to determine the cause of damage. Overall, SHM has an advantage compared to the traditional visual inspection in terms of monitoring and assessment of structural health in real-time. In addition, having a wireless or "smart" structure monitoring system, would mean that data gathered by the sensors installed on site would be easily accessible to engineers, maintenance managers, and other government offices that may have researches concerned with acceleration readings on a bridge.

2. Sensor Technology

The most common sensors used in SHM Systems are accelerometers and strain gauges. SHM systems monitors the health and behavior of a bridge, by embedding sensors during the construction of the bridge. Theoretically, embedded sensors are designed to gather data continuously throughout the lifespan of the bridge. Wired accelerometers are the most commonly used sensors in SHM Systems to monitor the structural health status and behavior of a bridge. Magalhaes et al. executed monitoring on a long span arch bridge in Portugal. After installing 12 wired accelerometers that were placed of equal spacing on the deck, they determined the first 12 mode shapes using Frequency Domain Decomposition (FDD) Method. These mode shapes, along with modal frequencies, were used to develop a numerical model to be utilized for the generation of events wherein damages are present [1]. However, one major disadvantage of these type of sensors is the wiring system that links all these sensors to a central server. Cabling numerous sensors is labor-intensive, and requires planning beforehand to avoid complications. Also, all wired sensors require a connection to a power supply [2].
In improving the sensor technology, studies such as the vibration monitoring experiment executed by Yu et. al. utilized wireless sensor technology, and compared it to wired sensors [3]. While, Hoult et al. used three wireless inclinometers and three wireless displacement transducers of a bridge in the United Kingdom, Ferriby Bridge, to assess the potential of WSSN for use in bridge management [4]. Wireless sensors are portable, and its reusability gives it a great advantage over wired sensors. Wireless sensors may also provide flexibility and reduce the number of required sensors compared to when using wired sensors, as they can be installed instantly or with ease, and removed in a shorter amount of time. Due to the easy installation and reinstallation of wireless sensors, a lot of tests can be done at other parts or sections of the same bridge. In conclusion, wireless sensors are more economical and effective when implemented to perform the structural health monitoring of a bridge. Due to the wireless feature, the sensors need to have batteries as power source. Thus, a disadvantage of this system is having a power supply that is limited, since batteries run out and needs to be replaced periodically. However, with advantages such as reusability and ease of installation, wireless sensors are clearly currently the best, reliable, and cost-effective option for sensors to use in a SHM system.
Numerous other SHM strategies needs a lot of sensors to be installed in the structure. With this, the SHM strategy shall be scalable, depending on the number of sensors. Scalability refers to the ability to

add number of sensors in a certain network while maintaining manageable increases in cost and preserving efficiency of data collection [5].
The SmartBridge Sensor Node used in this study, is an accelerometer taking advantage of the wireless technology. It meets the minimum requirement for accelerometers set by the local standards as indicated in Section 105.2 of the National Structural Code of the Philippines 2010 [6] and Section 102 of the National Building Code of the Philippines, otherwise known as P.D. 1096. The specifications were also approved by the DPWH as part of the Guidelines and Implementing Rules on Earthquake Recording Instrumentation for Buildings [7] of the National Building Code. Note that the SmartBridge Sensor node can be modified to satisfy other requirements or codes in other countries. The SmartBridge Sensor Node stores acceleration readings in the x, y, and z directions, with a sensitivity range of 2g to 16g and natural frequency of 100Hz. The recording ability complies with the minimum requirement of 100 samples per second, and has a memory card built-in with a 32GB storage capacity. Its power supply comes from an AC source, but in the event of a power outage, the sensor has a battery which can power the device for at least 8 hours. Data are stored in each individual sensor and then uploaded to a central server via the internet. Figure 1 below shows the IP67 Polycarbonate enclosure of the sensor, in compliance with the required specifications for weather-proofing purposes.

Figure 1. Sensor Casing.

Due to the AC and battery power source, SHM systems which utilizes the SmartBridge Sensor nodes are better for continuous SHM than periodic SHM.

3. SHM System

For this study, the SmartBridgeSensor Nodes were installed on an existing concrete bridge with American Association of State Highway and Transportation Officials (AASHTO) Type IV Girders. The bridge is a reinforced concrete bridge built in the 1960's and became operational in the year 1963. Its length is about 410 meters, linking Pandacan and Paco, Manila. It is shown in Figure 2.

Figure 2. Padre Jacinto Zamora Bridge.

Sensors will be placed on the concrete girders and columns of the bridge. As shown in Figure 2, the bridge was installed with four (4) AASHTO Type IV Girders. The sensors were placed on the interior face of the interior girders as shown in Figure 3.

Figure 3. Installed Sensor on Girder.

SmartBridge Sensor Nodes for this study were installed with stainless steel bracings for theft-proofing purposes. Considering a single span of the bridge, the interior girder was installed with three (3) sensors: one at the mid-span of the girder, and two sensors, each at a distance of 2H from the support. Wherein H is the overall height of the AASHTO Type IV Girder. Sensors were installed at 2H from the support since it is the location wherein the girder is expected to undergo flexural yielding and is also the location with the highest shear. The sensor is placed between the structural supports (the theoretical weakest point) for a worst-case measurement. There are two major sensor configurations for wired and wireless accelerometer that were commonly used in most of the SHM projects. Firstly, it is very common to place the accelerometers near the mid-span, where highest vibration responses are likely to occur. This configuration is ideal to obtain the modal frequencies of the structure. The sensor configuration should be defined based on the project's interest. If mode shapes are required, the sensor should be equally distributed on the girder of interest. On other hand, if modal frequencies are the only interest, mid-span where highest vibration responses are likely to occur will be the ideal location to place the sensor. (Teng, 2012). For sensor placement on the piers, all shall be installed at the mid height, wherein, maximum deflection occurs, as shown in Figure 4. Inelastic Time History Analyses, these analyses predict that the maximum column demand moments occur at approximately mid-height of concrete structures. In addition, the structural model of the bridge yielded the highest value for deflections at the mid-height of the columns.

Figure 4. Installed Sensor on Column.

With the sensors installed, acceleration readings are continuously monitored and transmitted to a central server through the internet. The Theoretical Framework is shown in Figure 5.

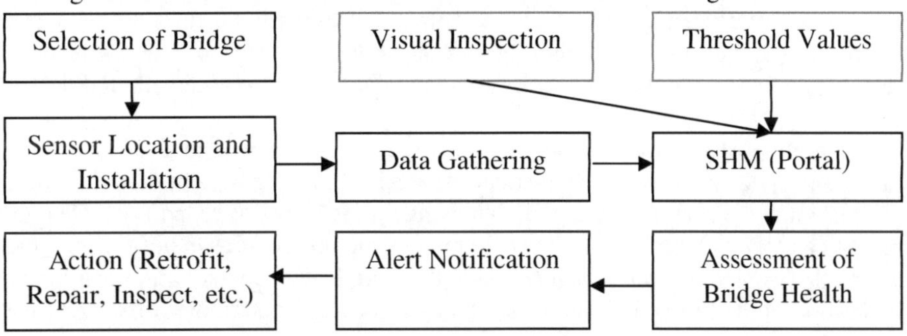

Figure 5. Theoretical Framework.

The SHM System designed is accessible through a portal or a webpage wherein data acquired is shown through a graph or a table, depending on the user. For the SmartBridge System, there are levels of access for the portal or the webpage, namely: a) Level I – Public access; b) Level II - Technical Staff; and c) Level III - Admin User.
Public access requires no account sign up for access, however, the page will only show basic information about the bridge, its maintenance managing party, and overall health of the bridge. For the data viewing and access to editing reports, the Level II access are included but not limited to the Maintenance Engineers that handles the visual inspection of the bridge. Level III access on the other hand are for the Maintenance Managers, Researchers, Administrators, and other Government Officials that may download and utilize data. Level III access allows the user to download visual inspection reports and data.
The Visual Inspection and Threshold Value generation are not included in this study. Visual Inspection Reports shall come from the Maintenance Engineers, whose frequency of inspections will be dependent upon the health of the bridge, as assessed by the SHM System. In the event of an acceleration reading reaching a certain threshold, the system triggers an alert and sends an e-mail notification to the Maintenance Managers that has authority over the bridge.

4. Results

With the installation procedures set, one span of the bridge would have a total of five (5) sensors installed. These sensors in one span is called a cluster. The configuration indicates that each SmartBridge Sensor Node has the ability to connect to the internet in order to transmit data to the central server, ready for remote access by the authorized users. The SHM System designed in this study would capture the actual acceleration readings on the bridge in real-time, and compare these readings to thresholds set through the structural analysis of the bridge.
The SHM System webpage or portal shall be designed to contain at least, but not limited to the following: Bridge Profile, Editable Inspection Report, Acceleration Graph, Downloadable Data, Levels of Access, and Alert/Notification System.
Structural Health Monitoring of Bridges is one of the interests in civil engineering that is quickly improving due to the fact that calamities such as earthquakes may damage lifeline structures such as a bridge. One of the most common SHM Methods of bridges revolves around the idea that structural damage in a structure is related to the modal properties and vibration of a structure. A Wireless Smart Sensor Network was installed on an existing Concrete Bridge with AASHTO Type IV Girder, and threshold values for each of the axes were used as limits for a Single Threshold Based SHM System. The SmartBridge Sensor Node was installed and utilized in this study to record acceleration responses in specific parts of the existing concrete bridge. Through Continuous SHM, each acceleration response is then monitored against the set thresholds values in g. A simple integration and double integration algorithm was used in order to derive velocity response and displacement response respectively. As part of the Portal/Webpage wherein continuous monitoring is done, the users of Level III access are able to receive a notification in the event of a threshold breach. Attached along with the notification e-mail and SMS, is a set of data recorded 20 seconds before and 30 seconds after the threshold breach. With this, the Level III users may react accordingly, such as being able to send inspectors in the particular portion of the bridge, wherein the threshold breach occurred. Due to the continuous aspect of the recording of acceleration responses, researchers may also find the gathered data in this study to be beneficial.
Using a different SHM method for the data gathered by the SmartBridge Sensor Nodes can also be used for further studies. Through Statistical Analysis, the normal vibration of the bridge can be determined through the use of data in which the structure is in normal condition. This normal vibration of the bridge can then be compared to the vibration of the bridge after the event of a threshold breach. In summary, the ambient or normal vibration of the bridge after an excitation can be compared to the normal vibration of the bridge before the excitation in order to determine if the response of the bridge has changed, hence determining a possible structural damage.

IOP Conf. Series: Materials Science and Engineering **216** (2017) 012050 doi:10.1088/1757-899X/216/1/012050

5. References

[1] Magalhaes F, Cunha A and Caetano E. (2008). "Dynamic Monitoring of a Long Span Arch Bridge." *Engineering Structures*, 30, 3034-3044.

[2] Teng C. (2012). "Structural Health Monitoring of a Bridge Structure Using Wireless Sensor Network". Master's Theses. Paper 79.

[3] Yu Y, Xie H, Ou J. (2010). "Vibration Monitoring Using Wireless Sensor Networks on Dongying Huanghe River Bridge." *Earth andSpace 2010: Engineering, Science, Construction, and Operations in Challenging Environments,* ASCE.

[4] Hoult, N.A., Fiddler, P. R. A., Hill, P.G., Middleton, C.R. (2010). "Long-Term Wireless Structural Health Monitoring of the Ferriby Road Bridge." Journal of Bridge Engineering, ASCE.

[5] Rice, J.A. And Spencer Jr., B.F. (2009). "Flexible Smart Sensor Framework for Autonomous Full-scale Structural Health Monitoring," NSEL Report Series, No. 18, University of Illinois at Urbana-Champaign. http://hdl.handle.net/2142/13635. Shankar P.M. (2002). Introduction to Wireless Systems. Wiley and Sons, New York.

[6] National Structural Code of the Philippines. (2010) (6th ed.). Pasay, Quezon City, Philippines: Association of Structural Engineers of the Philippines.

[7] Department of Public Works and Highways. (2015). Guidelines and Implementing Rules on Earthquake Recording Instrumentation for Buildings. 1- 17.

IOP Conf. Series: Materials Science and Engineering **216** (2017) 012051 doi:10.1088/1757-899X/216/1/012051

Vehicle-bridge Interaction Analysis Based on the ANCF Quasi-conforming Plate Technique

Bingjian Wang[a], **Hua He**[b]

[a] State Key Laboratory of Hydroscience and Engineering, Tsinghua University, Beijing 100084, China

[b] Guangxi Hong Wu Expressway Company, Ltd, Wu Zhou, GuangXi Province, 543000, China

Email: tyqh03@yeah.net;

Abstract. A new plate element is developed for analysis of plate structures in vehicle-bridge interaction analysis based on the combining of absolute nodal coordinate formulation (ANCF) and quasi-conforming technique (QCT). In order to simulate complex contact and large deformation during vehicle-bridge interaction (VBI) for the slender bridge, new curvature strains and explicit formulation of internal forces are developed for the shell elements of the bridge deck. The developed QCT_ANCF shell element is compared with the original ANCF element to verify its locking remedies. Compared with the original model, the new QCT_ANCF element shows better convergence and curvature continuity and is more accurate under the same number of elements. Numerical cases are analyzed using the QCT_ANCF element in comparison to analytical solutions and the original ANCF shell element. Meanwhile, there is less high frequency vibration in the velocity and acceleration curve by comparing with the original model. Furthermore, the vehicle-bridge interaction is parametrically analyzed using the new QCT_ANCF element under series of road roughness index and vehicle speeds. The impact factors based on the displacements and strains over the transverses of the bridge are investigated.

1. Introduction

Bridge-vehicle interaction (VBI) and road surface roughness are very essential to check the true capacity of existing bridges to heavier traffic and for proper design of new bridges. Plate is the most suitable element for these slender bridges among finite element method, which could not only simulate the 3D deformation contrast to the beam element but also save the computation contrast to the solid element. González et al [1] studied the structural deformation based on the Kirchhoff plate element. Kwasniewski et al [2] took 612000 solid elements on LS-DYNA to simulate the bridge on VBI analysis, and the computation had to conduct on a cloud platform. Nassif & Liu [3] simulated the bridge and vehicle interaction by beam element. During VBI analysis, the vehicle structural system and the bridge structural system deform separately, and the contact boundaries between the two systems are changing with time. Therefore, the VBI is a nonlinear procedure which could be taken as a multibody problem. So, it is a natural choice to take absolute nodal coordinate formulation (ANCF) to conduct the VBI analysis. The generalized coordinates of ANCF are absolute displacements and slopes and make no small deformation assumptions [4-6]. The mass matrix of the plate element keeps constant, which will reduce the degree of non-linearity of the dynamic equations [7]. Its discontinuities in bending strain would bring unrealistic strain energy to the element, resulting in an increased stiffness, element distortion, and locking [8, 9]. Many studies were carried out to take care of the locking, such as redefined polynomial expansion [10], selective integration procedure [11-13], modifying the constitutive relations [14], and using higher order geometry elements. Nowadays, selective integration is the primary method to solve the locking problems. However, it is constrained in boundary conditions. Tang et al. [15] presented the quasi-conforming technique. Instead of displacement fields traditionally, the strain fields are described by the polynomials [16]. As reported, the QCT elements are free from shear and membrane locking phenomena, and spurious kinematic modes.

This study, applying the quasi-conforming techniques, presents an ANCF element, namely QCT_ANCF element. Compared with the original ANCF element, this element matrix could be explicitly integrated or numerically applied by modifying the original code minimally. The element exhibits its computational efficiency and is robust to distortion.

2. ANCF Thin Plate Element

The coordinates at the node of the plate element are defined as rp:

$$r_p = S(x_p, y_p, z_p)e(t_p) \tag{1}$$

A lower order ANCF plate element (Figure 1) developed by Gerstmayr and Shabana is introduced in this section. For the freedom of each node, there are only three position vector and two gradient vector obtained by differentiating with respect to coordinates in element plane. The displacements are interpolated in x and y by polynomials with three orders.

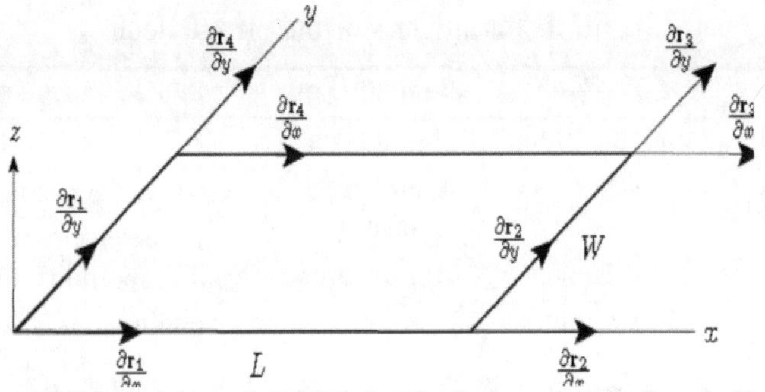

Figure 1. Kinematics of the 36 d.o.f. element by Dufva & Shabana

The axial curved strain in axis of element is as follows [17]: $\kappa = |r_x \times r_{xx}| / |r_x|^3$ (2)

3. Formulation of absolute nodal coordinate cable/beam element based on the quasi-conforming techniques

In Eq. (2), the curvatures r_{xx} are discontinuous, which could be the sources of lockings. The Quasi-Conforming technique is taken as the tool to take care of the problems. Described the strain fields by the polynomials, the formulation of Quasi-Conforming cable/beam element, named as QCT_ANCF, is briefly presented.

In Eq. (4) and Eq. (5), the strain resultant is as follows:

$$r_{xx} = \frac{\partial^2 r}{\partial x^2} = a^T P_\varepsilon,\ P_\varepsilon = [2 \quad 6x]^T, a = \begin{bmatrix} a_2 & a_3 \\ b_2 & b_3 \\ c_2 & c_3 \end{bmatrix}^T \tag{3}$$

Integral domain can be written as [x_1, x_2]. Using weighted integration with respect to x by weight P_ε, the curvature strain can be computed as:

$$\int_{x_1}^{x_2} \frac{\partial^2 r}{\partial x^2} P_\varepsilon dx = a\int_{x_1}^{x_2} P_\varepsilon P_\varepsilon^T dx, \quad \int_{x_1}^{x_2} P_\varepsilon P_\varepsilon^T dx = A \tag{4}$$

To evaluate the right hand side integral of Eq. (4) evaluated, the Green's theorem is used. As composed by polynomials on the left side of (4), it may be integrated easily.

$$\int_{x_1}^{x_2} \frac{\partial^2 r}{\partial x^2} P_\varepsilon dx = P_\varepsilon \frac{\partial r}{\partial x}\Big|_{x_1}^{x_2} - \int_{x_1}^{x_2} \frac{\partial r}{\partial x}\frac{\partial P_\varepsilon}{\partial x} dx = Ce \tag{5}$$

The parameters for the assumed strain as in Eq. (3) could be expressed as:

$$a = A^{-1}Ce \tag{6}$$

In order to obtain the strain discretization matrix, B, equations (6) to (3) is substituted into curvature strain term. The element strains could be expressed as:

$$r_{xx} = \mathrm{P}_\varepsilon^T a = \mathrm{P}_\varepsilon^T A^{-1}Ce \tag{7}$$

4. Engineering application

In the research, the bridge deck is simulated by QCT_ANCF plate element and the vehicle model is composed by mass-spring-damper. The plate is simple supported. The parameters are shown as Table 1 and Table 2.

Table 1. Parameters of the bridge deck

element type	**QCT_ANCF**	**element type**	**ANCF mass-spring-damper**
Dimensions (m)	24m*12m*0.8m	**Load type**	track
elastic modulus	3.0E9	**Geometric params**	axil distance in longitude 5.5m; axil distance in transverse 2m
passion ratio	0.15	**tire params**	elastic modulus 1E6; damper 3E3
density(m)	1500kg/m^3	**weight**	vehicle 165kN;front gear 30kN; back gear 60kN
dimension	1.2*1.0	**moment**	Ix: 4.5E5, Iy: 4.5E5

Time-displacement curve at the midpoint of the span under the speed of 60Km/h is shown in Figure 2. As can be seen, with the large of the road rough index, the displacement of the middle became larger.

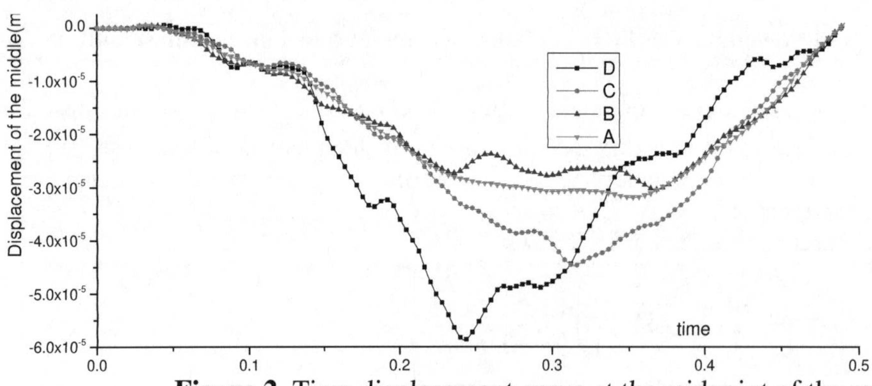

Figure 2. Time-displacement curve at the midpoint of the span

The impact factor at the right left and middle of the deck under the speed of 90Km/h is shown in Figure 3.

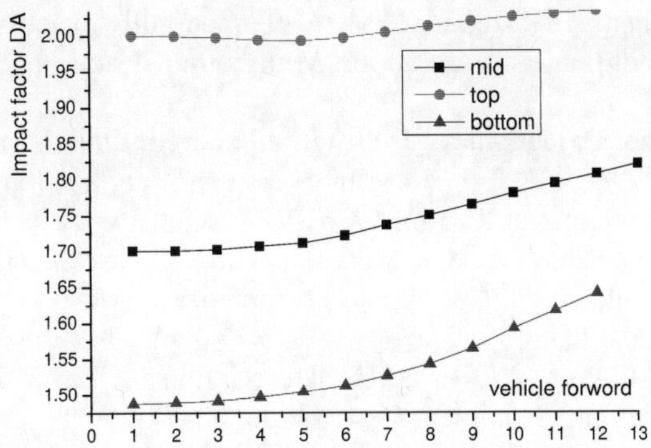

Figure 3. The impact factor at the right left and middle of the deck

5. Conclusion

1) This QCT_ANCF element does not depend on the higher order or assuming strain, and suffers less from locking and enjoys better convergence characteristics in simulating the flexible beams, etc. Therefore, the quasi-conforming technique could be a quick, easy, and cheap way to improve the computing performance for absolute nodal coordinate method.

2) With the element, the analysis suitable for the VBI are conduct under the united platform, and the convergent and accurate are promoted obviously.

6. Acknowledgments

Authors wishing to acknowledge the financial support from Science and Technology Program Project Funds of Guangxi Zhuang Autonomous Region (2016).

7. References

[1] Cantero, D., OBrien, E.J. & González, A. Modelling the vehicle in vehicle infrastructure dynamic interaction studies[P]. Proceedings of the Institution of Mechanical Engineers, Part K:Journal of Multi-body Dynamics, 2010, **224**: pp 243-248.

[2] Kwasniewski, L., Li, H., Wekezer, J., and Malachowski, J. Finite element analysis of vehicle-bridge interaction. Finite Elements in Analysis and Design, 2006, **42(11)**: pp 950–959.

[3] Nassif, H.H. & Liu, M. Analytical modelling of bridge-road-vehicle dynamic interaction system. Journal of Vibration and Control, 2004, **10**: p 215.

[4] Shabana AA. Flexible multibody dynamics: review of past and recent developments. Multibody System Dynamics, 1997, **1(2)**: pp 189-222

[5] Shabana AA. An absolute nodal coordinate formulation for the large rotation and large deformation analysis of flexible bodies. Technical Report No. MBS96-1-UIC, Department of Mechanical Engineering, University of Illinois at Chicago, 1996

[6] Shabana AA. Definition of the slopes and the finite element absolute nodal coordinate formulation. Multibody System Dynamics, 1997, **1(3)**: pp 339-348

[7] Shabana AA. Definition of ANCF finite elements. Journal of Computational and Nonlinear Dynamics, 2015, 10(5): 054506

[8] Sopanen JT, Mikkola AM. Description of elastic forces in absolute nodal coordinate formulation. Nonlinear Dynamics, 2003, **34(1-2)**: pp 53-74

[9] Schwab AL, Meijaard JP. Comparison of three-dimensional flexible beam elements for dynamic analysis: finite element method and absolute nodal coordinate formulation. ASME, 2005: pp 1341-1349

[10] García-Vallejo D, Mikkola AM. A new locking-free shear deformable finite element based on absolute nodal coordinates. Nonlinear Dynamics, 2007, **50(1-2)**: pp 249-264

[11] Gerstmayr J, Matikainen MK, Mikkola AM. A geometrically exact beam element based on the absolute nodal coordinate formulation. Multibody System Dynamics, 2008, **20(4)**: pp 359-384

[12] Nachbagauer K, Gruber P, Gerstmayr J. Structural and continuum mechanics approaches for a 3D shear deformable ANCF beam finite element: application to static and linearized dynamic examples. J Comput NonlinJ COMPUT NONLIN DYN, 2013, **8(2)**

[13] Kerkka¨nen, K. S., Sopanen, J. T., and Mikkola, A. M., 2005, "A Linear Beam Finite Element Based on the Absolute Nodal Coordinate Formulation," ASME J. Mech.Des., **127**, pp. 621–630.

[14] Sopanen, J. T., and Mikkola, A. M., 2003, "Description of Elastic Forces in Absolute Nodal Coordinate Formulation," Nonlinear Dyn. **34**, pp. 53–74.

[15] Dong-sheng, H. & Li-min, T. The displacement function of quasi-conforming element and its node error, Appl Math Mech (2002) **23**: p 127.

[16] Gilson Rescober Lomboy, Nonlinear Formulations of a Four-Node Quasi-Conforming Shell Element [J], Arch Comput Methods Eng (2009) **16**: pp 189–250

[17] Rade,L. and Westergren, B., Mathematics Handbook for Science and Engineering, 4th edn., Springer-Verlag, Berlin Heidelberg, 1999

IOP Conf. Series: Materials Science and Engineering **216** (2017) 012052 doi:10.1088/1757-899X/216/1/012052

A Reliability analysis on the applicability range of Caltrans-SDC method to address the P-Delta effects

P Heydarpour

Department of Civil and Environmental Engineering, The George Washington University, Washington, DC 20057, USA

pheydarpour@gwu.edu

Abstract. The main objective of this research is to investigate the applicability range on the Caltrans Seismic Design Criteria (SDC) to address the P-Delta effect. Caltrans SDC defines a maximum ratio for bending moment induced by P-Delta effects over the yielding moment capacity of the column. For columns satisfying this criterion predefined target ductility is used to design the columns. Columns with higher P-Delta induced bending moments should be subjected to nonlinear time history analysis to verify their performance. Typically, designers to prevent performing time consuming, nonlinear time history analyses resize the column to remain in the safe margin. This study through rigorous push over analyses identifies the applicability range for the Caltrans-SDC method.

1. Introduction

P-Delta effects can have a detrimental impact on the seismic response of bridges because of a reduction in both the shear capacity and initial stiffness of RC bridge columns [1], [2]. The reduction in the initial stiffness imposes an increase in the natural period of the system, and a likely surge in the design displacement demand. In cases that it is not allowed to neglect the P-Delta effects, it is required to consider the dynamic effect of gravity loads acting through lateral displacements in design[3], [4]. Wei et al. [5] provides a detailed description on methodologies to compensate for P-Delta effects on structures. As design codes are progressing towards performance-based metrics, there is an additional need to quantify the destabilizing effect of gravity loads and its effect on the seismic response of bridge columns [6], [7].

2. Background

Caltrans SDC [3] provides a procedure that can be used to evaluate whether P-Delta effects can be ignored in design. In design circumstances, not requiring to consider P-Delta effects, structural components can be designed based on predefined ductility demands. This study intends to find the range which (is valid. In cases which (is not satisfied, increasing the section size or reinforcement ratio can be used to increase the yielding moment capacity of the colum, Howevre, Caltrans SDC reccommends to perform nonlinear time-history analysis to verify that the column is capable to resist the P-Delta effects.

$$P \times \Delta_r = 0.2 \times M_p^{col} \quad (1)$$

Where, Δ_r is the lateral offset between the point of contra-flexure and the base of the plastic hinge, and $\mathbf{M_p^{col}}$ is the idealized plastic moment capacity of a column calculated by M-φ analysis. If (is satisfied, predefined ductility demands limits the design of structural components. According to Caltrans SDC target displacement ductility for single column and multi-column bents are as follows.

Single Column Bents supported on fixed foundation $\mu_D \leq 4$

Multi-Column Bents supported on fixed or pinned footings $\mu_D \leq 5$

3. Method

In this research using OpenSees, nonlinear static analysis (Pushover) is performed to obtain the moment-curvature and subsequently the load-deformation for various columns. The main interest is to obtain the value for the applied load and displacement at yielding and ultimate capacity of the column. The value for yielding displacement and load can be obtained by linear interpolation between the yielding and the ultimate point.

Table 1 shows the Column Properties subjected to this study.

Table 1. Column Properties

Parameters	
Concrete Strength, f'c (MPa, ksi)	27.5 (4)
yield Strength, fy (MPa, ksi)	413 (60.0)
Modulus of elasticity, Es (MPa, ksi)	2×105 (29,000)
Longitudinal reinforcing steel: yield strain, εy	0.0015
Column diameter, L (m, ft)	1.21 (4)
Column aspect ratio, CAR	4 to12
Cover concrete (cm, in)	5 (2)
Axial load (kips)	389,589,778,973,1168

3.1 Finite element model

Open System for Earthquake Engineering Simulation (OpenSees) [8] is used to investigate the nonlinear load-deformation response of RC bridge columns. The circular cross-section was represented by a fiber-based model and the concrete cover and core sections were modelled with the "*Concrete07*" uniaxial concrete material class.

4. Results

This section looks at nine different column height ratios ranging between four to twelve. longitudinal reinforcement ration is changing between one to four percent. Five different levels of axial load is applied on top of the column. Target displacement ductility level used to design the columns is obtained from the Caltrans SDC, which suggests single column bents supported on fixed foundation can be designed for ductility level 4.

Table 2. Yielding displacement

ALR	LRR	M_y (kip.in)	4	5	6	7	8	9	10	11	12
389kips	*1*	26253	1.92	2.19	2.51	2.96	3.51	4.14	4.80	5.57	6.39
	2	45670	2.96	3.27	3.71	4.31	5.05	5.88	6.79	7.88	9.04
	3	62245	3.76	4.09	4.64	5.32	6.18	7.21	8.37	9.58	10.94
	4	80691	4.63	4.98	5.60	6.39	7.38	8.55	9.90	11.38	13.01
589 kips	*1*	28445	2.03	2.27	2.61	3.08	3.65	4.23	4.99	5.79	6.65
	2	47400	3.04	3.34	3.80	4.43	5.13	5.98	6.95	8.01	9.20
	3	64628	3.84	4.17	4.73	5.43	6.32	7.31	8.43	9.76	11.14
	4	82067	4.69	5.05	5.66	6.47	7.47	8.65	10.02	11.48	13.20
778 kips	*1*	30480	2.08	2.35	2.78	3.19	3.73	4.40	5.10	5.93	6.80
	2	49664	3.09	3.42	3.86	4.51	5.22	6.08	7.06	8.19	9.33
	3	66127	3.91	4.25	4.81	5.52	6.43	7.44	8.63	9.89	11.30
	4	83392	4.74	5.10	5.72	6.54	7.57	8.76	10.10	11.62	13.29
973kips	*1*	32727	2.18	2.43	2.89	3.32	3.89	4.58	5.31	6.17	7.08
	2	51245	3.16	3.46	3.93	4.61	5.37	6.22	7.17	8.31	9.48
	3	67557	3.97	4.31	4.87	5.61	6.48	7.50	8.71	10.03	11.46
	4	84676	4.81	5.17	5.81	6.63	7.67	8.85	10.25	11.75	13.44
1168 kips	*1*	34636	2.30	2.53	2.92	3.41	4.05	4.71	5.46	6.35	7.29
	2	52799	3.23	3.55	4.05	4.65	5.42	6.32	7.35	8.45	9.72
	3	68956	4.04	4.39	4.95	5.70	6.60	7.65	8.88	10.18	11.63
	4	85942	4.87	5.24	5.86	6.72	7.77	8.96	10.34	11.90	13.62

(All results are in inch)

The P-Delta induced bending moment at target ductility level of four is equal to four times of the yielding displacement (**Table 2**) multiplied by the load at ductility level four (**Table 3**). According to Caltrans SDC if the ratio of bending moment induced by P-Delta effects to the yielding moment capacity of column is less the twenty percent, then structural components shall be designed based on predefined displacement ductility demands.

Table 3. Load at ductility level 4(interpolated from idealized bilinear force-displacement diagram)

		Column height ratio								
	ρ_L	4	5	6	7	8	9	10	11	12
389kips	1	154.04	122.51	101.75	87.00	76.04	67.53	60.77	55.19	50.60
	2	278.58	218.26	179.73	153.07	133.57	118.34	106.27	96.61	88.45
	3	399.05	307.80	251.79	213.08	185.14	163.90	147.16	133.42	122.05
	4	540.04	411.48	333.88	281.25	243.59	215.27	192.93	174.78	159.88

589 kips	1	164.32	130.58	108.35	92.65	80.99	71.89	64.73	58.79	53.90
	2	288.45	225.76	185.90	158.35	137.89	122.18	109.77	99.73	91.31
	3	408.59	315.34	257.77	218.28	189.72	167.82	150.45	136.56	124.93
	4	NA	418.89	339.46	286.05	247.81	218.76	196.11	177.66	162.64
778 kips	1	173.79	138.00	114.60	97.90	85.54	75.97	68.30	62.10	56.88
	2	298.07	233.15	191.69	163.18	142.06	125.99	113.09	102.70	94.06
	3	417.80	322.38	263.42	223.09	193.76	171.41	153.77	139.41	127.54
	4	NA	425.26	344.71	290.50	251.68	222.19	199.02	180.33	164.96
973kips	1	183.71	145.76	121.00	103.38	90.32	80.22	72.12	65.52	60.06
	2	307.56	240.01	197.39	168.05	146.28	129.63	116.30	105.62	96.74
	3	426.88	328.95	268.59	227.49	197.58	174.62	156.81	142.25	130.15
	4	NA	431.84	349.99	294.91	255.54	225.51	202.02	183.07	167.44
1168 kips	1	193.53	153.33	127.14	108.69	95.00	84.34	75.83	68.90	63.16
	2	316.22	246.74	202.85	172.35	150.17	133.05	119.56	108.42	99.37
	3	435.61	335.78	274.10	232.10	201.57	178.22	159.91	145.00	132.66
	4	NA	438.12	354.98	299.23	259.16	228.70	204.87	185.66	169.81

Table 4 presents the results obtained for the Caltrans SDC criterion for ignoring the P-Delta effects. Regions colored in green are where Caltrans allows ignoring the P-Delta effects, while the red region requires performing nonlinear time-history analysis to study the structural behavior with inclusion of the P-Delta effects. Column with high column height ratio and high axial load ratio are the most susceptible columns to P-Delta effects. Results indicate that for these columns Caltrans SDC equation is not satisfied and nonlinear time-history analysis should be implemented to make any decision regarding the adequacy of the design. Typically under these circumstances design firms in order to prevent performing time consuming and highly advanced nonlinear time history analysis with a wide range of earthquake records they prefer to resize the column in order to satisfy the Caltrans equation.

Table 4. Ratio of P-Delta induced bending moment to plastic moment capacity

		Column height ratio								
	ρ_L	4	5	6	7	8	9	10	11	12
389kips	1	0.11	0.13	0.15	0.18	0.21	0.25	0.28	0.33	0.38
	2	0.10	0.11	0.13	0.15	0.17	0.20	0.23	0.27	0.31
	3	0.09	0.10	0.12	0.13	0.15	0.18	0.21	0.24	0.27
	4	0.09	0.10	0.11	0.12	0.14	0.17	0.19	0.22	0.25
589 kips	1	0.17	0.19	0.21	0.25	0.30	0.35	0.41	0.48	0.55
	2	0.15	0.16	0.19	0.22	0.25	0.29	0.34	0.39	0.45
	3	0.14	0.15	0.17	0.20	0.23	0.26	0.30	0.35	0.40
	4	0.13	0.14	0.16	0.18	0.21	0.25	0.29	0.33	0.38
778 kips	1	0.21	0.24	0.28	0.33	0.38	0.45	0.52	0.61	0.70
	2	0.19	0.21	0.24	0.28	0.33	0.38	0.44	0.51	0.59
	3	0.18	0.20	0.23	0.26	0.30	0.35	0.41	0.47	0.53
	4	0.18	0.19	0.21	0.24	0.28	0.33	0.38	0.43	0.50
973kips	1	0.26	0.29	0.34	0.40	0.46	0.55	0.63	0.73	0.84
	2	0.24	0.26	0.30	0.35	0.41	0.47	0.54	0.63	0.72
	3	0.23	0.25	0.28	0.32	0.37	0.43	0.50	0.58	0.66
	4	0.22	0.24	0.27	0.30	0.35	0.41	0.47	0.54	0.62
1168 kips	1	0.31	0.34	0.39	0.46	0.55	0.64	0.74	0.86	0.98
	2	0.29	0.31	0.36	0.41	0.48	0.56	0.65	0.75	0.86

3	0.27	0.30	0.34	0.39	0.45	0.52	0.60	0.69	0.79
4	NA	0.28	0.32	0.37	0.42	0.49	0.56	0.65	0.74

5. Result evaluation

In this section a sample column with height of 32 ft with 1 % reinforcement ratio, and 389 kips applied load has been studied under a set of earthquake records. This column fails to satisfy the Caltrans SDC criterion.

Throughout this research ATC Far-Field, ground motion record set is used. The ground motion set is collected from Pacific Earthquake Engineering Research Center (PEER-NGA) database. Table 5 and Table 6 tabulate the characteristics of the ground motion set.

Table 5. Ground motion properties

Distance R	R > 10 km
Large Magnitude Events	M > 6.5
Equal Weighting of Events	≤ 2 records per event
Strong Ground Shaking	PGA > 0.2g /PGV > 15 cm/sec
Source Type	Both Strike-Slip and Thrust Fault Sources
Site Conditions	Rock or Stiff Soil Sites Vs > 180 m/s
Record Quality	Lowest Useable Frequency < 0.25 Hz

Magnitude and PGA for the Far-Field earthquake records are tabulated in Table 6.

Table 6. Ground motion records

EQ ID	Earthquake			PGA	EQ	Earthquake			PGA
	M	Year	Name	(g)	ID	M	Year	Name	(g)
12011	6.7	1994	Northridge	0.52	12092	7.3	1992	Landers	0.42
12012	6.7	1994	Northridge	0.48	12101	6.9	1989	Loma Prieta	0.53
12041	7.1	1999	Duzce, Turkey	0.82	12102	6.9	1989	Loma Prieta	0.56
12052	7.1	1999	Hector Mine	0.34	12111	7.4	1990	Manjil, Iran	0.51
12061	6.5	1979	Imperial Valley	0.35	12121	6.5	1987	Superstition Hills	0.36
12062	6.5	1979	Imperial Valley	0.38	12122	6.5	1987	Superstition Hills	0.45
12071	6.9	1995	Kobe, Japan	0.51	12132	7.0	1992	Cape Mendocino	0.55
12072	6.9	1995	Kobe, Japan	0.24	12141	7.6	1999	Chi-Chi, Taiwan	0.44
12081	7.5	1999	Kocaeli, Turkey	0.36	12142	7.6	1999	Chi-Chi, Taiwan	0.51
12082	7.5	1999	Kocaeli, Turkey	0.22	12151	6.6	1971	San Fernando	0.21
12091	7.3	1992	Landers	0.24	12171	6.5	1976	Friuli, Italy	0.35

The procedure is as the following steps:

Step1: Perform pushover analysis and obtain load-deformation diagram

IOP Conf. Series: Materials Science and Engineering **216** (2017) 012052 doi:10.1088/1757-899X/216/1/012052

Figure 1 shows the result obtained from pushover analysis.

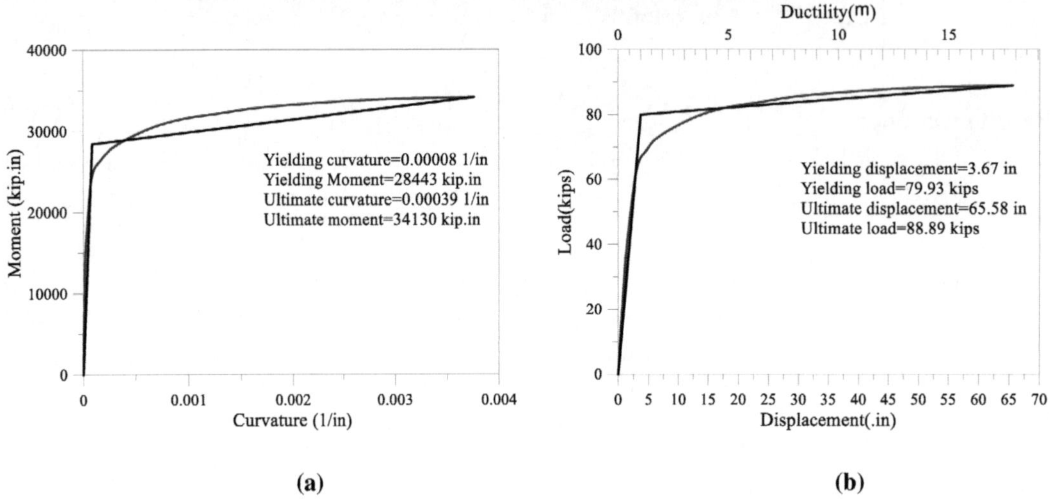

Figure 1. Moment-curvature and pushover analysis (Col spec: Col Height=32 ft., Axial load=389 kips, ρ_L=1%).

Step2: Compute the target spectral displacement

The target displacement at the target ductility can be obtained using (2. Design target ductility of four is suggested by Caltrans SDC for single column bent supported on fixed foundation.

$$SD_{cap} = \mu.\Delta_y = 14.68 \text{ in} \tag{2}$$

$$\mu=4$$

Figure 2 shows the procedure to compute the effective stiffness at target ductility. Force-displacement analysis provides the load and displacement values at yielding and ultimate points.

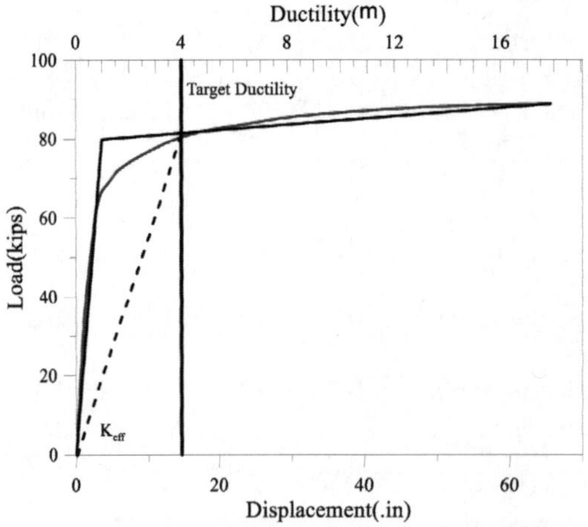

Figure 2. Effective stiffness at target ductility of 4.

Step3: Compute the effective period of the column at the target ductility

The lateral load at the target ductility can be obtained using (.

$$F_{target} = F_y + \frac{F_u - F_y}{\Delta_u - \Delta_y}(\Delta_y(\mu - 1)) = 81.52 kips \tag{3}$$

Effective stiffness at target ductility can be obtained using (.

$$K_{eff} = \frac{F_{target}}{\mu\, \Delta_y} = 5.55\ kips/inch \tag{4}$$

The effective period of structure can be obtained using

$$T_{eff} = 2\pi \sqrt{\frac{W}{g\, K_{eff}}} = 2.67\ sec \tag{5}$$

Step4: Compute the earthquake record scale factor

Expression proposed by Chopra and Goel (1999), $\hat{\zeta_{eq}} = \zeta + \zeta_{eq}$ was used for total viscous damping ratio.

$$\zeta_{eq} = \frac{1}{\pi}(1 - \frac{0.95}{\sqrt{\mu}} - 0.05\sqrt{\mu}) \tag{6}$$

$$\mu = 4, \zeta_{eq} = 0.185$$

Where, μ is the target ductility level and ζ is the elastic viscous damping and ζ_{eq} is equivalent viscous damping.
Newmark integration method is used to develop spectral displacement and spectral acceleration versus period of structure. Figure 3(A) shows the spectral displacement versus period of the structure for the earthquake record with EQID=120111 with 18.5% damping ratio. Figure 3 (B) shows the spectral acceleration versus time. For a structure with the period of 6.14 sec the spectral displacement is 8.02 inch.

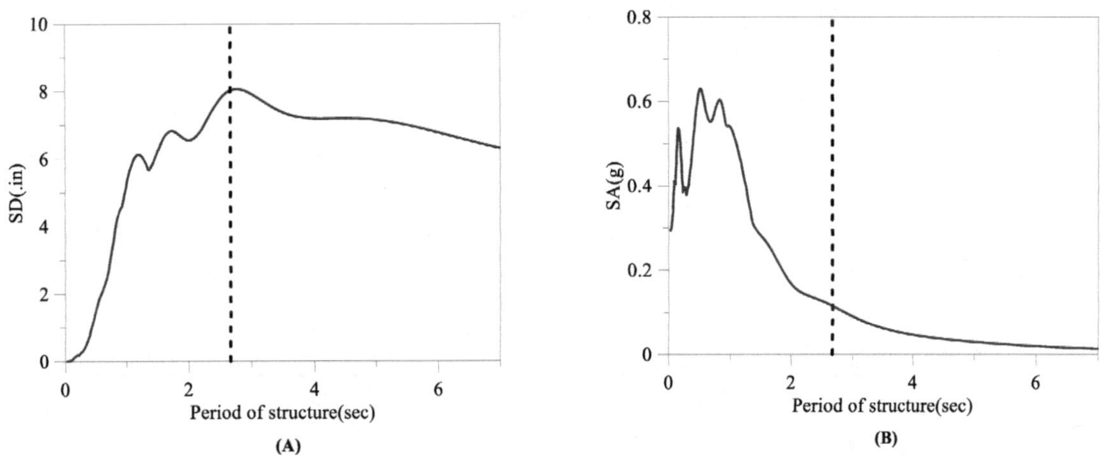

Figure 3. Response spectrum (EQID=120111, ζ_{eq} = 0.185).

The earthquake scale factor is the ratio of spectral displacement computed in previous step to the spectral displacement for the non-scaled earthquake record (at the effective period).

$$scalefactor = \frac{SD_{capacity}}{SD_{demand}} = \frac{14.68}{8.02} = 1.83 \quad (7)$$

Step5: Perform nonlinear time-history analysis

In this stage the earthquake record is amplified by the by the earthquake scale factor. Nonlinear time history response for the structure is presented.

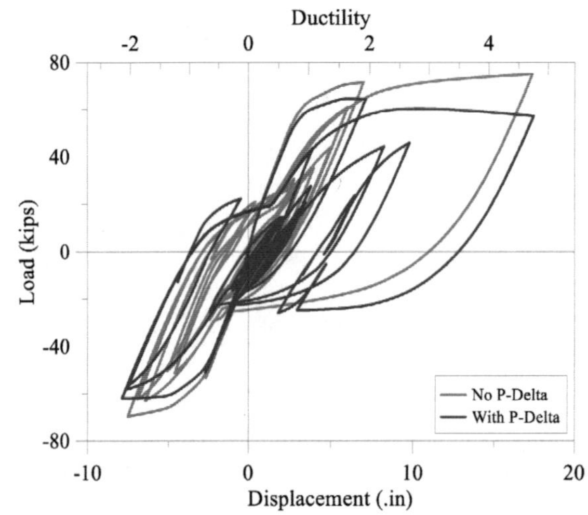

Figure 4. Dynamic inelastic time history analysis (EQID=120111)

Same analysis is performed on all the earthquake records introduced in Table 6. Figure 5 shows the ratio between the displacement ductility with P-Delta effects over displacement ductility ignoring the P-Delta effects for all earthquake records. The result from nonlinear time-history analysis indicates that although this column does not comply with the Caltrans SDC criterion for ignoring the P-Delta effects, the column is stable and the damage ductility is close to target ductility of four.

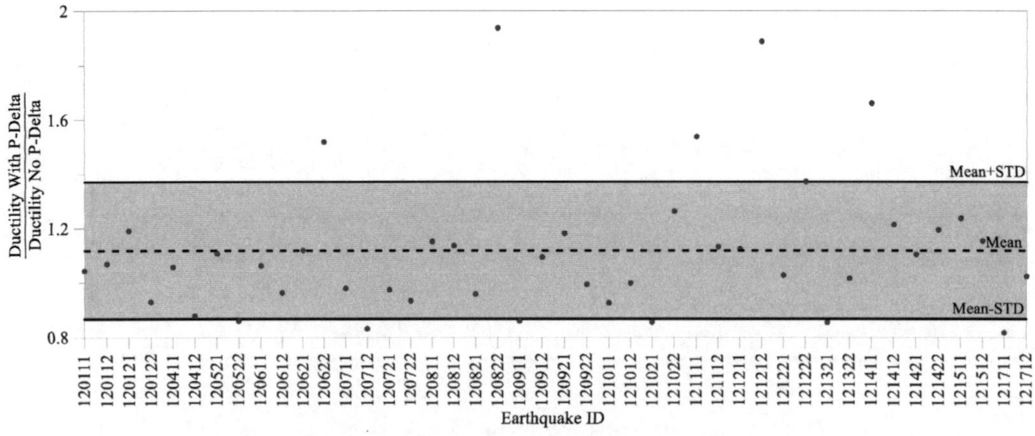

Figure 5. Obtained results for all earthquake records.

The mean value for ductility with P-Delta effects over the ductility without P-Delta effects for the studied column is 1.119 with standard deviation of 0.25.

6. Conclusion

Following observations have been made as the result of this research.

- Caltrans SDC method for columns with high aspect ratios(10-12) was not applicable, although these columns are most susceptible to P-Delta effects.
- Caltrans SDC encourages designers to use higher longitudinal reinforcement ratio or bigger section sizes instead of performing nonlinear time-history analysis.
- This method does not explicitly address the instability of the column.

References

[1] S. Mahin and R. Boroschek, "Background Report Influence of Geometric Nonlinearities on the Seismic Response and Design of Bridge Structures," no. October, 1991.

[2] D. Bernal, "Instability of buildings during seismic response," *Eng. Struct.*, vol. 296, no. April, 1998.

[3] *Caltrans SDC*. 2013.

[4] paul c Jennings and R. Husid, "Collapse of yielding structures during earthquake." 1968.

[5] B. Wei, Y. Xu, and J. Li, "Treatment of P-Δ Effects in Displacement-Based Seismic Design for SDOF Systems," *ASEC J. Bridg. Eng.*, vol. 17:3, pp. 509–518, 2012.

[6] D. Vamvatsikos, "SEISMIC PERFORMANCE, CAPACITY AND RELIABILITY OF STRUCTURES AS SEEN THROUGH INCREMENTAL DYNAMIC ANALYSIS," 2002.

[7] D. Vamvatsikos, "Seismic Performance Uncertainty Estimation via IDA with Progressive Accelerogram-wise Latin Hypercube Sampling 1."

[8] S. Mazzoni, F. McKenna, M. H. Scott, G. L. Fenves, and others, "OpenSees command language manual," *Pacific Earthq. Eng. Res. Cent.*, 2006.

IOP Conf. Series: Materials Science and Engineering **216** (2017) 012053 doi:10.1088/1757-899X/216/1/012053

Performance-Based Seismic Design of Bitlis River Viaduct Based on Damage Control Using Seismic Isolation and Energy Dissipation Devices

Murat Dicleli[1, a *] and Ali Salem Milani[2]
[1]Department of Engineering Sciences at the Middle East Technical University, Ankara, Turkey
[2]Middle East Technical University, Ankara, Turkey
[a]E-mail: mdicleli@metu.edu.tr

Abstract. This paper presents a sample application of seismic isolation techniques in performance-based design of a major viaduct. The Bitlis River viaduct is located in a seismically active region. The targeted performance goal required no damage at 475-year return period earthquake and repairable damage at 2475-year return period earthquake. The bridge is designed with a seismic isolation system composed of spherical bearings and MRSD (Multidirectional Re-centering steel Damper) hysteretic dampers. The MRSD is a recently-developed hysteretic damper with a controllable post-elastic stiffness. To keep the dampers from being activated during the thermal displacements, the attachment of the dampers to the deck are made through elongated holes oriented in the longitudinal direction of the bridge. The gaps are sized based on the amount of expected maximum thermal displacement in each pier. The gap length is thus different for different piers. This means that the number of the dampers to be engaged during an earthquake will depend on the intensity of the displacements. The distinct feature in this design is how it achieves double purpose: (i)preventing the dampers from engagement during service life as a result of thermal displacements and (ii) sequential engagement of the dampers depending on the level of seismically-induced displacements. The paper presents the basic design features of this seismically isolated bridge designed based on performance-based principles, a brief description of the newly-developed damper and a summary of analyses results.

1. Introduction

Special seismic protection, usually in the form of isolation or energy dissipation devices or combination of both, is often required for seismic protection of important structures located in areas of high risk of seismic activity to satisfy design objectives of controlled structural response and minimal or no damage. For structures subjected to earthquakes with intense long duration acceleration pulses, although seismic isolation technology may be used to reduce and control the magnitude of the forces, such a system alone may not be adequate to reduce the displacement demand to practical ranges of application [1]. In such cases, a combination of seismic isolation and energy dissipation devices or dampers is used to reduce and control both forces and displacements. Use of seismic isolation combined with energy dissipaters in bridges is as widespread. This paper is meant to be a demonstration of application of modern seismic isolation techniques to achieve a performance-based design of a bridge.

2. Description of the bridge and the construction site

Fig. 1 shows the satellite view and a perspective view of the Bitlis River Viaduct. The viaduct spans the Bitlis River in eastern Turkey with a total length of 1390 meters. Part of this bridge with a length of 801 meters containing 17 spans is designed as a post-tensioned box girder bridge, 19.6 meter width and girder depth of 3.0 meters, which is to be constructed using the incremental launching method. The height of the piers in this part of the bridge vary between 14 to 37 meters. Longitudinal view and a typical cross section of the bridge and the deck are shown in Fig. 2.

3. Seismicity of the site and seismic performance goals

Fig. 3 shows the site seismicity and the site-specific design spectrum, as obtained from a probabilistic seismic hazard analysis [2]. The site is located in a seismically active zone where seven potential sources of seismic activity were identified, two of which are capable of producing earthquakes with maximum magnitude of 7.6. The performance goals set for this viaduct are defined as follows:

- For 72-year return period earthquake (i.e., during construction): No damage
- For 475-year return period earthquake (design-basis earthquake): No damage
- For 2475-year return period earthquake (maximum considered earthquake): Repairable damage

Figure 1. Satellite and perspective views of the viaduct.

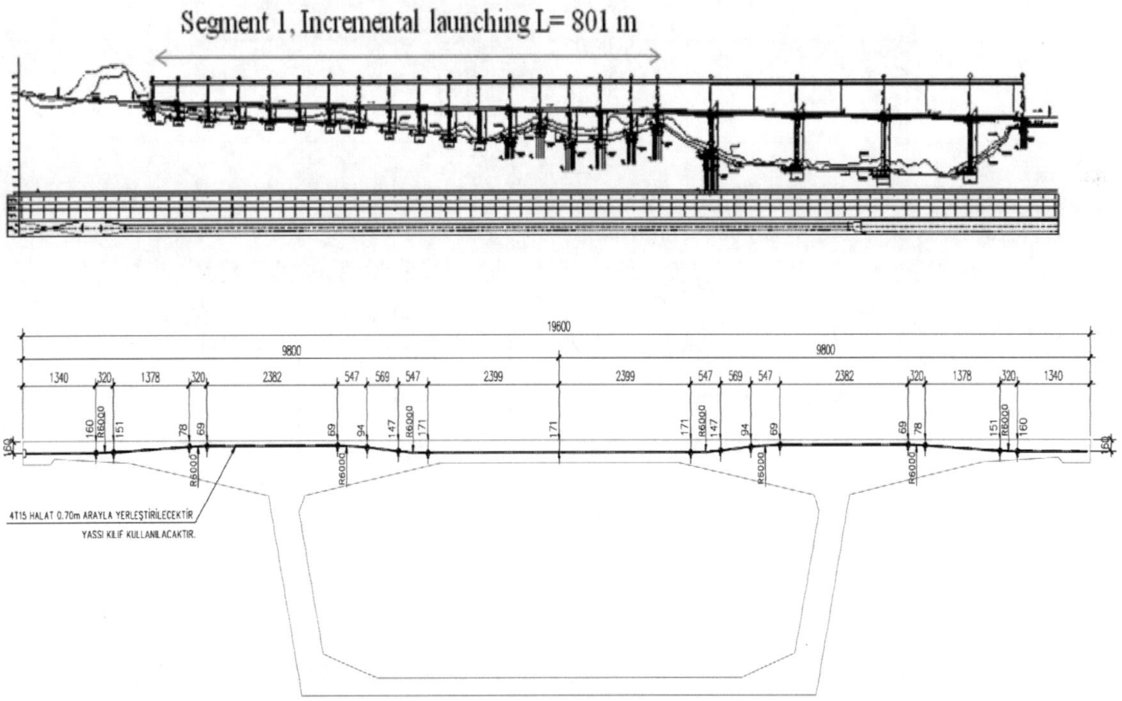

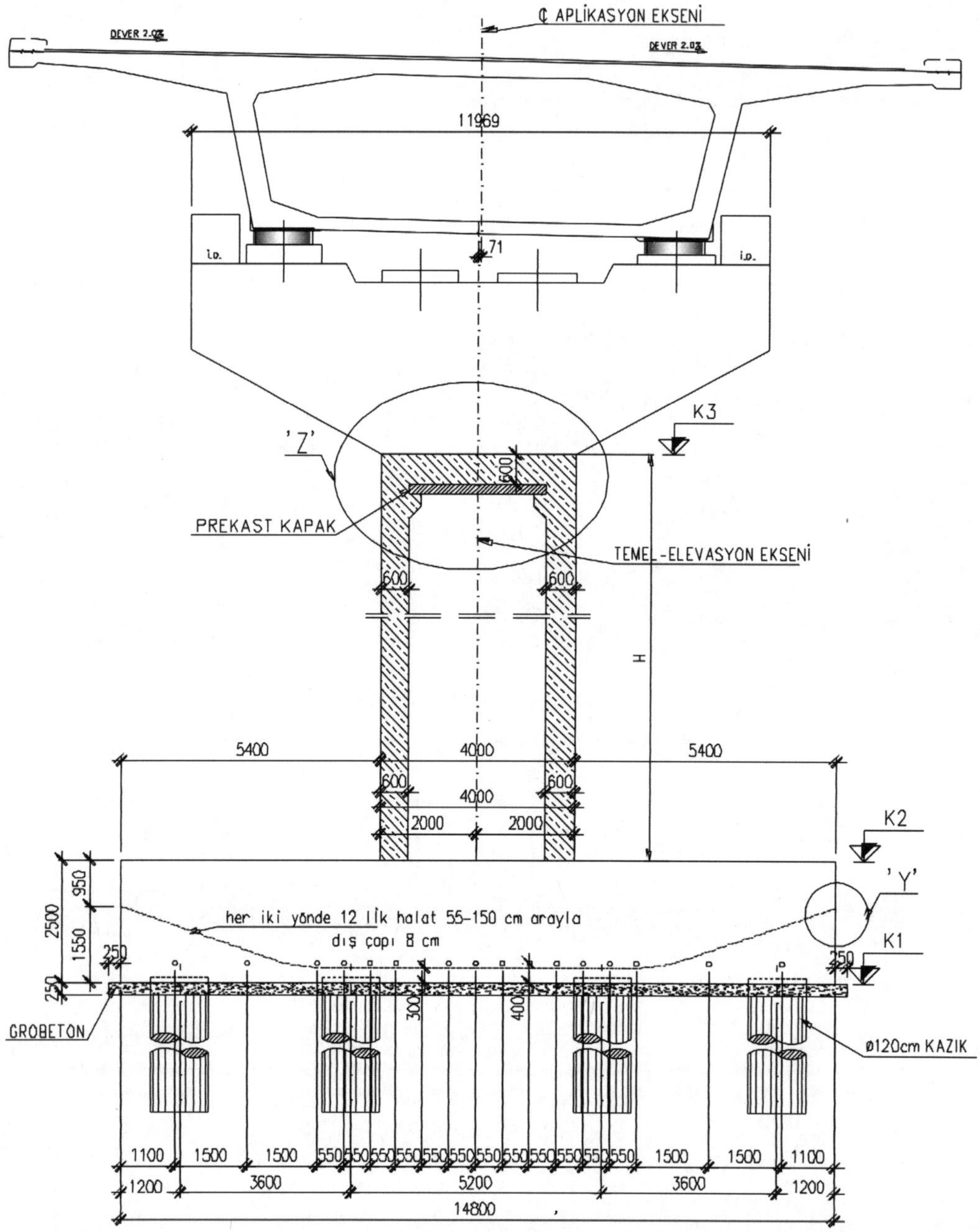

A-A KESİTİ
ÖLÇEK:1/100

Figure 2. Longitudinal and cross-section views of the bridge and the deck

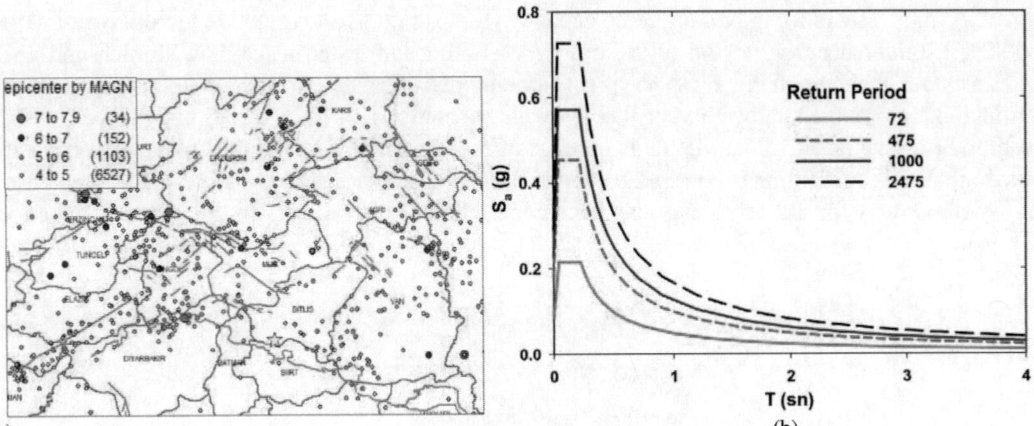

(a) (b)

Figure 3. (a) Site seismicity (the site is shown by a yellow star); (b) Site-specific design spectrum.

4. Seismic isolation system selected for design

The viaduct is located in a very cold area where the temperature can reach -22°C, a seismic isolation system that performs reliably in cold temperatures is needed. Accordingly, spherical bearings coupled with steel hysteretic dampers with re centering capability; MRSD (Multidirectional Re-centering Steel Damper) are chosen. A brief description of the newly-developed MRSD damper is given in the following.

MRSD (Multidirectional Re-centering Steel Damper) is designed to dissipate energy through yielding and plasticization of cylindrical energy dissipaters under torsion. Eight of these identical energy dissipaters each attached to a torsion arm are arranged in a symmetric configuration to create the MRSD. Fig 4(a) and (b) show perspective and side views of MRSD, respectively. Fig. 4(c) shows a section view of MRSD with a typical energy dissipation unit of MRSD. All of the parts which compose the damper are named in Fig 4(a) and Fig 4(c). However, for the sake of brevity, detailed description of the device is not given here and the interested reader is referred to reference [3]. To convert translational motion of the structure to twisting in the cylindrical energy dissipaters, each arm is coupled with a guiding rail which through a low-friction slider block guides the motion of the arm. Schematic top-views of MRSD at un-displaced and displaced positions are shown in Fig. 4(d). A distinguishing feature in force-displacement response of MRSD is the geometric hardening behavior which is the outcome of translation-to-rotation motion conversion mechanism in the energy dissipation units of MRSD, as schematized in Fig. 5(a). This mechanism, magnifies the reaction force required to balance the torque in energy dissipaters. Fig. 5(a) shows a typical energy dissipation unit with arm length L, subjected to displacements d_1 and d_2, where $d_2 > d_1$. Let the rotation angle of the arm, the reaction force and the torque are denoted by θ, f and T, respectively and let the numerical index indicate the corresponding state of displacement. Given that $T = f.L.\cos\theta$, and since increasing displacement leads to reduction in $\cos\theta$ without any reduction in T, it can be easily shown the $f_2 > f_1$, that is:

$$\frac{T_2}{L\cos\theta_2} > \frac{T_1}{L\cos\theta_1} \Rightarrow f_2 > f_1 \tag{1}$$

Note that the reaction force of the device is the sum of projections of all eight forces coming from eight energy dissipation units. Thus, the hardening behavior at eight energy dissipation units directly leads to similar behavior in global response of the device. The described mechanism also offers the possibility of controlling the desired level of hardening in force-displacement response, through adjustment of the arm length to maximum displacement ratio. Varying levels of hardening obtained as such, leads to hysteresis loops of different shapes as shown in Fig. 5(b). As indicated on these graphs, the parameter used to characterize hardening in the MRSD is called 'Hardening Index', defined as;

$$HI = \frac{F_{max}}{F_Y} \tag{2}$$

Where F_{max} and F_Y stand for maximum force capacity (force at D_{max}) and effective yield force of MRSD. A 200kN, 120mm-capacity version of the device was built and tested in UniBw/Munich and also at METU/Ankara, as shown in Fig. 6(a). A typical force-displacement response loop obtained from tests is given in Fig. 6(b), which shows a very stable cyclic response with little variation in force levels not exceeding %4.0 the mean value. MRSD is capable of reaching high force and displacement capacities, shows high levels of damping, controllable post-elastic stiffness and very stable cyclic response. A design methodology for the device has also been completed. Further details on this device can be found in [3], [4].

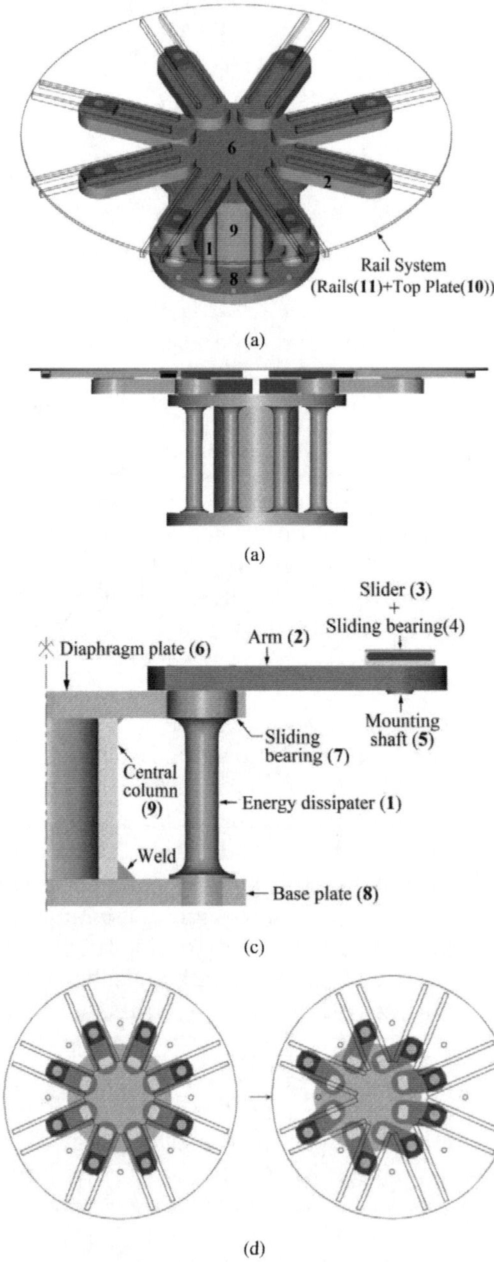

Figure 4. Multidirectional Re-centering Steel Damper (MRSD): (a) Isometric view showing the rail system and base device underneath; (b) side view; (c) Section view showing a single energy dissipation unit of MRSD; (d) Schematic top-view of MRSD at un-displaced and displaced positions.

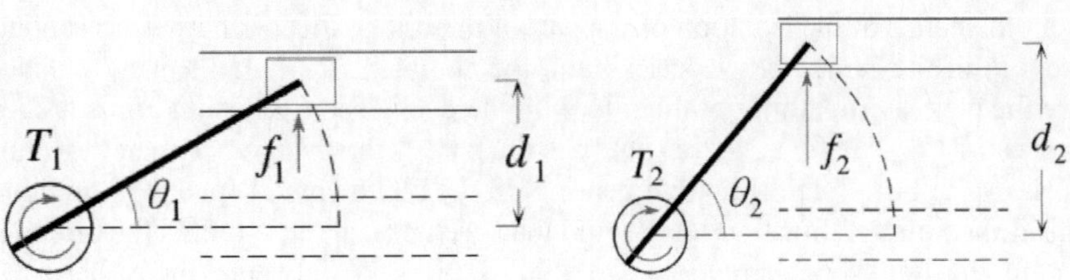

(a)

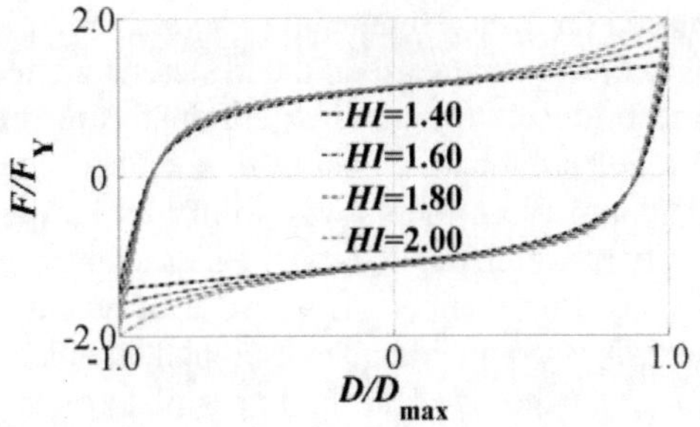

(b)

Figure 5. (a) Working mechanism of MRSD responsible for geometric hardening; (b) MRSD response for different design hardening indices.

(a)

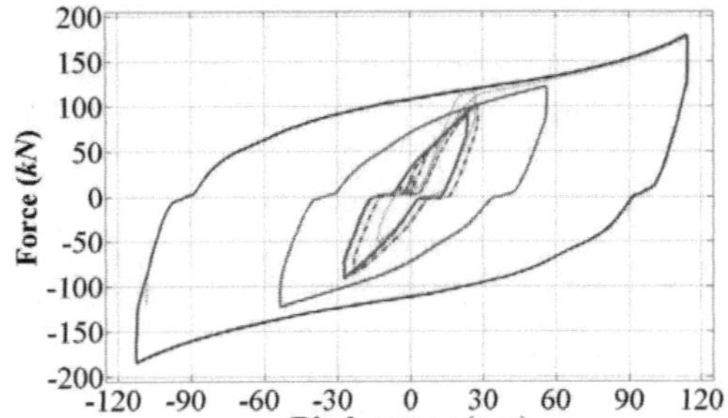

(b)

Figure 6. (a) 200kN, 120mm-capacity prototype MRSD; (b) Cyclic response obtained from tests.

5. The isolation scheme of the bridge

Fig. 7 shows a schematics of the isolation of the part of the bridge located between the abutment A1 and pier P17. The part of the bridge between P17 and the abutment A2 is built using balanced cantilever method where the piers are built monolithically with the deck. The expansion joints are located on two abutments and pier P17. The focus in this study is the part of the bridge between the abutment A1 and pier P17. As shown in Fig. 7, the spherical bearing on the abutment A1, pier P17 and three additional piers on each side are unidirectional to resist wind loads. That is, at these points the bridge is fixed to the abutment/pier in the transverse direction. On the 10 piers in between, the spherical bearings are multidirectional and the bridge is free to move both laterally and longitudinally during an earthquake. Under wind loads, the dampers provide the required resistance in the transverse direction within their elastic limit. The MRSDs are placed on these 10 piers, two on each pier, as shown in the cross-section view in Fig. 8(a). An issue to be tackled with the use of the hysteretic dampers is the presence of thermal movements in certain piers. The bridge is designed to eliminate the thermal movements at its two abutments where the expansion joints are located. That is, thermal action expands or contracts the deck from the middle. Therefore, the dampers on piers away from the middle pier(s) will be subjected to thermal displacements, the intensity of which depends on the pier's distance from the middle point of the deck. To prevent the low-cycle fatigue in dampers as a results of repeated thermal displacements, the attachment of the dampers to the deck is designed to be via elongated holes (slots), as shown in Fig. 8(b). This way, a gap is left between the anchorage and the upper plate of the MRSD, in the longitudinal direction of the bridge. The gap is sized to accommodate the maximum probable thermal displacement per each pier. The amount of gap provided for MRSDs on each pier is indicated in Fig. 7. An alternative solution would be to use shock transmission units (luck-up devices) to connect the dampers to the deck. However, this solution entails increased cost and reduced reliability since shock transmission units are both expensive and require maintenance. The design of the MRSDs with gaps, as described above, was also meant to serve a second objective. Presence of gaps in connections of certain MRSDs means that the engagement of these dampers during an earthquake depends on whether the intensity of the displacements are large enough for the gaps to close. That is, at very low-intensity events, only dampers on middle piers (P8 and P9) are engaged; thus, preventing both the unnecessary increase in base shear on the other piers and also damage to the MRSDs on those piers. With increasing intensity of the ground motion more number of MRSDs on piers come into action. This sequential engagement of the dampers is the performance-based oriented feature in this design. In addition the dampers, which are connected without slots on Piers 8 and 9 provide the required resistance within their elastic range against breaking forces.

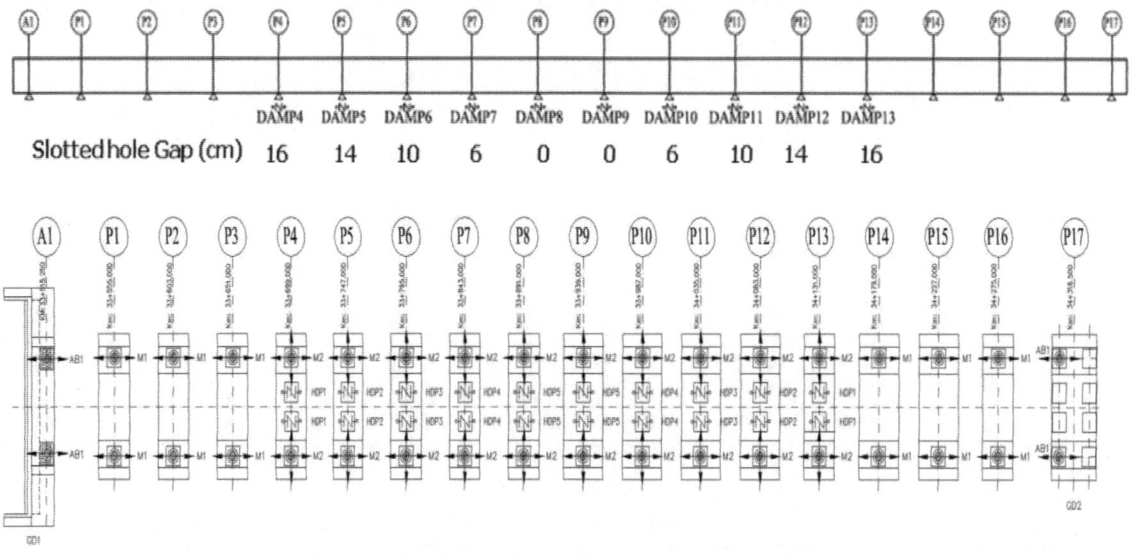

Figure 7. Schematics of the isolation system.

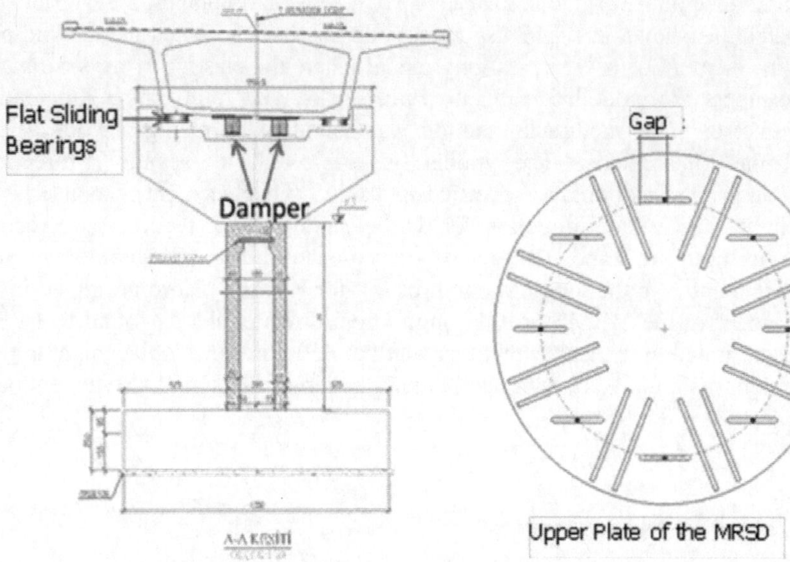

(a) (b)

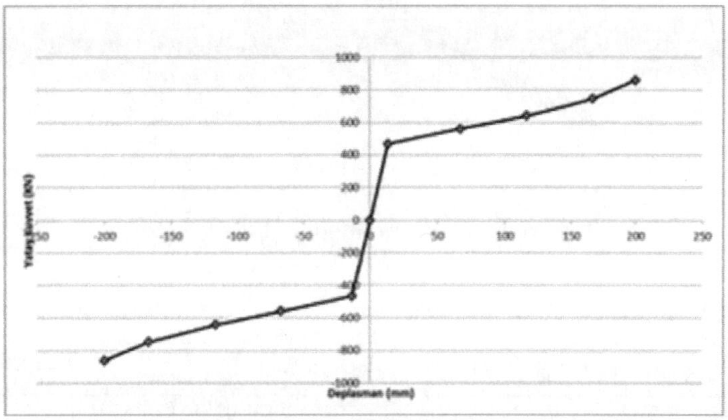

(c)

Figure 8. (a) Installation of two MRSDs on a typical pier; (b) The provided gaps (longitudinal direction) on the upper plate of the MRSDs where the device is mounted to the deck; (c) Force-displacement response of MRSD.

6. Seismic performance criteria for the mrsd-equipped viaduct

Based on the arrangement of the MRSDs, as laid out in previous section, the seismic performance goals of no damage at DBE and repairable damage in MCE are redefined as follows: at DBE (475 years return period), limited number of dampers (those with zero and 6 cm slot gaps) will be engaged during the earthquake in the longitudinal direction. No damage will be inflicted in the substructure members. If needed, energy dissipaters of few dampers could be replaced after the earthquake. At MCE (2475 years return period), dampers with larger slot gaps will also be engaged sequentially as the intensity of the ground shaking increases. The central piers may yield after the damper reach a certain force level. In the preliminary design stage, equivalent linear analysis method was used to determine the required surface coefficient of the spherical bearings and force/displacement capacity of the MRSDs. Following the initial design, a series of time-history response analyses are performed to assess whether the proposed design meets the performance objectives. The results are presented in the following section.

7. Time-history response analyses

To assess the performance of the bridge in DBE and MCE-level earthquakes, a 3D model of the bridge was built in SAP2000, as shown in Fig. 9. Spherical bearings are modelled using rigid plastic model (using nonlinear link element with Wen plasticity model where the elastic stiffness is taken very high) and the MRSD dampers are modelled using nonlinear links with multi-linear kinematic hardening behaviour. The effect of soil-structure interaction was found to be negligible due to the stiff soil condition under the foundations. The model is subjected to seven bi-directional design spectrum-compatible ground motions, as specified in Table I. Displacement response histories of the MRSD dampers in the longitudinal direction for Kocaeli and Landers records (two records with the largest magnitude) for both MCE and DBE events are presented in Fig. 10. Sample hysteresis loops of damper on piers No. 6 and 9 in the longitudinal direction for Kocaeli record are given in Fig. 11. The maximum displacement of the MRSDs in the longitudinal directional was found to be 330mm. The dampers were thus designed for a displacement capacity of 350mm. The displacement in the transverse direction of the bridge was found to be small and did not govern the design displacement of the damper.

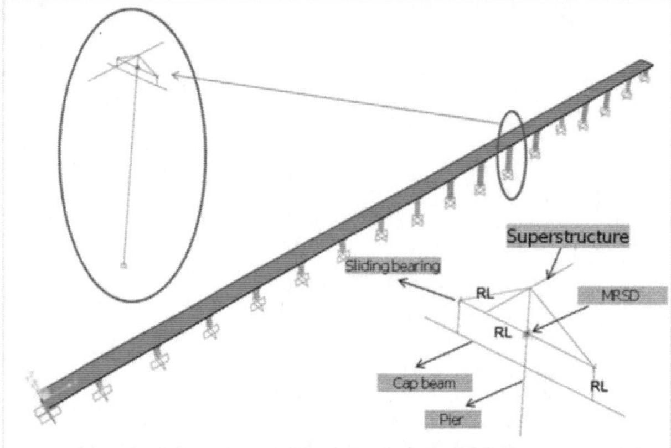

Figure 9. Structural model of the bridge.

Table 1. Specification of the design spectrum-compatible ground motions.

No.	Event	M_w	Fault Type	Station	Record	Scale Factor %10/50yr	%5/50yr	%2/50yr
1	Imperial Valley, 1979	6.5	Strike Slip	6604 Cerro Prieto	IMPVALL/H-CPE147 (001) IMPVALL/H-CPE237 (002)	1.00	1.25	1.40
2	Landers, 1992	7.3	Strike Slip	21081 Amboy	LANDERS/ABY000 (003) LANDERS/ABY090 (004)	0.75	0.85	0.90
3	Kocaeli, 1999	7.4	Strike Slip	Gebze	KOCAELI/GBZ000 (005) KOCAELI/GBZ270 (006)	0.75	0.90	1.00
4	Duzce, 1999	7.1	Strike Slip	531 Lamont	DUZCE/531-N (007) DUZCE/531-E (008)	1.40	1.70	1.90
5	Nahanni, 19856	6.8	Reverse Oblique	6099 Site 3	NAHANNI/S2330 (009) NAHANNI/S2240 (010)	0.75	0.95	1.10
6	Spitak, 1988	6.8	Reverse Oblique	12 Gukasian	SPITAK/GUK090 (011) SPITAK/GUK000 (012)	0.90	1.20	1.40
7	Loma Prieta, 1989	6.9	Reverse Oblique	58378 APEEL 7	LOMAP/A07000 (013) LOMAP/A07090 (014)	1.00	1.15	1.25

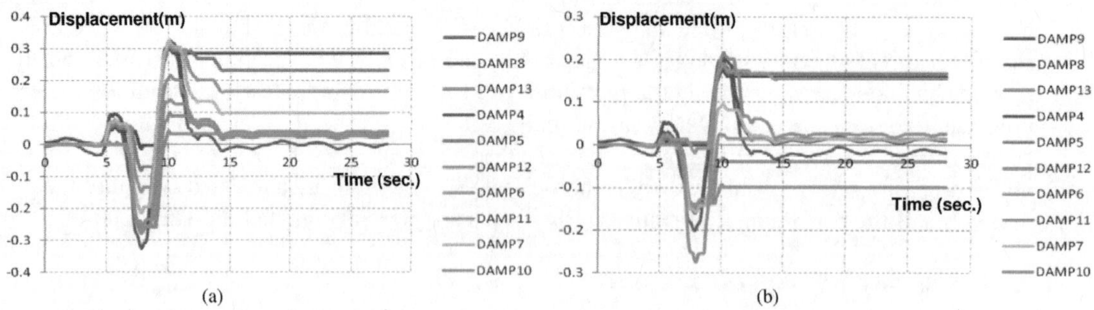

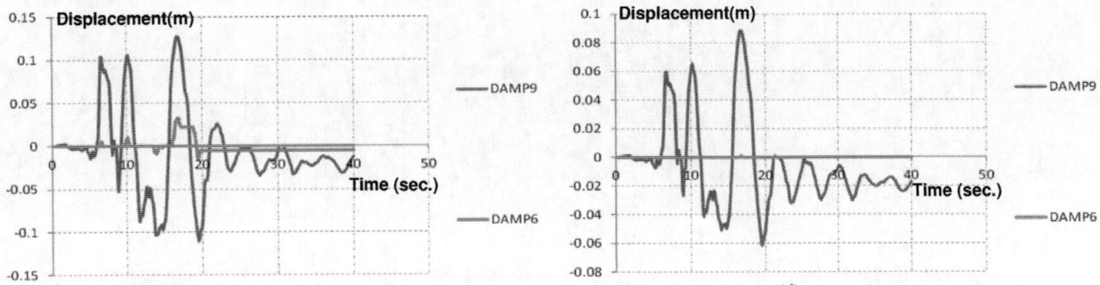

(c) (d)

Figure 10. Displacement response time history of the dampers in the longitudinal direction for (a) Kocaeli MCE; (b) Kocaeli DBE; (c) Landers MCE and (d) Landers DBE, earthquake records.

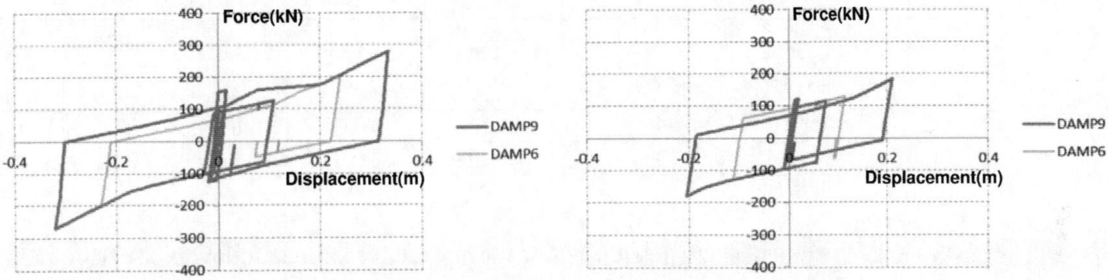

Figure 11. Sample hysteresis loops of damper on piers No. 6 and 9 in the longitudinal direction for (a) Kocaeli, MCE and (b) Kocaeli, DBE records.

To evaluate the performance of the bridge, the state of damage in the piers should be assessed. The damage Model of Hindi and Sexsmith [5] is used for this purpose. The damage model takes as a reference the monotonic energy dissipation capacity of a structure in the undamaged virgin state, which is defined as the area, A_o, under the static pushover curve up to the point of failure (Fig. 12(a)). With the actual "n" cycles of load-displacement history applied on the structure due to a potential earthquake, the remaining monotonic energy dissipation capacity of the structure, compared to that in its virgin state, defines the extent of damage. The remaining monotonic energy dissipation capacity of the structure is defined as the area, A_n, under the static pushover curve obtained from the end of the last cycle, n, to the failure point (Fig. 12(b)). Accordingly, the damage index is the ratio:

$$DI_n = \frac{A_o - A_n}{A_o} \quad (3)$$

A damage index of 0.0 (A_n=A_o) is indicative of no damage, whereas a damage index of 1.0 (A_n=0) is indicative of complete damage or collapse. The damage index is correlated with the physical state of damage, according to the following scale:

DI<0.2: Minor damage–light cracking–very easy to repair.

0.2≤DI<0.4: Moderate damage–severe cracking, cover spalling–repairable.

0.4≤DI<0.6: Severe damage-extensive cracking, reinforcement exposed–repairable with difficulties.

0.6≤DI<1.0: Severe damage–concrete crushing, reinforcement buckling–irreparable.

DI=1.0: Complete collapse.

The calculated damage indices are given in Table II. For DBE-level earthquakes, the DI values are all below 0.2, indicating that the objective of no damage at DBE is met. Likewise, in case of MCE-level earthquakes, all DI values fall below 0.4 indicating that the objective of repairable damage at MCE is met.

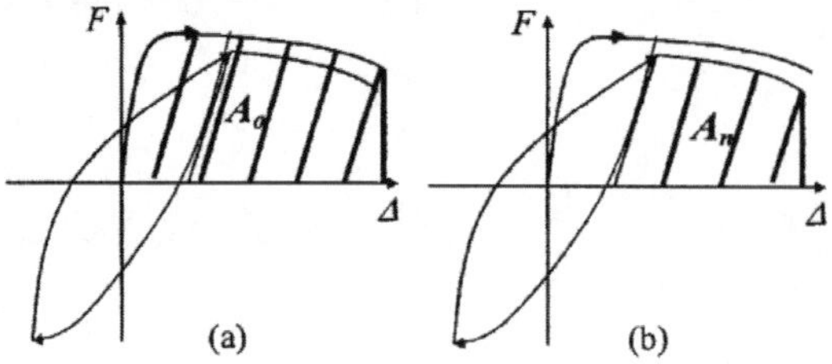

Figure 12. Definition of damage equation parameters of the model by Hindi and Seximith [5]: (a) monotonic energy in the virgin state; (b) monotonic energy after the application of load-displacement cycles.

Table 2. Calculated damage indices.

Earthquake	Imperial Valley, 1979	Lander s, 1992	Kocaeli , 1999	Duzce, 1999	Nahanni, 1985	Spitak, 1988	Loma Prieta, 1989	Averag e
DBE	0.06	0.09	0.12	0.10	0.03	0.05	0.08	0.076
MCE	0.11	0.19	0.27	0.21	0.07	0.13	0.22	0.171

8. Conclusions

The paper presents a practical application of seismic isolation technique following a performance-based design approach for a bridge. The bridge is designed with a seismic isolation system composed of spherical bearings and Multidirectional Re-centering Steel Damper (MRSD). The MRSD is a recently-developed hysteretic damper with a controllable post-elastic stiffness. To keep the dampers from being activated during the thermal displacements, the attachment of the dampers to the deck is made through elongated holes oriented in the longitudinal direction of the bridge. The size of these gaps depend on the amount of expected maximum thermal displacement in each pier and is thus different for different piers. This means that the number of the dampers to be engaged during an earthquake will depend on the intensity of the displacements. The slotted connections of MRSD ensures a progressive energy dissipation that is a function of the intensity of the earthquake in the longitudinal direction where the piers are weaker. The progressive design solution ensures minimal or no damage of substructure at small intensity, more frequent earthquakes while damage progressively increases in response to less frequent, larger earthquakes. The progressive / adaptive solution used in the design balanced the damage and risk producing an economical design solution.

9. References

[1] M. Dicleli, "Performance of seismic-isolated bridges with and without elastic-gap devices in near-fault zones," *Earthquake Engineering and Structural Dynamics*, vol. 37, Issue 6, pp. 935-954, May 2008.

[2] K. Ö. Çetin, "Probabilistic seismic hazard analysis of Bitlis River Viaduct", Technical Report, Middle East Technical University, METU Revolving funds Project No. 2010.03.03.01.03.24, Ankara, Turkey, 2010 (in Turkish).

[3] Salem Milani, A., and Dicleli, M. (2016) Systematic development of a new hysteretic damper based on torsional yielding: part I—design and development. *Earthquake Engineering and Structural Dynamics*, vol. 45, issue 6, pp. 845–867, May 2016. doi: 10.1002/eqe.2684.

[4] Salem Milani, A., and Dicleli, M. (2016) Systematic development of a new hysteretic damper based on torsional yielding: part II—experimental phase. *Earthquake Engineering and Structural Dynamics*, vol. 45, issue 5, pp. 779–796, April 2016. doi: 10.1002/eqe.2685.

[5] R. A. Hindi and R. G. Sexsmith, "A proposed damage for RC bridge columns under cyclic loading", *Earthquake Spectra*, vol. 17, no. 2, pp. 261–289, May 2001.

IOP Conf. Series: Materials Science and Engineering **216** (2017) 012054 doi:10.1088/1757-899X/216/1/012054

A PFC2D model of the interactions between the tire and the aggregate filled arrester bed on escape ramp

Pin-pin Qin[1], Chui-ce Chen[1], Shi-kang Pei[1] and Xin Li[1]
[1]College of Mechanical Engineering, Guangxi University, Nanning 530004, China

Email: qpinpin@gxu.edu.cn;1649989152@qq.com;pskmail@qq.com;1045375763@qq.com;

Abstract. The stopping distance of a runaway vehicle is determined by the entry speed, the design of aggregate-filled arrester bed and the longitudinal grade of escape ramp. Although numerous previous studies have been carried out on the influence of speed and grade on stopping distance, taking into account aggregate properties is rare. Firstly, this paper analyzes the interactions between the tire and the aggregate. The tire and the aggregate are abstracted into a big particle unit and a particle combination unit consisting of lots of aggregates, respectively. Secondly this paper proposes an assumption that this interaction is a kind of particle flow. Later, this paper uses some particle properties to describe the tire-particle unit and aggregate-particle unit respectively, then puts forward several simplified steps of modeling by particle flow code in 2 dimensions (PFC2D). Therefore, a PFC2D micro-simulation model of the interactions between the tire and the aggregate is proposed. The parameters of particle properties are then calibrated by three groups of numerical tests. The calibrated model is verified by eight full-scale arrester bed testing data to demonstrate its feasibility and accuracy. This model provides escape ramp designers a feasible simulation method not only for predicting the stopping distance but also considering the aggregate properties.

1. Introduction

Decades of traffic engineering practice all over the world shows that the escape ramp is the most effective engineering measure to improve highway safety on continuous downhill section [1]. Since the first escape ramp was constructed in California during the 1950s, the United States, Australia, South Africa and Japan have carried out extensive research in order to guide escape ramp design [2]. Unfortunately, China's research lags behind. The first escape ramp and related infrastructure were set up on the Beijing Badaling highway section in only 1998 [3]. Until now perfect design standard and specification of the escape ramp are not yet developed in China.

The length of arrester bed is the key part of escape ramp design, which is determined by the stopping distance [4]. Factors effecting stopping distance are the entry speed of runaway trucks, the aggregate filled arrester bed and the longitudinal grade of escape ramp [5]. Although Al-Qadi (1991) developed an arrester bed stopping model using gravel properties, the complexity of the model application in engineering practice becomes its weakness. Full-scale arrester bed testing is the most effective method to predict the stopping distance, but huge cost and safety limits its application. The traditional kinematic method has limitations such as the description of the interactions between the tire and the aggregate, but discrete element methods such as particle flow code in 2 dimensions (PFC2D) does not have such limitations. The PFC2D software can simulate the motion and interaction of circular granular media by the discrete element method. So the PFC2D has become an effective method to simulate solid mechanics and particles problems and often used as an alternative to very costly experiments [6].

2. The PFC2D micro-simulation model

2.1. The assumption of model

There are seven forces resisting the motion of the tire on the arrester bed [7]: (a) direct momentum transfer from the tire to the aggregate particles; (b) the rolling resistance; (c) air drag; (d) grade resistance; (e) compaction resistance; (f) bulldozing resistance; (g) side-shearing resistance. The arrester bed usually is filled with loose aggregates (such as gravel, broken stone, etc), rolling resistance generates by sinking into the aggregates. The more the tire sinkage, the more the rolling resistance and the less the stopping distance. Meanwhile, the tire has a compaction effect on the aggregate particles, consuming part of the kinetic energy of the tire. Because of compaction resistance, an aggregate height is created before and after the tire. The tire needs to consume a certain amount of kinetic energy to push aside the aggregates in front, so there are bulldozing resistance and side-shearing resistance. Because of the gravity, the tire on a certain grade produces grade resistance.

The runaway vehicle is assumed to be rigid and brakeless. Its weight is assumed to be equally distributed on each tire. The tires are assumed to be rigid free-rolling wheels. The aggregates are assumed to be rigid, spherical, homogeneous, and uniform. The tire and the aggregate are abstracted into a big particle unit and particle combination unit consisting of large amount of aggregates, respectively. This research proposes an assumption that the tire-aggregate interaction is a kind of particle flow.

2.2. The steps of modeling

Based on previous assumption, the steps of modeling using PFC2D are proposed. The steps are the following:

Step 1: Establish the wall. Generate a polygon with certain length and wide according to the actual arrester bed scale.

Step 2: Randomly generate a certain radius of particles. Let the particles fill the whole wall, and then compaction particles to achieve stability. Here is the stability condition: the ratio of the average contact force and the average unbalanced force is less than 1%.

Step 3: Initialize settings. It is mainly the setting of gravity acceleration and friction coefficient.

Step 4: Load experiments. Remove the wall and represent the tire with a big particle unit. Impose a downward force on the big particle, which is equivalent to the actual load of the tire.

Step 5: Run the program. Give the tire a running speed until it stops. The running speed is equal to entry speed of runaway trucks. Calculate the distance from the running to the stopping, which is called stopping distance.

The tire-particle unit properties were represented by particle radius, particle density, particle normal stiffness, particle tangential stiffness. The aggregate-particle unit properties were represented by particle radius, particle density, particle normal stiffness, particle tangential stiffness and friction coefficient between particles.

3. The calibration of model

There are two steps. Step1: The numerical simulation of calibration. Referencing to the static triaxial test results [8, 9], three groups of numerical simulation were carried out by PFC2D embedded fish language program. (a) The relative density of 25%, confining pressure of 69,138kpa; (b) The relative density of 50%, confining pressure of 69,207kpa; (c) The relative density of 75%, confining pressure of 138,207kpa. The results are shown in Table 1. Step2: The parameters of the tire-particle unit and the aggregate-particle unit are determined (see Table 2 to Table 5). Referencing to the data from the literature [10, 11], which are results of the full-scale arrester bed testing. The aggregate filled arrester bed on G139 highway section is river gravel with 4.75mm-53mm radius. The aggregate filled arrester bed on Zhanglong highway section is broken stone with 20mm-50mm radius.

4. The validation of model

4.1. Data

The stopping distance is chosen as an indicator of the validation. The validation of model is to compare indicator values to full-scale arrester bed testing data and to see whether simulation accuracy of model is acceptable. Only the verified model can be applied to various simulations. The results of predicting and full-scale measured stopping distance are shown in table 6. Test numbers 1 to 4 represent the data from G139 highway section [10]. Test numbers 5 to 8 represent the data from Zhanglong highway section [11].

4.2. Results and analysis

In table 6 the average error between predicting and full-scale measured stopping distance is 13.16% for gravel aggregate and 7.14% for broken stone aggregate respectively, which shows that the latter's predicting accuracy is better than the former. When the aggregate is gravel or broken stone and the vehicle weight is the same, the larger the entry speed, the longer the stopping distance, and the average error is decreased. The minimum difference between predicting and full-scale measured stopping distance is 0.4m among the 8 data. All the predicted stopping distances are less than the full-scale measured, which shows that the application of simulation model tends to be risky. This is probably because the state of aggregates-filled on arrester bed is ideal for simulation. Usually the ideal aggregates should be spherical, rigid, homogeneous, and uniform [8, 9].

Table 1. Calibration simulation results compared to static triaxial test results.

Relative Density/%	Confining Pressure/kpa	Peak press/kpa		Peak strain		Volume strain	
		Test	Simulation	Test	Simulation	Test	Simulation
25	69	284	220	0.073	0.03	+0.0032	-0.025
	138	459	410	0.077	0.05	-0.0046	-0.004
50	69	332	310	0.037	0.03	-0.0012	-0.02
	207	558	630	0.066	0.07	+0.002	-0.05
75	138	518	550	0.033	0.06	-0.0037	-0.045
	207	692	760	0.057	0.08	+0	——

Table 2. The parameters of tire-particle unit on G139 site.

radius（m）	normal stiffness（N/m）	shear stiffness （N/m）	friction parameters
0.6	2.5e5	2.5e5	0.25

Table 3. The parameters of gravel- particle on G139 site.

radius （mm）	density （kg/m^3）	number	Normal stiffness（N/m）	Shear stiffness （N/m）	friction parameters
4.75-53	2630	42000	4.8e6	2.4e7	30

Table 4. The parameters of tire-particle unit on Zhanglong site.

radius（m）	normal stiffness（N/m）	shear stiffness （N/m）	friction parameters
0.5	2.5e5	2.5e5	0.10

Table 5. The parameters of gravel- particle on Zhanglong site.

radius （mm）	density （kg/m^3）	number	Normal stiffness（N/m）	Shear stiffness （N/m）	friction parameters
20-50	2600	71500	1e8	1e8	22

Table 6. Predicting the stopping distance compared to full-scale measured values.

Num.	Aggregate type	vehicle weight (t)	Entry speed (km/h)	Stopping distance (m) Measured	Predicting	D-value	relative error(%)
1	gravel	12.4	32	11.7	10.3	1.4	11.97%
2		12.4	40	18.0	16.21	1.8	9.94%
3		22.46	33	14.6	11.2	3.4	23.29%
4		22.46	45	18.8	17.4	1.4	7.45%
average						2	13.16%
5	broken stone	18	44	29.3	24.13	5.2	17.65%
6		18	51	35.4	32.17	3.2	9.12%
7		18	61	45.3	44.8	0.5	1.10%
8		18	70	59.0	58.59	0.4	0.69%
average						2.3	7.14%

5. Conclusions

This study aims to propose a simulation model of the interactions between the tire and the aggregate and provides a feasible method to escape ramp designers for predicting the length of arrester bed through the stopping distance. A micro-simulation model of the interactions between the tire and the aggregate were done using PFC2D software. The calibration of model was done using numerical simulation data. The verification of model was done using full-scale arrester bed testing data. The study shows the accuracy of predicting stopping distance is acceptable, which is important in two respects. First, it confirmed that the assumption of the interactions between the tire and the aggregate is a kind of particle flow is right through PFC2D software. Second, it provided confidence that some dangerous and infeasible full-scale experiments would probably be feasible through PFC2D software at a low economic cost.

6. Acknowledgements

The authors gratefully acknowledge the support of the Guangxi Natural Science Found, Project 2015GXNSFAA139280, in supporting portions of this study.

7. References

[1] A Policy on Geometric Design of Highway and Streets (Green Books) 2001 *American Association of State Highway Transportation Officials* (AASHTO)

[2] Eck R W 1980 *Development of warrants for the use and location of truck escape ramps* Volume I - final report

[3] Research Institute of Highway of Ministry of Transport 2007 *The Study on the Safety Enhancement of the Continue, Long and Steep Downgrade* (Beijing: Research Institute of Highway, Ministry of Transport) p 49-51

[4] Truetlood M S 1984 *Design and Construction of a Truck Arrester Bed Research Facility* (The Pennsylvania State University)

[5] Wambold J C, Rivera-Ortiz L A, Wang M C and Wambold J C 1988 *A field and laboratory study to establish truck escape ramp design methodology* Volume II, Final report.

[6] 2014 *Itasca Consulting Group PFC Theory and Background* (Minneapolis)

[7] Bekker M G 1969 *Introduction to Terrain-Vehicle Systems* (Michigan: The University of Michigan Press)

[8] Al-Qadi I L and Rivera-Ortiz L A 1991 Laboratory Testing of River Gravel Used in Arrester Beds *J. Journal of Testing and Materials*, **19** (4) :280-91.

IOP Conf. Series: Materials Science and Engineering **216** (2017) 012054 doi:10.1088/1757-899X/216/1/012054

[9] Al-Qadi I L and Rivera-Ortiz L A 1991 Use of Gravel Properties to Develop Arrester Bed Stopping Model *J. Journal of Transportation Engineering*, **117**(5):566-84.
[10] Lai Shigui. The Experimental Study on Stopping Distance of Truck Escape Ramp in Longyan Mountainous Area *J. Traffic Engineering*, 2016.01.80-8 (in Chinese).
[11] Chen Bo 2007 Study on the Truck Escape Ramp Design Method of Mountainous Continue, Long and Steep Downgrade(Cheng Du：*Southwest Jiao Tong University*)

IOP Conf. Series: Materials Science and Engineering **216** (2017) 012055 doi:10.1088/1757-899X/216/1/012055

Basic Study on Term of Warranty Liability for Water Supply, Drainage, and Sanitation Arrangement Work Defect in Apartment Building

Junmo Park[1, a] and DeokSeok Seo[2,b *]
[1]Kyungsan Engineering, Seocho-gu, Seoul, 06646, Korea
[2]School of Architecture, Halla University, Wonju-si, Gangwon, 26404, Korea

E-mail: aadviser.cm@gmail.com; bseodk@halla.ac.kr;

Abstract. The defect lawsuit of the apartment which is the representative residential style of Korea continues and becomes a social problem. In the defect lawsuit, the term of warranty liability is a period that can demand the defect repair according to defect occurrence, and the exclusion period of the exercise of rights. However, the term of warranty liability stipulated in relevant laws such as Enforcement Decree of the Housing Act is being changed arbitrarily, without any established grounds. Therefore, a reasonable standard for establishing the term of warranty liability is required. In this study, the defects of water supply, drainage and sanitation arrangement work were studied. As a result of analyzing the number of defect occurrence in the apartment, it was shown that the defects in water supply, drainage and sanitation arrangement work occurred more than 80% in the 1st ~ 2nd year after completion. However, the occurrence of defects from the 3rd year was extremely slight. On the other hand, it was confirmed that the defect occurrence continued until fairly late point of time as the end point of time of the defects was in the 7th to 9th years.

1. Introduction

In Korea, the dominant forms of housing are chiefly apartment houses, especially apartments (more than five stories). These apartments are very popular in Korea because they are more advantageous than existing houses in terms of convenience, maintenance cost, safety, etc [1]. However, Korean apartments are expensive among housing, and the disputes surrounding them are accelerating due to the upgraded image by the subject of the business and the improvement of the divided owner's eye level [2]. This dispute is expanding into lawsuit, which is pointed out as a big social problem, and this is called a defect lawsuit.

The term of warranty liability for apartment stated in the Housing Act [3] is exclusion period of the exercise of rights as the period that can demand the defect repair according to defect occurrence. At the same time, it is a standard for preventive and smooth maintenance of apartment. In the case of a defect lawsuit, it is very important because it judges whether or not to request defect repair according to the term of warranty liability [4]. However, there is no clear ground for enacting the term of warranty liability. Although it was arbitrarily stipulated in the first Regulations on the Management of Apartment Buildings [5], the clauses were added, deleted, and the term was changed after the law revisions were carried out several times. Recently, with the revision of Apartment House Management Law [6], the term of warranty liability was changed again, but this was still done arbitrarily without any clear ground.

1.1. Purpose

In order to smoothly resolve defect lawsuit surrounding the apartment, objective and reasonable standard for establishing the term of warranty liability is required. To do this, it is important to provide an objective basis and interpret it logically. This study is the basis for reasonable establishment of the term of warranty liability.

1.2. Scope and Method

This study investigated the defects of water supply, drainage and sanitation arrangement work among various defects occurring in apartment housing. The water supply, drainage and sanitation arrangement work are the second most defects in the number of defect occurrence, and the most construction type of defect occurrence among the facility work of apartment housing.

In this study, we investigated the number of defects occurring in apartment housing in Korea. In addition, we examined the occurrence trends of defects in 18 construction types under the Housing Act [3] by the term of warranty liability.

2. Literature Review

2.1 Definition on Defect and Term of Warranty Liability

The defects of water supply, drainage and sanitation arrangement work in apartment buildings are caused by pipe leaks, poor drainage within households, defective pipes, poor fixation and dropouts of sanitary fixtures, etc. Water supply, drainage and sanitation arrangement work are mostly water leakage or clogging in piping. Therefore, it is an important defect because it directly affects the life of the apartment residents. Briefly in terms of the safety, function, and aesthetics, it is mainly related to functional matters. On the other hand, there is little direct influence in terms of safety or aesthetics, but if defects are neglected, it is likely to expand as safety and aesthetic problem.

2.2 History of Law Revision and Problem

The types and term of warranty liability of water supply, drainage and sanitation arrangement work have been changed several times according to the enactment and revision of relevant laws and regulations. Typical examples are as follows.

First, it was the Regulations on the Management of Apartment Buildings [7] in 1979 that first specified the defect repair period for major parts by construction type in relation to defect repair.

Second, it was the revision of the Management of Apartment Buildings in 1981 [5] that the term of warranty liability was first specified. However, there are no ground materials and purpose explanations for establishing the term of warranty liability. At the time when the relevant regulations were enacted, it seemed that it was not a topic because it was the beginning of the construction of apartment in Korea and there was no defect dispute. However, since the 2000s after more than 20 years later, the term of warranty liability has played a role as important determining basis of defect dispute. Through this, it can be confirmed that how necessary sufficient consultation on legal provisions is, and how important judgment considering the future impacts is. Meanwhile, at that time, there were five detailed construction types of water supply, drainage and sanitation arrangement work. For the term of warranty liability of each detailed construction type, water supply facilities, hot water supply facilities and drainage ventilation facilities were 2 years, and sanitation arrangement facilities and heat reserving work were 1 year.

Third, the provisions related to the period of liability for collateralized liability did not change much in 1981, even after several revisions of relevant laws and regulations. However, as the law was amended in 1998, the relevant contents were all transferred to the cooperative housing control ordinance [8]. However, there was no change in contents. In 2003, the apartment house control order was abolished and related contents were relocated to the Enforcement Decree of the Housing Act [9].

Fourth, with the revision of the Enforcement Decree of the Housing Act in 2007, the existing term of warranty liability extended from one to three years, to one to four years [10]. At the same time, special arrangement work was added to the water supply, drainage and sanitation arrangement work and the term of warranty liability was adjusted to two years in a lump. Especially, the revision in 2007 added a lot of major construction types of apartment, and the term of warranty liability was most extended. The reason for the revision of the law said that the extension of the term of warranty liability was caused by protection of tenants, development of construction-related technology, etc. However, in the case of revision of the law in Korea, no additional evidence was confirmed even in the official gazette which has announced the reason for the revision or the purpose of the revision, etc [11].

3. Review on Occurrence of Defect

3.1 Outline

As previously discussed, it was confirmed that there was no ground for establishing the term of warranty liability in the laws and regulations related to apartment. We intended to investigate the actual status of defects in apartment, and tried to identify to what extent defects occur, and at which time defects are concentrated, etc.

For this study, we investigated the number of defects that occurred in 177 apartment complexes constructed by 12 construction companies. The total size of each apartment complex was 125,816 households, and construction was completed from 2002 to 2011. It was difficult to identify other information such as the number of houses, the number of floors, and the area of households, etc. Therefore, it is noted that the analysis of these matters was excluded in this study.

3.2 Occurrence of Total Defect

A total of 1,449,745 defects were found in the apartment where the defects occurred. Based on the number of defect occurrence, the first place was finish work of 42.54%, and the second place was supply, drainage and sanitation arrangement work of 13.72%. The third place was window and door work of 12.66%, and the fourth place was wood work of 8.65%. The fifth place was miscellaneous work of 7.41% and the sixth was electricity and electric power facility work of 7.20%. The other defects occurred in 12 construction types were insignificant compared to those of the above 6 construction types (Figure 1).

3.3 Occurrence of Water Supply, Drainage, and Sanitation Arrangement Work Defect

The details of defects of water supply, drainage and sanitation arrangement work, which are direct subjects of this study, are as follow by detailed construction type.

Examined from the detailed types with the large number of defect occurrence, it appeared in order of drainage and ventilation work, sanitation arrangement work, water supply work, hot water supply work, and special arrangement work. Drainage and ventilation work, and sanitation arrangement work, accounted for 45% and 34%, respectively. On the other hand, special work was only 46 cases, and the defects of steel and heat reserving work did not occur at all (Figure 2).

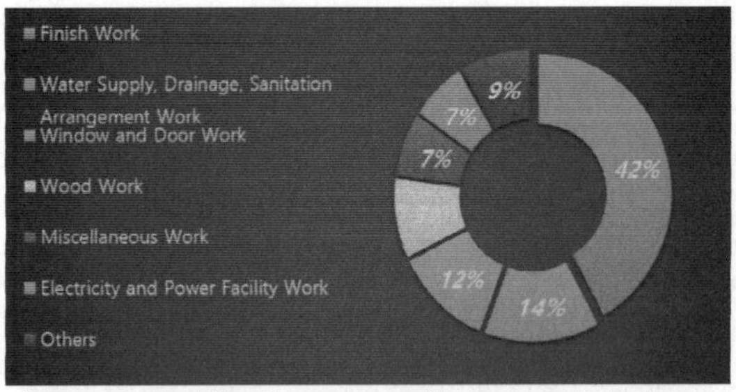

Figure 1. Whole status of occurred defect

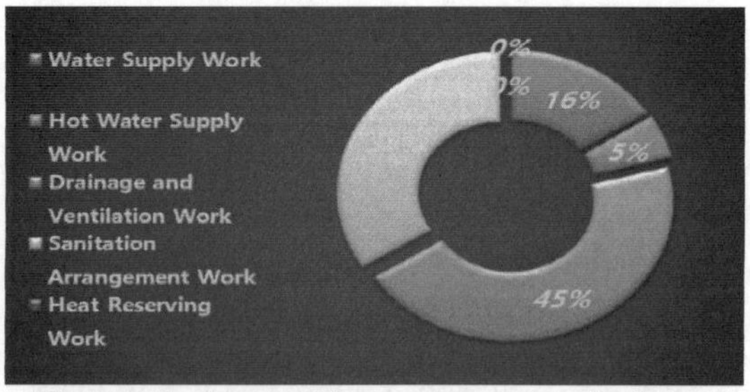

Figure 2. Status of water supply, drainage, and sanitation arrangement work defect

4. Occurrence Trend of Water Supply, Drainage, and Sanitation Arrangement Work Defect

In this chapter, we analyzed the cumulative trends of the number of defect occurrence by each construction type of water supply, drainage and sanitation arrangement work. The heat reserving work and special works with insignificant defects were excluded from this analysis.

4.1 Water Supply Work Defect

The occurrence of water supply work defect was 57% in the 1st year, and occurrence of cumulative defects was 83% in the 2nd year. It reached 95% in the 3rd year, increased 99% in the 5th year, and the occurrence of defect ended with 100% in the 9th year (Figure 3).

4.2 Hot Water Supply Work Defect

The occurrence of hot water supply work defect was 57% in the 1st year and cumulative defects up to 96% in the 2nd year. It was 99% in the 3rd year, and the defect occurrence ended with 100% in the 7th year (Figure 4).

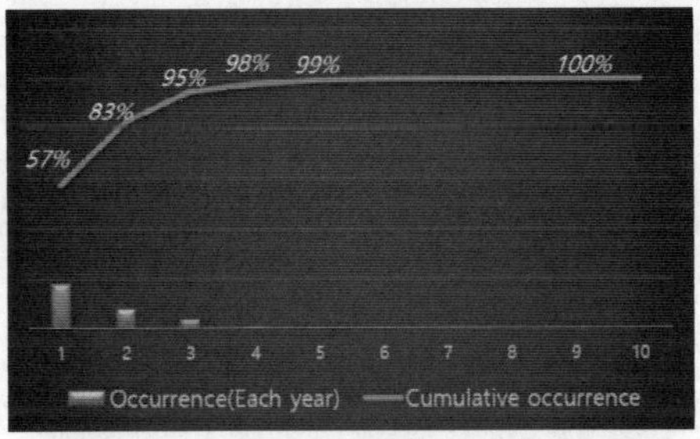

Figure 3. Water supply work defect

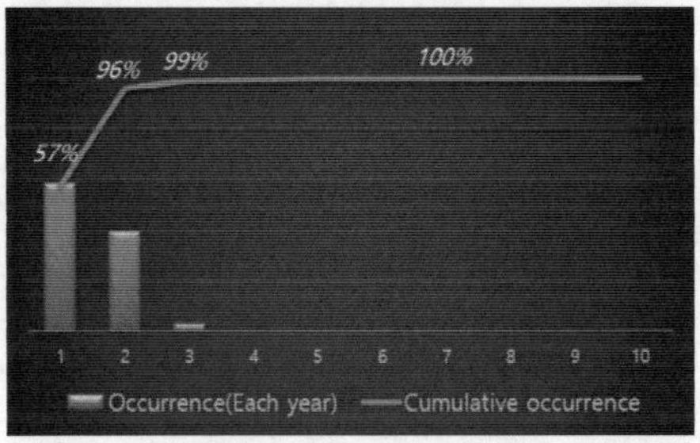

Figure 4. Hot water supply work defect

4.3 Drainage and Ventilation Work Defect

The occurrence of drainage and ventilation work defect was 59% in the 1st year and accumulated defect 86% in the 2nd year. It was 96% in the 3rd year, 99% in the 4th year, and the occurrence of defect ended in the 8th year (Figure 5).

4.4 Sanitation Arrangement Work Defect

The occurrence of sanitation arrangement work defect was 63% in the 1st year. It was 93% until the 2nd year after that, and 99% the 4th year, respectively, and the occurrence of defect ended in the 9th year (Figure 6).

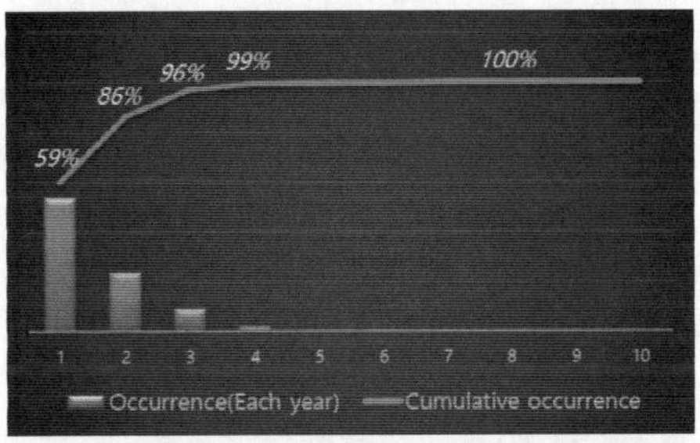

Figure 5. Drainage and ventilation work defect

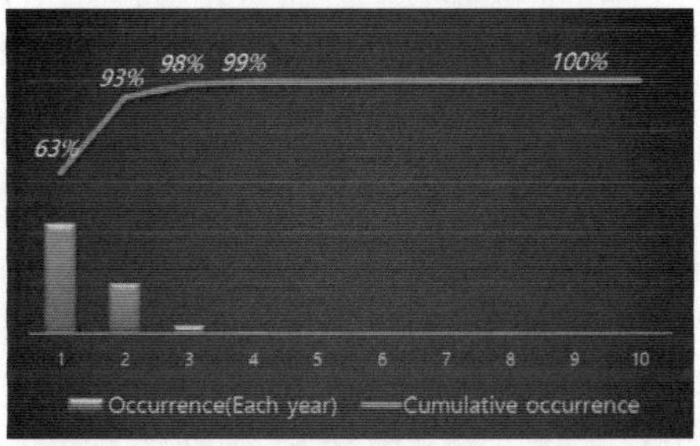

Figure 6. Sanitation arrangement work defect

4.5 Result and Implication

The results of analyzing the trend of the number of defect occurred in the water supply, drainage and sanitation arrangement work of the apartment house are summarized as follows. As a result of

checking the number of defect occurrence in the 1st and 2nd years, more than 80% of defects were concentrated in water supply work and drainage and ventilation work. On the other hand, in the case of hot water supply work and sanitary arrangement work, more than 90% of defects were concentrated. Through this, it can be seen that the occurrence of defect of the hot water supply work and the sanitary arrangement work were concentrated in the early stage.
Examining the overall trends suggests the following points. It can be seen that the defects of water supply, drainage and sanitation arrangement work are concentrated more than 80% in the first and second year. And the defects from the third year were extremely slight. On the other hand, it was confirmed that the point of time when the occurrence of defect ended was 7 ~ 9th year and the occurrence of defect continued until the considerably late time.

5. Conclusion

The lawsuits surrounding the defects of the apartment, which is the representative residential style of Korea, continues and becomes a social problem. In the defect lawsuit, the term of warranty liability is a period that can demand the defect repair according to defect occurrence, and the exclusion period of the exercise of rights. However, this important the term of warranty liability has been arbitrarily stipulated without establishing grounds and continued for about 40 years. Now that many apartments have been built and become the home of the nation, the term of warranty liability should be re-established on a reasonable basis through objective data and scientific analysis.
This study was conducted to investigate the defects of water supply, drainage and sanitation arrangement work among various defects occurring in apartment. As a result of the analysis, it was shown that the defects in water supply, drainage and sanitation arrangement work occurred more than 80% in the 1st ~ 2nd year after completion. However, the occurrence of defects from the 3rd year was extremely slight. On the other hand, it was confirmed that the defect occurrence continued until fairly late point of time as the end point of time of the defects was in the 7th to 9th years.

6. Acknowledgment

This work was supported by the National Research Foundation of Korea (NRF) grant funded by the Korea government (2016R1A2B4012485).

7. References

[1] J. Park and D. Seo, "Basic Study on Influence Factors for Defect Repairing Cost of Apartment Building in Korea," presented at the 2016 International Conference on Applied Mathematics and Mechanics, Bangkok, Thailand, Oct 29-30, 2016
[2] J. Park, D. Seo, J. Choi, O. Kim, K. Park, and J. Jo, (February 2012) Analysis on legal issue of lawsuits and subjective judgment on defects in apartment building. Journal of the Korea Institute of Building Construction. 12(1). Pp.42-53.
[3] Enforcement Decree of the Housing ACT, 2008
[4] J. Park and D. Seo, "Basic Study on Term of Warranty Liability for Finish Work Defect in Apartment Building," presented at the 6th International Conference on Studies in Architecture, Civil, Construction, and Environmental Engineering, Singapore, March 27-28, 2017
[5] Regulations on the Management of Apartment Buildings, 1981
[6] Enforcement Decree of Apartment House Management ACT, 2016
[7] Regulations on the Management of Apartment Buildings, 1979
[8] Decree on the Management of Apartment Houses, 1998
[9] Enforcement Decree of the Housing ACT, 2003
[10] Enforcement Decree of the Housing ACT, 2007
[11] Ministry of the Interior, the Official Gazette in Korea, 2007. 3. 16. No.16463, p.45

IOP Conf. Series: Materials Science and Engineering **216** (2017) 012056 doi:10.1088/1757-899X/216/1/012056

Stimulating a Sustainable Construction through Holistic BIM Adoption: The Root Causes of Recurring Low BIM Adoption in Malaysia

S Mamter[1], AR Abdul-Aziz[2], ME Mamat[3]
[1,3]Faculty of Architecture, Planning and Surveying, UiTM Perak, 32610 Seri Iskandar , Perak, Malaysia
[2]School of Housing, Building and Planning, Universiti Sains Malaysia, 11800 Pulau Pinang, Malaysia

Email: shahela8299@gmail.com; arashid@usm.my;eshamqs@gmail.com

Abstract. Fostering the Building Information Modelling (BIM) implementation is one of Malaysia sustainable strategies towards greener construction. Hence, the Eleventh Malaysia plan focuses on transforming construction industry through the increase of technology adoption in order to enhance construction productivity. Therefore, there is a growing and urgent demand to provide BIM competent. However, a significant number of parties are reluctant to develop and invest in BIM due to unsolved root causes. Scholars have identified barriers relating to the infancy stage of BIM adoption in Malaysia. Unfortunately, there is a lack of study to explore deeper the root causes of recurring for the barriers anticipate the low BIM adoption. This paper attempts to delve into the initiatives of BIM stake players in fostering BIM adoption and to determine the root causes of recurring barriers due to low BIM adoption. The study adopted the semi-structured interviews which involved BIM stake players as a sample population. From the findings, authors revealed four root causes of recurring barriers; absence of BIM policy and BIM compulsion, poor holistic readiness, software integration competition strategy, and reluctant in sharing knowledge. The findings espoused here are preliminary and more results are expected to emerge as the research progresses.

1. Introduction

Construction industry keeps contributing significantly to the Gross Domestic Product (GDP) of Malaysia economy. As reported in Construction Industry Transformation Plan (CITP) the construction industry will be expecting to contribute 5.5 per cent to the Malaysia GDP in 2020 [1]. Consequently, productivity in the construction sector will be focusing on increasing BIM adoption. BIM is a collaborative process and provides the environment with tools to manage and integrate information over the project lifecycle [2]. However, challenges faced by potential BIM adopters may result in poor adoption. Identifying the root causes to poor adoption on BIM will foster the holistic BIM adoption in Malaysia. Researchers [3], [4] defined root causes as the most basic reasons for an effect, which if corrected will prevent recurrence. Some scholars have identified the barriers to BIM adoption in Malaysia, unfortunately, there are still lack of studies to explore deeper the root causes of recurring barriers which accountable to the low BIM adoption

2. Fostering sustainable strategies

In Malaysia, sustainable strategies towards greener construction (CIDB Guidebook Green Practices, 2013) have focused on Involvement of stake players, Building Information Modelling, Green Building

IOP Conf. Series: Materials Science and Engineering **216** (2017) 012056 doi:10.1088/1757-899X/216/1/012056

Design, and Industrialised Building System. In achieving these strategies, the most recent Eleventh Malaysia plan has focused on increasing the construction productivity by increasing the BIM adoption amongst the construction players. Nevertheless, CITP reported that the adoption of Building Information Modelling (BIM) technology in Malaysia is still low, estimated at only 10 percent adoption amongst construction stake players. Since 2007, Malaysia BIM wave started with the publication of internal implementation by Public Work Department (PWD) or also known as Jabatan Kerja Raya (JKR) throughout BIM pilot projects. The wave of Malaysia adoption can be divided into five waves (Figure 1). Figure 1 shows PROKOM that was established in 2012 as a BIM unit project to handle JKR pilot project [5].

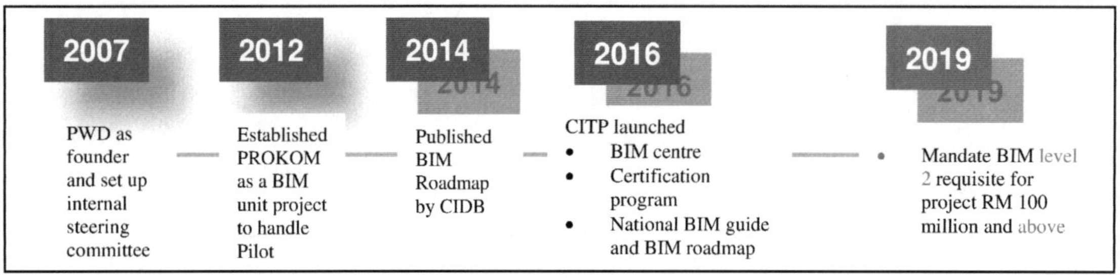

Figure 1. Malaysia BIM evolution. Authors' compilation from references [5]-[8]

As stated by one of the interviewees (R3), the adoptions of BIM in Malaysia actually took place first before the Singapore adoption. During that time JKR used UK BIM standard as a reference for their pilot projects. In 2012, in order to overcome the lack of awareness amongst potential BIM adopters, Construction Industry Development Board (CIDB) had set up National BIM committee to forge BIM guide and roadmap. Accordingly, Table 1 indicates a summary of a number of massive initiatives supported by construction stake players to foster the BIM adoption.

Table 1. BIM Stake Players Initiatives

Stake Players	Involvement Initiatives	
PWD (Public Work Department)	• Publishing JKR BIM Standard and Guideline • BIM pilot projects	• Adoption roadmap for in-house design for consultant/contractor [5]
CIDB (Construction Industry Development Board)	• Coordinating with all BIM principles and players • Provide high quality training [8]	• Set up the National BIM Committees for BIM guide and Roadmap • Establish BIM Centre • Subsidised BIM training programs • BIM transformation fund [8]
ACEM (Association of Consulting Engineers Malaysia)	• Provide awareness workshops and road shows [8]	
IEM (Institution of Engineers Malaysia)	• Provide awareness, buy-in program and internal trainings [8]	
PAM *(*Pertubuhan Akitek Malaysia)	• Provide awareness, buy-in program and internal trainings [8]	
RISM (Royal Institute Surveyor Malaysia)	• Formed 5 BIM sub-committee on Building, Civil, ICT, SMM and academic aspects • 5D BIM workshop	• BIM educational framework • BIM conference [RISM Annual Report 55th]

BIPC (Building Industry Presidents Council)	• Formed independent BIM committee to gather industry needs towards BIM implementation [8]	
MBAM (Master BuildersAssociation Malaysia)	• Provide awareness, buy-in program and internal trainings	
Developers	• Formed independent BIM adoption and implementation by mandatory requirements to their consultants/contractors Building Information Modelling (BIM) [8]	
Small Medium Enterprise	• Participate in seminar and workshop BIM [7]	• Formed strategic partnerships with multi-national companies (MNCs) and government-linked Companies(GLCs) [1]

To overcome the low BIM adoption, CIDB initiative strategies in 2016 were launched with BIM centre, certification program, national guide, roadmap, subsidised training programs and transformation fund. Since BIM was adopted, the entire professional bodies ACEM, IEM, PAM, RISM, BIPC, and MBAM have participated in all BIM initiative programs. Khairul [9] determined that most SME contractors do not have high capability to implement BIM. Thus, CITP involves High-performing small and medium enterprises (SMEs) to form strategic partnerships with multi-national companies (MNCs) and government-linked companies (GLCs) which will further drive the BIM adoption.

3. Methodology

Data collection for this paper is gained through a literature review and expert interviews conducted by authors. A literature review was conducted to explore on the initiatives to foster the BIM adoption in Malaysian construction industry. In addition, the preliminary semi-structured interview was conducted with construction experts with more than 10 years of experience and exceeding 5 years of experience in BIM projects in order to cater the second objective of this study i.e. to determine the root causes of recurring barriers due to low BIM adoption. Various samples of BIM stake players were approached by the authors for interviews through email, telephone calls, and face to face consultations, however at the time of writing this paper, a total of six semi-structured interviews were conducted. Although the sample size is slightly small, the in-depth nature and detailed explanations of the issues provided by experienced interviewees are in line with Romney [10] recommendation of samples size of four to six interviews when interviewees have an expert knowledge in the area of study. From Table 2 below, various position groups can be deciphered. These interviews lasted for an average of sixty (60) minutes each. The data collected from these interviews were transcribed and subsequently analyzed. During these interviews, the interviewees were asked to indicate their organisational position and years of experience in BIM practices.

Table 2. List of Interviewees

Interviewee's Coding Detail			
Position	***Industry Background***	***Philosophy adoption****	***ID***
Product Manager	Supplier/Vendors	Innovators	R1
Director of company	Quantity Surveyor	Innovators	R2
BIM Manager	Architect	Adopters	R3
BIM Engineer	Civil & Structural Engineering	Adopters	R4
Head of Unit	Building	Innovators	R5
BIM Engineer	Mechanical & Electrical Engineering	Adopters	R6

*System Dynamic Modelling Philosophy adoption classification[11]; Potential adopter, Innovator, Imitator, Adopter

IOP Conf. Series: Materials Science and Engineering **216** (2017) 012056 doi:10.1088/1757-899X/216/1/012056

Other questions asked were to include the best categories to describe their organisation's philosophy adoption. The classification of philosophy adoption is essential to analyse adoption rate using system dynamic modelling [12] for subsequent study. System dynamic model provides a holistic methodology hence BIM stake players may gain insights into the causal factors influencing potential BIM adopter in decision making processes, and thereby into the potential diffusion patterns resulting from those adoption decisions.

4. Finding

Due to limited pages, the following root causes of recurring barriers in BIM adoption were revealed and grouped under four themes:

4.1. Absence of BIM policy and compulsion

The absence of BIM policy is one of the root causes highlighted by the interviewees. Contrary to [13], who found an adequacy of Malaysian Government policies for BIM. The interviewees mentioned that the adoptions of BIM in Malaysia actually took place first before Singapore adoption. During that time Singapore adopted BIM only as a requisite for submission plan to the authority, but Malaysia adopted BIM at the preliminary stage until the production of construction drawings. As quoted by one of the interviewees; "Actually, BIM was already here since 1999 with the introduction of these software; ArchiCAD, Orion, and QTO as tools to expedite the construction processes. Unfortunately, no parties knew about the arrival of BIM and its integration inside software" (R3). In addition, "Since BIM was adopted in Malaysia, there had been no standard guideline being launched from Malaysian government. Currently, we have been using only client term and condition of BIM contract as a per-time basis" (All interviewees). All interviewees highlighted that the top managements of Malaysian Government are poorly educated in BIM and agreed that the future trend of BIM remains stable until there are compulsions from the government. Therefore, it is evident that the absence of a policy on the BIM implementation has led to the low adoption of BIM in Malaysia.

4.2. Poor Holistic Readiness

The interviewees suggested that the holistic readiness among construction stake players is the best way to fostering BIM process. Recently, BIM scholars found that the readiness among construction stake players is very positive. Contrary to [14], the seminar revealed that the land surveyor's roles are not formally recognised in BIM implementation based on the CIDB roadmap document. This is also supported by one of the interviewees; "In Malaysia, the key role of land surveyor is still missing in the BIM process" (R4, R5). Nevertheless, Ference Acs[15] found that the land surveying profession as an integral part of the AEC industry can substantially affected by the growing role and importance of the BIM environment.

4.3. Software Integration Competition Strategy

The nature of software development is constantly changing with the periodic hype of a new software invention [15]. Furthermore, any change or improvement of the BIM model is accessible in real-time and almost simultaneous for the stakeholders' utilisation. Moreover, various BIM authoring software can be used with a different software integration strategy to challenge with other competitors. The holistic BIM adoption therefore depends much on the integration of all BIM software. The collaboration of all the authoring software requires high cost to purchase by the BIM stake players. The interviewees opined that "Software is a tool for a successful BIM collaboration amongst construction team. Unfortunately, every year, BIM authoring software competition strategies are a burden to us because it involves high cost" (R3).

4.4. Reluctant to change & Reluctant to sharing knowledge

The interviewees suggest a good relation among stake players is the best way to foster the BIM adoption. On the other hand, the positive collaboration is capable to transfer knowledge and enhance the BIM knowledge among the team members. As observed by one interviewee; "The construction experts in Malaysia are reluctant to adopt BIM because they refuse to change. It is a tough age to educate when they are more than 40 years old." (R1). In addition, the interviewees opined that they are reluctant because of "The process coordination is much longer than preparing CAD drawings but for amendment works during construction stage; BIM can speed up the construction pace" (R4, R5, R6). Reported by [16], BIM requires a dedicated group of stakeholders to drive and facilitate the BIM implementation process. Interestingly, these isolated root causes arise in Malaysia are manifested from focal person who has high knowledge and expertise in software but reluctant to sharing was observed in the views espoused by two of the interviewees; "That focal person is the only one expert and competent with high capability in running the software to do concurrent analysis and prepare model in Malaysia. We need high knowledge and time to gain expertise on software which capable to do rigorous risk analysis in the infrastructure projects" (R4, R3).

5. Conclusion

The study has found that the absence of BIM policy and compulsion; poor holistic readiness; software integration competition strategy; and reluctant to sharing knowledge are the root causes of recurring barriers due to low BIM adoption in Malaysia. These findings will be further used to construct causal loops diagram in system dynamic model. As a limitation, the findings espoused here are preliminary and should be regarded with caution. It is expected that more results will emerge as the research progress.

6. Acknowledgment

The authors acknowledge the support from Universiti Teknologi MARA and Malaysia Ministry of Education.

7. References

[1] Construction Industry Development Board (CIDB) Malaysia, 2015 Construction Industry Transformation Programme (CITP) 2016-2020, no. 1.

[2] C. Eastman, P. Teicholz, R. Sacks, and K. Liston, 2011.BIM Handbook: A Guide to Building Information Modeling for Owners, Managers, Designers, Engineers and Contractors, vol. 2.

[3] J. Bhattacharya, M. P. Pharmaceutics, M. B. A. Hrm, and M. P. Management, 2014 "Root Cause Analysis – A Practice to Understanding and Control the Failure Management in Manufacturing Industry," vol. 3, no. 10, pp. 12–20.

[4] D. L.Gano, 2011 "Root Cause Analysis Compared," in Seven Steps To Effective Problem-Solving And Strategies For Personal Success, pp. 1–25.

[5] M. P. W. D. PWD, "Unit Building Information Modeling (BIM)," 2013. [Online]. Available: https://www.jkr.gov.my.

[6] Construction Industry Development Board (CIDB) Malaysia, 2015, Construction Industry Transformation Programme 2016-2020, no. 1.

[7] E. P. Unit, "Eleventh Malaysia Plan," 2015, pp. 1–77.

[8] CIDB, 2017, "CIDB BIM international Construction Weeks 2017 slide presentation.pdf".

[9] K. F. Anuar, M. Hezriq, and I. Zainal, 2015 "The Challenges in Implementing Building Information Model (BIM) for SME's Contractor in the Construction," Infrastruct. Univ. Kuala Lumpur Res. J., vol. Vol. 3 No., no. 1, pp. 40–49.

[10] A. K. Romney, S. C. Weller, and W. H. Batchelder, 2009 "Culture as Consensus : A Theory of Culture and Informant Accuracy," Blackwell Publ. behalf Am. Anthropol. Assoc., vol. Vol. 88, N, no. 2, pp. 313–338.

[11] B. & B. W. Harris, "System dynamics Methodology," 2005.

[12] J. Forrester, "System dynamics, systems thinking, and soft OR, 1994," Syst. Dyn. Rev., vol. 10, no. 2, pp. 1–14.

[13] C. P. John Rogers, Heap-Yih Chong, 2015 "Adoption of Building Information Modelling technology (BIM): Perspectives from Malaysian engineering consulting services firms," Emerald,.

[14] S. M. K. Adimin, 2016 "Roles and Readiness of Land Surveyors towards the Implementation of BIM in Malaysia." p. 60.

IOP Conf. Series: Materials Science and Engineering **216** (2017) 012056 doi:10.1088/1757-899X/216/1/012056

[15] F. Acs, 2015 "Building Information Modelling Impacts and Opportunities for Land surveying and the Cadastre."
[16] Construction Industry Development Board (CIDB) Malaysia; 2011workshop report, "Industrialised Building System (IBS): Mechanisation through Building Information Modelling (BIM)."

Development of an Adjustable board and a Rotational Board for Scaffold

Myunghoun Jang
Department of Architectural Engineering, Jeju National University, Jeju, 63243, Republic of Korea

Email: jangmh@jejunu.ac.kr

Abstract. Scaffold is widely used in high work-places inside and outside of a building construction site. It is inexpensive and is installed and dismantled easily. Although standards and ledgers of a steel tube and coupler scaffold are installed in a regular distance, the distances of transoms are not equal in some places. Sometimes a working platform or a board is absent in the corner of scaffold. This may cause safety accidents because a foothold is not stable on the transoms. An adjustable safety board and a rotational safety board are suggested in this paper. The adjustable board consists of two footholds. The small one is inserted into the large one. The rotational board covers not only right angle but also acute or obtuse angles. These safety boards for scaffold help to decrease safety accidents in construction sites.

1. Introduction

Curtain walls or stone panels are used for the exterior of a building. And, finishes with paint or tiles are often used in flat houses or a small house. These finish works require scaffold because the most works are performed in high-place.

A tube and coupler scaffold [1], [2] has steel tubes and clamps. Vertical tubes (standards) are connected to horizontal tubes (ledgers) via right angle clamps. Diagonal tubes (transoms) are periodically connected to the scaffold via swivel clamps in order to stabilize the scaffold. It is common in construction.

Planking or walking boards are placed between transoms installed periodically. In some case, the distance between transoms is not equal, so the walking board is not fixed strongly with the transoms. Fig. 1(a) shows a board installed unsafely that is longer than the distance between the transoms. This board may fall and cause a safety accident. Sometimes scaffold boards are not connected in a corner as shown in Fig. 1(b). The disconnected scaffold is very dangerous to work.

(a) A board longer than the distance between transoms

(b) Boards disconnected in a corner

Figure 1. Scaffold boards installed unsafely

IOP Conf. Series: Materials Science and Engineering **216** (2017) 012057 doi:10.1088/1757-899X/216/1/012057

2. Safety Accidents

According to KOSHA [3], the number of workers and fatal injuries of all industries in 2015 are shown Table 1. The fatal injuries in construction industry are 437 persons (45.8%) and they are very larger than those of manufacture industry (251 persons, 26.3%). FRTTY (number of fatality rate per 10,000 workers by year) of construction industry is 1.30, and it is larger than all industry (0.53) and manufacture (0.60).

Table 2 shows the accident types of non-fatal injuries and fatalities in all industries. Various types of safety accident happened. Among them, the number of slip and tip (15,632 persons) is the highest, the second is fall and drop (14,126 persons), and the third is getting caught-in (13,467 persons). The fatalities are 339 persons in fall and drop, 121 persons in getting caught-in, and 96 persons in collision except traffic accidents. Fall and drop accidents accounted for over 35% of all fatalities.

Table 1. Fatalities in 2015 (South Korea)

Industry	No. of workers (persons)	No. of fatalities		FRTTY*
		Persons	Ratio (%)	
Total	17,968,931	955	100.0	0.53
Construction	3,358,813	437	45.8	1.30
Mining	11,615	4	0.4	3.44
Manufacture	4,161,536	251	26.3	0.60
Electricity, Gas & Water supply	64,244	3	0.3	0.47
Transportation, storage, & communication	805,403	88	9.2	1.09
Forestry	86,565	15	1.6	1.73
Agriculture, Financing & Insurance	8,701,072	149	15.6	0.17
Etc	779,68.	8	0.8	0.10

* FRTTY: Fatality rate per 10,000 workers by year

Table 2. Injuries/Fatalities and Accident Types in 2015

Type	Injuries (persons)	Ratio (%)	Fatalities (persons)	Ratio (%)
Total	82,210	100.0	955	100.0
Fall & Drop	14,126	17.2	339	35.5
Slip & Trip	15,632	19.0	16	1.7
Crush & Overturn	2,321	2.8	69	7.2
Collision	6,658	8.1	96	10.1
Flying object	7,151	8.7	59	6.2
Collapse	476	0.6	34	3.6
Getting caught-in	13,467	16.4	121	12.7
Cut, Puncture	8,743	10.6	4	0.4
Explosion	589	0.7	50	5.2
Traffic accident	4,342	5.3	97	10.2
Improper action	3,090	3.8	0	0.0
Etc	5,615	6.8	70	7.3

2.1. Cases of Scaffold Foothold

The sliding type safety foothold [4] has two boards as shown in Fig. 2. The one is larger than the other. The small board is inserted into the big board to adjust its length. The big board has guide rails and supporting bars across the rails. The big and small boards have so little difference of height that a worker does not slip or trip. But, the safety hooks in the end of the board are too weak to connect the board and a transom.

Fig. 3 shows the portable working way for construction site [5] which has a main platform and a sub platform. The two platforms are fixed to each other with pins. The main platform's tip is slightly thin for a worker to walk without any obstacle. A fixing material which fixates the board with scaffold bearers is separable and is connected with a connecting material.

The safety board with a sliding foothold [6] has fixed length as shown in Fig. 4. The sliding plate is not only a working platform but also a door to move downward when it is open. A safety accident may happen when the sliding foothold is close incompletely.

Fig. 5 shows a foothold [7] developed for a curved wall and corners. The foothold has a main board with two holes and a rotational head with two plugs. The main board and the head are linked using nut and bolt. One foothold is connected other foothold putting the plugs into the holes. The pin connection enables the head to rotate, so several footholds are installed to scaffold along a curved wall. But, the gap between the main board and the triangular head may make a worker's foot be trapped or may cause material drop.

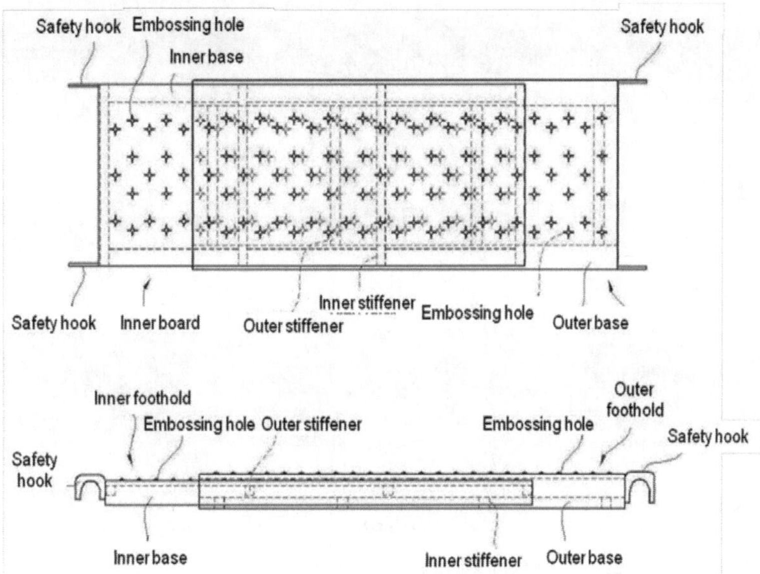

Figure 2. A sliding type safety foothold

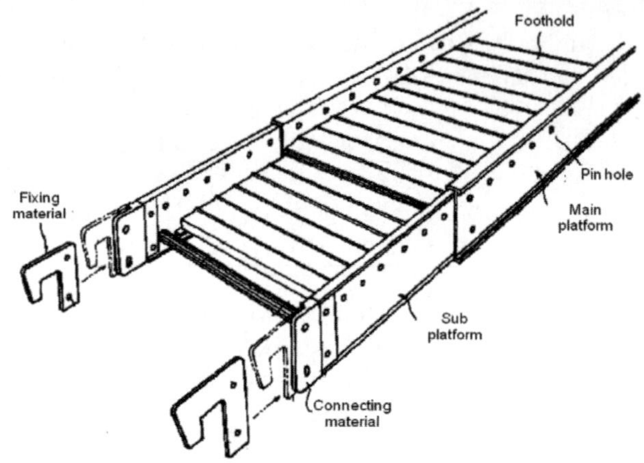

Figure 3. Portable working way

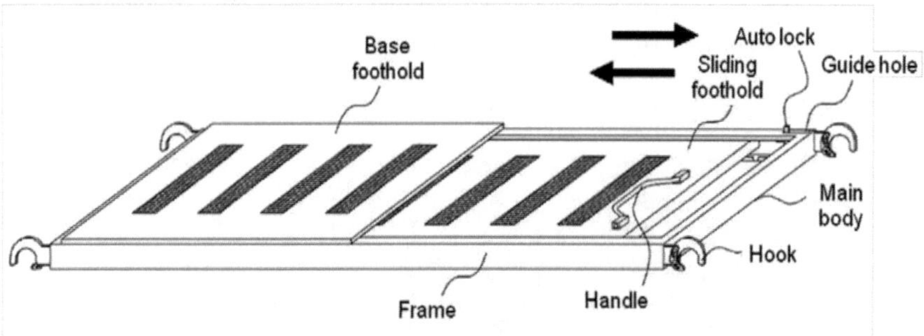

Figure 4. Safety board with a sliding foothold

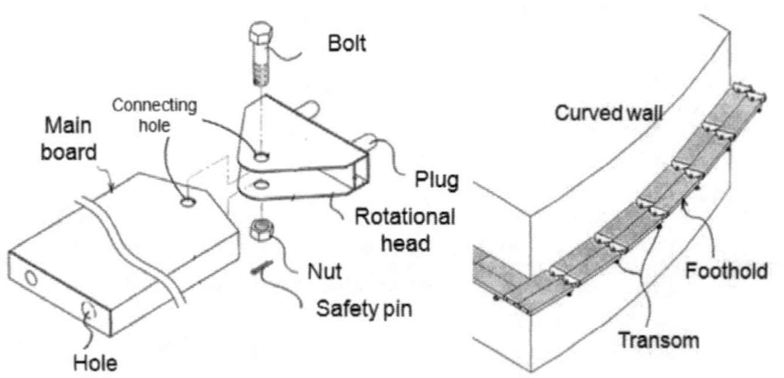

Figure 5. Foothold for curved wall

2.2. Development of an Adjustable Safety Board

It is important to find a method to lengthen and shorten the length of the board and to fix the board with the scaffold in order to develop a scaffold board with adjustable length. The previous cases use two boards or platforms different in size. This research also uses the idea of sliding style, but add an idea to fix the board with scaffold's transoms.

An adjustable board for scaffold [8] consists of two footholds as shown in Fig. 6. The small foothold is inserted into the large foothold to adjust the length of the board, and the small one is pull and pushed in order to adapt the distance between transoms. The length of the board is limited by a stopper which prevents the small frame and the large one from being separated apart.

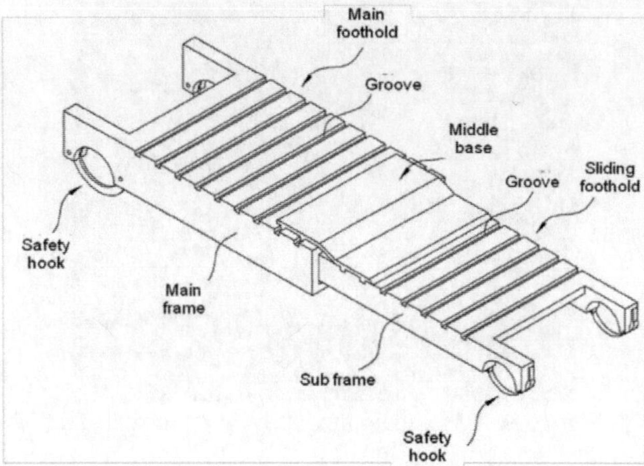

Figure 6. The adjustable safety board

There are linear grooves on the surface of the board not to be slippery. The middle cover is fixed into the grooves, and has a small slope that prevents stumbling owing to the height difference between the small frame and the large frame.

A circle-like hook is added at each end of the safety board as shown in Fig. 7. It holds a transom to install the board to the scaffold and closes by a pin. The board is safe from wind, live and dead load, or other impact because the board cannot be separated without removing the pin.

Fig. 8 shows a full-scale model of adjustable safety board installed to scaffold. The scaffold consists of standards, ledgers, and transoms which are fixed by clamps with each other. The full-scale model shows that it is easy to install and dismantle the adjustable safety board.

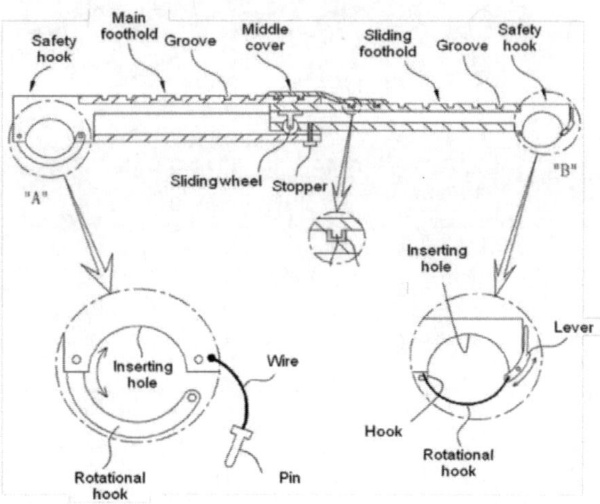

Figure 7. Section of the adjustable safety board

Figure 8. A full-scale model installed to scaffolds

2.3. Development of a Rotational Safety Board

A rotational safety board [9] is composed of a main platform and a rotational platform as shown in Fig. 9. The main board has two hooks to install to a transom, but the rotational board has fixing frames to place over a custom board without any hook. The custom board is a linear and fixed length working platform with hooks.

The rotational board rotates freely on the rotational axis, thus it is used not only at a right angle but also at an acute or obtuse angle. Fig. 10 shows installation examples. The hooks of the main platform grasp a transom and the fixing frames lock the rotational platform over a custom board installed to other transoms. The rotational platform has sloped edges in order not to trip on the rotational platform.

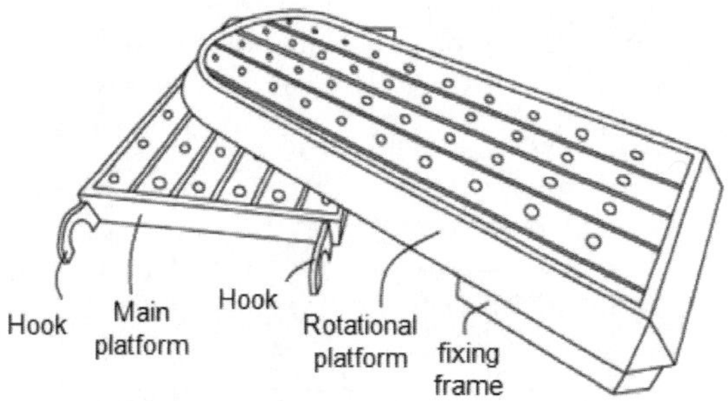

Figure 9. A rotational safety board

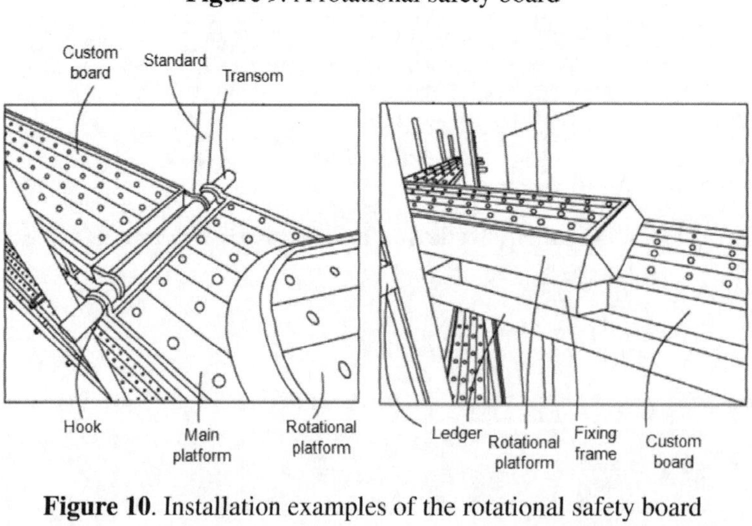

Figure 10. Installation examples of the rotational safety board

IOP Conf. Series: Materials Science and Engineering **216** (2017) 012057 doi:10.1088/1757-899X/216/1/012057

2.4. Structural Safety Analysis

A rotational -- The adjustable safety board and the rotational safety board are composed of two parts connected with each other. The connected parts are so weak for bending or deflection. Thus, structural analysis is performed to review the safety of the boards.

Sections of the safety board are designed shown in Fig. 11 according to Korean Standards [10]. The Standards regulate that the bending performance should resist the bending power on the center of the board and the deflection of the center should be less than 10mm. The overlapped part of the components of boards is also included in the analysis.

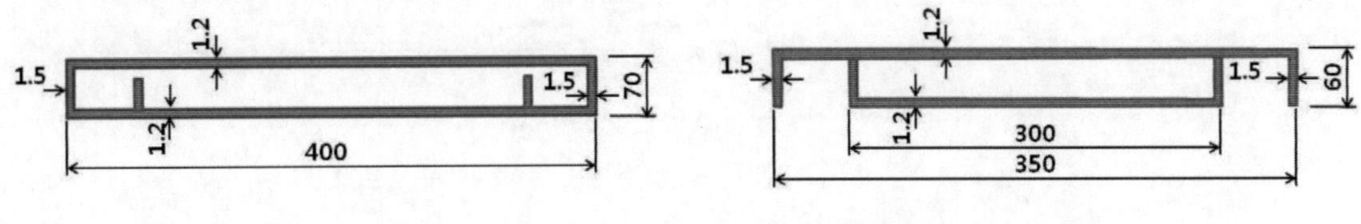

(a) Dimensions of the sub frame (b) Dimensions of the main frame

Figure 11. Design of the adjustable safety board

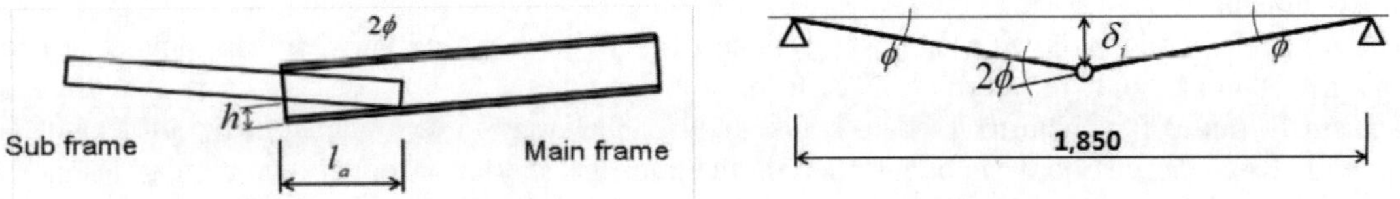

(a) Rotation on the board center (b) The instant deflection

Figure 12. Design of the adjustable safety board

Comparing the stress on the section over the maximum bending moment (2.04 kNm) with the allowable stress assuming the weight (4,400 N) is loaded on the center of the board (1,850 mm), the maximum stress of the main and sub platforms are 58.8 MPa and 64.8 MPa respectively. Because each maximum stress is less than the allowable stress (156 MPa) of SS400, the bending performance is structurally safe.

Rotation (2ϕ) and instant deflection (δi) occur on the board center shown in Fig. 12, and are calculated by the gap (h) between the main and sub frame and the overlapped length (la) in the equation (1) and (2).

$$2\phi = \mathrm{Tan}^{-1}\,(h / l_a) \tag{1}$$

$$\delta_i = \tan\phi \times (l / 2) \tag{2}$$

The flexible deflection is 3.0mm calculated by MIDAG GEN when the weight of 4,400 N is loaded on the center of the board (1,850mm long). The total deflections are the sum of the instant deflection and the flexible deflection. Fig. 13 shows that the overlapped length should be more than 650mm so that the total deflection of the board center is less than the allowable deflection in Korean Standards.

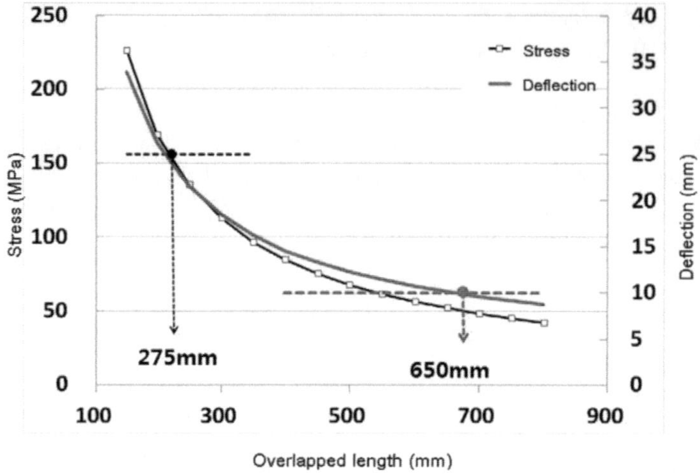

Figure 13. Deflection and overlapped length

3. Conclusions

The research develops the adjustable safety board for irregular period between transoms of scaffold and the rotational board for a corner of scaffold.

The small frame of the adjustable board is inserted into the large frame to adjust the board length. The shorted or lengthened board can be installed on the transoms shorter or longer than regular distance. A full-scale model is made to review the proposed safety boards to investigate the function of adjustable length and the hooking materials.

The rotational board can be used at any angle of a corner because the rotational platform rotates freely. It is installed over a custom board and is fixed strongly with fixing frames.

Structural safety is analyzed according to Korean Standards. The result shows the stress and deflection on the center of the board is allowable. But, mock-up test is necessary for full-scale models to ensure safety.

The proposed safety boards can reduce safety accidents in scaffold. But, the verification for reusability and production cost decreasing is necessary to use them on a building construction. The combination of adjustable and rotational functions is developed in further studies.

4. Acknowledgments

This research was supported by the 2017 scientific promotion program funded by Jeju National University.

5. References

[1] Unites States Department of Labor 2017 *OSHA (Occupational Safety and Health Agency)* http://www.osha.gov/SLTC/etools/scaffolding/supported/tubeandcoupler.html

[2] Wikipedia 2017 *Tube and clamp scaffold* https://en.wikipedia.org/wiki/Scaffolding

[3] KOSHA (Korea Occupational Safety and Health Agency) 2016 *Occupational injuries and illness status in 2015* www.kosha.or.kr/www/boardView.do?contentId=363787&menuId=554&boardType=A2

[4] Kim J 2015 *A sliding type safety foothold* Korea Patent 20-0391348

[5] Takahiro E et al. 1994 *Portable working way for construction site* Japan Patent 1992-255499

[6] Lee C 2013 *Safety foot board having sliding foot board for scaffold* Korea Patent 10-1251353

[7] Kim H 2011 *Foothold for scaffold corners* Korea Patent 10-1098844

[8] Jang M, Kim B, Jeong Y, Park S, and Lee S 2013 *Adjustable safety foothold for scaffold* Korea Patent 10-1289563

[9] Jang M, Lee H, Moon S, Youn J, and Lee W 2013 *Plate for control of angle with scaffold* Korea Patent 10-1347491

[10] Korea Agency for Technology and Standards 2009 *KS (Korean Standards) F 8012 Working boards*

IOP Conf. Series: Materials Science and Engineering **216** (2017) 012058 doi:10.1088/1757-899X/216/1/012058

Robotic Form-Finding and Construction Based on the Architectural Projection Logic

Sun Zexin, Hongyuan Mei
School of Architecture, HIT, China

E-mail: 736087330@qq.com

Abstract. In this article we analyze the relationship between the architectural drawings and form-finding, indicate that architects should reuse and redefine the traditional architectural drawings as a from-finding tool. Explain the projection systems and analyze how these systems affected the architectural design. Use robotic arm to do the experiment and establish a cylindrical projection form-finding system.

1. Introduction

According to the working mechanism of architecture, the operations of architectural design are always based on a series of geometric operations; these operations are usually functioning inside the abstract space of architecture and described by a series of drawings, such as plans, sections, elevations, perspectives and so on. In the context of contemporary architecture, design methods can be characterized by two types, the first one is translating the drawings from 2D medium into 3D objects according to the traditional projection method; the other one is purely 3D operation in digital environment. It is an undeniable fact that the introduction of computational technology has brought a huge design freedom for the latter, and the architects are able to work in a 3D atmosphere to create complex spatial forms; however, with the isolation of 2D drawings and 3D forms during the design process, the architectural drawings are just as additions of the design, representing the design result by traditional projection methods. The form-finding ability of the architectural drawings is being gradually neglected. [1]

2. Architectural Drawings and Form-finding

“Form-Finding” has a long history in architecture. As early as before the "Computer Age", the Spanish architect Antoni Gaudí had used upside down force model to explore the geometrical forms of catenary arches, such forms had been used in his projects such as Basilica and Expiatory Church and the Holy Family and Church of Colònia Güell (Fig. 1). Through using this completely new method – which is based on motion dynamics and gravity, Gaudí invented his own architecture languages and achieved his greatest architectural goal - to perfect and go beyond Gothic style. [2]
The term “form-finding” was brought by the German architect Frei Otto for the first time; due to his contribution of using the physical models as design engine and form-finding tool, Otto is recognized as pioneer of the parametric design in architecture. [3] Otto was specialized in lightweight tensile and membrane structures, the forms were derived from a series of soap bubble experiments which used the surface tension of the liquid to find the minimum surface to cover a closed shape (Fig. 2). [4]

From the "nonlinear system theory" to the "parametric design", the computational technology has brought a completely new medium to the architectural design, and it's even altering the fundamental theories and thinking processes behind it. With the popularization of digital software, the mechanics of how architects work have changed; less and less are using traditional architectural drawings as a from-finding tool. However, the traditional architectural drawings' ability of form-finding has played an important role in the development of the architecture with the following three aspects:

First, human beings live in 3-D world; human brain has a better ability of awareness and controllability for lower dimensions (1-D & 2-D). Through using traditional architectural drawings, such as plans, elevations, and sections and so on, architects are able to operate 2-D geometries on paper to obtain 3-D architectural forms. It simplifies the process of design and improves the creativity greatly.

Second, the design process will be stretched when use drawings as design medium, thoughts and ideas of the architects can be delivered to paper directly. During the design, 2-D drawings are much easier to perceive and operate, thus making a better condition for architects to compare and adjust, providing an intuitive means of visualized information to make decisions.

Third, the biggest contribution of the drawings is introducing the projection methods to the architecture, so that the work of architects is based on information transmission between dimensions rather than operating 3-D objects directly. Since the Renaissance, architects were no longer involved in the construction directly, but rather using the drawings to connect design and construction. [5] So that the projection methods began to affect the architecture forms. [6] The transmission between dimensions will inevitably be accompanied with adding or cutting information in some specific dimension, which makes the design results some kind of unpredictable.

There seems to be a paradox, as mentioned before, since the traditional architectural drawings can't meet the contemporary demands of creating complex forms in 3-D atmosphere due to the restriction of dimension, and also medium function of drawings to connect design and construction is being gradually replaced by computer, then how can we redefine and reuse the traditional architectural drawings as from-finding tools in contemporary architectural works? In order to answer this problem, we need to analyze the evolution of projections and drawings in the history of architecture.

Figure 1. An upside down force model of the Colònia Güell

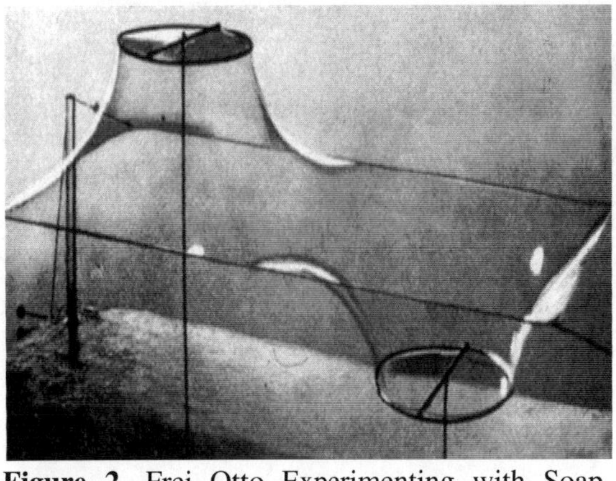

Figure 2. Frei Otto Experimenting with Soap Bubbles.

3. Redefinition of the Architectural Drawings Based on the Architectural Projection Logic

The functions of plans, sections and elevations, as three of the most important architectural drawings, are representing 3-D buildings with 2-D drawings based on parallel projection methods. Parallel projections are functioning with projection lines which are parallel both in reality and in projection planes, and the work can be described as a series of parallel lines projecting the building's geometry on drawing planes. [7] For different types of drawings, the projection lines are pointing to different directions, if the projection lines are orthogonal relations, it will carry out the orthogonal architectural work planes - plans, sections and elevations. Under such circumstance, the architecture can be viewed as in a cube full of parallel projection lines, the architectural drawings are presented on the surfaces of the cube, and thus this cube could be viewed as the workspace of architecture (Fig. 3). On one hand, the representations of the architecture are based on the drawings within this workspace; on the other hand, as an important design medium, the drawings can also affect design through the workspace. For example, if architect use plans as design medium, the work will tend to present richness and ductility in horizontal direction; if architect use sections as design medium, the work will tend to be more recognizable on the vertical dimension. All of these reflect that the forms are affected by the types of drawings and projection methods.

Painting and architecture are both based on projection systems and then presenting the results on papers. Looking back, the shift of projection methods affected the way of expression and had revolutionary impacts on innovations of painting and architecture. [8] For example, both abandoned perspective projection method which had been used for hundreds of years in painting, Futurism adopted a dynamic projection method and tried to add time as the fourth dimension on 2-D images; while Cubism tried to overlap multiple projection systems with different vanishing points together rather than use the traditional perspective with single vanishing point to describe 3-D objects. Similarly, in architecture, architects found it's difficult to describe and create nonorthogonal complicated forms through the traditional workspace with parallel projection system, such as the shapes of the different stones in dome, and then the stereotomy - a more effective graphic description of nonorthogonal form – was invented. Stereotomy abandoned the traditional cube workspace, combining the parallel projection with perspective projection to achieve higher degrees of freedom to operate nonorthogonal forms (Fig. 4).

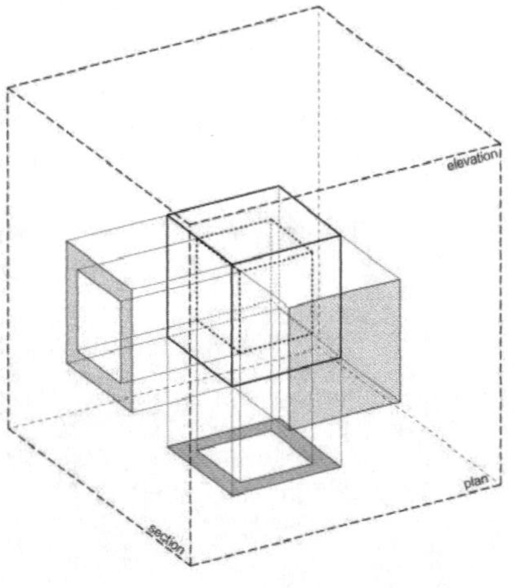

Figure 3. Traditional architectural workspace.

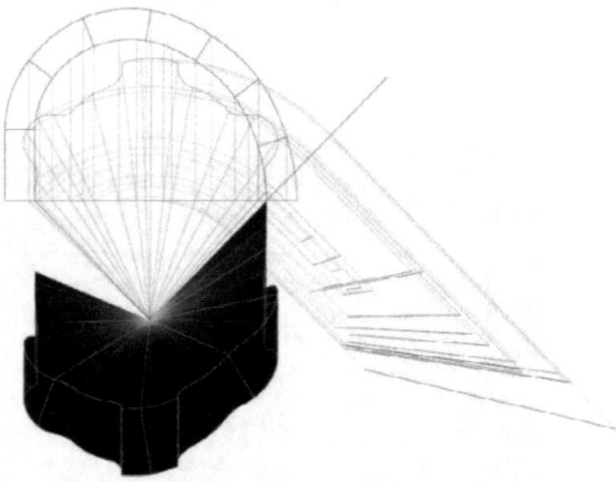

Figure 4. Stereotomy combine the parallel projection with perspective projection

In the context of contemporary architecture, architects tend to only care about the change of graphical languages, while ignoring the innovation power of drawings and projections as source of form-finding since it is difficult to create complex forms in the traditional architecture workspace which is based on the X - Y - Z orthogonal parallel projection system. On one hand, it is much easier for the architects to operate complex shapes with the help of computers, so the architectural drawings are gradually independent from the design process and only be regarded as additions to represent certain degrees of the results; on the other hand, the direct operation of 3-D shapes in digital is missing the unpredictability of forms due to the lack of conversions between dimensions. [9]

From perspective drawings in the Renaissance to the computer graphics of today, the working mechanism of the drawings as the design medium has shifted from the 2-D orthogonal projection to a multidimensional composite. The meaning of the word "drawings" becomes rich, more pointing to the diagrams with both logicality and fuzziness. Just as Neil Lynch said: "Computer Age is not giving a new style, but a new design technique. We apply computer technology in evolutionary and emerging systems, establishing an implementation and testing system make the diagrams to reality, reality to diagrams. In this new field, form is not important at all. We should explore the potential of "algorithm technology", and focus more on intelligent and logical processes from design to construction, logic is the new form." [10] So we can see that the combination of the architectural drawings with computer technology based on the architectural projection logic is the redefinition of the drawings as design medium. At the same time, with the development of CAD (Computer Aided Design) and CAM (Computer Aided Manufacturing), the contemporary architectural form-finding is not only the from top to bottom style – which is providing inspirations and references to architects during the one-way design process, but also the from bottom to top style, that set up the basis for material control and construction.[11] So, if we change the projection modes of the traditional architecture workspace, in which design is forming and operating, we will be able to explore multiple possibilities of drawings. Meanwhile, we introduce the robotic arm to connect materialization and projection, in the end, through combining the two methods' advantages, a new design medium, with both form-finding ability of drawings and the controllability for nonorthogonal geometries, will be set up.

4. Robotic Form-finding Based on the Architectural Projection Logic
In this study, we use the KUKA six axis articulated robotic arm as the bridge to connect drawings with construction, thus establishing a medium between design and materiality, which is totally different from the working mechanism of today's architectural design. The robotic arm is widely applied in areas of aerospace and automobile manufacturing, and it is gradually replacing the manpower by achieving precise positioning and point-to-point operations in 3-D space. The movement of the robotic arm follows the computer program instructions according to the principle of inverse kinematics, which is reaching the desired position through rotating the joints with certain parameters. [12] In recent years, with the popularity of utilizing the CAM, there are more and more robotic applications in architecture. In 2008 the Venice Biennale of Architecture, the ETH Zurich professor Fabio Mario and Matthias Koehler explored robotic construction by using bricks - one of the most traditional materials. They used six axis robotic arm to lay bricks and built a parametric curved wall, yield unusually brilliant result. ICD and ITKE of the University of Stuttgart were inspired by the inner structures of sand dollar and sea urchin, they applied seven axis robotic arm for precast productions, developed a new method of wood shell construction and finished the Research Pavilion 2011 (Fig. 5). [13]

Figure 5. ICD-ITKE Research Pavilion 2011.

Figure 6. The robotic arm is equipped with a hot-wire cutter to cut an EPS foam cube.

In our research, the robotic arm is equipped with a hot-wire cutter to cut an EPS foam cube (Fig. 1), the mechanism is controlling the path of hot-wire to get a desired shape. In geometry the smooth surface we got is a ruled surface, as the parametric representation of a curved surface is

$$r(u, v) = a(u) + v * l(u) \tag{1}$$

In ruled surface, l (u) is unit-length vector, the v curve is a straight line, and ruled surface is swept by this moving straight line which is referred as the generatrix. Ruled surface had been used in architecture for long due to the clear construction logic and diverse forms. Gaudi applied isometric mapping operation method to build ruled surface like hyperboloid and parabolic, which were widespread used in his projects. [14].

In traditional, the construction of ruled surface is the bottom to top style, which starts with material and ends with construction, the process is dominated by the materiality logic. While, the contemporary construction is usually the top to bottom style based on the computational technology, the process is dominated by form. So we imagine operating the robotic arm with the architectural projection logic so that we can break the unidirectional transmission between digital and physical, set up a form-finding method with the construction of the ruled surface.

At the beginning, we simulate the movement of the robotic arm by using Grasshopper plug-in of Rhino (Fig. 7), the simulation shows that:

1. The trajectory of hot-wire make a ruled surface, hot-wire is the generatrix of the surface.
2. The movement of the two endpoints of the hot-wire line generate two separate space curves.
3. The ruled surface can be described as a straight line sweeping through the two space curves.

Therefore, it's obvious to see that how to manipulate the two space curves is the key to cut the ruled surface. Though we can modeling the two space curves in digital precisely, however, the curves cannot be intuitively controlled since the mechanism of the digital software is using projection methods to simulate a 3-D workspace and then project shapes on screen, it will inevitably increase the difficulty of control.

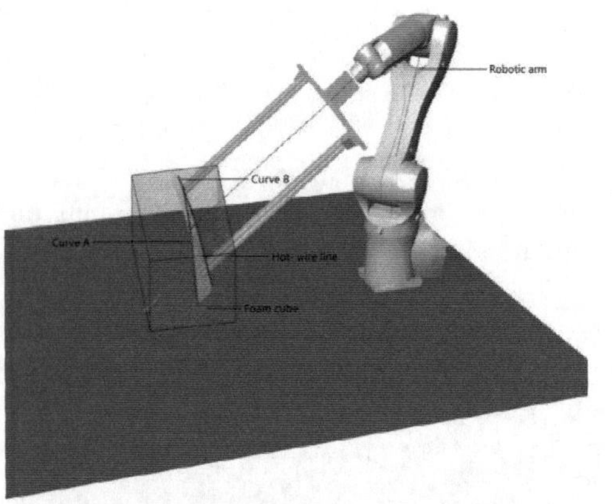

Figure 7. Simulation of the movement in Rhino

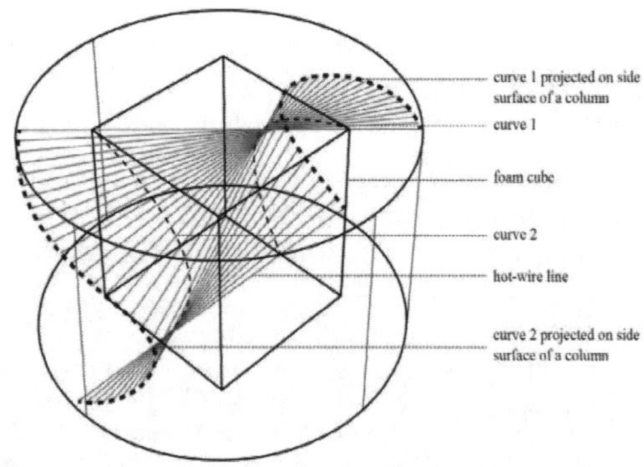

Figure 8. The two space curves can be presented and projected onto the side surface of a column around the foam cube.

In order to solve this problem, we need to reduce the dimension and create a corresponding workspace. It is apparently that the two space curves can always be presented and projected onto the side surface of a column around the foam cube (Fig. 8), so we introduce cylindrical projection to our system. The cylindrical projection is one kind of map projections and it is widely used in mapping systems around the world which is known as Mercator projection. In a cylindrical projection map, the mapping of meridians to vertical lines can be visualized by imagining a cylinder whose axis coincides with the

IOP Conf. Series: Materials Science and Engineering **216** (2017) 012058 doi:10.1088/1757-899X/216/1/012058

Earth's axis of rotation. This cylinder is wrapped around the Earth, projected onto, and then unrolled. [15] In our system, the two space curves can be presented on the unrolled cylindrical screen first (Fig. 9), and then position in 3-D space by wrapping the screen into cylinder. It is a reversed operation of cylindrical projection and the unrolled screen is within the Cartesian coordinate system and also it could be described as the new workspace with the ability of form-finding.

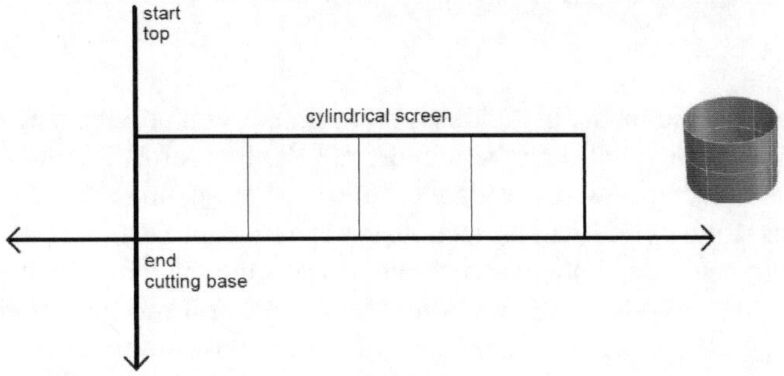

Figure 9. The unrolled cylindrical screen.

In this Cartesian coordinate system, hot-wire moves along with the y axis from the top edge to the bottom x axis. So we set two curves along the movement direction, which are rail a, and rail b, then we wrapped the screen into a cylinder, so that the two curves rail a, and rail b will be positioned in 3-D (Fig. 10). By using this workspace with the cylindrical projection method, it is able to describe any kind of movement of hot-wire under such circumstance.

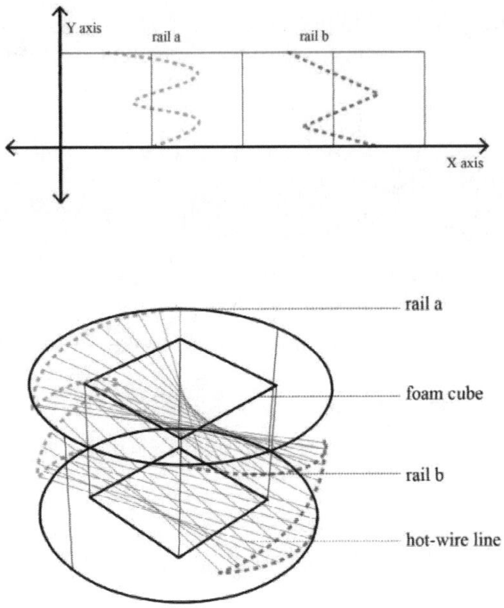

Figure 10. Position rail a, and rail b in 3D.

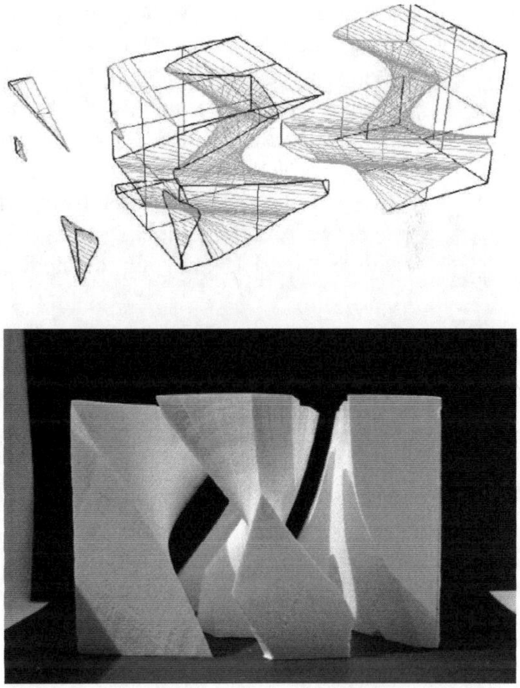

Figure 11. Cut the foam with the logic of architectural projection

As we control the trajectory of the robotic arm with the cylinder projection method, which is a 2-D workspace, so we didn't make any predictions about the forms of the ruled surfaces. The working process is based on the architectural projection logic, multiple possibilities and unpredictable forms are generated. During the construction, we control and simulate the movement of the robotic arm in digital first, so that we can correct the operations in time and represent the machining process intuitively, paperless construction will be realized. Finally, we evaluate the system by cutting the foams and feedback the evaluation results to the digital, so that we are able to establish a circuit fusion of digital and physical and cut the foam with logic of architectural projection (Fig. 11).

5. Conclusion

Drawings and projections are the most basic media in architectural design, it will provide a new workspace and new way of form-finding by redefining them in digital. Explore form-finding possibilities through changing the architectural representing mechanism of projections rather than operating the forms directly, such as using cylindrical projection and spherical projection instead of the traditional orthogonal parallel projection, so under such circumstances, the drawings are both form-finding media and representation methods. Besides, control the robotic arm and by applying different projection logics, we will eventually achieve the reconstruction of form-finding from drawings to constructions.

References

[1] Leach, Neil, C. Williams, and D. Turnbull. "Digital Tectonics." Wiley (2004).
[2] Han Pei, and Mei Hongyuan. “A Studyon Dynamic Design Strategy of Cold Regions Architecture Based on Photokinetic Mechanism of Plant” [J]. New Architecture, 2016, 03:64-68, ISSN：1000-3959
[3] Schumacher, Patrik, S. Zhang, and D. U. Luo. "Parametricism the Parametric Paradigm and the Formation of a New Style." Time Architecture (2012).
[4] Lopes, João V, A. C. Paio, and J. P. Sousa. “Parametric urban models based on frei otto’s generative form-finding processes." Rethinking Comprehensive Design: Speculative Counterculture, Proceedings of the, International Conference on Computer-Aided Architectural Design Research in Asia 2014.
[5] Carpo, Mario. "The Art of Drawing." Architectural Design 83.5(2013):128–133.
[6] Evans, Robin. "Translations from Drawing to Building and Other Essays." Journal of the Society of Architectural Historians 57.2(1986):211-212.
[7] Maynard, Patrick. Drawing distinctions: the varieties of graphic expression. Cornell University Press, 2005.
[8] Evans, Robin. "The Projective Cast: Architecture and Its Three Geometries." Renaissance Quarterly 50.3(2000).
[9] Frascari, Marco, Jonathan Hale, and Bradley Starkey. From models to drawings: imagination and representation in architecture. Routledge, 2013.
[10] Leach, Neil, David Turnbull, and Chris JK Williams. "Digital tectonics." (2004).
[11] Adriaenssens, Sigrid, et al. Shell structures for architecture: form finding and optimization. Routledge, 2014.
[12] Helm, et al. "In-Situ Robotic Construction: Extending the Digital Fabrication Chain in Architecture." 2012.
[13] ITKE, Stuttgart, and U. O. Stuttgart. "ICD | ITKE Research Pavilion 2011 / ICD / ITKE University of Stuttgart." Archdaily.
[14] Santiago Huerta. "Structural Design in the Work of Gaudí." Architectural Science Review volume 49.4(2006):324-339.
[15] Projection, Map. "Cylindrical projection." Dictionary Geotechnical Engineering/wörterbuch Geotechnik (2014):329-329.

IOP Conf. Series: Materials Science and Engineering **216** (2017) 012059 doi:10.1088/1757-899X/216/1/012059

Green Architecture Approach on Mosque Design in Cipendawa Village, Cianjur, West Java, Indonesia

Rahma Purisari[1a], Ratna Safitri[1b], Eka Permanasari[1c], Feby Hendola[1d]
[1]University of Pembangunan Jaya, South Tangerang, Indonesia
[a]E-mail: rahma.purisari@upj.ac.id; [b]E-mail: ratna.safitri@upj.ac.id; [c]E-mail: eka.permanasari@upj.ac.id; [d]E-mail: feby.kaluara@upj.ac.id

Abstract. Climate change has been a global concern as many buildings are constructed and maintained without considering environmental impacts. Global warming, flood and environmental damage have become everyday phenomena. Architecture plays a vital role to in shifting design paradigm green architecture that is sustainable and environmental friendly. This research-based design investigates the implementation of green architectural approach on a mosque design at a waqf land in Cipendawa Village, Cianjur. The site is located on the hill where water catchment is required. Green architecture approach is expected to minimize negative environmental impacts and provide solution for low budget mosque.

1. Introduction

The focus of this research-based design is to identify problems and potential features on the site and propose green architectural concept which enhances energy efficiency, effective building maintenance and at the same time provides comfort for users. Therefore, the approach is contextual.

In Indonesian society, a mosque serves both as a religious and as a social venue. People congregate on a mosque to pray and to have social agenda such as community meeting and learning activities. The needs for providing a mosque requires good knowledge on what the community needs. Majority of the villagers in Cipendawa Village, Cianjur West Java are moslem but the mosque are rarely to be found due to the budget constraint in constructing it. There are only two mosques found within the area and in far distance. This research-based design took place on a piece of waqf land in this village which serves community for religious activities. The closest mosques to this land are within 3 km (Al-Misbah mosque) and 5 km (Al Qossimiah Mosque). Considering the moslems should pray five times a day, these distance are not reachable in walking distance, especially for Friday or Eid prayings.

There is Mrs. Hj. Yuslida Hamid, who agrees to give away her 270 m2 land on Hanjawar road to build a mosque serving the community. However, she has some budget constraint in constructing the building. Understanding this constraint, the team agreed to take the challenge and propose the design.

The team surveyed the location and identified key potential points. The land is primarily located on the main road. It has beautiful scenery, and slightly cold weather. The development of the mosque should incorporate this potential weather condition. However, the restricted budget and the beautiful scenery require the low-cost budget and minimum impact to the environment. Therefore, this research-based design is aimed to carefully identifying problem on the site to correctly propose the design which accommodate people's need for religious activity and for reducing the carbon footprints. By closely analyze the environment, we propose passive design strategy to reduce energy consumption for the benefit of the occupants [1]. Furthermore, the developed green building concept is inline with the economically viable considerations, whereby the entire building life cycle – from concept to planning stage, from construction to operation and then back to denaturation – is taken into account. Green buildings, therefore, are based on an integrated and future-oriented approach [2].

IOP Conf. Series: Materials Science and Engineering **216** (2017) 012059 doi:10.1088/1757-899X/216/1/012059

Figure 1: The location of the proposed mosque design

Figure 2: The site condition

2. Research Method

In order to propose green architectural design, the team uses research design method of imaging, presenting and testing (Zeisel, 1984) [4].

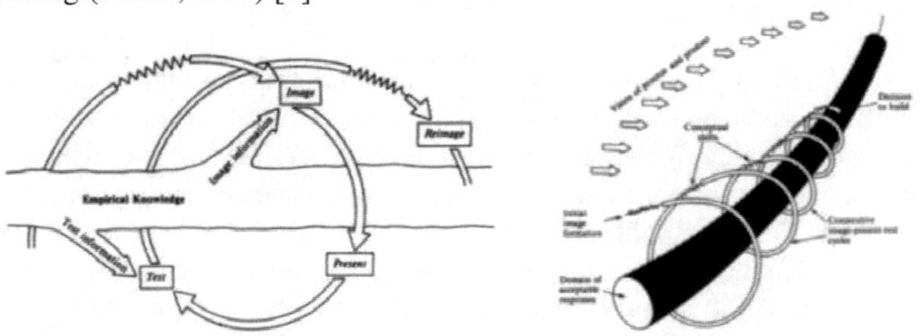

Figure 3**:** Design Development Spiral (Zeisel, 1984)

Imaging is proposing a new idea which illustrates criteria of best architectural practice in the future. Imaging includes problem identification, data gathering, analysis and synthesis. The goal is to propose a low-cost design serving the community and at the same time is environmentally friendly. There are four approaches in data gathering: literature study, precedent, site analysis and interview.
Site analysis starts with the macro and micro conditions to identify the problems on the site located on the hills. This site is required to function as a water catchment. The interview with Mrs. Hj. Yuslida Hamid was aimed to get ideas on the society's needs and religion-based activities on the surrounded area in Cipendawa village. After that, the team is analyzing design criteria to propose the space required and green architectural concerns to be applied onto the site, space, building form, interior and building facade.

IOP Conf. Series: Materials Science and Engineering **216** (2017) 012059 doi:10.1088/1757-899X/216/1/012059

Presenting is the result of applied criteria on the design using these principles: conserving energy, working with climate, respect for the site, respect for users, limiting new resources and holistic.
Testing is to evaluate on the proposed design based on the criteria applied. Testing is done by verifying the interview with the literature study and comparing the mosque design with the precedents.

3. Theoretical Framework

In the attempt to keep the environment sustainable and to reduce the environmental damage after building the mosque, the team uses green architectural approaches introduced by Brenda and Robert Vale (1991) [3] in the book, Green Architecture Design for Sustainable Future. The principles and criteria to achieve this green design is through the following ways:

3.1. Conserving energy

This approach is to operate the building with the minimum unrenewable energy. This is through the following methods:
3.1.1. The building is designed in elongated form in order to bring the light inside. Other attempt is to use skylight in areas which do not have direct sunlight.
3.1.2. The building uses sunlight as thermal energy to produce electricity by using photovoltaic devices. In addition, the uses of sunlight can also be used by tilting the roof facing the east-west walls or in line with the sun path.
3.1.3. The building uses low level electric devices and controls the high uses through automatic devices.
3.1.4. The building applies sunscreen on windows which can adjust the light intensity getting through the interior.
3.1.5. Interior uses light colors to enhance light intensity.
3.1.6. The building should avoid artificial heater. The heat should come from the inhabitants and the sunlight entering the building through ventilation.
3.1.7. The users should minimize the use of air conditioning to reduce the energy consumption.

3.2. Working with climate

This approach can be achieved through these following ways:
3.2.1. To locate the building orientation against the sun path.
3.2.2. To apply air pump system and cross ventilation.
3.2.3. To use plants and water as climate adjuster.
3.2.4. To use windows and roof which can be opened and shut depending on the users' need.

3.3. Respect for site

This approach requires interaction between the building and its site. This can be achieved through the following ways:
3.3.1. To minimize site alteration by deploying design following the site condition.
3.3.2. To use minimum building coverage ratio and to favor vertical building approach.
3.3.3. To apply local materials available from the surrounding area

3.4. Respect for users

This approach plays an important role as building has a close relationship with the users, activity and needs.

3.5. Limiting new resources

This approach is to limit the use of new resources and use the current available resources.

3.6. Holistic approach

IOP Conf. Series: Materials Science and Engineering **216** (2017) 012059 doi:10.1088/1757-899X/216/1/012059

This approach is aimed to design a building by applying these five principles. These green architecture design are linked and cannot be separated.

4. Analysis

4.1. Site Analysis

The proposed land is surrounded by the main Hanjarwar road, which functions as an alternative route from Puncak to Cianjur, West Java, on the north east side and walls on the south side (image 4). With the beautiful scenery and tranquility, this land has a very high quality for religious activities. To respond such quality the team deeply analyze the site to map the contour, potential points and environment.

The site contour is not that extreme, in fact it can be said as flat. The proposed land currently has some areas for planting sweet potatoes and peanuts. On the side near the main road, the area is currently being used for selling flowers. The rest of area are clear with grass covering most of it. This flat contour benefits the team in designing the mosque as there is no cut and fill approach involved.

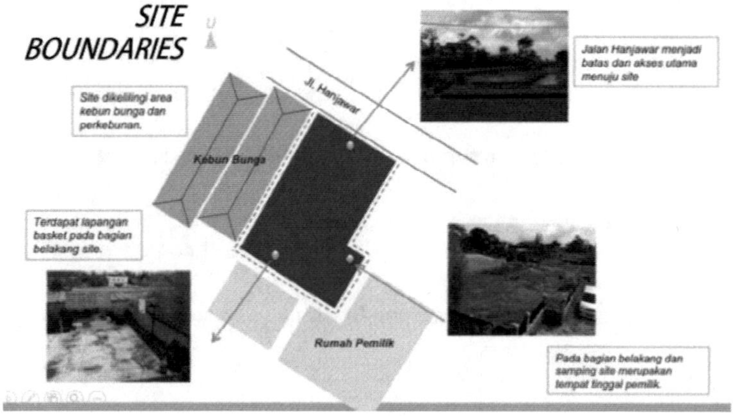

Figure 4: Site Boundaries

The beautiful scenery and slightly cold weather are the other two main pints to be included in the design. The scene of mountain at the back of the land and will be captured and be part of the mosque design.

The noise level is practically low with the high intensity is towards the main road. When designing the mosque, there should be a treatment responding this noise level (Figure 5).

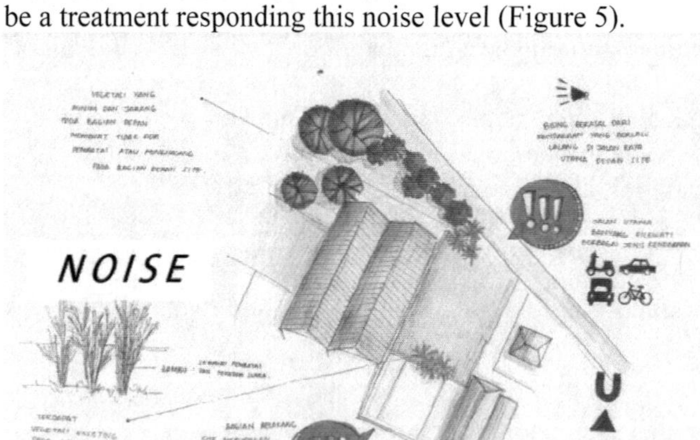

Figure 5: Noise Analysis

IOP Conf. Series: Materials Science and Engineering **216** (2017) 012059 doi:10.1088/1757-899X/216/1/012059

During the day, the weather on the site is comfortable because the wind speed is around 0,2 – 0,6 km/hr. The location of the site which is near the mountain makes the weather drops drastically during the night (figure 6)

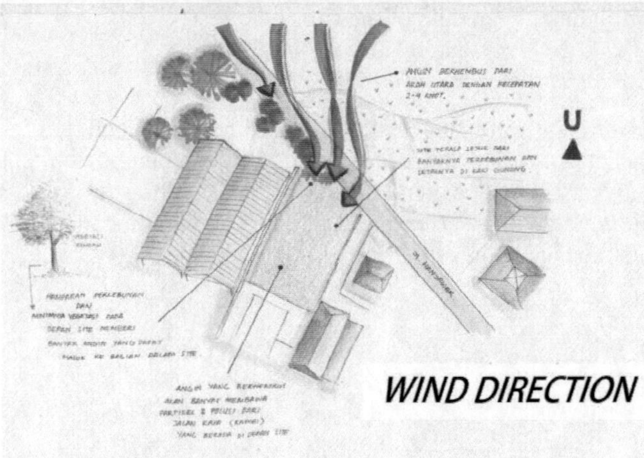

Figure 6: Wind Direction Analysis

The site receives abundance natural lighting, especially the area near the Hanjarwar road. The site has nearly 12 hours full daylight during the day as not many of high shading trees around (Figure 7)

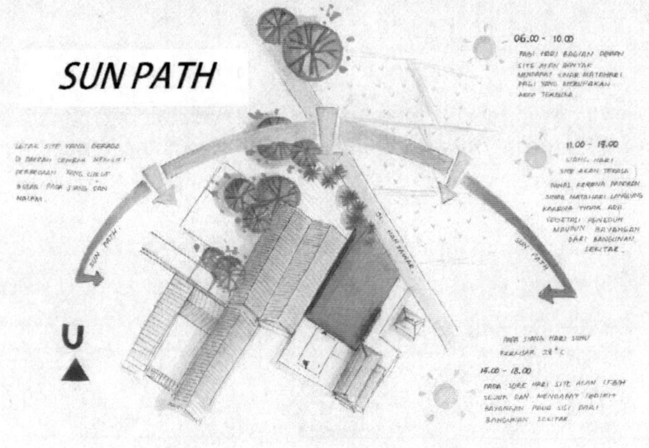

Figure 7: Sun Path Diagram

Apart from contour, temperature, wind direction and natural light, the team also analyses existing drainage system on the site. The team found that there is a drainage system right in the middle of the site with 1 m width and run through the city drainage system located on Hanjawar road. The water is categorized as grey water as it came from overflow rain and the land next to the site (Figure 8)

Figure 8: Existing drainage on the site

IOP Conf. Series: Materials Science and Engineering **216** (2017) 012059 doi:10.1088/1757-899X/216/1/012059

From the explanation above, the existing site has strengths points such as comfortable weather, beautiful scenery and significant amount of plantation. These strengths will be incorporated in the design. However, the team also need to anticipate the temperature drop during the night and the current drainage system right in the middle of the site.

4.2. *Building response towards site analysis*

Based on the site analysis, the team first respond to the weather condition. Since there are no big plantation surround the mosque, and the site receives direct sunlight, the team carefully treat this slightly cold weather by directing the wind inside of the building to create cross ventilation and minimize the use of air conditioning during the day. The ventilation is also required to anticipate the temperature drop during the night (Figure 9 and 10)

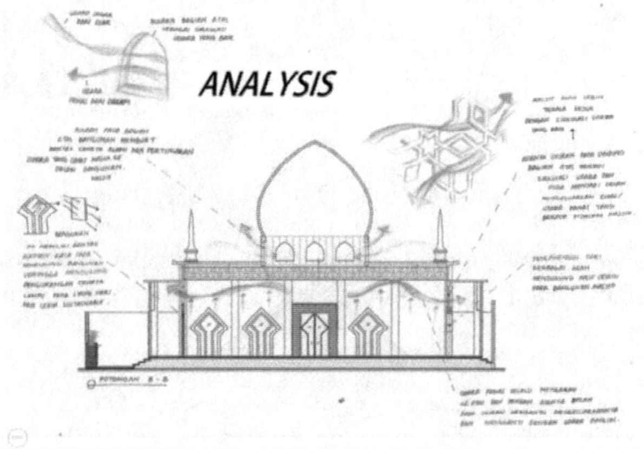

Figure 9: Building's Responses towards the Wind

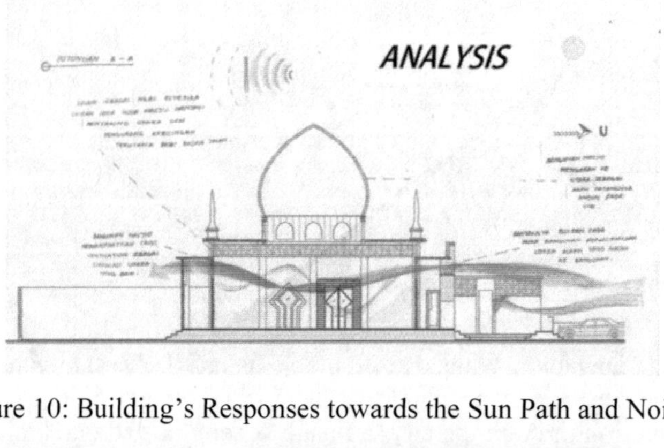

Figure 10: Building's Responses towards the Sun Path and Noise

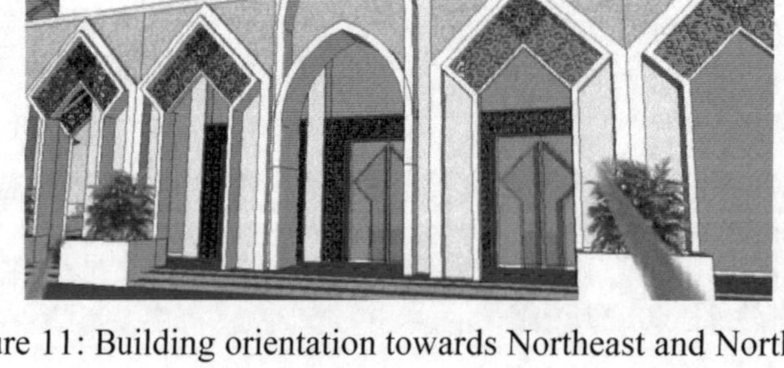

Figure 11: Building orientation towards Northeast and Northwest

Table 1: Conserving Energy Principle

Conserving Energy	
PRINCIPLES	APPLICATION
To operate the building with the minimum unrenewable energy.	To position the building openings towards the northeast and northwest (minimizing the opening towards the west)
	To provide absorbtion ion well and biopori system to catch ground water.
	To maximize cross ventilation and passive cooling by providing green areas surrounding the mosque in order to avoid the use of Air Conditioner.

Figure 12. The building is designed in elongated form to bring the light inside facing the Southeast and Southwest.

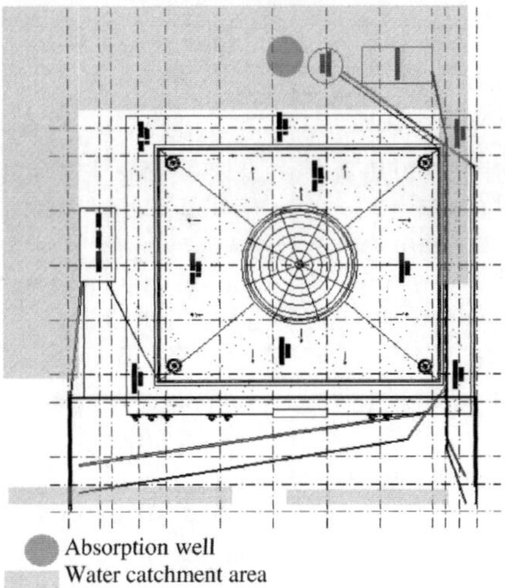

Figure 13. Absorption well and water catchment area

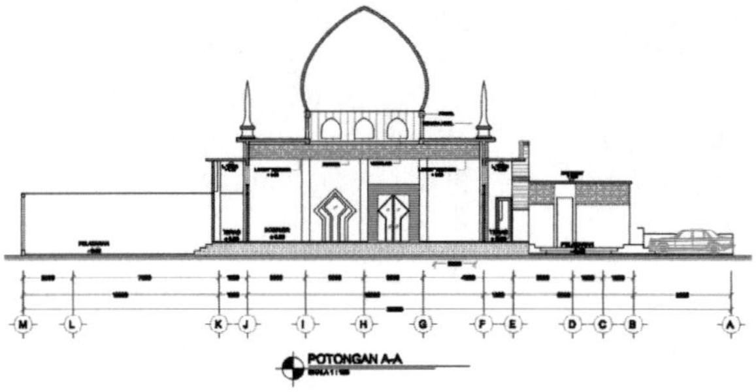

Figure 14. Green area surrounding the Mosque

To respond to the drainage in the middle of the site, the planning is to divert this drainage so that it will not be in the middle. Since the water came from the neighboring site, the team decided to move the drainage to the neighboring site so it will have direct access to the city drainage system. The site has a good capacity to catch the water and reduce the amount of water that goes to the city drainage system.

The team also considers the input from the land owner. Ibu Hj. Yuslida Hamid wants the design to be contemporary but does not exclude the symbol of a mosque as a religious building. Specifically she insists to have a lower level open court before the main entrance, a space for marbot (the person who cleans up the mosque), ablution space on the west and southeast sides, as well as parking space at the front facing the Hanjarwar road.

4.3. Green building approach in the building.

Based on the building response towards the site analysis, the application of the green building approach is as followed:

Table 2: Working with Climate Principle

Working with Climate	
PRINCIPLE	APPLICATION
To adjust and use the existing weather, climate and environment	To locate the building orientation against the sun path.
	To use plants and water as climate adjuster
	To use windows and roof which can be opened and shut depending on the users' need.
	To design terrace as buffer/ shading area.

Figure 15: Building orientation and vegetation to adjust the climate

IOP Conf. Series: Materials Science and Engineering **216** (2017) 012059 doi:10.1088/1757-899X/216/1/012059

Figure 16: Roof ventilation enabling light and wind circulation

Table 3: Respect for Site Principle

Respect for Site	
PRINCIPLE	APPLICATION
This approach requires interaction between the building and its site	To minimize site alteration by deploying design following the site condition.
	To apply local materials available from the surrounding area
	To use absorption well to catch rain water.
	To apply biopori holes in green areas surrounding the mosque.

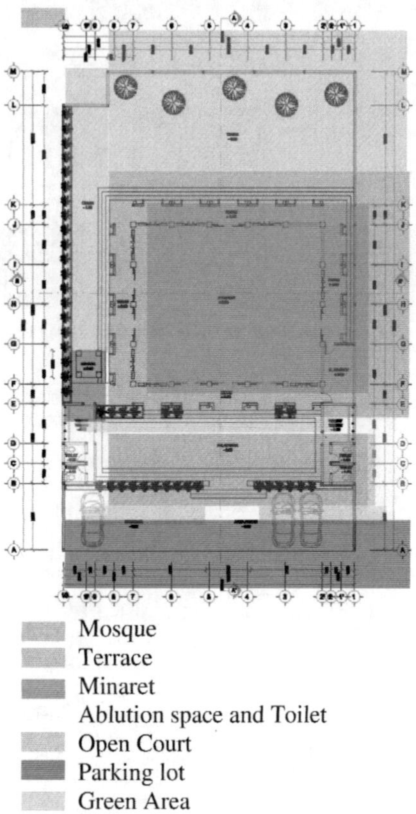

Figure 17: Site Plan Mosque

IOP Conf. Series: Materials Science and Engineering **216** (2017) 012059 doi:10.1088/1757-899X/216/1/012059

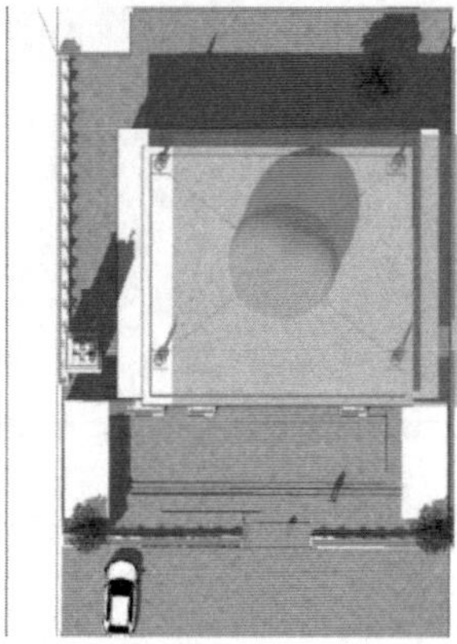

Figure 18: Building mass following the site

Figure 19: The use of local roster ornament in the facade

Figure 20: The use of local timber in the Mosque doors

Table 4: Respect for Users Principle

Respect for Users	
PRINCIPLE	APPLICATION
To pay attention on the users' needs, social condition and building maintenance.	To design a contemporary mosque which attracts and welcome visitors
	To provide parking space in front of the mosque to enable travelers have comfortable break in their journey..

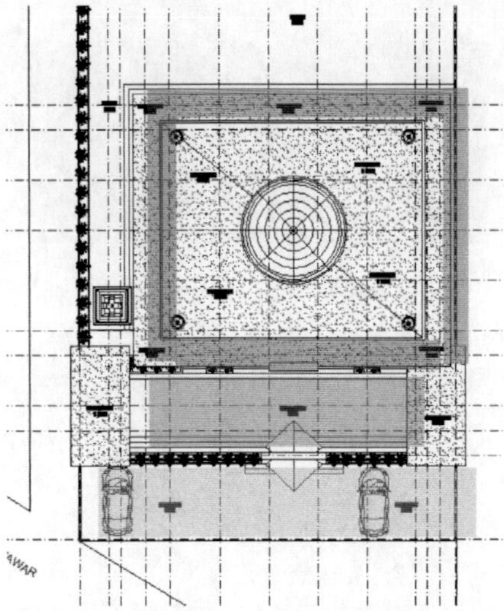

Open court to welcome visitors and function as transitional space between inside and outside
Parking Lot
Open terrace

Figure 21: Floor Plan Mosque

Figure 22: View from Northwest

Figure 23: View from Northeast

Figure 24: Front Building Façade

Figure 25: Side Building Facade

Table 5: Limiting New Resources Principle

Limiting New Resources	
PRINCIPLE	APPLICATION
to limit the use of new resources and use the current available resources	To use low-cost material and construction but environmentally friendly.

Figure 26: Open court using Paving Block

Figure 27: Building facade using granite

Figure 28: Paint finish on walls and minaret

Built environment is an important factor in supporting users' activities. Through this design, green architecture approach shows to be the best solution in emitting carbon emission.

Table 6: Holistic Approach Principle

Holistic approach	
PRINCIPLE	APPLICATION
To design a building by applying these five principles	To ensure and evaluate the approaches according to the principles.

5. Conclusion

The analysis and the design approach of the mosque show that designing a building requires comprehensive strategy in formulating the concept based on the existing site condition. Green architectural approach is then applied to minimize the environmental damage, reduce the building cost yet accommodating user's need. Applying the green approach enables users to maintain the building easily and at the same time reducing the running cost.

The Mosque in Cipendawa Village is designed following the green principles which carefully looks at the site, space, form and facade as well as interior. This can be achieved through the following ways: (1) to position the building orientation against the sun path and wind direction; (2) to provide absorption well and biopori system to catch ground water; (3) To maximize cross ventilation and passive cooling by providing green areas surrounding the mosque in order to avoid the use of Air Conditioner; (4) to use plant as climate adjuster; (5) to use low-cost local material but environmentally friendly; (6) To provide open court as transitional space between inside and outside; (7) to provide parking lot and green areas as much as possible.

By applying this, the users will have a different perception while having activities in the mosque. In this case, green architecture does not only deal with the physical appearance only but can also provide a positive psychological effect to the users.

6. Acknowledment
We would like to thank University of Pembangunan Jaya for providing us with internal grant to conduct this research-based design. Also, we would like to thank Mrs. Hj. Yuslida Hamid, who agrees to give away her 270 m2 land to build the mosque.

7. References
[1] Bauer, M. *Green Building-Guide Book for Sustainable Architecture.* Heidelberg: Springer. 2010.
[2] Sahid, Larasati, D., Victoria, P. *Towards Implementation of Green Criteria in Indonesia; Needs dan Reality*. I-Dwell International Conference, 10th-13th August 2015, Lombok, Indonesia. 2015.
[3] Vale, R., Vale, B. *Green Architecture: Design for a Sustainable Future*. London: Thames and Hudson Ltd., p70-186. 1991.
[4] Zeisel, John. *Inquiry by Design: Tools for Environment-Behavior Research*, Cambridge: Cambridge University Press. 1984.

IOP Conf. Series: Materials Science and Engineering **216** (2017) 012060 doi:10.1088/1757-899X/216/1/012060

The Role of Sketch in Architecture Design

Yanjin Li[1], Wen Ning[2]

[1]Southwest Jiaotong university, Chengdu, Sichuan, China
[2]Southwest Jiaotong university, Chengdu, Sichuan, China

[1]449058351@qq.com, [2]522107258@qq.com

Abstract. With the continuous development of computer technology, we rely more and more on the computer and pay more and more attention to the final design results, so that we ignore the importance of the sketch. However, the sketch is the most basic and effective way of architecture design. Based on the study of the sketch of Tjibao Cultural Center of sketch , the paper explores the role of sketch in architecture design .

1. Introduction

Sketch is a kind of carrier of the architecture designers to express their own design intention and idea in an intuitive form. It is a process that the designers quickly record their own design. However a designer of hand drawing skills, understanding of architecture and engineering technology is reflected in the rapid expression of the process [1]-[2]. For example, a picture of ingenious design sketch can be used as a perspective renderings to convey understanding of the designer of the space, structure and shape .At the same time, sketch also plays a important role in guiding of the construction. However, with rapid development of software technology,we gradually neglect the importance of architecture sketch. This paper elaborates the role of sketch in architecture design illustrated by the example of sketch in architecture design.

a. architecture form [3]

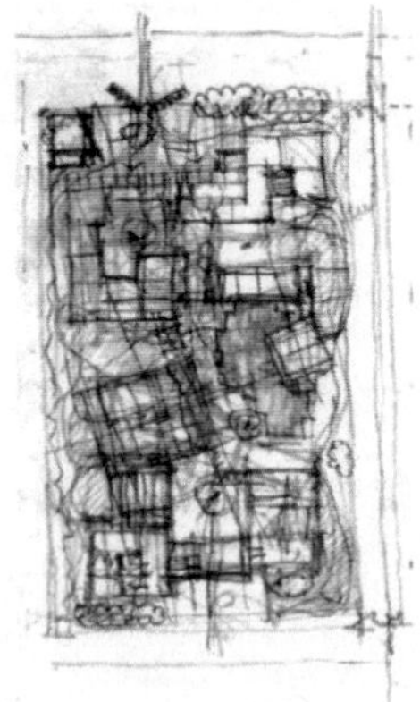

b. architecture layout[4]

IOP Conf. Series: Materials Science and Engineering **216** (2017) 012060 doi:10.1088/1757-899X/216/1/012060

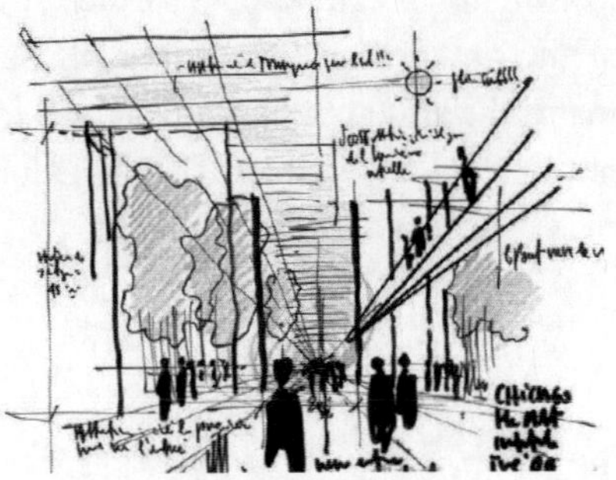

c. section and space [5]

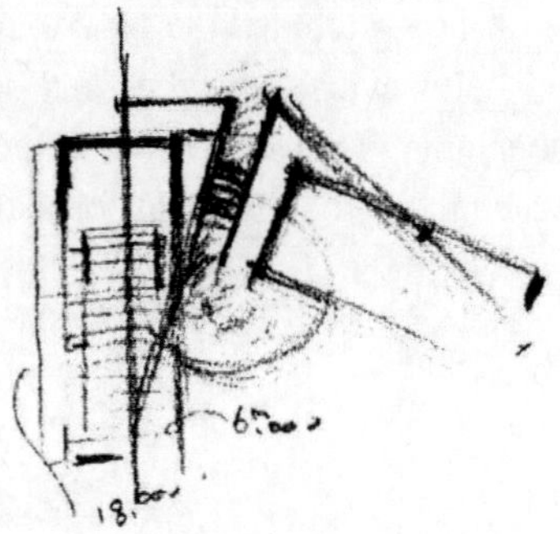

d. Plan [6]

Figure 1. Sketch of Architecture Design

(Source:webside)

2. Analysis of sketch of Tjibao cultural Center

2.1. Introduction of Tjibao Cultural Center

Tjibao Cultural Center was completed in 1994, located in New Caledonia Noumea. Renzo Piano took pains to understand the local customs, beliefs and aesthetic taste, when he designed the architecture.He won the aboriginal elders' support who are with wary of the Western colonial control [7]. Tjibao Cultural Center successfully drew lessons from other local traditional building "shed" and wood frame structure. These double braided facade of "shed" can withstand hurricanes and it also can use the sea breeze in the upper part of the formation of pressure from the lower part of the air to play the role of ventilation [8]. The shutter can also play a role in regulating the indoor light. He pays attention to the combination of architectural art, technology and the environment around the building. His architectural thinking is rigorous and lyrical on the traditional inheritance and transformation.

Figure 2. local traditional building "shed"
(Author self drawing)

Figure 3. Tjibao Culture Center
(Author self drawing).

2.2. Contrast between the architecture and sketch

From the expression of Piano's sketch, it skillfully expresses facade, section and structure. Section space can be thought as the half space of "the shed" section. In addition, in order to merge into the surrounding woods Piano learn from the local wood rib structure of the building to make architecture merge into the environment. The final sketch of and the architecture is basically in agreement with the architecture. But from the sketch and final architecture section, there is a difference in the curvature of the outer frame of the sketch and the final appearance of the architecture. The surface structure of the sketch is nimble. However, from the final section of the architecture, the curve of the structure is somewhat cumbersome. This shows that Piano coordinates the structure of the wood rib frame and the sketch of the structure, so that the final construction and sketches to minimize the deviation.

Figure 4. Sketch of Piano (author painting copy).

The wood rib of the culture center of the sesame is double layer, which is for ventilation and lighting. However, In order to strengthen the link between the two tier structures, the transverse members are connected. Horizontal linkage members draw lessons from veins of palm trees fan-shaped distribution of enlightenment and it also reflects the local architectural materials extracted architectural vocabulary. They are firmly anchored in a horizontal way to replace the weaving or tying of the folk and to increase the wind resistance of the building,

From a whole point of view, the structure of the sketch is similar to the structure of the final architecture. The angle of the component in the sketch is different from final architecture because the structure needs to achieve the best force transmission system through the computer technology to calculate and adjust the angle between the components .Secondly, Piano pays attention to the regulation of the transverse component location. It can ensure the reasonable structure system and make less influence on structure curve. This difference can be weaken by the foil of surrounding.

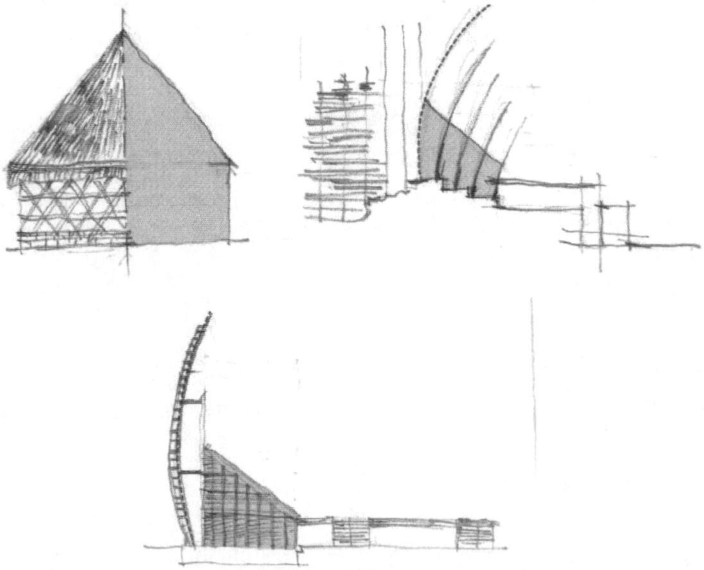

Figure 5. Comparison of shed skectch, sketch and final architecture section
(Author self drawing).

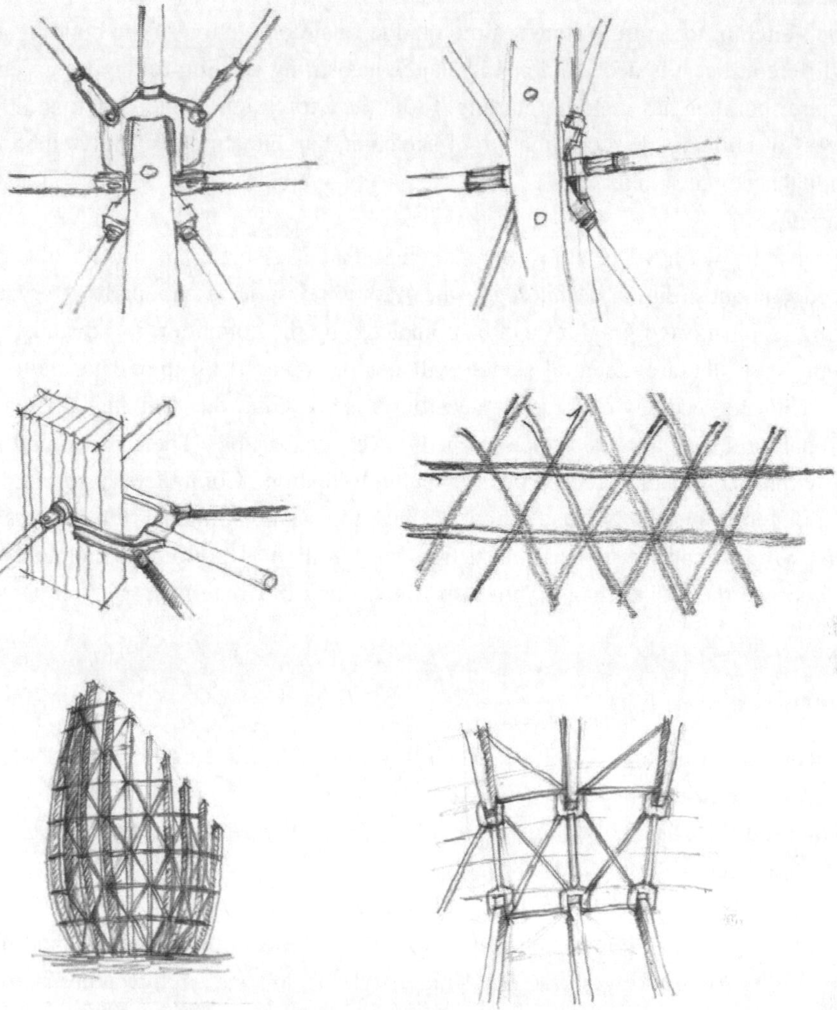

Figure 6. The main frame of the final architecture (Author self drawing).

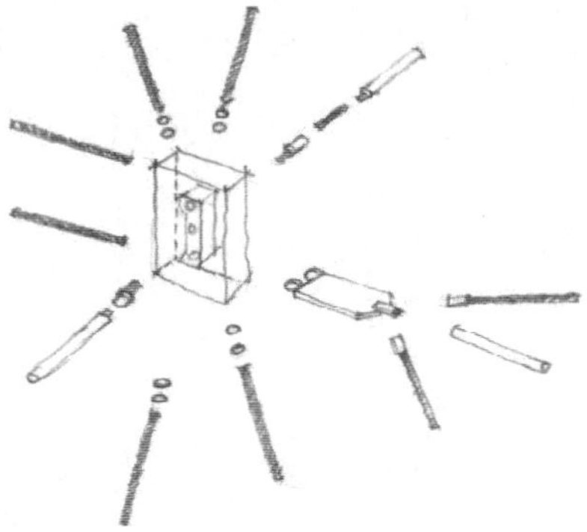

Figure 7. Sketch of Piano's structural components (Author self drawing).

2.3. Summary
From the sketch of the culture center, although the final completion of the building and the sketch are a little different, but it is acceptable. And sketch has strong guiding to the design and construction of architecture. building has a strong guiding. From the expression of the sketch and the construction of the process of building. the coordination of sketch and architecture also fully embodies the architect's professional accomplishment.

3. Conclusion
The reason why we have to study the sketch is that it is better to use it. In today's increasingly developed computer digital technology, the traditional style is gradually overwhelmed. Complex computing, sophisticated graphics can be completed by the computer, its advantage is beyond doubt. But in any case, the architectural sketch will not be replaced by digital technology. When we are obsessed with the accuracy of the computer, but it is also limit our thinking. So some people say that professionals still have to look at the sketch. It is very reasonable. The architectural sketch is the most spiritual expression in the process of architectural creation. Computers have also been designed to bring a lot of convenience. No matter how we have the way to stimulate thinking, the expression of the sketch is the most concise, concise, fast, free and intuitive. The training and analysis of sketch will help to improve the design quality, but also it is lacking in Contemporary Architectural Education in China.

References
[1] Xiaoming Hu,Jian Qiu,Jun zhu.Study on Architectural Sketches[J].Architecture & Culture,2012,(11):92-93.

[2] Dongzhu Chu.The Power of Flexbility:Study on Architecturahes[J].New Architecture:2008(3):114-119.

[3] http://www.jsdwy.cn/a/2016/0623/1437912.html

[4] Qing Lin,Rui Liao.Integration of Ecological and Innovation Space: Design of Tianfu Research Institute of Southwest Jiaotong University[J].Huazhong Architecuture,2016(1):65-69.

[5] http://www.soujianzhu.cn/ispace/profile/docrizi.aspx?u=152&id=2668

[6] http://blog.sina.com.cn/s/blog_5d3bfd340100mn5d.html

[7] http://www.sj33.cn/architecture/jzsj/200607/9170.html

[8] Kan Yao.Application of Traditional Wood Materials in Modern Architecture Design[J].Journal of Hefei University of Technology(Social Sciences) ,2007, 21(5):122-124

Charring Rate for Fire Exposed X-Lam

Bernice VY Wong[1], Kong Fah Tee[2]

[1] Ramboll Fire, 240 Blackfriars Road, London SE1 8NW, UK
[2] Department of Engineering Science, University of Greenwich, Kent ME4 4TB, UK

E-mail: Bernice.Wong@ramboll.co.uk

Abstract. Design of timber structures has been outlined in Eurocode 5. Notional charring rate for softwood and hardwood timber is given. For the performance of X-LAM panels in fire, only little information on charring is available and whether the fire behaviour of X-LAM is similar to homogenous timber panels has not yet been systematically analysed. This paper presents an overview of fire performance of X-LAM and evaluation of its resistance to elevated temperature as an element of structure in comparison to homogeneous timber panels. Numerical study has been carried out based on available experimental results. Charring rates for X-LAM panels obtained from experimental results are compared with those obtained from Eurocode 5 and proposed simplified model.

1. Introduction

Timber frame is one of the fastest growing modern method of construction in the United Kingdom, as wood is effectively carbon neutral, with timber frame benefiting from being the only organic, non-toxic and naturally renewable building material, whilst minimising energy consumption. In recent years many developments have been made in relation to timber technology and construction products. For heavy timber construction, cross-laminated timber (X-Lam) panel load bearing wall and floor assemblies system is becoming increasingly common.

X-Lam panels, with thickness varies from 50mm to 300mm in 3 to 8 layers, are cut to size in the factory according to the structural drawings taking into account openings such as doors, windows and stairwells, using layers of softwood planks glued together with each layer arranged at right angles to one another (figure 1), and are delivered to site ready for immediate erection of walls, floors and roofs. The benefits of this huge building system are the good structural performance, thermal and acoustic insulation characteristics, the great prefabrication, and the rapidity of erection [1]. The use of large solid timber panels is also favorable case of fire, as the risk of fire spread through void cavities is reduced in comparison to light timber frame constructions. Although, large solid timber panels increase the fire load in the room.

Advanced and simplified calculation methods have been recommended for analysing how timber behaves when exposed to elevated temperatures [2]. Experimental investigations on small and large scale tests of unloaded and loaded X-Lam specimens manufactured by various producers with different characteristics have been carried out [3-5]. The outcome of the tests revealed that the fire behavior of X-Lam elements were mostly linked to the thickness of the layers and the type of adhesive used to produce the timber panels [1]. Analytical and numerical methods are effective ways to study the mechanical and thermal performance of cross-laminated timber elements exposed to fire without performing experimental tests which are hazardous and expensive [1].

IOP Conf. Series: Materials Science and Engineering **216** (2017) 012061 doi:10.1088/1757-899X/216/1/012061

This paper presents an overview of fire performance of X-Lam and evaluation of its resistance to elevated temperature as an element of structure in comparison to homogeneous timber panels. Charring rates for X-Lam panels obtained from experimental results are compared with those obtained from Eurocode 5 [2] simplified calculation methods and proposed simplified model.

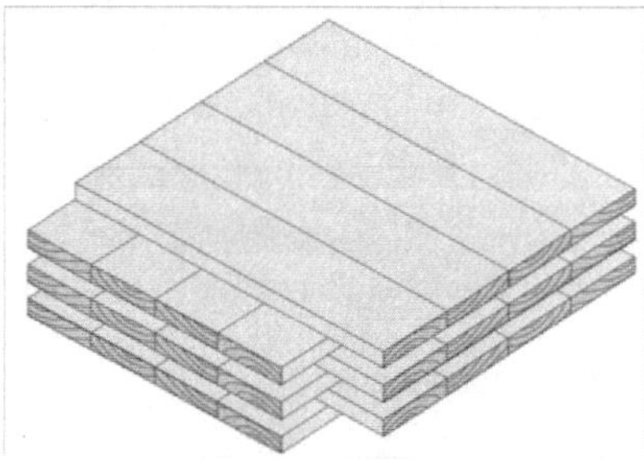

Figure 1. Individual timber panels used to make a typical X-Lam [6].

2. Performance of timber in fire

All buildings must be constructed to meet national building regulations, which requirements for fire resistance. Fire-resistance-rated construction requires structural components and assemblies to meet defined fire performance standards. The fire resistance rating of a building assembly has traditionally been assessed by subjecting a replicate of the assembly to ISO 834 standard fire-resistance test [7]. In term of fire resistance, a structural element should not collapse or deflect beyond the permitted levels when subjected to the applied load throughout the fire test. Combustible building materials like timber burn on their surface, release energy and thus contribute to fire propagation and the development of smoke in case of fire. It is often perceived that timber will perform unsatisfactorily during a fire situation; this is not necessarily the case as timber burns slowly and resists heat penetration by the formation of a self-insulating char.

When exposed to heat of a fire, timber undergoes a thermal breakdown (pyrolysis) into combustible gases. Pyrolysis takes place in the timber when they are exposed to elevated temperatures between 150°C to 200°C. A layer of charcoal forms on the burning surface, the char layer grows in thickness as the fire grows, reducing the cross-sectional dimensions of the wood element. The char layer is a poor thermal conductor and protects the un-charred remaining residual cross-section from fire. When large timber members are subjected to fire, the uncharred inner portion maintains its strength, giving the structure a higher survival factor and more importantly, it retains its structural integrity being one of the reasons why large timber sections can often be used in unprotected situations where non-combustible materials such as steel would require special fire protection.

The rate at which timber chars varies between species and is predominately dependent on density and moisture content. When timber burns, it occurs at a predictable speed known as the charring rate. For unprotected wood surfaces throughout the duration of fire exposure, the charring rate for softwood is 0.65 mm/min as given in Eurocode 5 [2].

For protected timber surfaces, a simplified model has been suggested [2] taken into consideration the fact that different charring rates should be applied during different phases of the fire exposure, where an increased charring is expected after failure of the protection materials. It was recommended double the rate of initially unprotected surfaces, until the charring depth has reached 25mm [2].

3. Performance of X-Lam in fire

The interest for X-Lam as a buildings system is increasing whilst for X-Lam panels in fire only little information on charring is available. A few experimental investigations are available [3-5] on the fire behavior of X-Lam panels. It has been revealed that, the fire behavior of X-Lam panels is characterised by the behavior of the single layers. In the case of fire, if the charred layers of X-Lam panel remain in place, the charcoal protects the remaining uncharred layers of the X-Lam solid panels

against heat, behaves similarly to homogenous timber panels in fire [3]. An increased charring is expected if the charred layers fell off after the panel system. This appearance is similar to the increased charring observed for protected timber surfaces after failure of the protection materials.
The thickness of the X-Lam layers as well as the behavior of the bonding adhesive at elevated temperature can influence the falling of the charred layers [3]. X-Lam with thicker panel layers behave better in fire than that of thinner, due to the fact that X-Lam panels with thicker panel layers expected behave similar to homogenous timber panels and therefore the one-dimensional charring rate of 0.65 mm/min can be assumed [2], while X-Lam with thinner panel layers should behaves more like plywood, where charring rate of 1.0 mm/min is assumed for plywood [2].
Observation from the fire tests on X-Lam panels under ISO-fire exposure [4,5], has suggested that at elevated temperatures falling off of the charred layers leading to increase charring rates in comparison to homogeneous timber panels. Similar effect to a protected timber member after the fire protection has fallen off. For X-Lam panels where no falling off of the charred layers was occurred, the fire behavior of the X-Lam panels is similar to that of homogeneous timber panels.
The individual layers forming the X-Lam members are bonded and pressed together using with durable, moisture-resistant structural adhesives. For X-Lam panels bonding system, Melamine-urea-formaldehyde (MUF) and one-component polyurethane adhesives (1K-PUR) are frequently used. At high temperature the performance of the assembled timber member depends highly on the resistance of laminating adhesives to elevated temperatures. It was demonstrated that in fire, the behavior of adhesive that used in the bond-line between the lamellas has little influence on the resistance of the glued laminated timber beam and anticipated that the fire resistance will be governed by the bending resistance rather than shear resistance [8]. A series of tests were conducted to study the shear behavior of different adhesives at high temperatures, has demonstrated that the fire behavior of cross-laminated timber panels is strongly influenced by the behavior of the adhesive that used for the X-Lam panels bonding system [5,9]. X-Lam panels manufactured with a less temperature-sensitive adhesive the charred layers almost remained in place throughout the fire tests and the panels behave just similar to homogenous timber panels exposed to ISO 834 [7] on one side. Test result has demonstrated that the measured charring rate of these timber panels has lower rate than the one-dimensional charring rate of 0.65 mm/min that assumed for solid timber according to Eurocode 5 [2]. Moreover, the thermal behavior of one component polyurethane systems can be greatly varied by modifying their chemical structure, test results that were based on 1K-PUR are therefore not valid for other polyurethane adhesives [5].

4. Simplified Calculation for X-Lam charring rate

To date, there has been limited experimental investigation into X-Lam in fire, and no published literature has been found regarding the charring rate of X-Lam panel. Based on the available fire tests on X-Lam panels [5, 10], a simplified charring model to determine the residual cross-section for X-Lam has been proposed [11]. Figure 2 illustrates the char depth vs. time plots, compared with the experimental results [5, 10], alongside the plot calculated based on the charring model according to Eurocode 5 [2]
The charring rate for the X-Lam panels studied in this investigation was calculated taken into account the subsequent fire-exposed layers, it is necessary to mathematically estimate an increased charring rate until the formation of a new 25mm thick char layer, the charring rate is suggested as below;

- 0.65 mm/min if only one layer is affected by exposure to fire;
- 1.30 mm/min for any additional layers affected by exposure to fire until charring or the formation of a 25mm thick char layer;
- a charring rate of 0.65 mm/min may be applied up to the next bonded joint.

From figure 2a it can be seen that, the best agreement between calculated and measured charring depth is observed from the fire test performed with X-Lam panel formed by two layers of 30mm panel. The effect of the increase of the charring rate can be seen from the test, the differences between calculated and measured charring depth are relatively small.

The specimen with three layers of 20mm panel (figure 2b) showed reasonable good agreement with the calculated values for the first and second layer. It should be note that the charring rate of the first layer measured from the fire test is higher than the value of 0.65 mm/min which has been used in the calculation as according to Eurocode 5 [2]. Test result has shown that after second charred layer has fall off, there was no significant change in term of the charring rate, although the calculated values for the simplified model has conservatively assumed a charring rate of 1.30 mm/min to be used for any additional layers affected by exposure to fire until charring or the formation of a 25mm thick char layer. The calculated values for Eurocode 5 assumed a 0.65 mm/min to the third layer give non-conservative estimation.

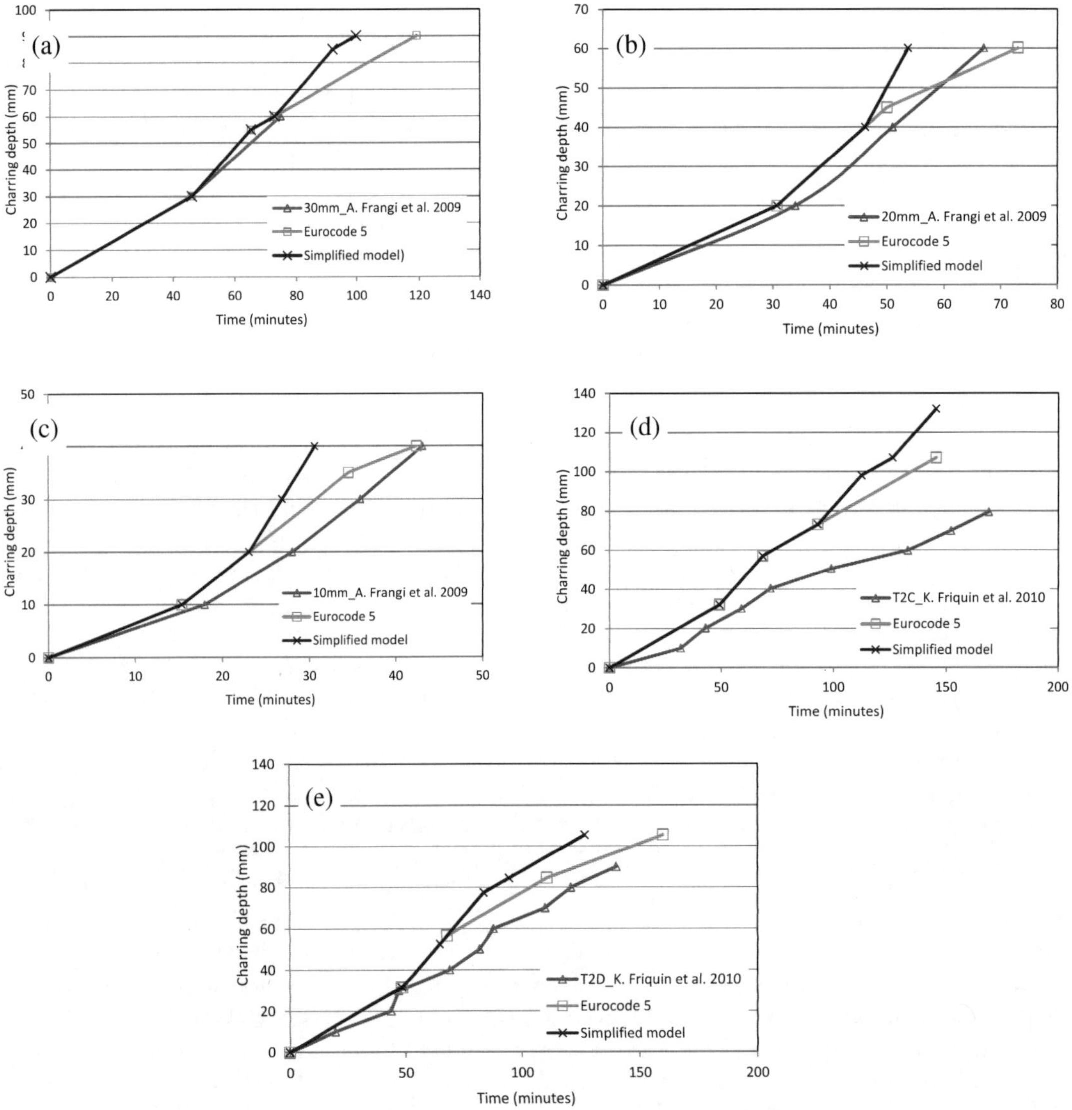

Figure 2. Charring depth for X-Lam panels [5, 10] compared with Eurocode 5 [2] and simplified model.

The effect of the increase of the charring rate can be seen from the test specimen with four layers of 10mm panel. The comparison between fire test and calculated values (figure 2c) showed that the simplified models lead to safe results. The specimen with three layers of 20mm panel showed reasonable good agreement with the calculated values for the first and second layer. Tests T2C and T2D (figure 2d and 2e) [10] were tested on X-Lam panel with different panel thicknesses. The average charring rate obtained from the test specimen T2C was 0.47 mm/min [10] which is lower than the value of 0.65 mm/min a charring rate for softwood [2] thus comparison of the test result against the simplified model does not agree. Whereas for fire test on specimen T2D, the test result shows that both the Eurocode 5 charring rate and the simplified charring models lead to safe results.

5. Conclusions

From the overview of available experimental results [3-5, 10] and comparisons made between test results and simplified models the following conclusions can be drawn:

- The thickness of a single layer of the X-Lam panel plays a significant role on the charring rate;
- The calculations of the charring depth of X-Lam panel should take into account the influence of the falling of charred layers;
- Eurocode 5 provides simplified model for homogenous timber but not overly conservative design methods for X-Lam.
- Fire behavior of X-Lam panel is being studied by using finite-element numerical approach, in order to extend and confirm the results of the experimental and simplified models performed. The studies however at the preliminary stage are still under investigation, the work will be presenting in the next stage.

6. References

[1] Fragiacomo M, A. Menis A, Clemente I, Bochicchio G and Tessadri B 2012 Experimental and Numerical Behaviour of Cross Laminated Timber Floors in Fire Conditions, *Proc. of World Conference on Timber Engineering (Auckland).*

[2] CEN (European Committee for Standardization) 2004 BS EN 1995-1-2 Eurocode 5: Design of timber structures - Part 1-2: General - Structural fire design (Brussels: BSI).

[3] Frangi A, Bochicchio G, and Ceccotti A 2008 Natural full-scale fire test on a 3 storey xlam timber building *Proc. of 10th World Conference on Timber Engineering (Japan).*

[4] Frangi A, Fontana M, Knobloch M and Bochicchio G 2009 Fire Behaviour of Cross-Laminated Solid Timber Panels *Fire Safety Science* vol. 9 pp. 1279-1290.

[5] Frangi A, Fontana M, Hugi E, and Jöbstl R 2009 Experimental analysis of cross-laminated timber panels in fire *Fire Safety Journal* vol. 44 (8) pp. 1078–1087.

[6] Karacabeyli E and Douglas B 2013 Cross Laminated Timber (CLT) Handbook: Introduction to cross-laminated timber *FPInnovations*.

[7] ISO 1999 ISO 834-1: Fire-resistance Tests—Elements of Building Construction—Part 1: General Requirements (Geneva: International Organization for Standardization).

[8] Klippel M, Frangi A and Fontana M 2011 Influence of the Adhesive on the Load Carrying Capacity of Glued Laminated Timber Members in Fire *Proc. of the 10th International Symposium of the International Association for Fire Safety Science (Maryland)* pp. 1219-1232.

[9] Frangi A, Fontana M and Mischler A 2004 Shear behaviour of bond lines in glued laminated timber beams at high temperatures *Wood Science and Technology* **38** pp.119–126.

[10] Friquin KL, Grimsbu M and Hovde PJ 2010 Charring rates for cross-laminated timber panels exposed to standard and parametric fires *Proc of the 11th World Conference on Timber Engineering*.

[11] Wong BVY and Tee KF 2014 The Fire Performance of Exposed Timber Panels *World Academy of Science, Engineering and Technology, International Journal of Civil, Environmental, Structural, Construction and Architectural Engineering* 8(10) pp 1045-1051